BASIC BIOMECHANICS OF THE MUSCULOSKELETAL SYSTEM

MARGARETA NORDIN, R.P.T., Dr. Sci.
Director, Occupational and Industrial Orthopaedic Center, Hospital for Joint Diseases Orthopaedic Institute;
Program Director, Program of Ergonomics and Occupational Biomechanics, New York University, New York, New York

VICTOR H. FRANKEL, M.D., Ph.D.
President, Hospital for Joint Diseases Orthopaedic Institute, Chairman, Department of Orthopaedic Surgery;
Professor of Orthopaedic Surgery, New York University School of Medicine, New York, New York

KAJSA FORSSÉN and HUGH NACHAMIE, Illustrators
LAURIE YELLE, Editorial Associate

BASIC BIOMECHANICS OF THE MUSCULOSKELETAL SYSTEM

Second Edition

LEA & FEBIGER
PHILADELPHIA, LONDON

Lea & Febiger
200 Chester Field Parkway
Malvern, Pennsylvania 19355-9725
U.S.A.
(215) 251-2230
1-800-444-1785

Lea & Febiger (UK) Ltd.
145a Croydon Road
Beckenham, Kent BR3 3RB
U.K.

Library of Congress Cataloging-in-Publication Data

Basic biomechanics of the musculoskeletal system / [edited by]
Margareta Nordin, Victor H. Frankel; Kajsa Forssén, illustrator;
Hugh Nachamie, illustrator; Laurie Yelle, editorial associate. —
2nd ed.
p. cm.
Rev. ed. of: Basic biomechanics of the skeletal system / Victor H.
Frankel, Margareta Nordin. 1980.
Includes bibliographies and index.
ISBN 0-8121-1227-X
1. Human mechanics. I. Nordin, Margareta. II. Frankel, Victor
H. (Victor Hirsch), 1925- . III. Frankel, Victor H. (Victor
Hirsch), 1925- Basic biomechanics of the skeletal system.
[DNLM: 1. Biomechanics. WE 103 B311]
QP303.B375 1989
612'.76—dc19
DNLM/DLC
for Library of Congress

89-2455
CIP

PRINTED IN THE UNITED STATES OF AMERICA

Print Number 3 2

The authors wish to dedicate this book to

Ingrid Forssén-Nordin,
Estelle H. Frankel,
and
Ruth E. Frankel

FOREWORD TO THE FIRST EDITION

Progress in restoring function has paralleled our gain in knowledge of the musculoskeletal system and the nature of its disabilities. Observation, thought, and experience were sufficient to solve some problems. Further answers evolved as we became more cognizant of underlying pathology and physiology. Many of the remaining problems await the application of biomechanical principles.

A major determinant of effective function is the force tolerance and mobility of the skeletal system. Increases in the intensity of activity of all age groups and extended human longevity are presenting new demands on these physical qualities. To develop an appropriate clinical response, biomechanical information must be acquired and the knowledge translated into therapeutic guidelines.

Victor Frankel has been a leader in both these endeavors. His contribution to biomechanical research is attested to by his extensive publications. This text represents his second effort to introduce biomechanical knowledge into patient care. It is not a simple task. Margareta Nordin has joined him in this responsibility for producing a new text. Her involvement has contributed to the informative style used to interpret complex data. The value of this text also has been enhanced by the added breadth provided by the contributors.

The engineering profession has developed the means for identifying the forces, motions, and tissue responses in human function. A considerable amount of information has been provided, but this information is dispersed in a wide variety of publications and is often couched in engineering terminology unfamiliar to clinicians.

This book on the basic biomechanics of the musculoskeletal system brings together the current knowledge and presents it in a clear, conversational style. The use of engineering terminology is restricted to that which is necessary to express essential concepts. Presentation of the literature is both inclusive and selective. The major contributions to each topic are cited as specific references, and other sources are identified as recommended reading.

While the text is primarily a compendium of information and illustrations of biomechanical principles that are clinically pertinent, a small portion of the book describes the means for calculating joint and muscle forces. This is presented lucidly with minimal mathematical formulae and very clear diagrams. The techniques are described in a way that makes them usable by the interested clinician.

This book is a well conceived bridge between engineering concepts and clinical practice. It should provide orthopaedists, physical and occupational therapists, and other health professionals with a working knowledge of biomechanical principles useful in the evaluation and treatment of musculoskeletal dysfunction.

Jacquelin Perry, M.D.

PREFACE

This book has been written to serve as an introduction to biomechanics for those who deal with disorders of the musculoskeletal system. The work is the result of 25 years of experience in biomechanics education and many conferences with colleagues in the fields of orthopaedic surgery, physical therapy, occupational therapy, bioengineering, ergonomics, and other allied specialties.

Biomechanics uses laws of physics and engineering concepts to describe motion undergone by the various body segments and the forces acting on these body parts during normal daily activities. The interrelationship of force and motion is important, and must be understood if rational treatment programs are to be applied to musculoskeletal disorders. Deleterious effects may be produced if the forces acting on the areas with disorders rise to high levels during exercise or activity.

The purpose of this volume is to acquaint the readers with the force-motion relationships within the musculoskeletal system and the various techniques used to understand these relationships. The volume is intended to be used as a textbook, either in conjunction with a biomechanics course or for independent study. The references and suggested readings at the end of each chapter can be utilized to amplify the discussions in these chapters.

Although we undertook an extensive review of the world literature on biomechanics in preparing this book, it has not been our purpose to publish a review of the material. Rather, we have selected examples to illustrate the concepts needed for a basic knowledge of musculoskeletal biomechanics. In addition, we have developed important engineering concepts throughout the volume. The text will serve as a guide to a deeper understanding of musculoskeletal biomechanics gained through further reading, analysis of case material, and independent research. The information presented should also guide the reader in assessing the literature on biomechanics. No attempt has been made to discuss therapy, as it was not our purpose to cover this area. Rather, we have described the underlying basis for rational therapeutic programs.

This book has been strengthened by the work of many excellent contributing authors. An initial section on the international system of measurement serves as an introduction to the physical measurements used throughout the book. The reader needs no more than a basic knowledge of mathematics to fully comprehend the material in the book, but it is important to first review the section on the SI system and its application to biomechanics.

The five chapters in Part One of the book deal with the biomechanics of the tissues and structures of the musculoskeletal system: biomechanics of bone, biomechanics of articular cartilage, biomechanics of tendons and ligaments, biomechanics of peripheral nerves, and biomechanics of skeletal muscle. This section has been expanded from the first edition to give the reader an understanding of the differences in mechanical behavior among the principal tissues composing the musculoskeletal system.

The ten chapters in Part Two cover the biomechanics of the major joints of the musculoskeletal system. Each of these chapters contains an anatomic overview followed by a discussion of the biomechanics specific to the particular joint or joint complex. In this new edition chapters on the biomechanics of the cervical spine, the biomechanics of the wrist, and the biomechanics of the hand have been added.

We feel that our efforts in rewriting and expanding this book will be justified if it brings about an increased awareness of the importance of biomechanics and engenders discussion.

Margareta Nordin
Victor H. Frankel

ACKNOWLEDGMENTS

This book was made possible through the efforts of all the contributors. Their outstanding knowledge, their understanding of the basic concepts of biomechanics, and their wealth of experience have brought both breadth and depth to this work. We are honored and grateful for their contributions and for their devoted work in preparing and finalizing this new edition. We are deeply saddened by the recent death of one contributor, Carl Carlstedt, who finalized his chapter from his hospital bed, near the end of his valiant struggle with leukemia.

A book of this size, with its large number of figures, legends, and references, cannot be produced without editorial efforts. Laurie Yelle's intense efforts shine throughout the entire book. Not only did she function as a grammarian and stylist, but her logical pattern of thinking had a great deal to do with the final organization of the book.

Kajsa Forssén and Hugh Nachamie, illustrators, were vital, energetic, and never-failing members of our publication team. Their quick grasp of the concepts to be illustrated and the clarity of their art work were of the greatest importance in the production of this book.

Our colleagues in the Occupational and Industrial Orthopaedic Center of the Hospital for Joint Diseases Orthopaedic Institute functioned as critical reviewers of the chapters. A special thanks is extended to Mohamad Parnianpour for his invaluable assistance. Our sincerest thanks also goes to Shelly Bade, Oscar Cartas, Nadia Greenidge, Inna Gitsevitch, Manny Halpern, Joan Kahn, Ali Sheikhzadeh, Judy Trucios, Pat Turiello, Kathy Viola, and Sherri Weiser. In the Department of Orthopaedic Surgery, Irene Campbell, Abe Moshel, and Alisa Rivera deserve our greatest appreciation for their valuable help. We thank Judy Raymond and Hilary Winter for word processing assistance and constructive suggestions. We are also most grateful to Drs. Alex Norman, Paul Bisson, and Steven Lubin for supplying vital roentgenograms for the new edition.

This new edition of the book was supported throughout its production by the Research and Development Foundation of the Hospital for Joint Diseases Orthopaedic Institute and the hospital administration, to whom we forward our sincerest gratitude.

To all who helped, hjärtligt tack!

Margareta Nordin
Victor H. Frankel

CONTRIBUTORS

FADI JOSEPH BEJJANI, M.D., D.E.M., M.A., Ph.D.
Research Professor and Director
Human Performance Analysis Laboratory
New York University
New York, New York
and
Consultant in Occupational Orthopaedics,
Electrodiagnosis, Motion Analysis, and Arts Medicine

CARL A. CARLSTEDT, M.D., Ph.D.
Karolinska Institute
Huddinge Hospital
Stockholm, Sweden
and
Research Fellow
Department of Bioengineering
Hospital for Joint Diseases Orthopaedic Institute
New York, New York

DENNIS R. CARTER, Ph.D.
Associate Professor of Mechanical Engineering
Stanford University
Stanford, California

VICTOR H. FRANKEL, M.D., Ph.D.
President
Hospital for Joint Diseases Orthopaedic Institute
Chairman, Department of Orthopaedic Surgery
New York, New York
and
Professor of Orthopaedic Surgery
New York University School of Medicine
New York, New York

MICHAEL A. KELLY, M.D.
Assistant Professor of Orthopaedic Surgery
College of Physicians and Surgeons
Columbia University
New York, New York

JOHAN M. F. LANDSMEER, Prof. Dr.
Professor Emeritus
Department of Anatomy and Embryology
Rijks Universiteit
Leiden, Holland

MARGARETA LINDH, M.D., R.P.T., Ph.D.
Department of Rehabilitation Medicine
Sahlgren Hospital
University of Gothenburg
Gothenburg, Sweden

GÖRAN LUNDBORG, M.D., Ph.D.
Professor and Chairman
Division of Hand Surgery
Department of Orthopaedics
University of Lund
Lund, Sweden
and
Laboratory of Experimental Biology
Department of Anatomy
University of Gothenburg
Gothenburg, Sweden

FREDERICK A. MATSEN III, M.D.
Professor and Chairman
Department of Orthopaedics
University of Washington
Seattle, Washington

VAN C. MOW, Ph.D.
Director
New York Orthopaedic Hospital Research Laboratory
Professor of Mechanical Engineering and Orthopaedic Bioengineering
Columbia-Presbyterian Medical Center
Columbia University
New York, New York

MARGARETA NORDIN, R.P.T., Dr. Sci.
Director
Occupational and Industrial Orthopaedic Center
Hospital for Joint Diseases Orthopaedic Institute
New York, New York
and
Program Director
Program of Ergonomics and Occupational Biomechanics
New York University
New York, New York

LARS PETERSON, M.D., Ph.D.
Associate Professor
Department of Orthopaedic Surgery
East Hospital
Gothenburg, Sweden

MARK I. PITMAN, M.D.
Chief of Sports Medicine
Attending Physician, Department of Orthopaedic Surgery
Hospital for Joint Diseases Orthopaedic Institute
New York, New York
and
Assistant Clinical Professor of Orthopaedic Surgery
New York University School of Medicine
New York, New York

CHRISTOPHER S. PROCTOR, M.D.
Research Fellow
New York Orthopaedic Hospital Research Laboratory
Columbia-Presbyterian Medical Center
New York, New York

BJÖRN RYDEVIK, M.D., Ph.D.
Associate Professor
Department of Orthopaedic Surgery I
Laboratory of Experimental Biology
Department of Anatomy
University of Gothenburg
Gothenburg, Sweden

G. JAMES SAMMARCO, M.D., F.A.C.S.
Associate Clinical Professor
Department of Orthopaedic Surgery
University of Cincinnati Medical Center
Cincinnati, Ohio
and
Director of Foot and Ankle Center
Good Samaritan Hospital
Cincinnati, Ohio

ILAN SHAPIRO, M.D., M.S.
Orthopaedic Surgery
Harwood Medical Associates
Wauwatosa, Wisconsin

RICHARD SKALAK, Ph.D.
Director, Bioengineering Institute
Department of Civil Engineering and Engineering Mechanics
Columbia University
New York, New York

STEVEN STUCHIN, M.D.
Chief of Arthritis Management Service
Assistant Chief of Hand Surgery
Hospital for Joint Diseases Orthopaedic Institute
New York, New York
and
Assistant Professor of Orthopaedic Surgery
Department of Orthopaedic Surgery
New York University School of Medicine
New York, New York

JOSEPH D. ZUCKERMAN, M.D.
Chief, Shoulder Clinic
Hospital for Joint Diseases Orthopaedic Institute
New York, New York
and
Assistant Professor of Orthopaedic Surgery
Department of Orthopaedic Surgery
New York University School of Medicine
New York, New York

CONTENTS

SI: THE INTERNATIONAL SYSTEM OF UNITS

Dennis R. Carter

The need for establishing systems of weights and measures for use in building, making clothes, and engaging in simple trade and commerce was recognized by primitive societies. The Bible and early records from Babylonian and Egyptian civilizations indicate that man first used parts of his body and common elements in his environment to establish simple systems of measurement. Time was commonly measured by periods of the sun. At the time of Noah, length was measured in terms of the cubit, which was equivalent to the distance from the elbow to the tip of the middle finger. A span, equivalent to one half a cubit, was taken as the distance from the tip of the thumb to the tip of the little finger with the hand fully spread. For smaller measurements the digit, or width of the thumb, was used. For measurements of volumes or fluid capacities, containers were filled with seeds, which were then poured out and counted to establish the volume of the container in terms of the number of seeds it could hold.

Primitive societies found that they could determine if one object was heavier than another by placing the two objects on a balance. The Babylonians refined this technique by balancing an object with a set of standard, well-polished stones. These stones were probably the first standards, and the measurement later evolved into the English legal stone, which in the imperial system is equivalent to 14 pounds (62.27 newtons).

The Egyptians and the Greeks both used the wheat seed as the standard for the smallest unit of weight. This concept evolved into the unit of the grain, which is still used today in limited applications. The Arabs established a standard of weights for precious stones and metals based on the weight of a small bean called a carob, a measurement which has evolved into the present unit of the carat.

Increasing trade among tribes and nations caused the measuring systems developed by early civilizations to become intermixed. Measuring standards were widely disseminated by the Romans. As the Roman soldiers marched in their conquests, distances were measured in terms of the pace, which was the distance between the points where one foot struck the ground on successive steps. A pace was therefore equivalent to two steps.

THE ENGLISH SYSTEM

The English system of weights and measures evolved from measuring systems used by the Babylonians, Egyptians, Romans, Anglo-Saxons, and Norman French. Body measurements such as the digit, palm, span, and cubit were replaced by the inch, foot, and yard. The Romans contributed the use of the base unit 12 to the English system, as the Roman foot, *pes,* had 12 divisions called *unciae.* The English words *inch* and *ounce* were derived from this Latin word.

The early Saxon kings wore a sash or girdle around their waists which could be removed conveniently and used for linear measurement. The word *yard* was derived from the Saxon word *gird,* meaning the circumference of a person's waist. In the twelfth century, King Henry I decreed the standard yard to be equivalent to the distance between his nose and the end of his thumb with his arm extended. In the thirteenth century, King Edward I took an important step forward in standardizing linear measurement by ordering a permanent measuring stick made of iron to be used as the standard yard for the kingdom. This measurement stick was coined the "iron ulna" in reference to the bone of the forearm. The measurement of a foot was established as one third the length of the measuring stick, and an inch was established as one-thirty-sixth of the standard yard. King Edward II apparently found this new measuring system to be confusing and passed a statute decreeing an inch to be equivalent to "three barleycorns, round and dry."

The English system of measurement evolved and became refined primarily through royal decrees. The early Tudor rulers established 220 yards as equivalent to 1 furlong. In the sixteenth century, Queen Elizabeth I changed the traditional Roman mile from 5,000 to 5,280 feet, making the mile equivalent to exactly 8 furlongs. In 1824 the English Parliament legalized the new standard yard, which was based on the measurements taken from a brass bar with a gold button near each end. A single dot was engraved on each button, and the yard was taken to be the distance between the two engraved dots.

The English colonization and dominance of world commerce in the seventeenth, eighteenth, and nineteenth centuries served to spread the English system of weights and measures throughout the world, including the American colonies. Although refined in some ways to meet the demands of commerce, the system has remained essentially unchanged.

THE METRIC SYSTEM

In the early eighteenth century no uniform system of weights and measures existed on the European continent. Measurements not only differed from country to country, but also from town to town. During the French Revolution in 1790, King Louis XVI and the National Assembly of France passed a decree calling on the French Academy of Sciences and the Royal Society of London to "deduce an invariable standard for all of the measures and all weights." The English, however, did not participate in this undertaking, and so the French alone set out to establish a new and uniform system of measures. The result was the metric system.

In 1793 the French government adopted a standard system of measurement based on the meter, which was defined as one ten-millionth of the distance from the North Pole to the equator on a line passing through Paris. All other linear measurements were established as decimal fractions of the meter. The metric unit of mass was established as the gram, which was defined as the mass of 1 cubic centimeter of water at its temperature of maximum density. The unit of fluid capacity was named the "litre" and defined as the volume contained by a cube which measured one tenth of a meter on each side.

Most Frenchmen thought that the new metric system of measurement was confusing and expended a great deal of effort converting from the new system to the familiar measuring system of yards and feet that had been used previously. Widespread resistance to the new system forced Napoleon to renounce the metric system in 1812. In 1837, however, France returned to the metric system in the hope that it would spread throughout the world.

In the nineteenth century, the metric system found favor among scientists because (1) it was an international system, (2) the units were independently reproducible, and (3) its use of a decimal system greatly simplified calculations. The British system had been designed for trade and commerce, but the metric system appeared to have significant advantages in the areas of engineering and science.

In the latter half of the nineteenth century, scientific advances necessitated the development of better metric standards. In 1875 an international treaty, the Treaty of the Meter, established well-defined metric standards for length and mass. In addition, the permanent machinery for further refinements of the metric system was established. This treaty was signed by 17 countries. By 1900, 35 nations had officially adopted the metric system.

At the time the Treaty of the Meter was signed, a permanent secretariat, the International Bureau of Weights and Measures, was established in Sèvres, France, to coordinate the exchange of information about the metric system. The diplomatic organization concerned with the metric system is the General Conference of Weights and Measures, which meets periodically to ratify improvements in the system and the standards. In 1960, the general conference adopted an extensive revision and simplification of the system. This modernized metric system was named Le Systeme International d'Unites (International System of Units), and was given the international abbreviation SI. The general conference adopted improvements and additions to the SI system in 1964, 1968, 1971, and 1975.

Between 1965 and 1972, the United Kingdom, Australia, Canada, New Zealand, and many other English-speaking countries adopted the metric system. In 1975, the United States Congress passed the Metric Conversion Act, providing for the adoption of the SI metric system as the predominant system of measurement units. Today, nearly the entire world either is using the metric system or is committed to its adoption.

THE SI METRIC SYSTEM

The Systeme International d'Unites (SI), the modern metric system, has evolved into the most exacting system of measures devised. In this section, the SI units of measurement used in the science of mechanics are described. SI units used in electrical and light sciences have been omitted for the sake of simplicity.

The SI units can be considered in three groups: (1) the base units; (2) the supplementary units; and (3) the derived units (Fig. 1). The base units are a small group of standard measurements which have been arbitrarily defined. The base unit for length is the meter (m), and the base unit of mass is the kilogram (kg). The base units for time and temperature are the second (s) and the kelvin (K), respectively. Definitions of the base units have become increasingly sophisticated in response to the expanding needs and capabilities of the scientific community (Table 1). For example, the meter is now defined in terms of the wavelength of radiation emitted from the krypton-86 atom.

The radian (rad) is a supplementary unit to measure plane angles. This unit, like the base units, is arbitrarily defined (Table 1). Although the radian is the SI unit for plane angle, the unit of the degree has been retained for general use, since it is firmly established and widely used around the world. A degree is equivalent to $\pi/180$ rad.

Most units of the SI system are derived units, meaning that they are established from the base units in accordance with fundamental physical principles. Some of these units are expressed in terms of the base units from which they are derived. Examples are area, speed, and acceleration, which are expressed in the SI units of square meters (m^2), meters per second (m/s), and meters per second squared (m/s^2), respectively.

Other derived units are similarly established from the base units but have been given special names (see Fig. 1 and Table 1). These units are defined through the use of fundamental equations of physical laws in conjunction with the arbitrarily defined SI base units. For example, Newton's second law of motion states that when a body which is free to move is subjected to a force, it will experience an acceleration proportional to that force and inversely proportional to its own mass. Mathematically this principle can be expressed as

$$\text{force} = \text{mass} \times \text{acceleration}.$$

The SI unit of force, the newton (N), is therefore defined in terms of the base SI units as

$$1 \text{ N} = 1 \text{ kg} \times 1 \text{ m/s}^2.$$

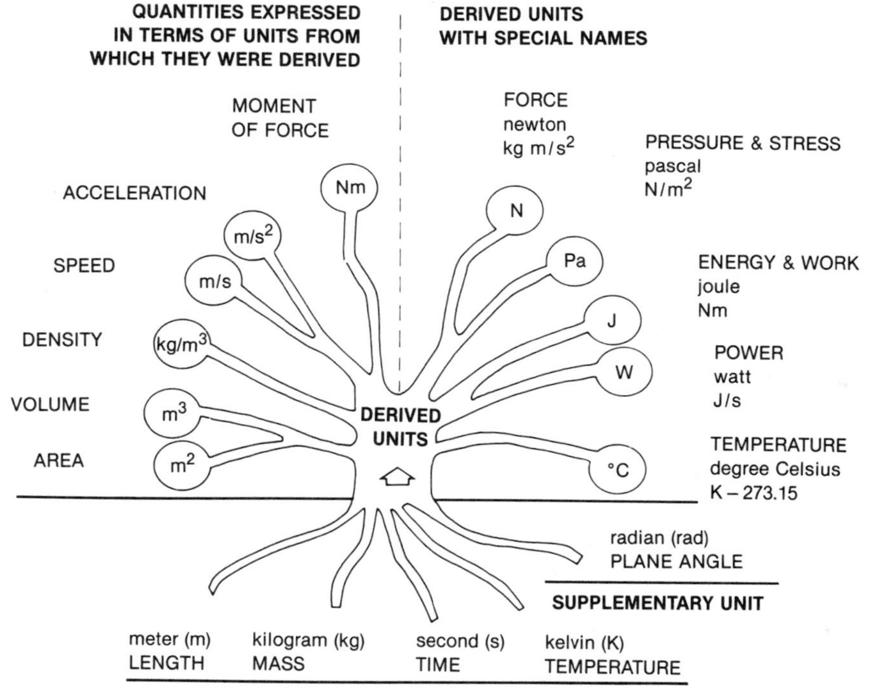

FIG. 1.
The International System of Units.

TABLE 1

DEFINITIONS OF SI UNITS

	BASE SI UNITS
meter (m)	The meter is the length equal to 1,650,763.73 wavelengths in vacuum of the radiation corresponding to the transition between the levels $2p_{10}$ and $5d_5$ of the krypton-86 atom.
kilogram (kg)	The kilogram is the unit of mass and is equal to the mass of the international prototype of the kilogram.
second (s)	The second is the duration of 9,192,631,770 periods of the radiation corresponding to the transition between the two hyperfine levels of the ground state of the cesium-133 atom.
kelvin (K)	The kelvin, a unit of thermodynamic temperature, is the fraction 1/273.16 of the thermodynamic temperature of the triple point of water.
	SUPPLEMENTARY SI UNIT
radian (rad)	The radian is the plane angle between two radii of a circle which subtend on the circumference an arc equal in length to the radius.
	DERIVED SI UNITS WITH SPECIAL NAMES
newton (N)	The newton is that force which, when applied to a mass of one kilogram, gives it an acceleration of one meter per second squared. 1 N = 1 kg m/s^2.
pascal (Pa)	The pascal is the pressure produced by a force of one newton applied, with uniform distribution, over an area of one square meter. 1 Pa = 1 N/m^2.
joule (J)	The joule is the work done when the point of application of a force of one newton is displaced through a distance of one meter in the direction of the force. 1 J = 1 Nm.
watt (W)	The watt is the power which in one second gives rise to the energy of one joule. 1 W = 1 J/s.
degree Celsius (°C)	The degree Celsius is a unit of thermodynamic temperature and is equivalent to K − 273.15.

The SI unit of pressure and stress is the pascal (Pa). Pressure is defined in hydrostatics as the force divided by the area of force application. Mathematically this can be expressed as

$$\text{pressure} = \frac{\text{force}}{\text{area}}.$$

The SI unit of pressure, the pascal (Pa), is therefore defined in terms of the base SI units as

$$1\ \text{Pa} = \frac{1\ \text{N}}{1\ \text{m}^2}.$$

Although the SI base unit of temperature is the kelvin, the derived unit of degree Celsius (°C or c) is much more commonly used. The degree Celsius is equivalent to the kelvin in magnitude, but the absolute value of the Celsius scale differs from that of the Kelvin scale such that °C = K − 273.15.

When the SI system is used in a wide variety of measurements, the quantities expressed in terms of the base, supplemental, or derived units may be either very large or very small. For example, the area on the head of a pin is an extremely small number when expressed in terms of square meters (m^2). On the other hand, the weight of a whale is an extremely large number when expressed in terms of newtons (N). To accommodate the convenient representation of small or large quantities, a system of prefixes has been incorporated into the SI system (Table 2). Each prefix has a fixed meaning and can be used with all SI units. When used with the name of the unit, the prefix indicates that the quantity described is being expressed in some multiple of ten times the unit used. For example, the millimeter (mm) is used to represent one thousandth (10^{-3}) of a meter and a gigapascal (GPa) is used to denote one billion (10^9) pascals.

One of the more interesting aspects of the SI system is its use of the names of famous scientists as standard units. In each case, the unit was named after a scientist in recognition of his contribution to the field in which that unit plays a major role. Table 3 lists a number of SI units and the scientist for which each was named.

The unit of force, the newton, was named in honor of the English scientist Sir Isaac Newton (1624–1727). He was educated at Trinity College at Cambridge and later returned to Trinity College as a professor of mathematics. Early in his career he made fundamental contributions to mathematics which formed the basis of differential and integral calculus. His other major discoveries were in the fields of optics, astronomy, gravitation, and mechanics. His work in gravitation was purportedly spurred by being hit on the head by an apple falling from a tree. It is perhaps poetic justice that the SI unit of one newton is approximately equivalent to the weight of a medium-sized apple. Newton was knighted in 1705 by Queen Mary for his monumental contributions to science.

The unit of pressure and stress, the pascal, was named after the French physicist, mathematician, and philosopher Blaise Pascal (1623–1662). Pascal conducted important investigations on the characteristics of vacuums and barometers and also invented a machine which would make mathematical calculations. His work in the area of hydrostatics and hydrodynamics helped lay the foundation for the later development of these scientific fields. In addition to his scientific pursuits, Pascal was passionately interested in religion and philosophy and thus wrote extensively on a wide range of subjects.

The base unit of temperature, the kelvin, was named in honor of Lord William Thomson Kelvin (1824–1907). He was of Scottish-Irish descent, and his given name was William Thomson. He was educated at the University of Glasgow and Cam-

TABLE 2

SI PREFIXES

PREFIX	SYMBOL	MEANING	EXPONENT OR POWER
exa	E	quintillion	10^{18}
peta	P	quadrillion	10^{15}
tera	T	trillion	10^{12}
giga	G	billion	10^{9}
mega	M	million	10^{6}
kilo	k	thousand	10^{3}
hecto	h	hundred	10^{2}
deka	da	ten	10^{1}
	BASE UNIT		
deci	d	tenth	10^{-1}
centi	c	hundredth	10^{-2}
milli	m	thousandth	10^{-3}
micro	μ	millionth	10^{-6}
nano	n	billionth	10^{-9}
pico	p	trillionth	10^{-12}
femto	f	quadrillionth	10^{-15}
atto	a	quintrillionth	10^{-18}

TABLE 3

SI UNITS NAMED AFTER SCIENTISTS

SYMBOL	UNIT	QUANTITY	SCIENTIST	COUNTRY OF BIRTH	DATES
A	ampere	electric current	Ampere, Andre-Marie	France	1775–1836
C	coulomb	electric charge	Coulomb, Charles Augustin de	France	1736–1806
°C	degree Celsius	temperature	Celsius, Anders	Sweden	1701–1744
F	farad	electric capacity	Faraday, Michael	England	1791–1867
H	henry	inductive resistance	Henry, Joseph	United States	1797–1878
Hz	hertz	frequency	Hertz, Heinrich Rudolph	Germany	1857–1894
J	joule	energy	Joule, James Prescott	England	1818–1889
K	kelvin	temperature	Thomson, William later Lord Kelvin	England	1824–1907
N	newton	force	Newton, Sir Isaac	England	1642–1727
Ω	ohm	electric resistance	Ohm, Georg Simon	Germany	1787–1854
Pa	pascal	pressure/stress	Pascal, Blaise	France	1623–1662
S	siemens	electric conductance	Siemens, Karl Wilhelm, later Sir William	Germany (England)	1823–1883
T	tesla	magnetic flux density	Tesla, Nikola	Croatia (United States)	1856–1943
V	volt	electrical potential	Volta, Count Alessandro	Italy	1745–1827
W	watt	power	Watt, James	Scotland	1736–1819
Wb	weber	magnetic flux	Weber, Wilhelm Eduard	Germany	1804–1891

bridge University and early in his career investigated the thermal properties of steam at a scientific laboratory in Paris. At the age of 32 he returned to the University of Glasgow to accept the chair of Natural Philosophy. His meeting with James Joule in 1847 stimulated interesting discussions on the nature of heat which eventually led to the establishment of Thomson's absolute scale of temperature, the Kelvin scale. In recognition of Thomson's contributions to the field of thermodynamics, King Edward VII conferred on him the title of Lord Kelvin.

The commonly used unit of temperature, the degree Celsius, was named after the Swedish astronomer and inventor Anders Celsius (1701–1744). Celsius was appointed professor of astronomy at the University of Uppsala at the age of 29 and remained at the university until his death 14 years later. In 1742 he described the centigrade thermometer in a paper prepared for the Swedish Academy of Sciences. The name of the centigrade temperature scale was officially changed to Celsius in 1948.

CONVERTING TO SI FROM OTHER UNITS OF MEASUREMENT

Table 4 has been provided to allow conversion of measurements expressed in English and non-SI metric units into SI units. One fundamental source of confusion in converting from one system to another is that two basic types of measurement systems exist. In the "physical" system (such as SI) the units of length, time, and *mass* are arbitrarily defined, and other units (including force) are derived from these base units. In "technical" or "gravitational" systems (such as the English system) the units of length, time, and *force* are arbitrarily defined, and other units (including mass) are derived from these base units. Since the units of force in gravitational systems are in

TABLE 4

CONVERTING TO SI FROM OTHER UNITS

LENGTH			
To convert to meters (m)			
multiply	in	by	0.0254
	ft		0.3048
	yd		0.9144
	mile		1609.3
AREA			
To convert to square meters (m^2)			
multiply	in^2	by	6.4516×10^{-4}
	ft^2		0.0929
	yd^2		0.8361
VOLUME			
To convert to cubic meters (m^3)			
multiply	in^3	by	1.6387×10^{-5}
	ft^3		0.0283
	yd^3		0.7646
(Note: A liter is equivalent to 1 $m^3 \times 10^{-3}$.)			
MASS			
To convert to kilograms (kg)			
multiply	lb	by	0.4536
	slug		14.594
	kgf s^2 m^{-1}		9.8067
DENSITY			
To convert to kg/m^3			
multiply	lb/in^3	by	27,680
	lb/ft^3		16.018
MOMENT OF INERTIA			
To convert to kg m^2 or, equivalently, to Nm sec^2			
multiply	lb ft^2	by	0.0421
	lb in^2		2.9264×10^{-4}
	kgf s^2 m		9.8067
FORCE			
To convert to newtons (N)			
multiply	kgf (=kp)	by	9.8067
	lbf		4.4482
PRESSURE AND STRESS			
To convert to pascals (Pa) or, equivalently, to N/m^2			
multiply	lbf/in^2	by	6894.8
	kgf/m^2		9.8067
	mm Hg		133.32
ENERGY, WORK, AND MOMENT OF FORCE			
To convert to joules (J) or, equivalently, to Nm			
multiply	ft lbf	by	1.3558
	in lbf		0.1130
	kgf m		9.8067
	Btu		1055.1
	cal		4.1868
TIME			
To convert to seconds (s)			
multiply	min	by	60
	h		3600
	d		86,400
SPEED			
To convert to m/s			
multiply	ft/s	by	0.3048
	km/h		0.2778
	mile/h		0.4470
ACCELERATION			
To convert to m/s^2			
multiply	ft/s^2	by	0.3048
TEMPERATURE			
To convert to degrees Celsius			
from deg K:	deg C = deg K − 273.15		
from deg F:	deg C = 5/9 (deg F − 32)		
PLANE ANGLE			
To convert to radians (rad)			
multiply	deg	by	π/180 (or 0.01745)
	rev		2π (or 6.2832)

fact the *weights* of standard masses, conversion to SI is dependent upon the acceleration of mass due to the Earth's gravity. By international agreement the acceleration due to gravity is 9.806650 m/s^2. This value has been used in establishing some of the conversion factors in Table 4.

BIBLIOGRAPHY

Feirer, J. L.: SI Metric Handbook. New York, Charles Scribner's Sons, 1977.

Pennycuick, C. J.: Handy Matrices of Unit Conversion Factors for Biology and Mechanics. New York, John Wiley and Sons, 1974.

World Health Organization. The SI for the Health Professions. Geneva, World Health Organization, 1977.

part one

BIOMECHANICS OF TISSUES AND STRUCTURES OF THE MUSCULOSKELETAL SYSTEM

1

BIOMECHANICS OF BONE

Margareta Nordin
Victor H. Frankel

The purpose of the skeletal system is to protect internal organs, provide rigid kinematic links and muscle attachment sites, and facilitate muscle action and body movement. Bone has unique structural and mechanical properties that allow it to carry out these roles. Bone is among the body's hardest structures, only dentin and enamel in the teeth being harder. It is one of the most dynamic and metabolically active tissues in the body and remains active throughout life. A highly vascular tissue, it has an excellent capacity for self-repair and can alter its properties and configuration in response to changes in mechanical demand. For example, changes in bone density are commonly observed after periods of disuse and of greatly increased use; changes in bone shape are noted during fracture healing and after certain operations. Thus, bone adapts to the mechanical demands placed on it.

This chapter describes the composition and structure of bone tissue, the mechanical properties of bone, and the behavior of bone under different loading conditions. Various factors that affect the mechanical behavior of bone in vitro and in vivo are also discussed.

BONE COMPOSITION AND STRUCTURE

Bone tissue is a specialized connective tissue whose solid composition suits it for its supportive and protective roles. Like other connective tissues, it consists of cells and an organic extracellular matrix of fibers and ground substance produced by the cells. The distinguishing feature of bone is its high content of inorganic materials, in the form of mineral salts, that combine intimately with the organic matrix. The inorganic component of bone makes the tissue hard and rigid, while the organic component gives bone its flexibility and resilience.

The mineral portion of bone consists primarily of calcium and phosphate, mainly in the form of small crystals resembling synthetic hydroxyapatite crystals with the composition $Ca_{10}(PO_4)_6(OH)_2$. These minerals, which account for 65 to 70% of the bone's dry weight, give bone its solid consistency. Bone serves as a reservoir for essential minerals in the body, particularly calcium.

Bone mineral is embedded in variously oriented fibers of the protein collagen, the fibrous portion of the extracellular matrix. Collagen fibers are tough and pliable, yet they resist stretching and have little extensibility. Collagen composes approximately 95% of the extracellular matrix and accounts for about 25 to 30% of the dry weight of bone. A universal building block of the body, collagen is also the chief fibrous component of other skeletal structures. (A detailed description of the microstructure and mechanical behavior of collagen is provided in Chapters 2 and 3.)

The gelatinous ground substance surrounding the mineralized collagen fibers consists mainly of protein polysaccharides, or glycosaminoglycans (GAGs), primarily in the form of complex macromolecules called proteoglycans (PGs). The GAGs serve as a cementing substance between layers of mineralized collagen fibers. These GAGs, along with various noncollagenous glycoproteins, constitute about 5% of the extracellular matrix. (The structure of PGs, which are vital components of articular cartilage, is described in detail in Chapter 2.)

Water is fairly abundant in live bone, accounting for up to 25% of its total weight. About 85% of the water is found in the organic matrix, around the collagen fibers and ground substance, and in the hydration shells surrounding the bone crystals. The other 15% is located in canals and cavities that house bone cells and carry nutrients to the bone tissue.

At the microscopic level, the fundamental structural unit of bone is the osteon, or haversian system (Fig. 1–1). At the center of each osteon is a small channel, called a haversian canal, that contains blood vessels and nerve fibers. The osteon itself consists of a concentric series of layers (lamellae) of mineralized matrix surrounding the central canal, a configuration similar to growth rings in a tree trunk.

Along the boundaries of each layer, or lamella, are small cavities known as lacunae, each containing one bone cell, or osteocyte (see Fig. 1–1C). Numerous small channels, called canaliculi, radiate from each lacuna, connecting the lacunae of adjacent lamellae and ultimately reaching the haversian canal. Cell processes extend from the osteocytes into the canaliculi, allowing nutrients from the blood vessels in the haversian canal to reach the osteocytes.

At the periphery of each osteon is a cement line, a narrow area of cementlike ground substance composed primarily of glycosaminoglycans. The canaliculi of the osteon do not pass this cement line. Like the canaliculi, the collagen fibers in the bone matrix interconnect from one lamella to another within an osteon but do not cross the cement line. This intertwining of collagen fibers within the osteon undoubtedly increases the bone's resistance to mechanical stress and probably explains why the cement line is the weakest portion of the bone's microstructure (Dempster and Coleman, 1960; Evans and Bang, 1967).

A typical osteon is about 200 micrometers (μm) in diameter. Hence, every point in the osteon is no more than 100 μm from the centrally located blood supply. In the long bones, the osteons usually run longitudi

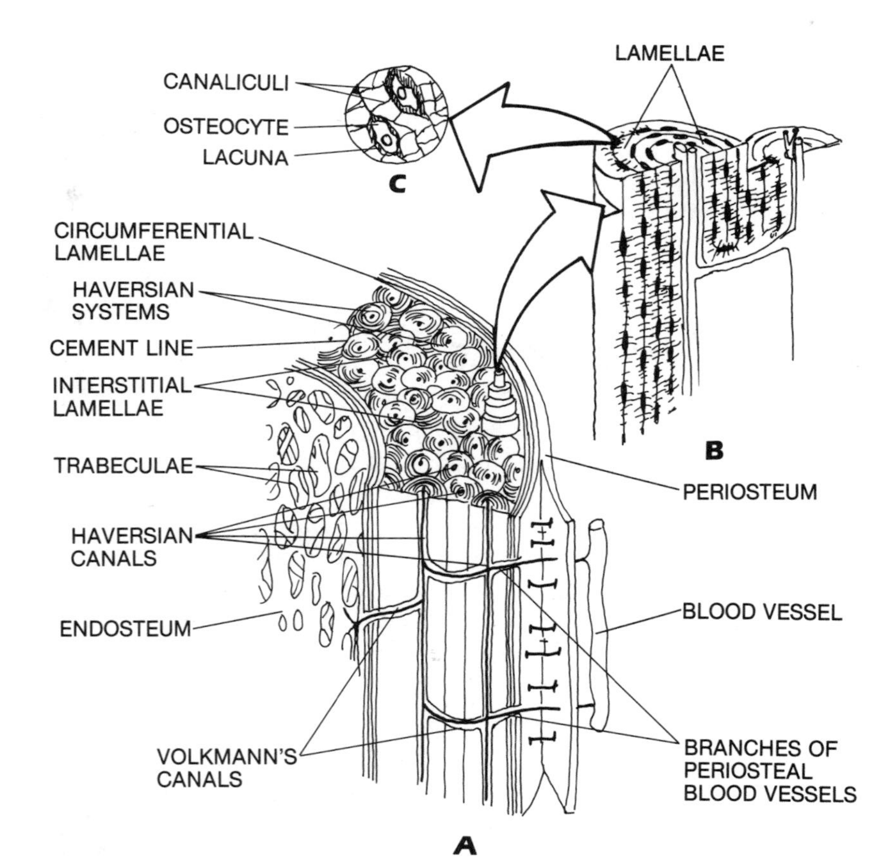

FIG. 1–1
A. The fine structure of bone is illustrated schematically in a section of the shaft of a long bone depicted without inner marrow. The osteons, or haversian systems, are apparent as the structural units of bone. In the center of the osteons are the haversian canals, which form the main branches of the circulatory network in bone. Each osteon is bounded by a cement line. One osteon is shown extending from the bone (20×). (Adapted from Bassett, 1965.) **B.** Each osteon consists of lamellae, concentric rings composed of mineral matrix surrounding the haversian canal. (Adapted from Tortora and Anagnostakos, 1984.) **C.** Along the boundaries of the lamellae are small cavities known as lacunae, each of which contains a single bone cell, or osteocyte. Radiating from the lacunae are tiny canals, or canaliculi, into which the cytoplasmic processes of the osteocytes extend. (Adapted from Tortora and Anagnostakos, 1984.)

nally, but they branch frequently and anastomose extensively with each other.

Interstitial lamellae span the regions between complete osteons (see Fig. 1–1A). They are continuous with the osteons and are just the same material in a different geometric configuration. As in the osteons, no point in the interstitial lamellae is farther than 100 μm from its blood supply. The interfaces between these lamellae contain an array of lacunae in which osteocytes lie and from which canaliculi extend.

At the macroscopic level, all bones are composed of two types of osseous tissue: cortical, or compact, bone and cancellous, or trabecular, bone (Fig. 1–2). Cortical bone forms the outer shell, or cortex, of the bone and has a dense structure similar to that of ivory. Cancellous bone within this shell is composed of thin plates, or trabeculae, in a loose mesh structure; the interstices between the trabeculae are filled with red marrow. Cancellous bone tissue is arranged in concentric lacunae-containing lamellae, but it does not contain haversian canals. The osteocytes receive nutrients through canaliculi from blood vessels passing through the red marrow. Cortical bone always surrounds cancellous bone, but the relative quantity of each type varies among bones and within individual bones according to functional requirements.

Since the lamellar pattern and material composition of cancellous and cortical bone appear identical, the basic distinction between the two is the degree of porosity. Biomechanically, the two bone types can be considered as one material whose porosity and density vary over a wide range (Carter and Hayes, 1977b). The difference in the porosity of cortical and cancellous bone can be seen in cross sections from human tibiae (Fig. 1–3). The porosity ranges from 5 to 30% in cortical bone and from 30 to over 90% in cancellous bone. The distinction between porous cortical bone and dense cancellous bone is somewhat arbitrary.

All bones are surrounded by a dense fibrous membrane called the periosteum (see Fig. 1–1A). Its outer layer is permeated by blood vessels and nerve fibers that pass into the cortex via Volkmann's canals, connecting with the haversian canals and extending to the cancellous bone. An inner, osteogenic layer contains bone cells responsible for generating new bone during growth and repair (osteoblasts). The periosteum covers the entire bone except for the joint surfaces, which are covered with articular cartilage. In the long bones, a thinner membrane, the endosteum, lines the central (medullary) cavity, which is filled with yellow fatty marrow. The endosteum

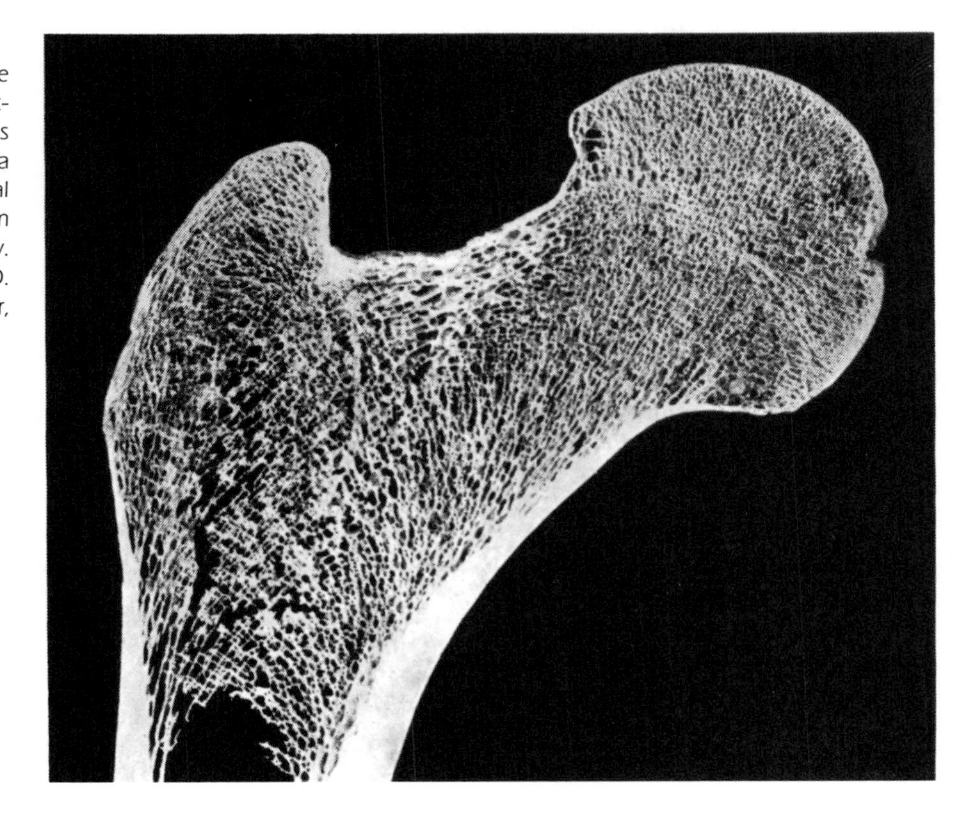

FIG. 1–2
Frontal longitudinal section through the head, neck, greater trochanter, and proximal shaft of an adult femur. Cancellous bone, with its trabeculae oriented in a lattice, lies within the shell of cortical bone. (Reprinted with permission from Gray, H.: Anatomy of the Human Body. 13th American Ed. Edited by C. D. Clemente. Philadelphia, Lea & Febiger, 1985.)

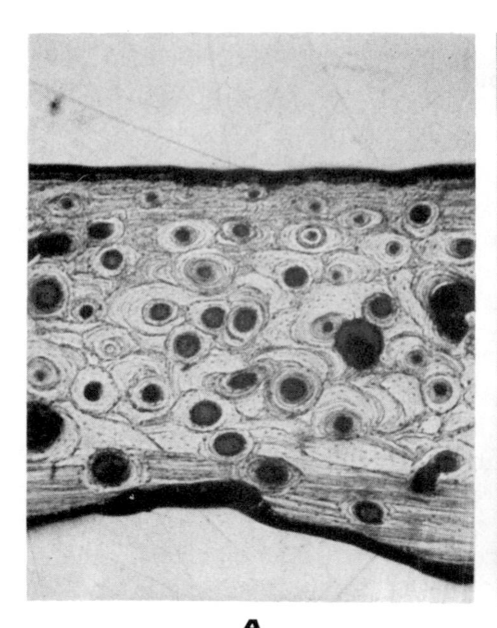

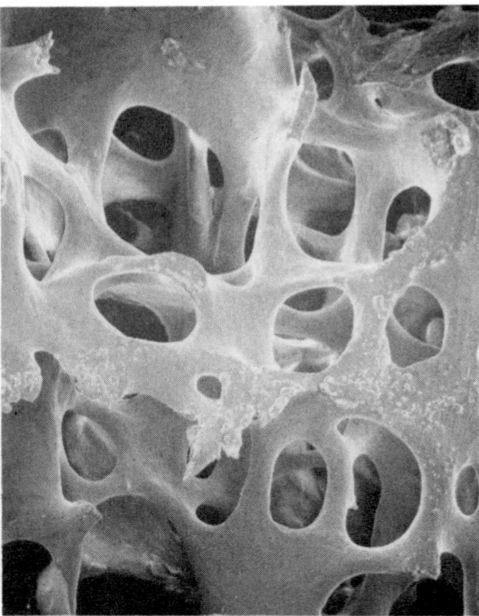

FIG. 1–3
A. Reflected-light photomicrograph of cortical bone from a human tibia (40×). (Courtesy of Dennis R. Carter, Ph.D.) **B.** Scanning electron photomicrograph of cancellous bone from a human tibia (30×). (Courtesy of Dennis R. Carter, Ph.D.)

contains osteoblasts and also giant multinucleated bone cells called osteoclasts, which play a role in the resorption of bone.

BIOMECHANICAL PROPERTIES OF BONE

Biomechanically, bone tissue may be regarded as a two-phase (biphasic) composite material, with the mineral as one phase and the collagen and ground substance as the other. In such materials (a nonbiologic example is fiberglass)— in which a strong, brittle material is embedded in a weaker, more flexible one— the combined substances are stronger for their weight than either substance alone (Bassett, 1965).

Functionally, the most important mechanical properties of bone are its strength and stiffness. These and other characteristics can best be understood for bone, or any other structure, by examining its behavior under loading, i.e., under the influence of externally applied forces. Loading causes a deformation, or a change in the dimensions, of the structure. When a load in a known direction is imposed on a structure, the deformation of that structure can be measured and plotted on a load-deformation curve. Much information about the strength, stiffness, and other mechanical properties of the structure can be gained by examining this curve.

A hypothetical load-deformation curve for a somewhat pliable fibrous structure, such as a long bone, is shown in Figure 1–4. The initial (straight line) portion of the curve, the elastic region, reveals the elasticity of the structure, i.e., its capacity for return-

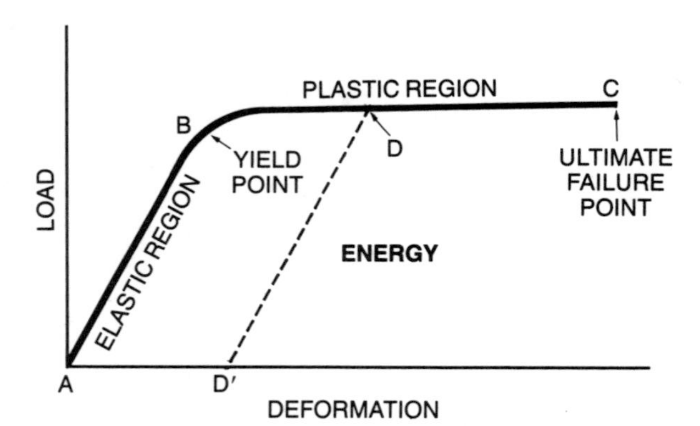

FIG. 1–4
Load-deformation curve for a structure composed of a somewhat pliable material. If a load is applied within the elastic range of the structure (A to B on the curve) and is then released, no permanent deformation occurs. If loading is continued past the yield point (B) and into the structure's plastic range (B to C on the curve) and the load is then released, permanent deformation results. The amount of permanent deformation that occurs if the structure is loaded to point D in the plastic region and then unloaded is represented by the distance between A and D′. If loading continues within the plastic range, an ultimate failure point (C) is reached.

ing to its original shape after the load is removed. As the load is applied, deformation occurs but is not permanent; the structure recovers its original shape when unloaded. As loading continues, the outermost fibers of the structure begin to yield at some point. This yield point signals the elastic limit of the structure. As the load exceeds this limit, the structure exhibits plastic behavior, reflected in the second (curved) portion of the curve, the plastic region. The structure will no longer return to its original dimensions when the load has been released; some residual deformation will be permanent. If loading is progressively increased, the structure will fail at some point (bone will fracture). This point is indicated by the ultimate failure point on the curve.

Three parameters for determining the strength of a structure are reflected on the load-deformation curve: (1) the load that the structure can sustain before failing, (2) the deformation that it can sustain before failing, and (3) the energy that it can store before failing. The strength in terms of load and deformation, or ultimate strength, is indicated on the curve by the ultimate failure point. The strength in terms of energy storage is indicated by the size of the area under the entire curve. The larger the area is, the greater the energy that builds up in the structure as the load is applied. The stiffness of the structure is indicated by the slope of the curve in the elastic region. The steeper the slope is, the stiffer the material.

The load-deformation curve is useful for determining the mechanical properties of whole structures such as a whole bone, an entire ligament or tendon, or a metal implant. This knowledge is helpful in the study of fracture behavior and repair, the response of a structure to physical stress, or the effect of various treatment programs; however, characterizing a bone or other structure in terms of the material that composes it, independent of its geometry, requires standardization of the testing conditions and the size and shape of the test specimens. Such standardized testing is useful for comparing the mechanical properties of two or more materials, such as the relative strength of bone and tendon tissue or the relative stiffness of various materials used in prosthetic implants. More precise units of measure can be used when standardized samples are tested, i.e., the load per unit of area of the sample (stress) and the amount of deformation in terms of the percentage of change in the sample's dimensions (strain). The curve generated is a stress-strain curve.

Stress is the load, or force, per unit area that develops on a plane surface within a structure in response to externally applied loads. The three units most commonly used for measuring stress in standardized samples of bone are newtons per centimeter squared (N/cm^2); newtons per meter squared, or pascals (N/m^2, Pa); and meganewtons per meter squared, or megapascals (MN/m^2, MPa).

Strain is the deformation (change in dimension) that develops within a structure in response to externally applied loads. The two basic types of strain are linear strain, which causes a change in the length of the specimen, and shear strain, which causes a change in the angular relationships within the structure. Linear strain is measured as the amount of linear deformation (lengthening or shortening) of the sample divided by the sample's original length. It is a nondimensional parameter expressed as a percentage (for example, centimeter per centimeter). Shear strain is measured as the amount of angular change (γ) in a right angle lying in the plane of interest in the sample. It is expressed in radians (one radian equals approximately 57.3 degrees) (International Society of Biomechanics, 1988).

Stress and strain values can be obtained for bone by placing a standardized specimen of bone tissue in a testing jig and loading it to failure (Fig. 1–5). These values can then be plotted on a stress-strain curve

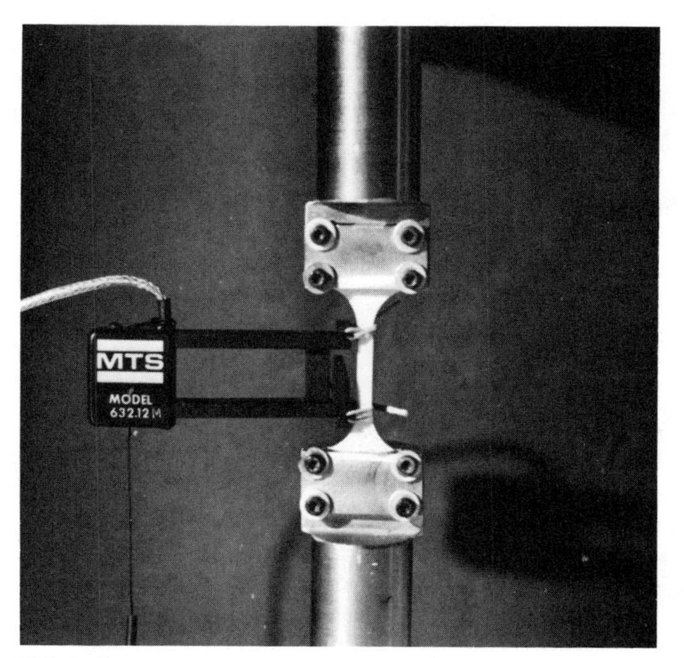

FIG. 1–5
Standardized bone specimen in a testing machine. The strain in the segment of bone between the two gauge arms is measured with a strain gauge. The stress is calculated from the total load measured. (Courtesy of Dennis R. Carter, Ph.D.)

(Fig. 1–6). The regions of this curve are similar to those of the load-deformation curve. Loads in the elastic region do not cause permanent deformation, but once the yield point is exceeded, some deformation is permanent. The strength of the material in terms of energy storage is represented by the area under the entire curve. The stiffness is represented by the slope of the curve in the elastic region. A value for stiffness is obtained by dividing the stress at any point in the elastic (straight line) portion of the curve by the strain at that point. This value is called the modulus of elasticity (Young's modulus). Stiffer materials have higher moduli.

Mechanical properties differ in the two bone types. Cortical bone is stiffer than cancellous bone, withstanding greater stress but less strain before failure. Cancellous bone in vitro does not fracture until the strain exceeds 75%, but cortical bone fractures when the strain exceeds 2%. Because of its porous structure, cancellous bone has a large capacity for energy storage (Carter and Hayes, 1976).

Stress-strain curves for cortical bone, metal, and glass illustrate the differences in mechanical behavior among these materials (Fig. 1–7). The variations in stiffness are reflected in the different slopes of the curves in the elastic region. Metal has the steepest slope and is thus the stiffest material.

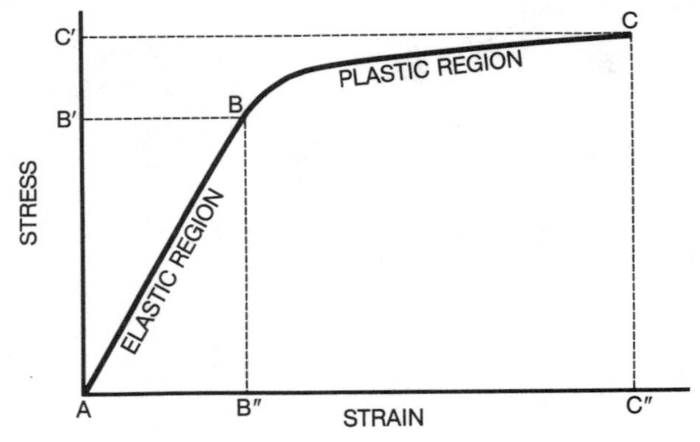

FIG. 1–6
Stress-strain curve for a cortical bone sample tested in tension (pulled). *Yield point (B):* point past which some permanent deformation of the bone sample occurred. *Yield stress (B'):* load per unit area sustained by the bone sample before plastic deformation took place. *Yield strain (B"):* amount of deformation withstood by the sample before plastic deformation occurred. The strain at any point in the elastic region of the curve is proportional to the stress at that point. *Ultimate failure point (C):* the point past which failure of the sample occurred. *Ultimate stress (C'):* load per unit area sustained by the sample before failure. *Ultimate strain (C"):* amount of deformation sustained by the sample before failure.

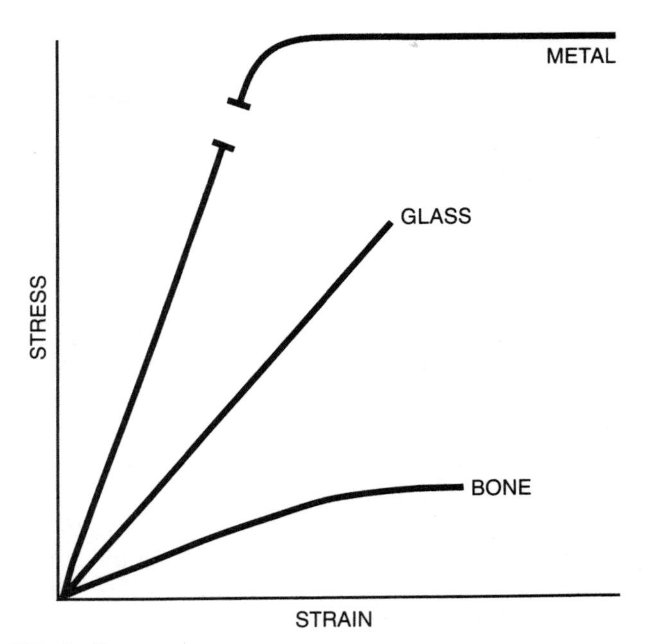

FIG. 1–7
Stress-strain curves for three materials. Metal has the steepest slope in the elastic region and is thus the stiffest material. The elastic portion of the curve for metal is a straight line, indicating linearly elastic behavior. The fact that metal has a long plastic region indicates that this typical ductile material deforms extensively before failure. Glass, a brittle material, exhibits linearly elastic behavior but fails abruptly with little deformation, as indicated by the lack of a plastic region on the stress-strain curve. Cortical bone, which possesses both ductile and brittle qualities, exhibits nonlinear elastic behavior. This behavior is demonstrated by a slight curve in the elastic region, which indicates some yielding during loading within this region. Cortical bone continues to deform before failure but to a lesser extent than does metal.

The elastic portion of the curve for glass and metal is a straight line, indicating linearly elastic behavior; virtually no yielding takes place before the yield point is reached. By comparison, precise testing of cortical bone has shown that the elastic portion of the curve is not straight but is slightly curved, indicating that bone is not linearly elastic in its behavior but yields somewhat during loading in the elastic region (Bonefield and Li, 1967).

After the yield point is reached, glass deforms very little before failing, as indicated by the absence of a plastic region on the stress-strain curve. By contrast, metal exhibits extensive deformation before failing, as indicated by a long plastic region on the curve. Bone also deforms before failing but to a much lesser extent than metal. The difference in the plastic behavior of metal and bone is due to differences in micromechanical events at yield. Yielding in metal (tested in tension, or pulled) is caused by plastic flow and formation of plastic slip lines; slip lines are formed when the molecules of the lattice structure of metal dislocate. Yielding in bone (tested in tension) is caused by debonding of the osteons at the cement lines and microfracture.

Materials are classified as brittle or ductile depending on the extent of deformation before failure. Glass is a typical brittle material, and soft metal is a typical ductile material. The difference in the amount of deformation is reflected in the fracture surfaces of the two materials (Fig. 1–8). When pieced together after fracture, the ductile material will not conform to its original shape whereas the brittle material will. Bone exhibits more brittle or more ductile behavior depending on its age (younger bone being more ductile) and the rate at which it is loaded (bone being more brittle at higher loading speeds).

Because the structure of bone is dissimilar in the transverse and the longitudinal directions, it exhibits different mechanical properties when loaded along different axes, a characteristic known as anisotropy. Figure 1–9 shows the variations in strength and stiffness for cortical bone samples from a human femoral shaft, tested in tension in four directions (Frankel and Burstein, 1970). The values for both parameters are highest for the samples loaded in the longitudinal direction. Although the relationship between loading patterns and the mechanical properties of bone throughout the skeleton is extremely complex, it can generally be said that bone strength and stiffness are greatest in the direction in which loads are most commonly imposed (Frankel and Burstein, 1970).

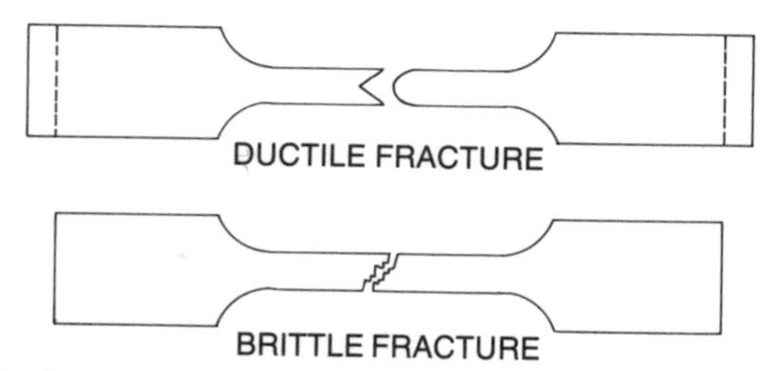

FIG. 1–8
Fracture surfaces of samples of a ductile and a brittle material. The broken lines on the ductile material indicate the original length of the sample, before it deformed. The brittle material deformed very little before fracture.

BIOMECHANICAL BEHAVIOR OF BONE

The mechanical behavior of bone—its behavior under the influence of forces and moments—is affected by its mechanical properties, its geometric characteristics, the loading mode applied, the rate of loading, and the frequency of loading.

BONE BEHAVIOR UNDER VARIOUS LOADING MODES

Forces and moments can be applied to a structure in various directions, producing tension, compression, bending, shear, torsion, and combined loading (Fig. 1–10). Bone in vivo is subjected to all of these loading modes. The following descriptions of these modes apply to structures in equilibrium (at rest or moving at a constant speed); loading produces an internal, deforming effect on the structure.

Tension

During tensile loading, equal and opposite loads are applied outward from the surface of the structure, and tensile stress and strain result inside the structure. Tensile stress can be thought of as many small forces directed away from the surface of the structure. Maximal tensile stress occurs on a plane perpendicular to the applied load (Fig. 1–11). Under tensile loading, the structure lengthens and narrows. At the microscopic level, the failure mechanism for bone tissue loaded in tension is mainly

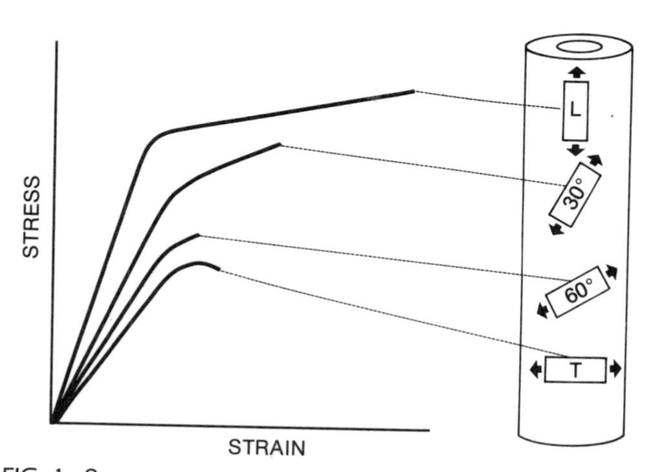

FIG. 1–9
Anisotropic behavior of cortical bone specimens from a human femoral shaft tested in tension (pulled) in four directions: longitudinal (L), tilted 30 degrees with respect to the neutral axis of the bone, tilted 60 degrees, and transverse (T). (Data from Frankel and Burstein, 1970.)

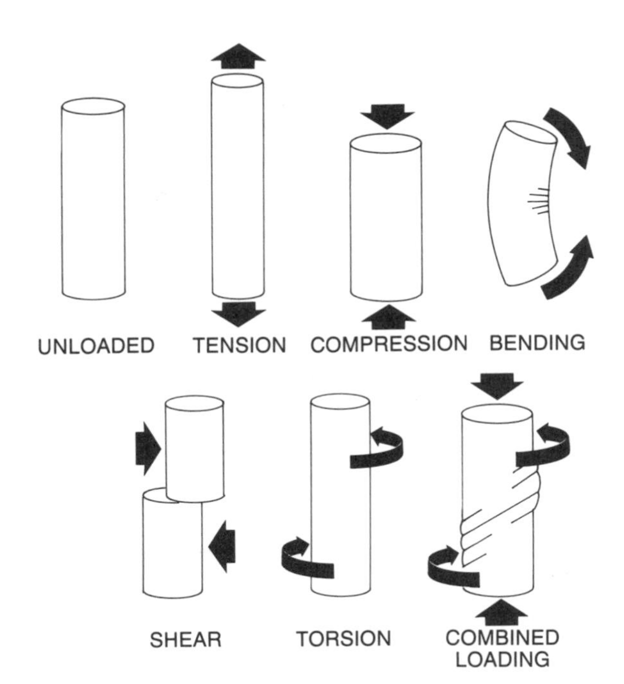

FIG. 1–10
Schematic representation of various loading modes.

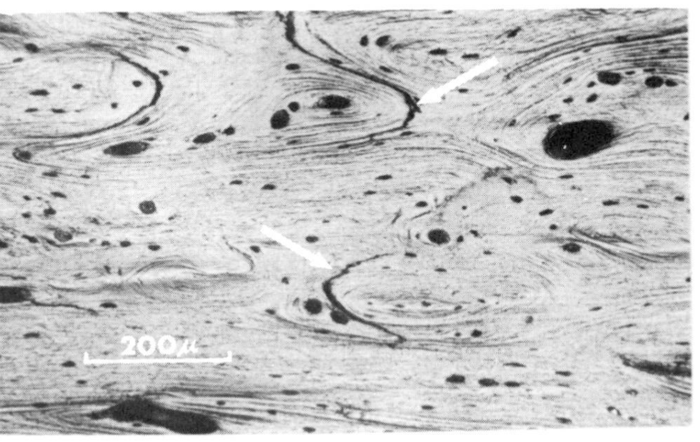

FIG. 1–12
Reflected light photomicrograph of a human cortical bone specimen tested in tension (30×). Arrows indicate debonding at the cement lines and pulling out of the osteons. (Courtesy of Dennis R. Carter, Ph.D.)

debonding at the cement lines and pulling out of the osteons (Fig. 1–12).

Clinically, fractures produced by tensile loading are usually seen in bones with a large proportion of cancellous bone. Examples are fractures of the base of the fifth metatarsal adjacent to the attachment of the peroneus brevis tendon and fractures of the calcaneus adjacent to the attachment of the Achilles tendon. Figure 1–13 shows a tensile fracture through the calcaneus; intense contraction of the triceps surae muscle produced abnormally high tensile loads on the bone.

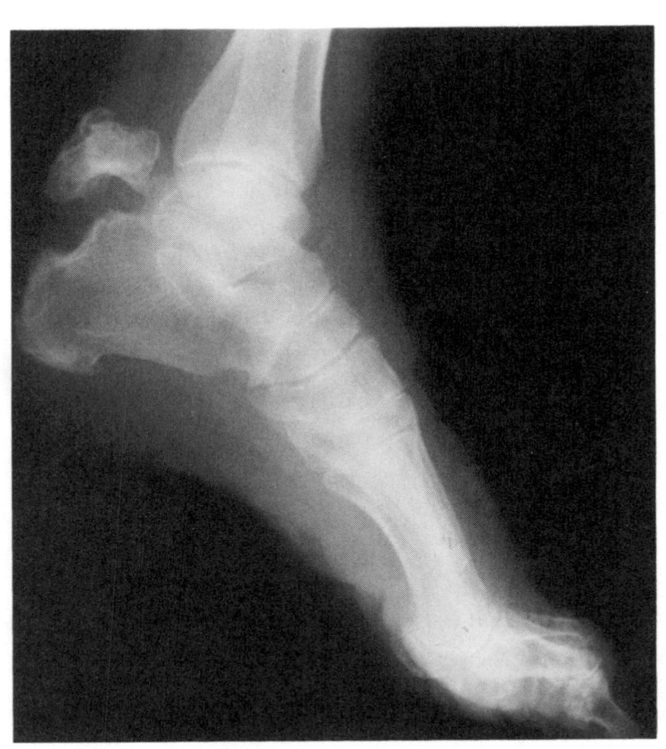

FIG. 1–13
Tensile fracture through the calcaneus produced by strong contraction of the triceps surae muscle during a tennis match. (Courtesy of Robert A. Winquist, M.D.)

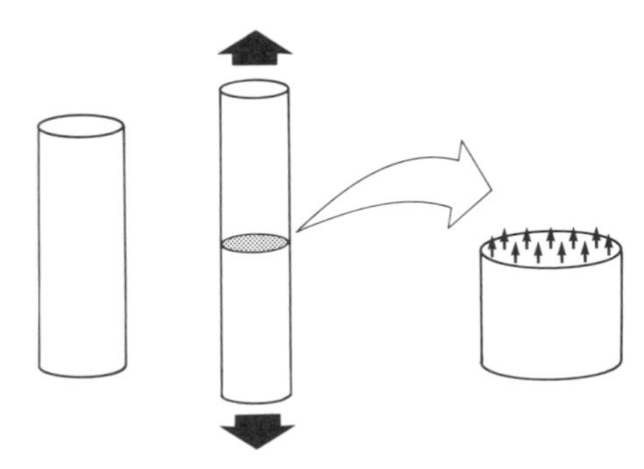

FIG. 1–11
Tensile loading.

Compression

During compressive loading, equal and opposite loads are applied toward the surface of the structure and compressive stress and strain result inside the structure. Compressive stress can be thought of as many small forces directed into the

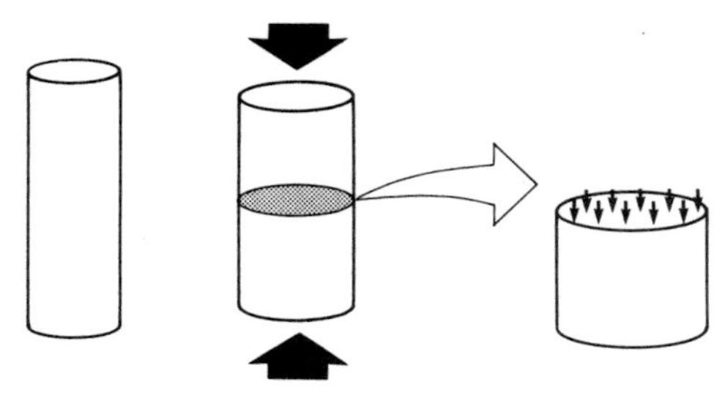

FIG. 1–14
Compressive loading.

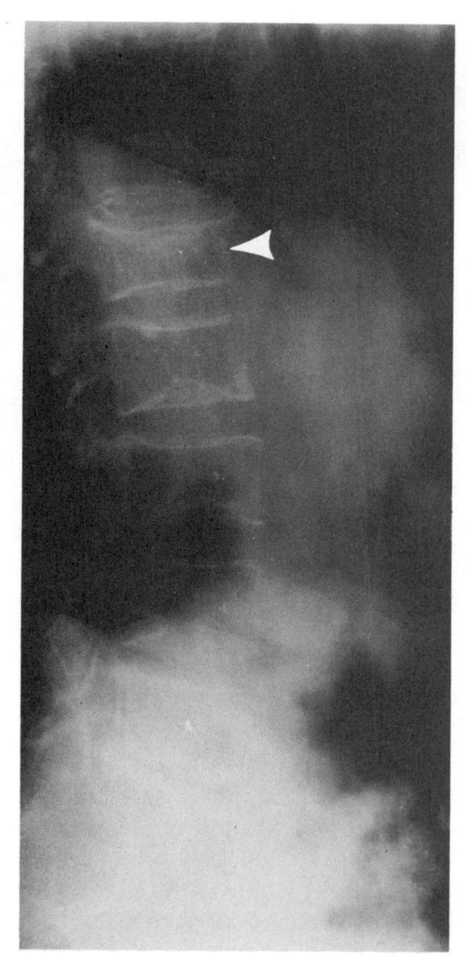

FIG. 1–16
Compression fracture of a human first lumbar vertebra. The vertebra has shortened and widened.

surface of the structure. Maximal compressive stress occurs on a plane perpendicular to the applied load (Fig. 1–14). Under compressive loading the structure shortens and widens. At the microscopic level, the failure mechanism for bone tissue loaded in compression is mainly oblique cracking of the osteons (Fig. 1–15).

Clinically, compression fractures are commonly found in the vertebrae, which are subjected to high compressive loads. These fractures are most often seen in the elderly, whose bones weaken as a function of aging. Figure 1–16 shows the shortening and widening that took place in a human vertebra subjected to a high compressive load. In a joint, compressive loading to failure can be produced by abnormally strong contraction of the surrounding muscles. An example of this effect is presented in Figure 1–17; bilateral subcapital fractures of the femoral neck were sustained by a patient undergoing electroconvulsive therapy; strong contractions of the muscles around the hip joint compressed the femoral head against the acetabulum.

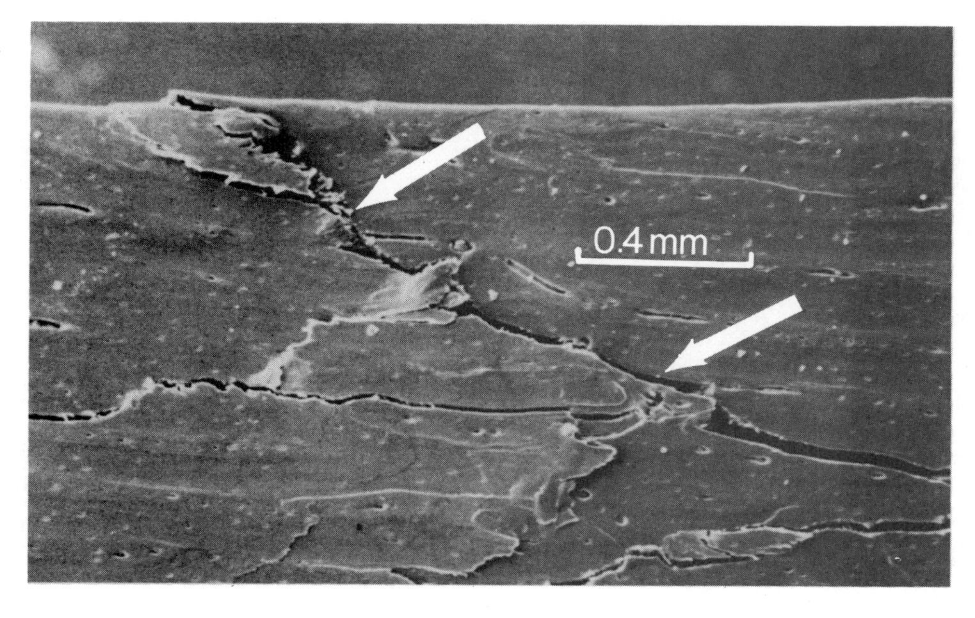

FIG. 1–15
Scanning electron photomicrograph of a human cortical bone specimen tested in compression (30×). Arrows indicate oblique cracking of the osteons. (Courtesy of Dennis R. Carter, Ph.D.)

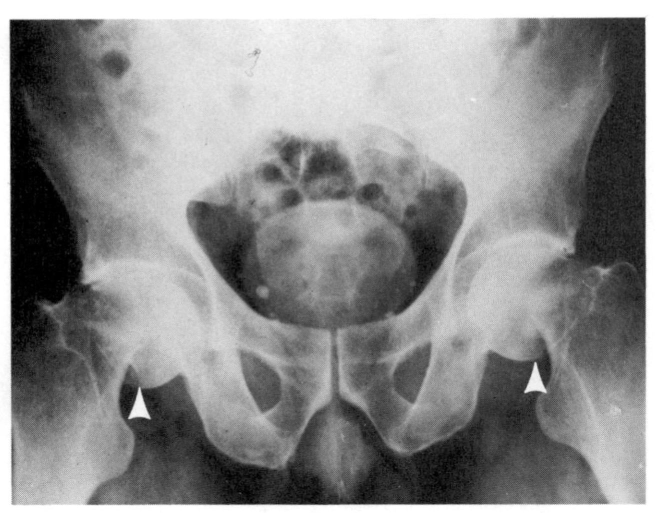

FIG. 1–17
Bilateral subcapital compression fractures of the femoral necks in a patient who underwent electroconvulsive therapy.

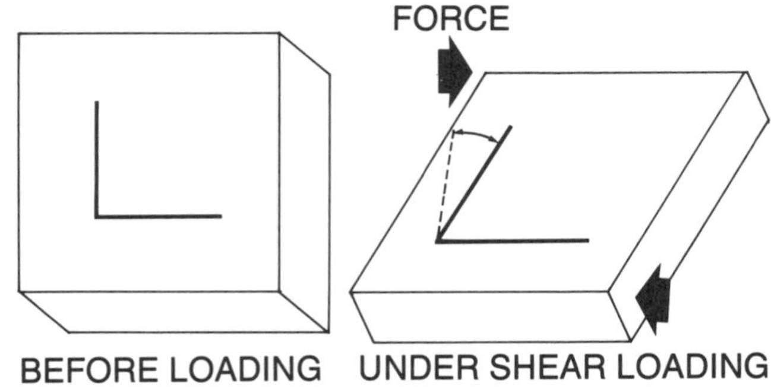

FIG. 1–19
When a structure is loaded in shear, lines originally at right angles on a plane surface within the structure change their orientation, and the angle becomes obtuse or acute. This angular deformation indicates shear strain. (Adapted from Frankel and Burstein, 1970.)

Shear

During shear loading, a load is applied parallel to the surface of the structure, and shear stress and strain result inside the structure. Shear stress can be thought of as many small forces acting on the surface of the structure on a plane parallel to the applied load (Fig. 1–18). A structure subjected to a shear load deforms internally in an angular manner; right angles on a plane surface within the structure become obtuse or acute (Fig. 1–19). Whenever a structure is subjected to tensile or compressive loading, shear stress is produced. Figure 1–20 illustrates angular deformation in structures subjected to these loading modes.

Clinically, shear fractures are most often seen in cancellous bone. Examples are fractures of the femoral condyles and the tibial plateau. A shear fracture of the tibial plateau is shown in Figure 1–21.

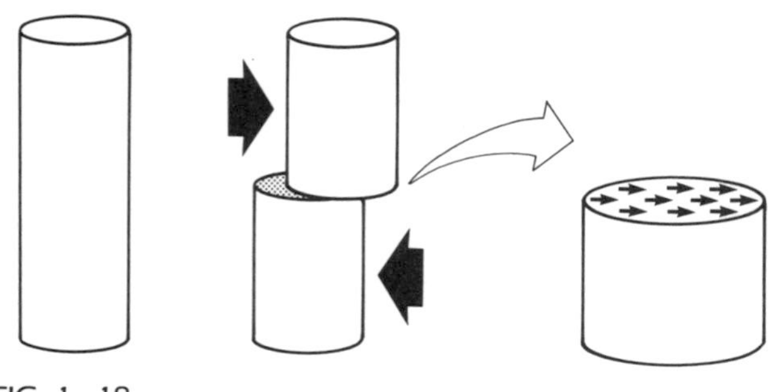

FIG. 1–18
Shear loading.

Human adult cortical bone exhibits different values for ultimate stress under compressive, tensile, and shear loading (Fig. 1–22). Cortical bone can withstand greater stress in compression than in tension and greater stress in tension than in shear (Reilly and Burstein, 1975). The value for the stiffness of a material under shear loading is known as the shear modulus rather than the modulus of elasticity.

Bending

In bending, loads are applied to a structure in a manner that causes it to bend about an axis. When a bone is loaded in bending, it is subjected to a combination of tension and compression. Tensile

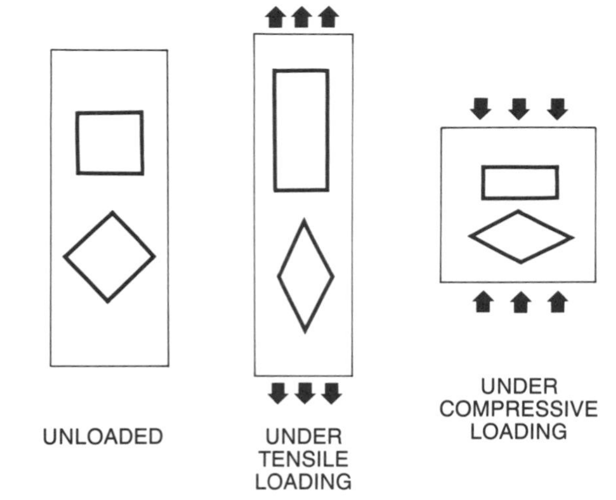

FIG. 1–20
The presence of shear strain in a structure loaded in tension and in compression is indicated by angular deformation. (Adapted from Frankel and Burstein, 1970.)

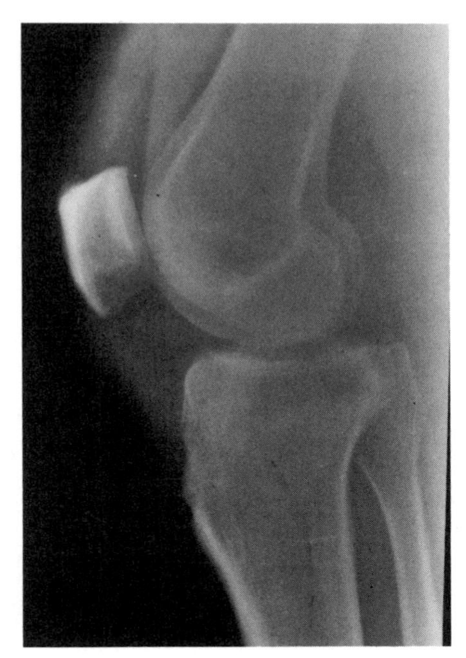

FIG. 1–21
Shear and compression fracture of the lateral tibial plateau. (Courtesy of Steven Lubin, M.D.)

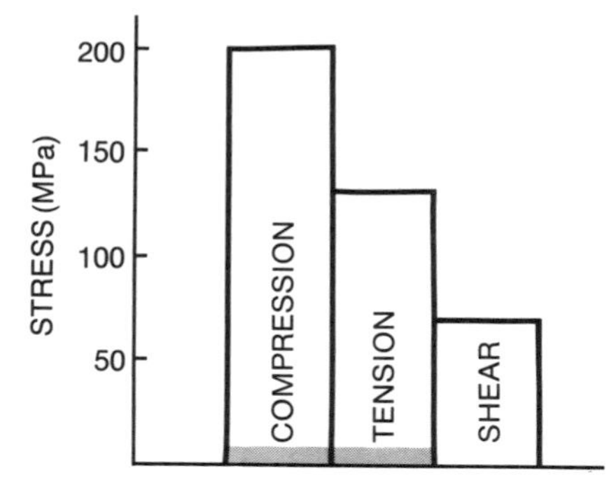

FIG. 1–22
Ultimate stress for human adult cortical bone specimens tested in compression, tension, and shear (average of data from Reilly and Burstein, 1975). Shaded area indicates ultimate stress for human adult cancellous bone with an apparent density of 35% tested in tension and compression (Carter, 1979).

stresses and strains act on one side of the neutral axis, and compressive stresses and strains act on the other side (Fig. 1–23); there are no stresses and strains along the neutral axis. The magnitude of the stresses is proportional to their distance from the neutral axis of the bone. The farther the stresses are from the neutral axis, the higher their magnitude. Because bone is asymmetrical, the tensile and compressive stresses may not be equal.

Bending may be produced by three forces (three-point bending) or four forces (four-point bending) (Fig. 1–24). Fractures produced by both types of bending are commonly observed clinically, particularly in long bones.

Three-point bending takes place when three forces acting on a structure produce two equal moments, each being the product of one of the two peripheral forces and its perpendicular distance from the axis of rotation (the point at which the middle force is applied) (see Fig. 1–24A). If loading continues to the yield point, the structure, if homogeneous and symmetrical, will break at the point of application of the middle force.

A typical three-point bending fracture is the "boot top" fracture sustained by skiers. In the "boot top" fracture shown in Figure 1–25, one bending moment acted on the proximal tibia as the skier fell forward over the top of the ski boot. An equal moment, produced by the fixed foot and ski, acted on the distal tibia. As the proximal tibia was bent forward, tensile stresses and strains acted on the posterior side of the bone and compressive stresses and strains acted on the anterior side. The tibia and fibula fractured at the top of the boot. Since adult bone is weaker in tension than in compression, failure begins on the side subjected to tension. Immature bone may fail first in compression, and a buckle fracture may result on the compressive side.

Four-point bending takes place when two force couples acting on a structure produce two equal moments. A force couple is formed when two parallel forces of equal magnitude but opposite direction are applied to a structure (see Fig. 1–24B). Because the

FIG. 1–23
Cross section of a bone subjected to bending, showing distribution of stresses around the neutral axis. Tensile stresses act on the superior side, and compressive stresses act on the inferior side. The stresses are highest at the periphery of the bone and lowest near the neutral axis. The tensile and compressive stresses are unequal because the bone is asymmetrical.

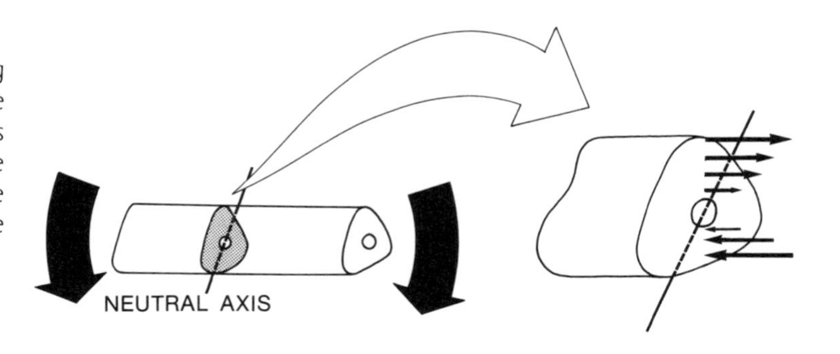

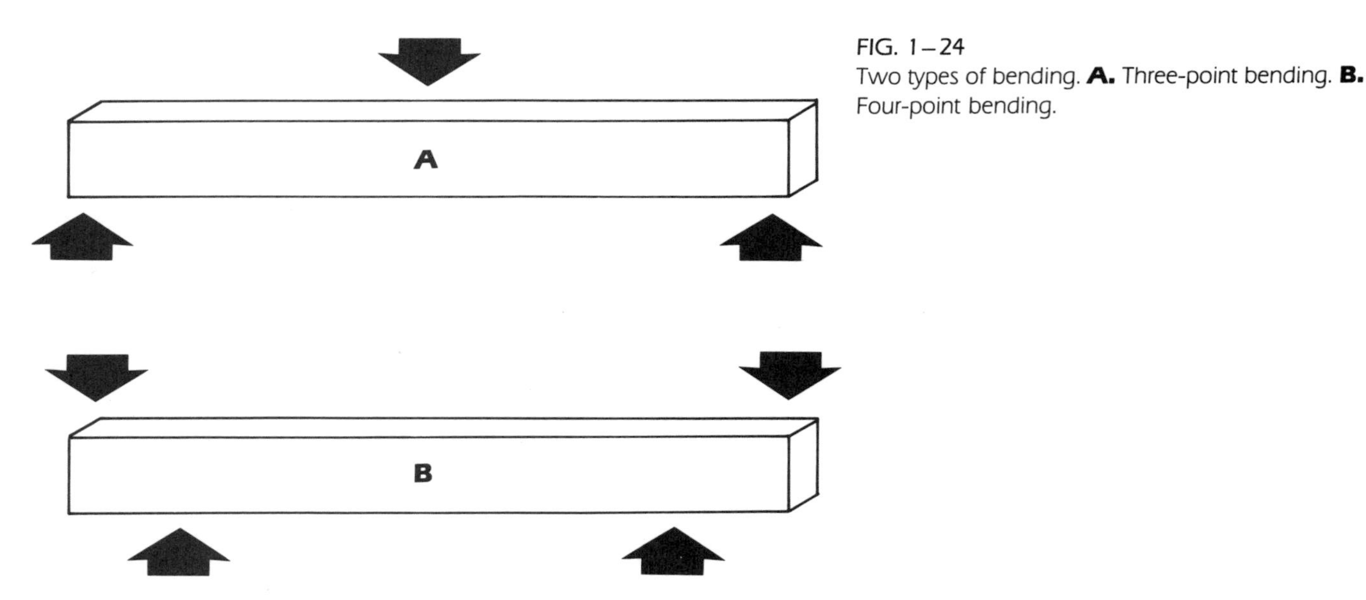

FIG. 1–24
Two types of bending. **A.** Three-point bending. **B.** Four-point bending.

magnitude of the bending moment is the same throughout the area between the two force couples, the structure breaks at its weakest point. An example of a four-point bending fracture is shown in Figure 1–26. A stiff knee joint was manipulated incorrectly during rehabilitation of a patient with a femoral fracture. During the manipulation, the posterior knee joint capsule and tibia formed one force couple and the femoral head and hip joint capsule formed the other. As a bending moment was applied to the femur, the bone failed at its weakest point, the original fracture site.

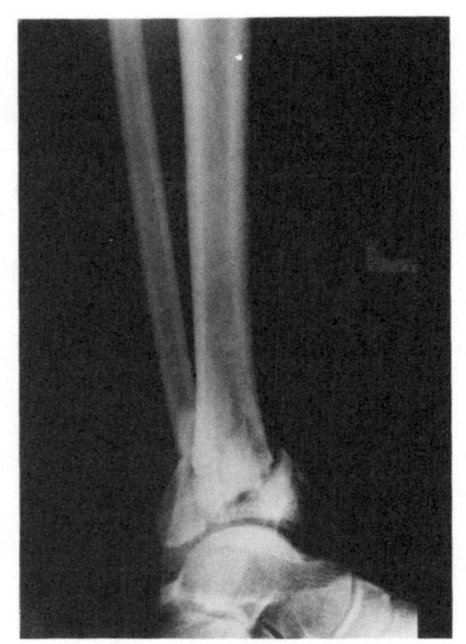

FIG. 1–25
Lateral roentgenogram of a "boot top" fracture produced by three-point bending. (Courtesy of Robert A. Winquist, M.D.)

Torsion

In torsion, a load is applied to a structure in a manner that causes it to twist about an axis, and a torque (or moment) is produced within the structure. When a structure is loaded in torsion, shear stresses are distributed over the entire structure. As in bending, the magnitude of these stresses is proportional to their distance from the neutral axis (Fig. 1–27). The farther the stresses are from the neutral axis, the higher their magnitude.

Under torsional loading, maximal shear stresses act on planes parallel and perpendicular to the neutral axis of the structure. In addition, maximal tensile and compressive stresses act on a plane diagonal to the neutral axis of the structure. Figure 1–28 illustrates these planes in a small segment of bone loaded in torsion.

The fracture pattern for bone loaded in torsion suggests that the bone fails first in shear, with the formation of an initial crack parallel to the neutral axis of the bone. A second crack usually forms along the plane of maximal tensile stress. Such a pattern can be seen in the experimentally produced torsional fracture of a canine femur shown in Figure 1–29.

Combined Loading

Although each loading mode has been considered separately, living bone is seldom loaded in one mode only. Loading of bone in vivo is complex for two principal reasons: bones are constantly subjected to multiple indeterminate loads, and their geometric structure is irregular. Measurement in vivo of the

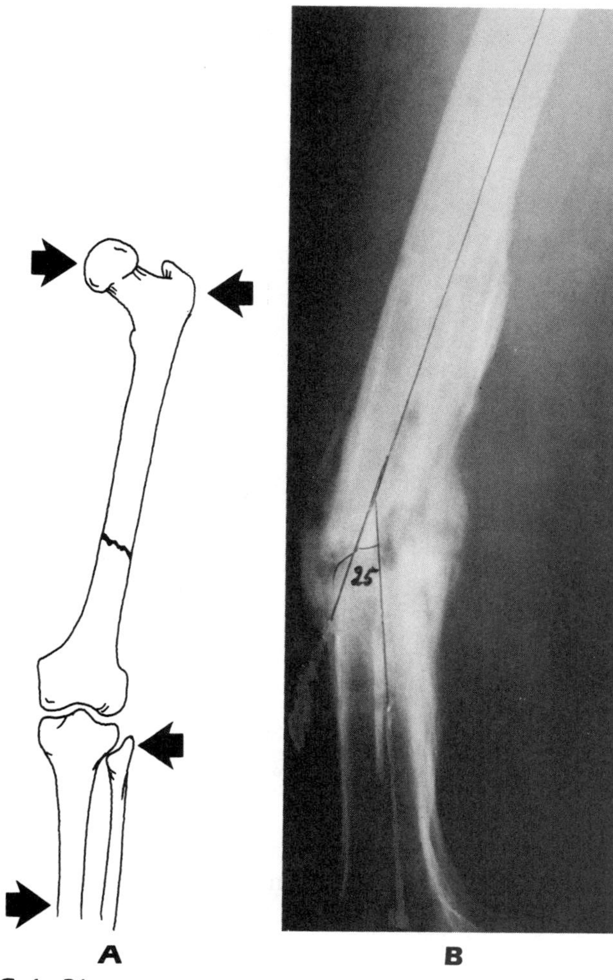

FIG. 1–26
A. During manipulation of a stiff knee during fracture rehabilitation, four-point bending caused the femur to fracture at its weakest point, the original fracture site. **B.** Lateral roentgenogram of the fractured femur. (Courtesy of Kaj Lundborg, M.D.)

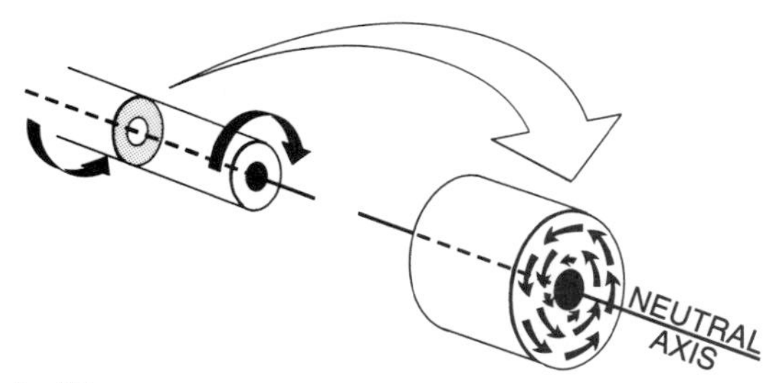

FIG. 1–27
Cross section of a cylinder loaded in torsion, showing the distribution of shear stresses around the neutral axis. The magnitude of the stresses is highest at the periphery of the cylinder and lowest near the neutral axis.

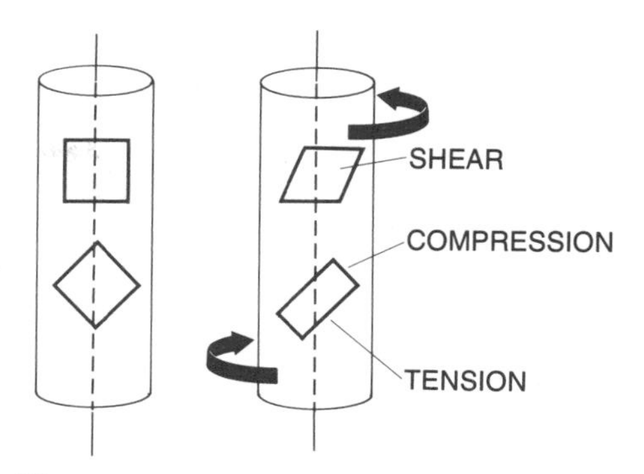

FIG. 1–28
Schematic representation of a small segment of bone loaded in torsion. Maximal shear stresses act on planes parallel and perpendicular to the neutral axis. Maximal tensile and compressive stresses act on planes diagonal to this axis.

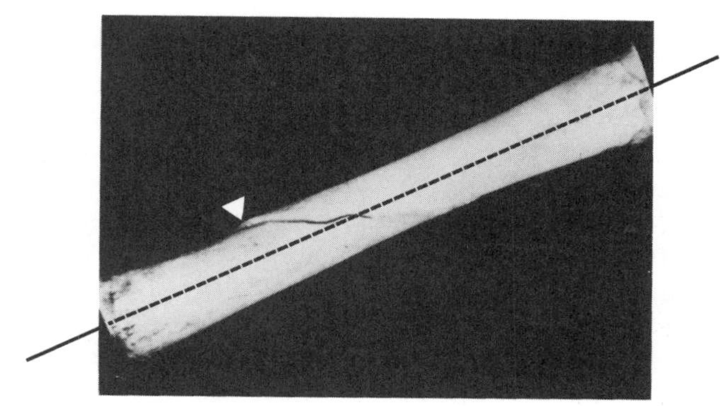

FIG. 1–29
Experimentally produced torsional fracture of a canine femur. The short crack (arrow) parallel to the neutral axis represents shear failure; the fracture line at a 30-degree angle to the neutral axis represents the plane of maximal tensile stress.

strains on the anteromedial surface of a human adult tibia during walking and jogging demonstrated the complexity of the loading patterns during these common physiologic activities (Lanyon et al., 1975). Stress values calculated from these strain measurements by Carter (1978) showed that during normal walking the stresses were compressive during heel strike, tensile during the stance phase, and again compressive during push-off (Fig. 1–30A). Values for shear stress were relatively high in the later portion of the gait cycle, denoting significant torsional loading. This torsional loading was associated with external rotation of the tibia during stance and push-off.

During jogging the stress pattern was quite different (Fig. 1–30B). The compressive stress predominating at toe strike was followed by high tensile stress during push-off. The shear stress was low through-

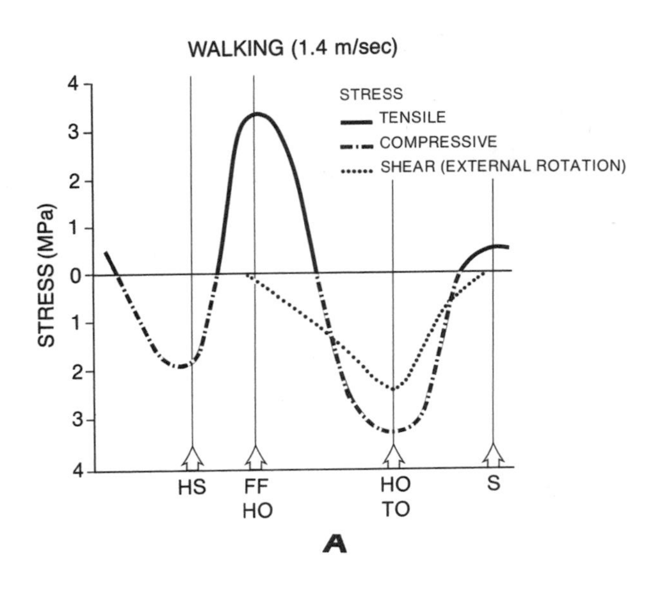

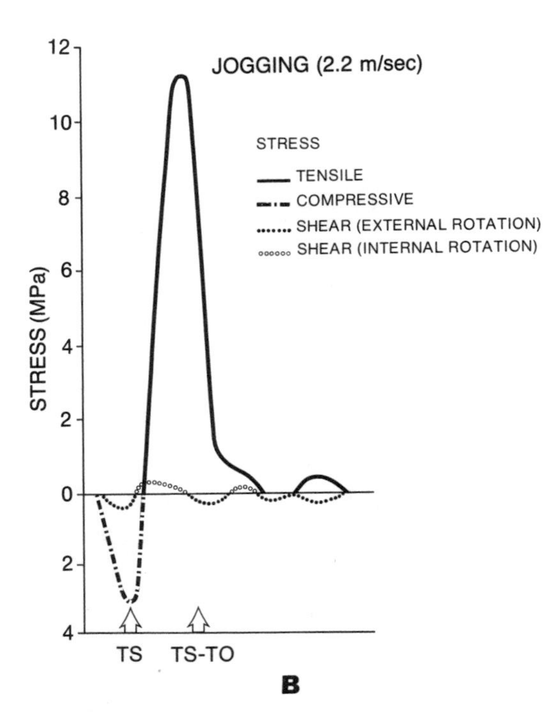

FIG. 1–30
A. Calculated stresses on the anteromedial cortex of a human adult tibia during walking. HS—heel strike; FF—foot flat; HO—heel-off; TO—toe-off; S—swing. (After Lanyon et al., 1975; courtesy of Dennis R. Carter, Ph.D.) **B.** Calculated stresses on the anteromedial cortex of an adult human tibia during jogging. TS—toe strike; TO—toe-off. (After Lanyon et al., 1975; courtesy of Dennis R. Carter, Ph.D.)

out the stride, denoting minimal torsional loading produced by slight external and internal rotation of the tibia in an alternating pattern. The increase in speed from slow walking to jogging increased both the stress and the strain on the tibia (Lanyon et al., 1975). This increase in strain with greater speed was confirmed in studies of locomotion in sheep, which demonstrated a fivefold increase in strain values from slow walking to fast trotting (Lanyon and Bourn, 1979).

Clinical examination of fracture patterns indicates that few fractures are produced by one loading mode or even by two modes. Indeed, most fractures are produced by a combination of several loading modes.

INFLUENCE OF MUSCLE ACTIVITY ON STRESS DISTRIBUTION IN BONE

When bone is loaded in vivo, contraction of the muscles attached to the bone alters the stress distribution in the bone. This muscle contraction decreases or eliminates tensile stress on the bone by producing compressive stress that neutralizes it either partially or totally.

The effect of muscle contraction can be illustrated in a tibia subjected to three-point bending. Figure 1–31A represents the leg of a skier who is falling forward, subjecting the tibia to a bending moment. High tensile stress is produced on the posterior aspect of the tibia, and high compressive stress acts on the anterior aspect. Contraction of the triceps surae muscle produces great compressive stress on the posterior aspect (Fig. 1–31B), neutralizing the great tensile stress and thereby protecting the tibia from failure in tension. This muscle contraction may result in higher compressive stress on the anterior surface of the tibia. Adult bone can usually withstand this stress, but immature bone, which is weaker, may fail in compression.

Muscle contraction produces a similar effect in the hip joint (Fig. 1–32). During locomotion, bending moments are applied to the femoral neck and tensile stress is produced on the superior cortex. Contraction

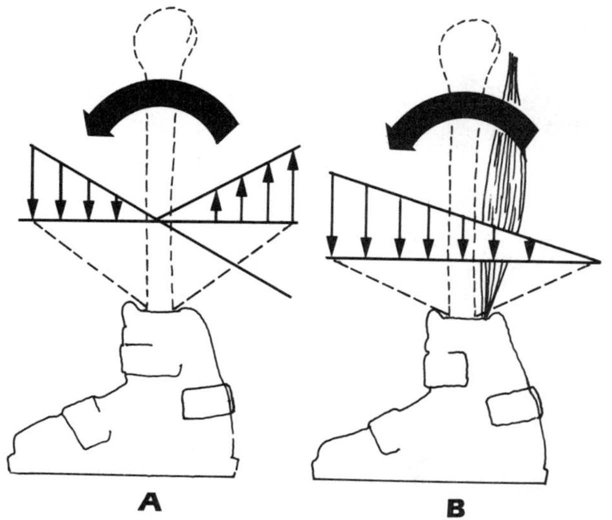

FIG. 1–31
A. Distribution of compressive and tensile stresses in a tibia subjected to three-point bending. **B.** Contraction of the triceps surae muscle produces high compressive stress on the posterior aspect, neutralizing the high tensile stress.

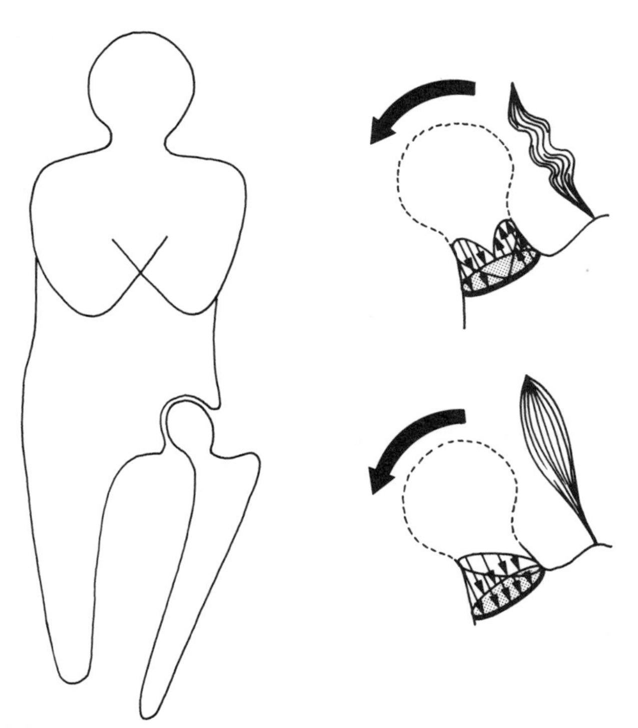
FIG. 1–32
Stress distribution in a femoral neck subjected to bending. When the gluteus medius muscle is relaxed (top), tensile stress acts on the superior cortex and compressive stress acts on the inferior cortex. Contraction of this muscle (bottom) neutralizes the tensile stress.

of the gluteus medius muscle produces compressive stress that neutralizes this tensile stress, with the net result that neither compressive nor tensile stress acts on the superior cortex. Thus, the muscle contraction allows the femoral neck to sustain higher loads than would otherwise be possible.

RATE DEPENDENCY IN BONE

Because bone is a viscoelastic material, its biomechanical behavior varies with the rate at which the bone is loaded (i.e., the rate at which the load is applied and removed). Bone is stiffer and sustains a higher load to failure when loads are applied at higher rates. Bone also stores more energy before failure at higher loading rates, provided that these rates are within the physiologic range.

The load-deformation curves in Figure 1–33 show the difference in the mechanical properties of paired canine tibiae tested in vitro at a high and a very low loading rate, 0.01 second and 200 seconds, respectively (Sammarco et al., 1971). The amount of energy stored before failure approximately doubled at the higher loading rate. The load to failure almost doubled, but the deformation to failure did not change significantly. The bone was about 50% stiffer at the higher speed.

The loading rate is clinically significant because it influences both the fracture pattern and the amount of soft tissue damage at fracture. When a bone fractures, the stored energy is released. At a low loading rate, the energy can dissipate through the formation of a single crack; the bone and soft tissues

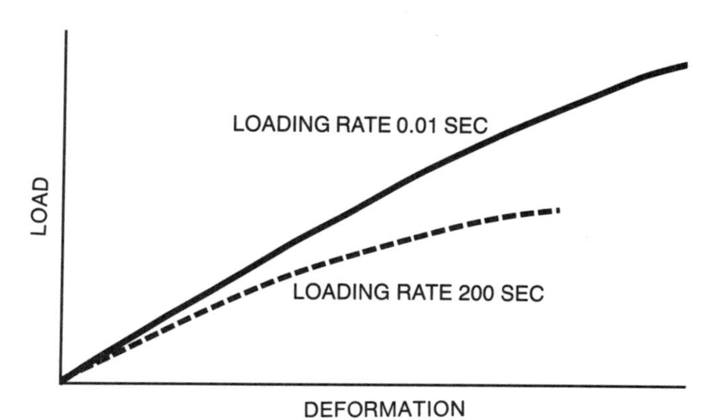

FIG. 1–33
Rate dependency of bone is demonstrated in paired canine tibiae tested at a high and a low loading rate. The load to failure and the energy stored to failure almost doubled at the high rate. (Adapted from Sammarco et al., 1971.)

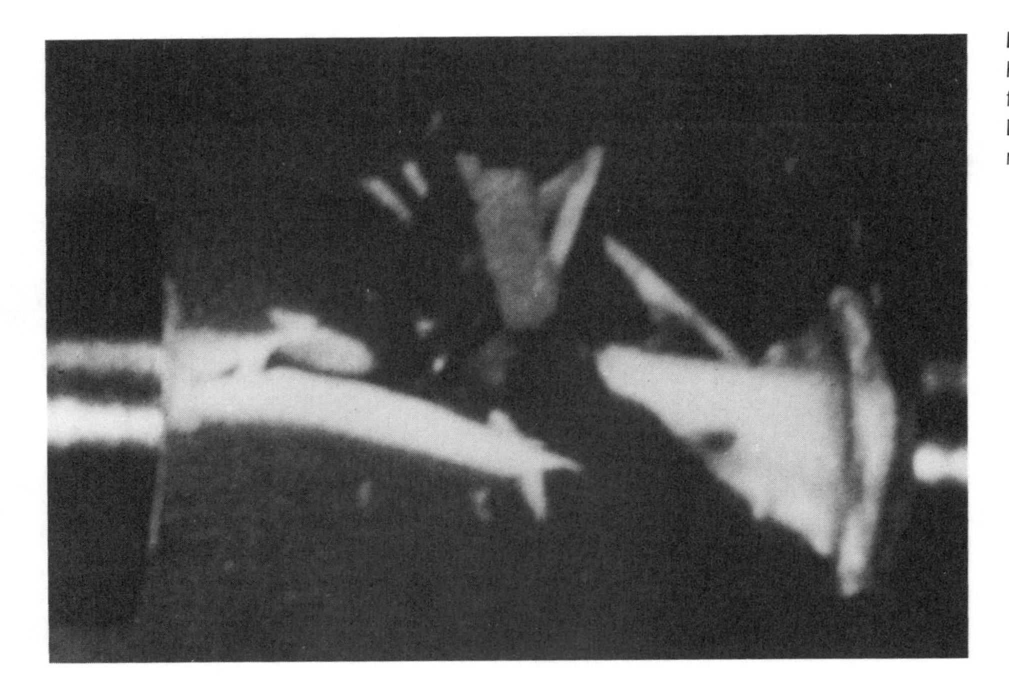

FIG. 1–34
Human tibia experimentally tested to failure in torsion at a high loading rate. Displacement of the numerous fragments was pronounced.

remain relatively intact, and there is little or no displacement of the bone fragments. At a high loading rate, however, the greater energy stored cannot dissipate rapidly enough through a single crack, and comminution of bone and extensive soft tissue damage result. Figure 1–34 shows a human tibia tested in vitro in torsion at a high loading rate; numerous bone fragments were produced, and displacement of the fragments was pronounced.

Clinically, bone fractures fall into three general categories based on the amount of energy released at fracture: low-energy, high-energy, and very high–energy. A low-energy fracture is exemplified by the simple torsional ski fracture; a high-energy fracture is often sustained during automobile accidents; and a very high–energy fracture is produced by very high–muzzle velocity gunshot.

Only a small proportion of the total energy storage capacity of bone is utilized during normal activity. Figure 1–35 illustrates just how little of this capacity is used during the normal physiologic activity of jogging.

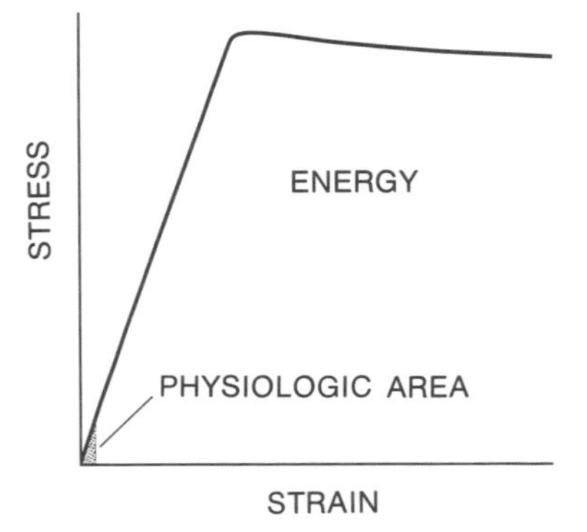

FIG. 1–35
Tensile strain values from a human adult tibia during jogging (Lanyon et al., 1975) have been plotted on a stress-strain curve for bone samples tested to failure in tension. A small proportion of the total energy storage capacity of the bone is utilized during this normal physiologic activity.

FATIGUE OF BONE UNDER REPETITIVE LOADING

Bone fractures can be produced by a single load that exceeds the ultimate strength of the bone or by repeated applications of a load of lower magnitude. A fracture caused by repeated applications of a lower load is called a fatigue fracture and is typically produced either by few repetitions of a high load or by many repetitions of a relatively normal load.

The interplay of load and repetition for any material can be plotted on a fatigue curve (Fig. 1–36). For some materials (some metals, for example), the fatigue curve is asymptotic, indicating that if the load

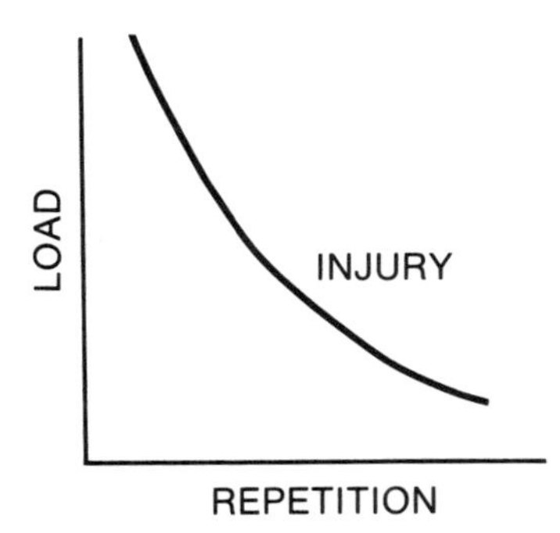

FIG. 1–36
The interplay of load and repetition is represented on a fatigue curve.

is kept below a certain level, theoretically, the material will remain intact, no matter how many repetitions. For bone tested in vitro, the curve is not asymptotic. When bone is subjected to repetitive low loads, it may sustain fatigue microfractures (Carter and Hayes, 1977a). Testing of bone in vitro also reveals that bone fatigues rapidly when load or deformation approaches the yield strength of the bone (Carter and Hayes, 1977a); that is, the number of repetitions needed to produce a fracture diminishes rapidly.

In repetitive loading of living bone, the fatigue process is affected not only by the amount of load and the number of repetitions but also by the number of applications of the load within a given time (frequency of loading). Since living bone is self-repairing, a fatigue fracture results only when the remodeling process is outpaced by the fatigue process, i.e., when loading is so frequent that it precludes the remodeling necessary to prevent failure.

Fatigue fractures are usually sustained during continuous strenuous physical activity, which causes the muscles to become fatigued and reduces their ability to contract. As a result they are less able to store energy and thus to neutralize the stresses imposed on the bone. The resulting alteration of the stress distribution in the bone causes abnormally high loads to be imposed, and a fatigue fracture may result. Bone may fail on the tensile side, the compressive side, or both sides. Failure on the tensile side results in a transverse crack, and the bone proceeds rapidly to complete fracture. Fatigue fractures on the compressive side appear to be produced more slowly; the remodeling is less easily outpaced by the fatigue process, and the bone may not proceed to complete fracture.

This theory of muscle fatigue as a cause of fatigue fracture in the lower extremities is outlined in the following schema:

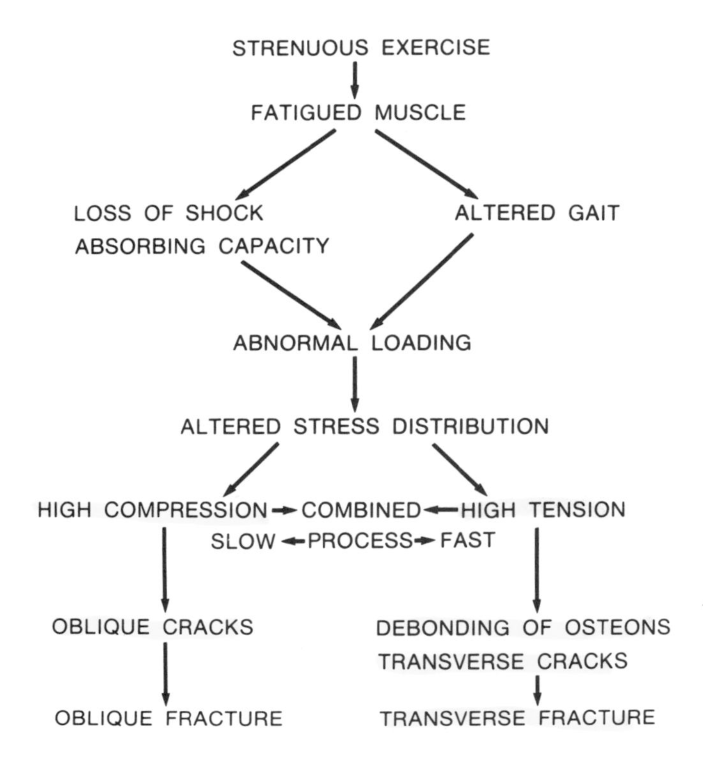

INFLUENCE OF BONE GEOMETRY ON BIOMECHANICAL BEHAVIOR

The geometry of a bone greatly influences its mechanical behavior. In tension and compression, the load to failure and the stiffness are proportional to the cross-sectional area of the bone. The larger the area is, the stronger and stiffer the bone. In bending, both the cross-sectional area and the distribution of bone tissue around a neutral axis affect the bone's mechanical behavior. The quantity that takes into account these two factors in bending is called the area moment of inertia. A larger area moment of inertia results in a stronger and stiffer bone.

Figure 1–37 shows the influence of the area moment of inertia on the load to failure and the stiffness of three rectangular structures that have the same area but different shapes. In bending, beam III is the stiffest of the three and can withstand the highest load, because the greatest amount of material is distributed at a distance from the neutral axis. For rectangular cross sections, the formula for the area

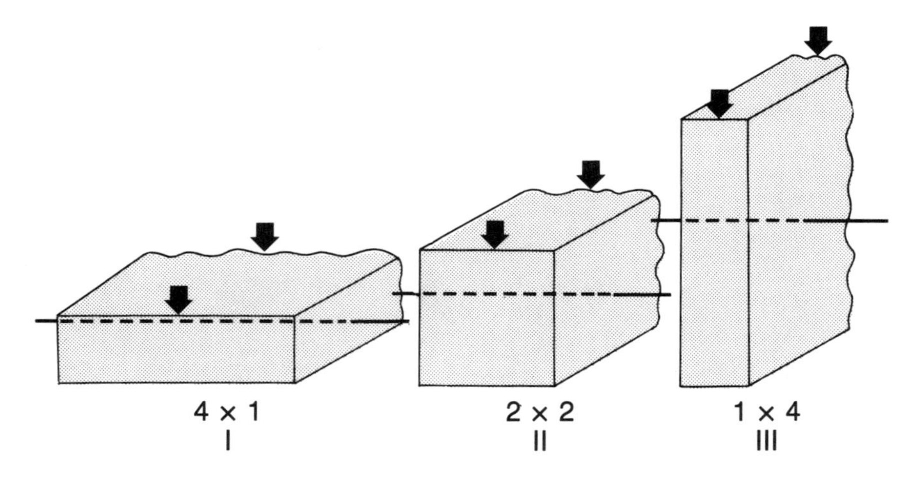

FIG. 1–37
Three beams of equal area but different shapes subjected to bending. For rectangular cross sections, the area moment of inertia is calculated by the formula $\frac{B \cdot H^3}{12}$, where B is the width and H, the height. The area moment of inertia for beam I is 4/12; for beam II, 16/12; and for beam III, 64/12. (Adapted from Frankel and Burstein, 1970.)

moment of inertia is the width (B) multiplied by the cube of the height (H^3) divided by 12:

$$\frac{B \cdot H^3}{12}$$

Because of its large area moment of inertia, beam III can withstand four times more load in bending than beam I.

A third factor, the length of the bone, influences the strength and stiffness in bending. The longer the bone is, the greater the magnitude of the bending moment caused by the application of a force. In a rectangular structure, the magnitude of the stresses produced at the point of application of the bending moment is proportional to the length of the structure. Figure 1–38 depicts the forces acting on two beams with the same width and height but different lengths: beam B is twice as long as beam A. The bending moment for the longer beam is twice that for the shorter beam; consequently, the stress magnitude throughout the beam is twice as high.

Because of their length, the long bones of the skeleton are subjected to high bending moments and, so, high tensile and compressive stresses. Their tubular

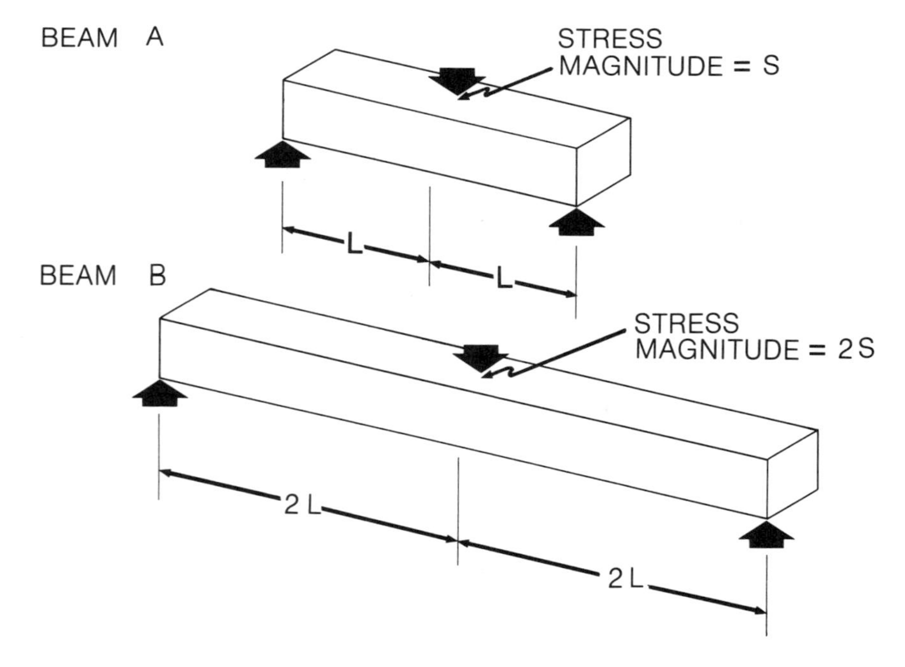

FIG. 1–38
Beam B is twice as long as beam A and sustains twice the bending moment. Hence, the stress magnitude throughout beam B is twice as high. (Adapted from Frankel and Burstein, 1970.)

shape gives them the ability to resist bending moments in all directions. These bones have a large area moment of inertia because much of the bone tissue is distributed at a distance from the neutral axis.

The factors that affect bone strength and stiffness in torsion are the same ones that operate in bending: the cross-sectional area and the distribution of bone tissue around a neutral axis. The quantity that takes into account these two factors in torsional loading is the polar moment of inertia. The larger the polar moment of inertia is, the stronger and stiffer the bone.

Figure 1–39 shows distal and proximal cross sections of a tibia subjected to torsional loading. Although the proximal section has a slightly smaller bony area than does the distal section, it has a much higher polar moment of inertia because much of the bone tissue is distributed at a distance from the neutral axis. The distal section, while it has a larger bony area, is subjected to much higher shear stress because much of the bone tissue is distributed close to the neutral axis. The magnitude of the shear stress in the distal section is approximately double that in the proximal section. Clinically, torsional fractures of the tibia commonly occur distally.

When bone begins to heal after fracture, blood vessels and connective tissue from the periosteum migrate into the region of the fracture, forming a cuff of dense fibrous tissue, or callus, around the fracture site, which stabilizes that area (Fig. 1–40A). The callus significantly increases the area and polar moments of inertia, thereby increasing the strength and stiffness of the bone in bending and torsion during the healing period. As the fracture heals and the bone gradually regains its normal strength, the callus cuff is progressively resorbed and the bone returns to as near its normal size and shape as possible (Fig. 1–40B).

Certain surgical procedures produce defects that greatly weaken the bone, particularly in torsion. These defects fall into two categories: those whose length is less than the diameter of the bone (stress raisers) and those whose length exceeds the bone diameter (open section defects).

A stress raiser is produced surgically when a small piece of bone is removed or a screw is inserted. Bone strength is reduced because the stresses imposed during loading are prevented from being distributed evenly throughout the bone and instead become concentrated around the defect. This defect is analogous to a rock in a stream, which diverts the water, producing high water turbulence around it (Fig. 1–41). The weakening effect of a stress raiser is particularly marked under torsional loading; the total decrease in bone strength in this loading mode can reach 60%.

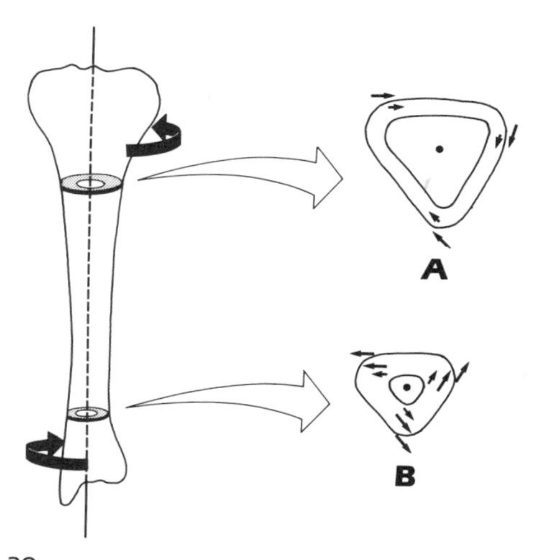

FIG. 1–39
Distribution of shear stress in two cross sections of a tibia subjected to torsional loading. The proximal section (A) has a higher moment of inertia than does the distal section (B), because more bony material is distributed away from the neutral axis. (Adapted from Frankel and Burstein, 1970.)

Burstein and associates (1972) showed the effect of stress raisers produced by screws and by empty screw holes on the energy storage capacity of rabbit bones tested in torsion at a high loading rate. The immediate effect of drilling a hole and inserting a screw in a rabbit femur was a 74% decrease in energy storage capacity. After 8 weeks, the stress raiser effect produced by the screws and by the holes without screws had disappeared completely because the bone had remodeled: bone had been laid down around the screws to stabilize them, and the empty screw holes had been filled in with bone. In femora from which the screws had been removed immediately before testing, however, the energy storage capacity of the bone decreased by 50%, mainly because the bone tissue around the screw sustained microdamage during screw removal (Fig. 1–42).

An open section defect is a discontinuity in the bone caused by surgical removal of a piece of bone longer than the bone's diameter (for example, by the cutting of a slot during a bone biopsy). Because the outer surface of the bone cross section is no longer continuous, the bone's ability to resist loads is altered, particularly in torsion.

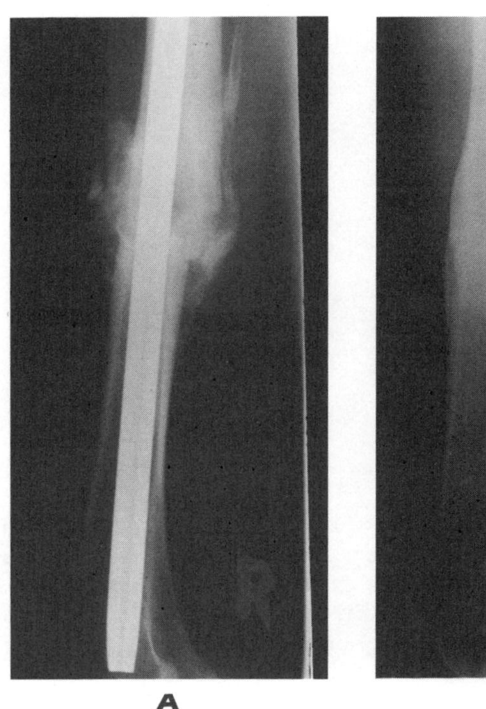

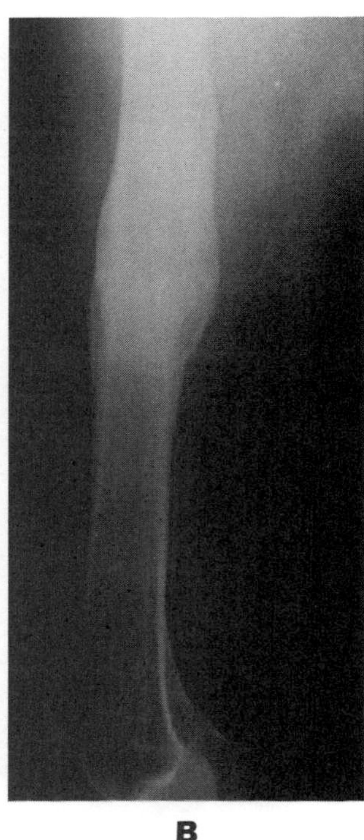

FIG. 1–40
A. Early callus formation in a femoral fracture fixed with an intramedullary nail. **B.** Nine months after injury the fracture has healed and most of the callus cuff has been resorbed. (Courtesy of Robert A. Winquist, M.D.)

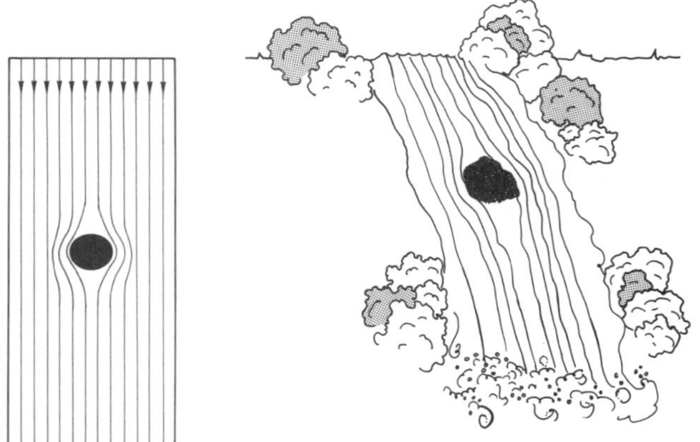
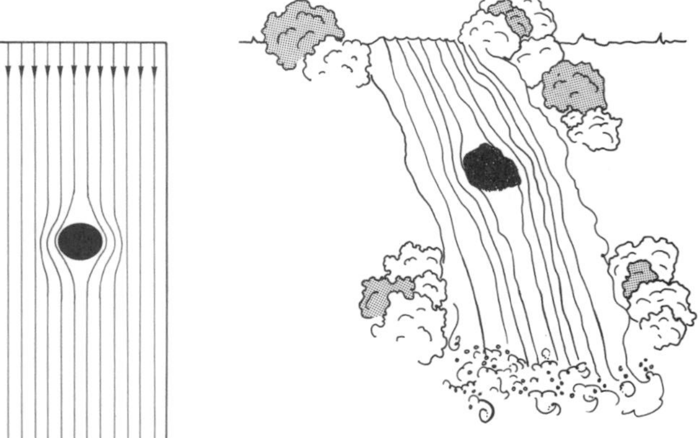

FIG. 1–41
Stress concentration around a defect; such a defect is analogous to a rock in a stream.

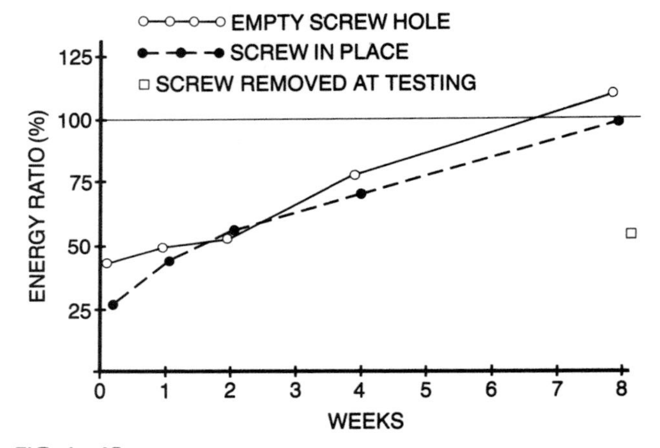

FIG. 1–42
Effect of screws and of empty screw holes on the energy storage capacity of rabbit femora. The energy storage for experimental animals is expressed as a percentage of the total energy storage capacity for control animals. When screws were removed immediately before testing, the energy storage capacity decreased by 50%. (Adapted from Burstein et al., 1972.)

In a normal bone subjected to torsion, the shear stress is distributed throughout the bone and acts to resist the torque. This stress pattern is illustrated in the cross section of a long bone shown in Figure 1–43A. (A cross section with a continuous outer surface is called a closed section.) In a bone with an open section defect, only the shear stress at the periphery of the bone resists the applied torque. As the shear stress encounters the discontinuity, it is forced to change direction (Fig. 1–43B). Through-

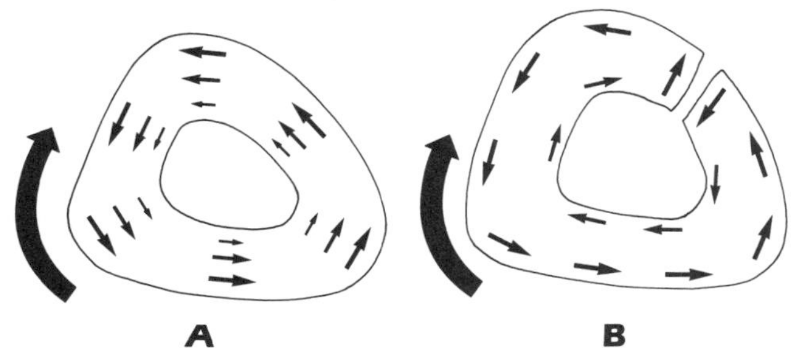

FIG. 1–43
Stress pattern in an open and closed section under torsional loading. **A.** In the closed section, all the shear stress resists the applied torque. **B.** In the open section, only the shear stress at the periphery of the bone resists the applied torque. (Adapted from Frankel and Burstein, 1970.)

out the interior of the bone, the stress runs parallel to to the applied torque, and the amount of bone tissue resisting the load is greatly decreased.

In torsion tests in vitro of human adult tibiae, an open section defect reduced the load to failure and energy storage to failure by as much as 90%. The deformation to failure was diminished by about 70% (Frankel and Burstein, 1970) (Fig. 1–44).

Clinically, surgical removal of a piece of bone can greatly weaken the bone, particularly in torsion. Figure 1–45 is a roentgenogram of a tibia from which a graft was removed for use in an arthrodesis of the hip. A few weeks after operation, the patient tripped while twisting in an attempt to rescue the Christmas ham from toppling onto the floor, and the bone fractured through the defect.

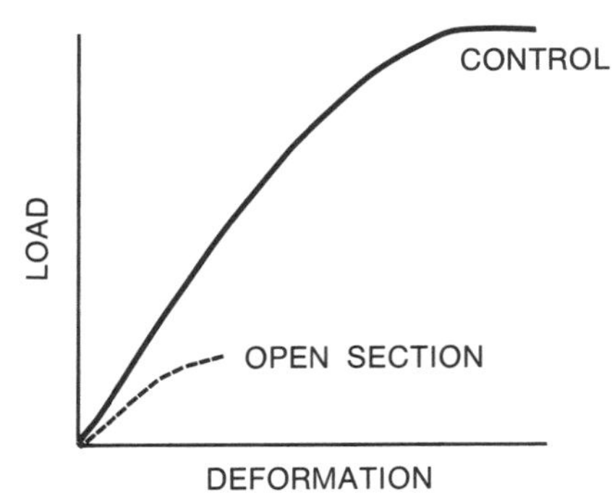

FIG. 1–44
Load-deformation curves for human adult tibiae tested in vitro under torsional loading. The control curve represents a tibia with no defect; the open section curve represents a tibia with an open section defect. (Adapted from Frankel and Burstein, 1970.)

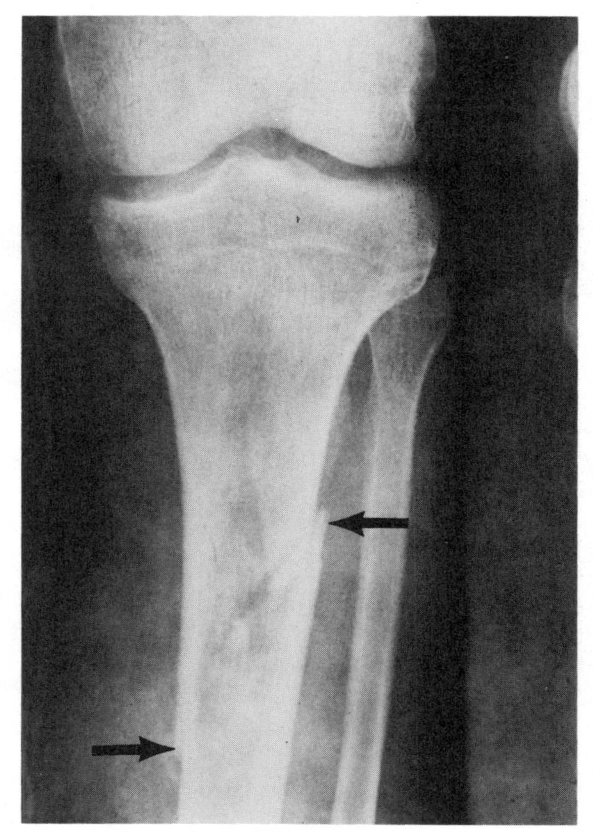

FIG. 1–45
A patient sustained a tibial fracture through a surgically produced open section defect when she tripped a few weeks after operation.

BONE REMODELING

Bone has the ability to remodel, by altering its size, shape, and structure, to meet the mechanical demands placed on it. This phenomenon, in which bone gains or loses cancellous and/or cortical bone in response to the level of stress sustained, is summarized as Wolff's law, which states that bone is laid down where needed and resorbed where not needed (Wolff, 1892).

If, because of partial or total immobilization, bone is not subjected to the usual mechanical stresses, periosteal and subperiosteal bone is resorbed (Jenkins and Cochran, 1969) and strength and stiffness decrease. This decrease in bone strength and stiffness was shown by Kazarian and Von Gierke (1969), who immobilized Rhesus monkeys in full-body casts for 60 days. Subsequent compressive testing in vitro of the vertebrae from the immobilized monkeys and from controls showed up to a threefold decrease in load to

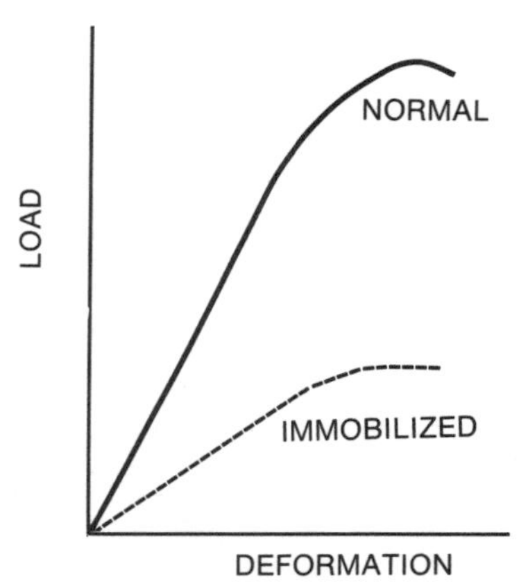

FIG. 1–46
Load-deformation curves for vertebral segments L5 to L7 from normal and immobilized Rhesus monkeys. (Adapted from Kazarian and Von Gierke, 1969.)

failure and energy storage capacity in the vertebrae that had been immobilized; stiffness was also significantly decreased (Fig. 1–46).

An implant that remains firmly attached to a bone after a fracture has healed may also diminish the strength and stiffness of the bone. In the case of a plate fixed to the bone with screws, the plate and the bone share the load in proportions determined by the geometry and material properties of each structure. A large plate, carrying high loads, unloads the bone to a great extent; the bone then atrophies in response to this diminished load. (The bone may hypertrophy at the bone-screw interface in an attempt to reduce micromotion of the screws.)

Bone resorption under a plate is illustrated in Figure 1–47. A compression plate made of a material approximately 10 times stiffer than the bone was applied to a fractured ulna and remained after the fracture had healed. The bone under the plate carried a lower load than normal; it was partially resorbed, and the diameter of the diaphysis became markedly smaller. A reduction in the size of the bone diameter greatly decreases bone strength, particularly in bending and torsion, as it reduces the area and polar moments of inertia. A 20% decrease in bone diameter may reduce the strength in torsion by 60%. Changes in bone size and shape illustrated in Figure 1–47 suggest that rigid plates should be removed shortly after a fracture has healed and before the bone has markedly diminished in size. Such a decrease in bone size is usually accompanied by secondary osteoporosis, which further weakens the bone (Slätis et al., 1980).

An implant may cause bone hypertrophy at its attachment sites. An example of bone hypertrophy around screws is illustrated in Figure 1–48. A nail plate was applied to a femoral neck fracture, and the bone hypertrophied around the screws in response to the increased load at these sites. Hypertrophy may also result if bone is repeatedly subjected to high mechanical stresses within the normal physiologic range. Hypertrophy of normal adult bone in response to strenuous exercise has been observed (Jones et al., 1977; Dalén and Olsson, 1974; Huddleston et al., 1980), as has an increase in bone density (Nilsson and Westlin, 1971).

A positive correlation exists between bone mass and body weight. A greater body weight has been associated with a larger bone mass (Exner et al., 1979). Conversely, a prolonged condition of weightlessness, such as that experienced during space travel, has

A
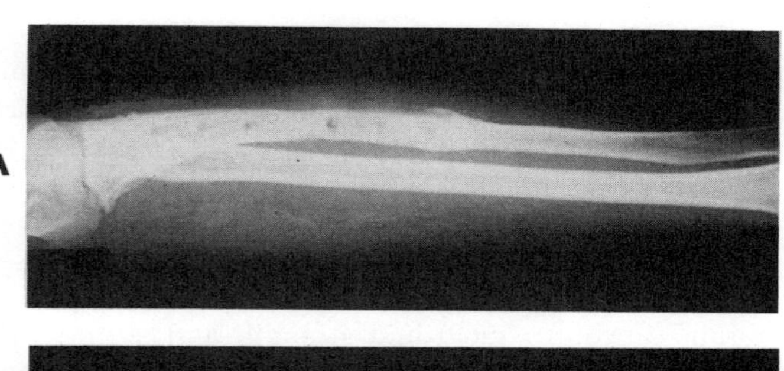

B
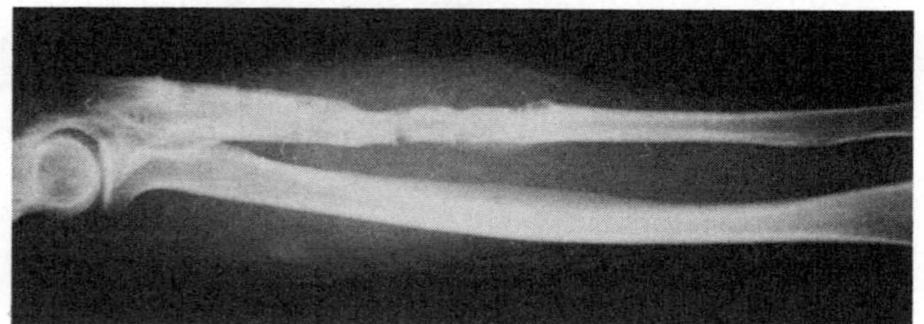

FIG. 1–47
Anteroposterior **(A)** and lateral **(B)** roentgenograms of an ulna after plate removal show a decreased bone diameter due to resorption of the bone under the plate. Cancellization of the cortex and the presence of screw holes also weaken the bone. (Courtesy of Marc Martens, M.D.)

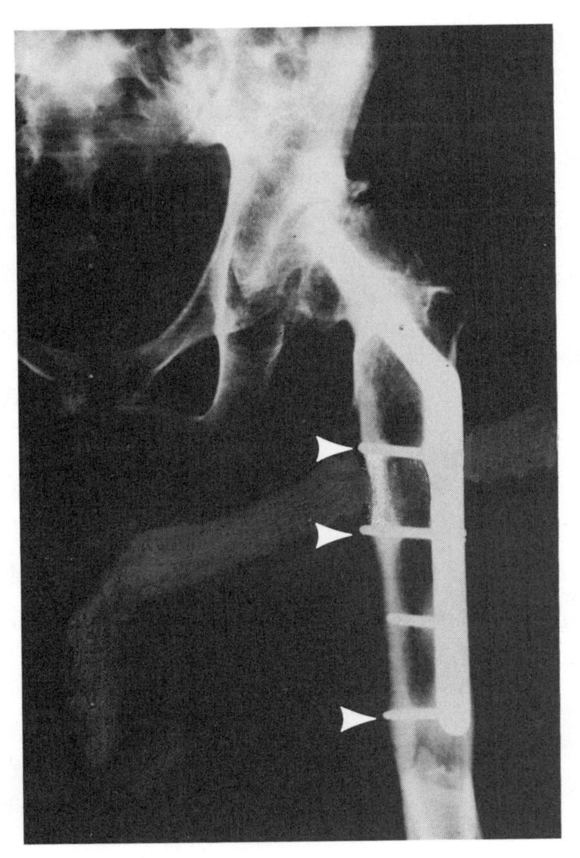

FIG. 1–48
Roentgenogram of a fractured femoral neck to which a nail plate was applied. Loads are transmitted from the plate to the bone via the screws. Bone has been laid down around the screws to bear these loads.

been found to result in a decreased bone mass in weight-bearing bones (Rambaut and Johnston, 1979; Gazenko et al., 1981).

DEGENERATIVE CHANGES IN BONE ASSOCIATED WITH AGING

A progressive loss of bone density has been observed as part of the normal aging process. The longitudinal trabeculae become thinner, and some of the transverse trabeculae are resorbed (Siffert and Levy, 1981) (Fig. 1–49). The result is marked reduction in the amount of cancellous bone and thinning of cortical bone. This decrease in the total amount of bone tissue and the slight decrease in the size of the bone reduce bone strength and stiffness.

Stress-strain curves for specimens from human adult tibiae of two widely differing ages tested in tension are shown in Figure 1–50. The ultimate stress was approximately the same for the young and the old bone. The old bone specimen could withstand only half the strain that the young bone could, indicating greater brittleness and a reduction in energy storage capacity.

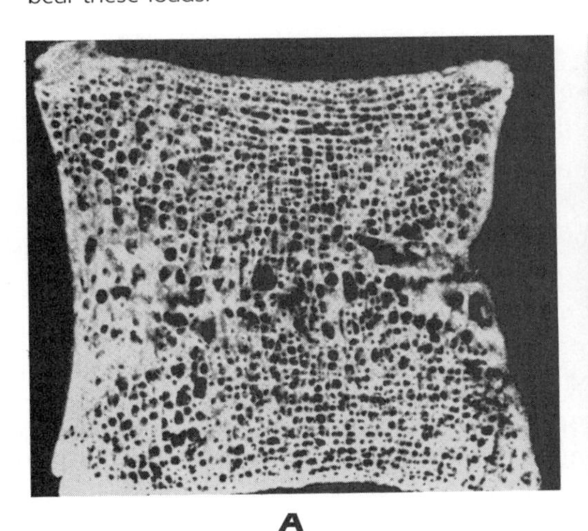

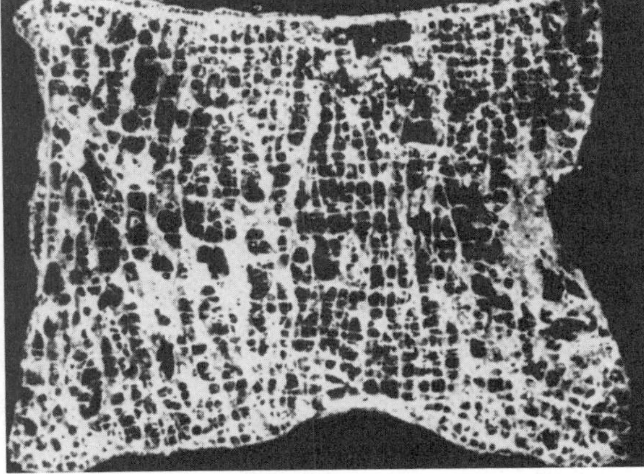

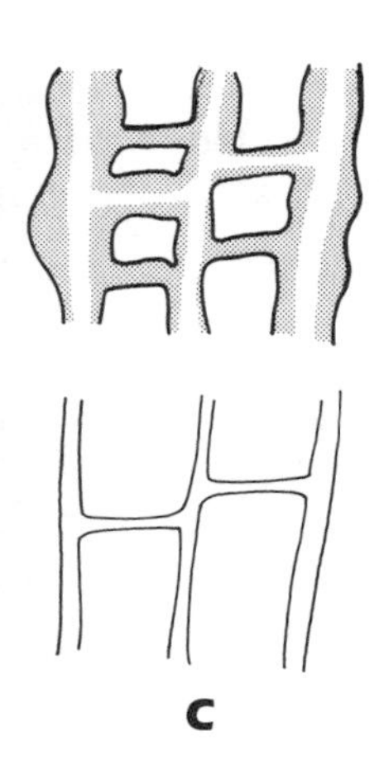

FIG. 1–49
Vertebral cross sections from autopsy specimens of young **(A)** and old **(B)** bone show a marked reduction in cancellous bone in the latter. (Reprinted with permission from Nordin, B.E.C.: Metabolic Bone and Stone Disease. Edinburgh, Churchill Livingstone, 1973.) **C.** Bone reduction with aging is schematically depicted. As normal bone (top) is subjected to absorption (shaded area) during the aging process, the longitudinal trabeculae become thinner and some transverse trabeculae disappear (bottom). (Adapted from Siffert and Levy, 1981.)

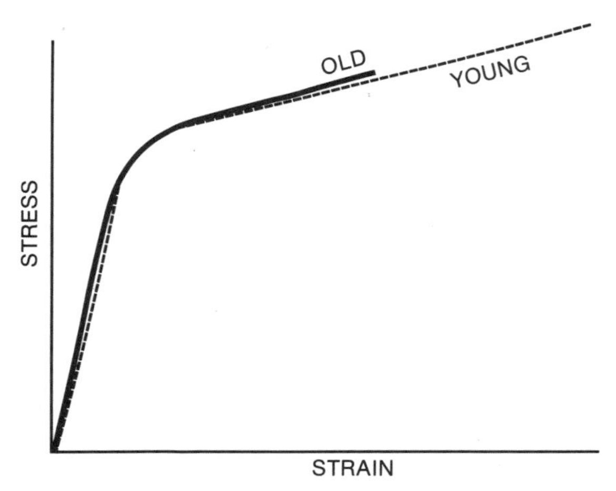

FIG. 1–50
Stress-strain curves for samples of adult human tibiae of two widely differing ages tested in tension. (Adapted from Burstein et al., 1976.)

SUMMARY

1. Bone is a two-phase composite material, inorganic mineral salts being one phase and an organic matrix of collagen and ground substance the other. The inorganic component makes bone hard and rigid, whereas the organic component gives bone its flexibility and resilience.
2. Microscopically, the fundamental structural unit of bone is the osteon, or haversian system, composed of concentric layers of mineralized matrix surrounding a central canal containing blood vessels and nerve fibers.
3. Macroscopically, the skeleton is composed of cortical and cancellous bone. Bone of both types can be considered as one material whose porosity and density vary over a wide range.
4. Bone is an anisotropic material, exhibiting different mechanical properties when loaded in different directions. Mature bone is strongest and stiffest in compression.
5. Bone is subjected to complex loading patterns during common physiologic activities such as walking and jogging. Most bone fractures are produced by a combination of several loading modes.
6. Muscle contraction affects stress patterns in bone by producing compressive stress that partially or totally neutralizes the tensile stress acting on the bone.
7. Bone is stiffer, sustains higher loads before failing, and stores more energy when loaded at higher rates.
8. Living bone fatigues when the frequency of loading precludes the remodeling necessary to prevent failure.
9. The mechanical behavior of a bone is influenced by its geometry (length, cross-sectional area, and distribution of bone tissue around the neutral axis).
10. Bone remodels in response to the mechanical demands placed on it; it is laid down where needed and resorbed where not needed.
11. With aging there is a marked reduction in the amount of cancellous bone and a decrease in the thickness of cortical bone. These changes diminish bone strength and stiffness.

REFERENCES

Bassett, C. A. L.: Electrical effects in bone. Sci. Am., *213*:18, 1965.

Bonefield, W., and Li, C. H.: Anisotropy of nonelastic flow in bone. J. Appl. Physics, *38*:2450, 1967.

Burstein, A. H., Reilly, D. T., and Martens, M.: Aging of bone tissue: Mechanical properties. J. Bone Joint Surg., *58A*: 82, 1976.

Burstein, A. H., et al.: Bone strength. The effect of screw holes. J. Bone Joint Surg., *54A*:1143, 1972.

Carter, D. R.: Anisotropic analysis of strain rosette information from cortical bone. J. Biomech., *11*:199, 1978.

Carter, D. R., and Hayes, W. C.: Bone compressive strength: The influence of density and strain rate. Science, *194*:1174, 1976.

Carter, D. R., and Hayes, W. C.: Compact bone fatigue damage. A microscopic examination. Clin. Orthop., *127*:265, 1977a.

Carter, D. R., and Hayes, W. C.: The compressive behavior of bone as a two-phase porous structure. J. Bone Joint Surg., *59A*:954, 1977b.

Carter, D. R., Schwab, G. H., and Spengler, D. M.: The effect of apparent density on the tensile and compressive properties of cancellous bone. Transactions of the 25th Annual Meeting, Orthopaedic Research Society, *4*:87, 1979.

Dalén, N., and Olsson, K. E.: Bone mineral content and physical activity. Acta Orthop. Scand., *45*:170, 1974.

Dempster, W. T., and Coleman, R. F.: Tensile strength of bone along and across the grain. J. Appl. Physiol., *16*:355, 1960.

Evans, F. G.: Bone and bones. J. Biomech. Eng., *104*:1, 1982.

Evans, F. G., and Bang, S.: Differences and relations between the physical properties and the microscopic structure of human femoral, tibial, and fibular cortical bone. Am. J. Anat., *120*:79, 1967.

Exner, G. U., et al.: Bone densitometry using computed tomography. Part I: Selective determination of trabecular bone density and other bone mineral parameters. Normal values in children and adults. Br. J. Radiol., *52*:14, 1979.

Frankel, V. H., and Burstein, A. H.: Orthopaedic Biomechanics. Philadelphia, Lea & Febiger, 1970.

Gazenko, O. G., Genin, A. M., and Yegorov, A. D.: Major medical results of the Salyut-6/Soyuz 185-day space flight. NASA NDB 2747. Proceedings of the XXXII Congress of the International Astronautical Federation, Rome, Italy, September 6–12, 1981.

Gray, H.: Anatomy of the Human Body. 13th American edition. Edited by C. D. Clemente. Philadelphia, Lea & Febiger, 1985.

Huddleston, A. L., Rockwell, D., Kulund, D. N., and Harrison, R. B.: Bone mass in lifetime tennis athletes. JAMA, *244*:1107, 1980.

International Society of Biomechanics: Quantities and Units of Measurement in Biomechanics, 1987 (unpublished).

Jenkins, D. P., and Cochran, T. H.: Osteoporosis: The dramatic effect of disuse of an extremity. Clin. Orthop., *64*:128, 1969.

Jones, H., Priest, J., Hayes, W., and Nagel, D.: Humeral hypertrophy in response to exercise. J. Bone Joint Surg., *59A*:204, 1977.

Kazarian, L. L., and Von Gierke, H. E.: Bone loss as a result of immobilization and chelation. Preliminary results in *Macaca mulatta*. Clin. Orthop., *65*:67, 1969.

Lanyon, L. E., and Bourn, S.: The influence of mechanical function on the development and remodeling of the tibia. An experimental study in sheep. J. Bone Joint Surg., *61A*:263, 1979.

Lanyon, L. E., Hampson, W. G. J., Goodship, A. E., and Shah, J. S.: Bone deformation recorded in vivo from strain gauges attached to the human tibial shaft. Acta Orthop. Scand., *46*:256, 1975.

Nilsson, B. E., and Westlin, N. E.: Bone density in athletes. Clin. Orthop., *77*:179, 1971.

Nordin, B. E. C.: Metabolic Bone and Stone Disease. Edinburgh, Churchill Livingstone, 1973.

Rambaut, P. C., and Johnston, R. S.: Prolonged weightlessness and calcium loss in man. Acta Astronautica, *6*:1113, 1979.

Reilly, D., and Burstein, A.: The elastic and ultimate properties of compact bone tissue. J. Biomech., *8*:393, 1975.

Sammarco, J., Burstein, A., Davis, W., and Frankel, V.: The biomechanics of torsional fractures: The effect of loading on ultimate properties. J. Biomech., *4*:113, 1971.

Siffert, R. S., and Levy, R. N.: Trabecular patterns and the internal architecture of bone. Mt. Sinai J. Med., *48*:221, 1981.

Slätis, P., Paavolainen, P., Karaharju, E., and Holmström, T.: Structural and biomechanical changes in bone after rigid plate fixation. Can. J. Surg., *23*:247, 1980.

Tortora, G. J., and Anagnostakos, N. P.: Principles of Anatomy and Physiology. 4th Ed. New York, Harper & Row, 1984.

Wolff, J.: Das Gesetz der Transformation der Knochen. Berlin, Hirschwald, 1892.

SUGGESTED READING

Whole Bones

Albright, J. A.: Bone: Physical properties. *In* The Scientific Basis of Orthopaedics. 2nd Ed. Edited by J. A. Albright and R. A. Brand. New York, Appleton-Century-Crofts, 1987, pp. 213–240.

Albright, J. A., and Skinner, H. C. W.: Bone: Structural organization and remodeling dynamics. *In* The Scientific Basis of Orthopaedics. 2nd Ed. Edited by J. A. Albright and R. A. Brand. New York, Appleton-Century-Crofts, 1987, pp. 161–198.

Asang, E.: Biomechanics of the human leg in alpine skiing. *In* Biomechanics IV. Edited by R. C. Nelson and C. A. Morehouse. Baltimore, University Park Press, 1974, pp. 236–242.

Asang, E.: Applied biomechanics of the human leg. A basis for individual protection from skiing injuries. Orthop. Clin. North Am., *7*:95–103, 1976.

Asang, E.: Experimental biomechanics of the human leg. A basis for interpreting typical skiing injury mechanisms. Orthop. Clin. North Am., *7*:63–73, 1976.

Bassett, C. A. L.: Electrical effects in bone. Sci. Am., *213*:18–25, 1965.

Burstein, A. H., and Frankel, V. H.: The viscoelastic properties of some biological materials. Ann. N.Y. Acad. Sci., *146*:158–165, 1968.

Burstein, A. H., et al.: Bone strength. The effect of screw holes. J. Bone Joint Surg., *54A*:1143–1156, 1972.

Carter, D. R.: Anisotropic analysis of strain rosette information from cortical bone. J. Biomech., *11*:199–202, 1978.

Currey, J.: The Mechanical Adaptations of Bones. Princeton, Princeton University Press, 1984, pp. 3–157.

Dalén, N., and Olsson, K.-E.: Bone mineral content and physical activity. Acta Orthop. Scand., *45*:170–174, 1974.

Dempster, W. T., and Coleman, R. F.: Tensile strength of bone along and across the grain. J. Appl. Physiol., *16*:355–360, 1960.

Evans, F. G.: Bone and bones. J. Biomech. Eng., *104*:1–5, 1982.

Evans, R. D. (ed.): Studies on the Anatomy and Function of Bone and Joints. New York, Springer-Verlag, 1966.

Exner, G. U., et al.: Bone densitometry using computed tomography. Part I: Selective determination of trabecular bone density and other bone mineral parameters. Normal values in children and adults. Br. J. Radiol., *52*:14–23, 1979.

Frankel, V. H.: The Femoral Neck: Function, Fracture Mechanisms, Internal Fixation. Springfield, Charles C Thomas, 1960.

Frankel, V. H., and Burstein, A. H.: Orthopaedic Biomechanics. Philadelphia, Lea & Febiger, 1970.

Frankel, V. H., and Burstein, A. H.: Biomechanics of the locomotor system. *In* Medical Engineering. Edited by C. D. Ray. Chicago, Year Book Medical Publisher, 1974, pp. 505–516.

Frost, H. M.: An Introduction to Biomechanics. Springfield, Charles C Thomas, 1976.

Hakim, N. S., and King, A. I.: Programmed replication of *in situ* (whole body) loading conditions during *in vitro* (substructure) testing of a vertebral column segment. J. Biomech., *9*:629–632, 1976.

Inman, V. T.: Functional aspects of the abductor muscles of the hip. J. Bone Joint Surg., *29A*:607–619, 1947.

Jenkins, D. P., and Cochran, T. H.: Osteoporosis: The dramatic effect of disuse of an extremity. Clin. Orthop., *64*:128–134, 1969.

Jensen, J. S., Hansen, F. W., and Johansen, J.: Tibial shaft fractures. A comparison of conservative treatment and internal fixation with conventional plates or AO compression plates. Acta Orthop. Scand., *48*:204–212, 1977.

Jones, H., Priest, J., Hayes, W., and Nagel, D.: Humeral hypertrophy in response to exercise. J. Bone Joint Surg., *59A*:204–208, 1977.

Kazarian, L., and Graves, G. A.: Compressive strength characteristics of the human vertebral centrum. Spine, *2*:1–14, 1977.

Kazarian, L. E., and Von Gierke, H. E.: Bone loss as a result of immobilization and chelation. Preliminary results in *Macaca mulatta*. Clin. Orthop., *65*:67–75, 1969.

Lakes, R., and Saha, S.: Cement line motion in bone. Science, *204*:501–503, 1979.

Lanyon, L. E., Hampson, W. G. J., Goodship, A. E., and Shah, J. S.: Bone deformation recorded in vivo from strain gauges attached to the human tibial shaft. Acta Orthop. Scand., *46*:256–268, 1975.

Murphy, E. F., and Burstein, A. H.: Physical properties of materials including solid mechanics. *In* Atlas of Orthotics. Biomechanical Principles and Application. St. Louis, C. V. Mosby Co., 1975, pp. 3–30.

Netz, P., Eriksson, K., and Strömberg, L.: Ultimate failure of diaphyseal bone. An experimental study on dogs. Acta Orthop. Scand., *51*:583–588, 1980.

Piziali, R. L., and Nagel, D. A.: Modeling of the human leg in ski injuries. Orthop. Clin. North Am., *7*:127–139, 1976.

Robinson, R. A.: Bone tissue: Composition and function. Johns Hopkins Med. J., *145*:10–24, 1979.

Rosse, C., and Clawson, D. K.: Introduction to the Musculoskeletal System. New York, Harper & Row, 1970.

Rybicki, E. F., Simonen, F. A., and Weis, E. B.: On the mathematical analysis of stress in the human femur. J. Biomech., *5*:203–215, 1972.

Sammarco, G. J., Burstein, A. H., Davis, W. L., and Frankel, V. H.: The biomechanics of torsional fractures: The effect of loading on ultimate properties. J. Biomech., *4*:113–117, 1971.

Sandler, R. B., and Herbert, D. L.: Quantitative bone assessments: Applications and expectations. J. Am. Geriatr. Soc., *29*:97–103, 1981.

Skinner, H. C. W.: Bone mineralization. *In* The Scientific Basis of Orthopaedics. 2nd Ed. Edited by J. A. Albright and R. A. Brand. New York, Appleton-Century-Crofts, 1987, pp. 199–212.

Slätis, P., Paavolainen, P., Karaharju, E., and Holmström, T.: Structural and biomechanical changes in bone after rigid plate fixation. Can. J. Surg., *23*:247–250, 1980.

Strömberg, L., and Dalén, N.: Experimental measurement of maximum torque capacity of long bones. Acta Orthop. Scand., *47*:257–263, 1976.

Strömberg, L., and Dalén, N.: The influence of freezing on the maximum torque capacity of long bones. An experimental study on dogs. Acta Orthop. Scand., *47*:254–256, 1976.

Toridis, T. G.: Stress analysis of the femur. J. Biomech., *2*:163–174, 1969.

Cortical Bone

Bonefield, W., and Li, C. H.: Anisotropy of nonelastic flow in bone. J. Appl. Physics, *38*:2450–2455, 1967.

Burstein, A. H., Reilly, D. T., and Martens, M.: Aging of bone tissue: Mechanical properties. J. Bone Joint Surg., *58A*:82–86, 1976.

Burstein, A. H., Currey, J. D., Frankel, V. H., and Reilly, D. T.: The ultimate properties of bone tissue: The effects of yielding. J. Biomech., *5*:35–44, 1972.

Carter, D. R., and Hayes, W. C.: Compact bone fatigue damage. A microscopic examination. Clin. Orthop., *127*:265–274, 1977.

Carter, D. R., and Hayes, W. C.: The compressive behavior of bone as a two-phase porous structure. J. Bone Joint Surg., *59A*:954–962, 1977.

Currey, J. D.: The mechanical properties of bone. Clin. Orthop., *73*:210–231, 1970.

Dalén, N., Hellström, L.-G., and Jacobson, B.: Bone mineral content and mechanical strength of the femoral neck. Acta Orthop. Scand., *47*:503–508, 1976.

Enneking, W.: Principles of Musculoskeletal Pathology. Gainesville, FL, Storter Printing Company, 1970.

Evans, F. G.: Mechanical Properties of Bone. Springfield, Charles C Thomas, 1973.

Evans, F. G., and Bang, S.: Differences and relations between the physical properties and the microscopic structure of human femoral, tibial, and fibular cortical bone. Am. J. Anat., *120*:79–88, 1967.

Fredensborg, N., and Nilsson, B. E.: The bone mineral content and cortical thickness in young women with femoral neck fracture. Clin. Orthop., *124*:161–164, 1977.

Frost, H. M.: The Laws of Bone Structure. Springfield, Charles C Thomas, 1964.

Lakes, R., and Saha, S.: Long-term torsional creep in compact bone. J. Biomech. Eng., *102*:178–180, 1980.

Mueller, K. H., Trias, A., and Ray, R. D.: Bone density and composition. Age-related and pathological changes in water and mineral content. J. Bone Joint Surg., *48A*:140–148, 1966.

Nilsson, B. E., and Westlin, N. E.: Bone density in athletes. Clin. Orthop., *77*:179–182, 1971.

Pope, M. H., and Outwater, J. O.: The fracture characteristics of bone substance. J. Biomech., *5*:457–465, 1972.

Reilly, D. T., and Burstein, A. H.: The mechanical properties of cortical bone. J. Bone Joint Surg., *56A*:1001–1002, 1974.

Reilly, D. T., and Burstein, A. H.: The elastic and ultimate properties of compact bone tissue. J. Biomech., *8*:393–405, 1975.

Reilly, D. T., Burstein, A. H., and Frankel, V. H.: The elastic modulus for bone. J. Biomech., *7*:271–275, 1974.

Sedlin, E. D.: A rheological model for cortical bone. A study of the physical properties of human femoral samples. Acta Orthop. Scand., Suppl. *83*:1–87, 1965.

Viano, D., Helfenstein, U., Anliker, M., and Rüegsegger, P.: Elastic properties of cortical bone in female human femurs. J. Biomech., *9*:703–710, 1976.

Weaver, J. K.: The microscopic hardness of bone. J. Bone Joint Surg., *48A*:273–288, 1966.

Cancellous Bone

Behrens, J. C., Walker, P. S., and Shoji, H.: Variations in strength and structure of cancellous bone at the knee. J. Biomech., *7*:201–207, 1974.

Carter, D. R., and Hayes, W. C.: Bone compressive strength: The influence of density and strain rate. Science, *194*:1174–1176, 1976.

Carter, D. R., and Hayes, W. C.: The compressive behavior of bone as a two-phase porous structure. J. Bone Joint Surg., *59A*:954–962, 1977.

Carter, D. R., Schwab, G. H., and Spengler, D. M.: The effect of apparent density on the tensile and compressive properties of cancellous bone. Transactions of the 25th Annual Meeting, Orthopaedic Research Society, *4*:87, 1979.

Chung, S. M. K., Batterman, S. C., and Brighton, C. T.: Shear strength of the human femoral capital epiphyseal plate. J. Bone Joint Surg., *58A*:94–103, 1976.

Enneking, W.: Principles of Musculoskeletal Pathology. Gainesville, FL, Storter Printing Company, 1970.

Evans, F. G.: Mechanical Properties of Bone. Springfield, Charles C Thomas, 1973.

Frankel, V. H., and Burstein, A. H.: Load capacity of tubular bone. *In* Biomechanics and Related Bio-engineering Topics. Edited by R. M. Kenedi. New York, Pergamon Press, 1965, pp. 381–396.

Frost, H. M.: The Laws of Bone Structure. Springfield, Charles C Thomas, 1964.

Galante, J., Rostoker, W., and Ray, R. D.: Physical properties of trabecular bone. Calcif. Tissue Res., *5*:236–246, 1970.

Hayes, W. C., and Carter, D. R.: Postyield behavior of subchondral trabecular bone. J. Biomed. Mater. Res., *7*:537–544, 1976.

Hayes, W. C., Boyle, D. J., and Velez, A.: Functional adaptation in the trabecular architecture of the human patella. Transactions of the 23rd Annual Meeting, Orthopaedic Research Society, 2:114, 1977.

Lindahl, O.: Mechanical properties of dried defatted spongy bone. Acta Orthop. Scand., *47*:11–19, 1976.

Pope, M. H., and Outwater, J. O.: The fracture characteristics of bone substance. J. Biomech., *5*:457–465, 1972.

Pope, M. H., and Outwater, J. O.: Mechanical properties of bone as a function of position and orientation. J. Biomech., *7*:61–66, 1974.

Siffert, R. S., and Levy, R. N.: Trabecular patterns and the internal architecture of bone. Mt. Sinai J. Med., *48*:221–229, 1981.

Weaver, J. K.: The microscopic hardness of bone. J. Bone Joint Surg., *48A*:273–288, 1966.

Weaver, J. K., and Chalmers, J.: Cancellous bone: Its strength and changes with aging and an evaluation of some methods for measuring its mineral content. I. Age changes in cancellous bone. J. Bone Joint Surg., *48A*:289–299, 1966.

Fatigue Fracture

Baker, J., Frankel, V. H., and Burstein, A. H.: Fatigue fractures: Biomechanical considerations. J. Bone Joint Surg., *54A*:1345–1346, 1972.

Burrows, H. J.: Fatigue infraction of the middle of the tibia in ballet dancers. J. Bone Joint Surg., *38B*:83–94, 1956.

Devas, M.: Stress Fractures. Edinburgh, London, and New York, Churchill Livingstone, 1975.

Frankel, V. H., and Hang, Y.-S.: Recent advances in the biomechanics of sport injuries. Acta Orthop. Scand., *46*:484–497, 1975.

Friedenberg, Z. B.: Fatigue fractures of the tibia. Clin. Orthop., *76*:111–115, 1971.

McBryde, A. M.: Stress fractures in athletes. J. Sports Med., *3*:212–217, 1976.

Walter, N. E., and Wolf, M. D.: Stress fractures in young athletes. Am. J. Sports Med., *5*:165–170, 1977.

Bone Remodeling

Abramson, A.: Bone disturbances in injuries to the spinal cord and cauda equina (paraplegia). Their prevention by ambulation. J. Bone Joint Surg., *30A*:982–987, 1948.

Bartley, M. H., Arnold, J. S., Haslam, R. K., and Jee, W. S. S.: The relationship of bone strength and bone quantity in health, disease, and aging. J. Gerontol., *21*:517–521, 1966.

Bassett, C. A. L.: Biologic significance of piezoelectricity. Calcif. Tissue Res., *1*:252–272, 1968.

Chamay, A., and Tschantz, P.: Mechanical influences in bone remodeling. Experimental research on Wolff's Law. J. Biomech., *5*:173–180, 1972.

Dalén, N., and Olsson, K.-E.: Bone mineral content and physical activity. Acta Orthop. Scand., *45*:170–174, 1974.

Franke, J., et al.: Physical properties of fluorosis bone. Acta Orthop. Scand., *47*:20–27, 1976.

Gjelsvik, A.: Bone remodeling and piezoelectricity—I. J. Biomech., *6*:69–77, 1973.

Horal, J., Nachemson, A., and Scheller, S.: Clinical and radiological long term follow-up of vertebral fractures in children. Acta Orthop. Scand., *43*:491–503, 1972.

Huddleston, A. L., Rockwell, D., Kulund, D. N., and Harrison, R. B.: Bone mass in lifetime tennis athletes. JAMA, *244*:1107–1109, 1980.

Jaworski, Z. F. G.: Physiology and pathology of bone remodeling. Cellular basis of bone structure in health and in osteoporosis. Orthop. Clin. North Am., *12*:485–512, 1981.

Jenkins, D. P., and Cochran, T. H.: Osteoporosis: The dramatic effect of disuse of an extremity. Clin. Orthop., *64*:128–134, 1969.

Jones, H. H., et al.: Humeral hypertrophy in response to exercise. J. Bone Joint Surg., *59A*:204–208, 1977.

Lanyon, L. E., and Bourn, S.: The influence of mechanical function on the development and remodeling of the tibia. An experimental study in sheep. J. Bone Joint Surg., *61A*:263–273, 1979.

Martin, R. B., Pickett, J. C., and Zinaich, S.: Studies of skeletal remodeling in aging men. Clin. Orthop., *149*:268–282, 1980.

Rambaut, P. C., and Johnston, R. S.: Prolonged weightlessness and calcium loss in man. Acta Astronautica, *6*:1113–1122, 1979.

Schock, C. C., Noyes, F. R., Mathews, C. H. E., and Crouch, M. M.: The effect of activity on surface remodelling in the Rhesus monkey rib and femur. Transactions of the 23rd Annual Meeting, Orthopaedic Research Society, *2*:36, 1977.

Sharpe, W. D.: Age changes in human bone: An overview. Bull. N.Y. Acad. Med., *55*:757–773, 1979.

2

BIOMECHANICS OF ARTICULAR CARTILAGE

Van C. Mow
Christopher S. Proctor
Michael A. Kelly

The human body has three types of joints: fibrous, cartilaginous, and synovial. Only one, the synovial, or diarthrodial, joint, allows a wide range of motion. The articulating bone ends of diarthrodial joints are covered by a thin (1 to 5 mm), dense white connective tissue called hyaline articular cartilage.* Articular cartilage is a highly specialized tissue precisely suited to withstand the rigorous joint environment without failing during an average person's lifetime. Physiologically, it is virtually an isolated tissue, devoid of blood vessels, lymph channels, and nerves. Furthermore, its cellular density is less than that of any other tissue.

Articular cartilage in diarthrodial joints has two primary functions: to distribute joint loads over a wide area, thus decreasing the stresses sustained by the contacting joint surfaces (Askew and Mow, 1978), and to allow relative movement of the opposing joint surfaces with minimal friction and wear (Armstrong and Mow, 1980). In this chapter we describe how the biomechanical properties of articular cartilage, as determined by its composition and structure, allow for optimal performance of these functions.

*A notable exception are the temporomandibular joints, synovial joints in which fibrocartilage covers the bone ends. Fibrocartilage and a third type of cartilage, elastic cartilage, are closely related to hyaline cartilage embryonically and histologically but have vastly different mechanical and biochemical properties. Fibrocartilage represents a transitional cartilage found at the margins of some joint cavities, in the joint capsules, and at the insertions of ligaments and tendons into bone. Fibrocartilage also forms the menisci interposed between the articular cartilages of some joints and composes the outer covering of the intervertebral discs, the annulus fibrosus. Elastic cartilage is found in the external ear, in the cartilage of the eustachian tube, in the epiglottis, and in certain parts of the larynx.

COMPOSITION AND STRUCTURE OF ARTICULAR CARTILAGE

Chondrocytes, the sparsely distributed cells in articular cartilage, account for less than 10% of the tissue's volume (Stockwell, 1979; Muir, 1980, 1983). The zonal arrangement of these cells is shown schematically in Figure 2–1. Despite their sparse distribution, chondrocytes manufacture, secrete, and maintain the organic component of the extracellular compartment, the matrix. The organic matrix is composed of a dense network of fine collagen (type II) fibrils enmeshed in a concentrated solution of proteoglycans (PGs). The collagen content of cartilage tissue ranges from 10 to 30% by net weight and the PG content from 3 to 10% by wet weight (Muir, 1980); the remaining 60 to 87% is water, inorganic salts, and small amounts of other matrix proteins, glycoproteins, and lipids (Linn and Sokoloff, 1965; Armstrong and Mow, 1982a). Collagen fibrils and PGs are the

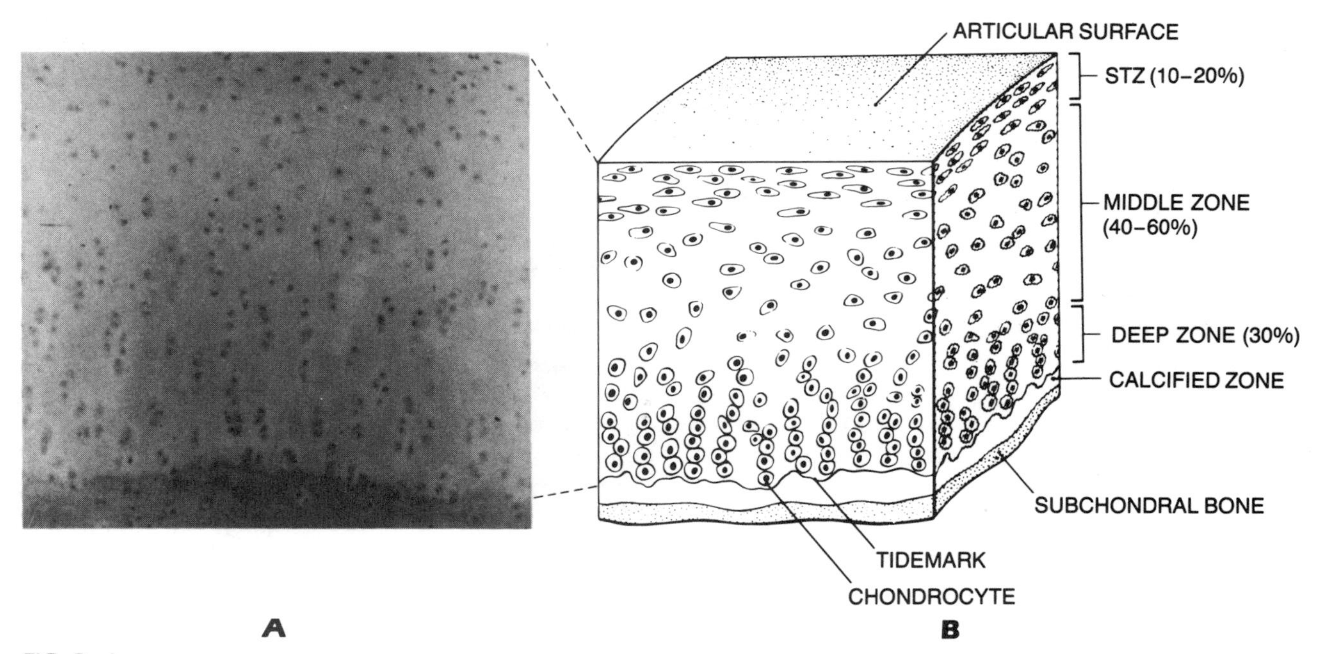

FIG. 2–1
Photomicrograph **(A)** and schematic representation **(B)** of the chondrocyte arrangement throughout the depth of noncalcified articular cartilage. In the superficial tangential zone, chondrocytes are oblong with their long axes aligned parallel to the articular surface. In the middle zone, the chondrocytes are "round" and randomly distributed. Chondrocytes in the deep zone are arranged in columnar fashion oriented perpendicular to the tidemark, the demarcation between the calcified and noncalcified tissue.

structural components supporting the internal mechanical stresses that result from loads being applied to the joint cartilage; it is these structural components, together with water, that determine the biomechanical behavior of this tissue (Kempson et al., 1976; Mow and Lai, 1980; Mow et al., 1984a).

COLLAGEN

Collagen is the most abundant protein in the body. In articular cartilage collagen has a high level of structural organization that provides a fibrous ultrastructure. The basic biologic unit of collagen is tropocollagen, a structure composed of three procollagen polypeptide chains (α chains) coiled into left-handed helices (Fig. 2–2A), which are further coiled about each other into a right-handed triple helix (Fig. 2–2B). These rodlike tropocollagen molecules (1.4 nm in diameter and 300 nm long) (Fig. 2–2C, D) polymerize into larger collagen fibrils (Eyre, 1980). In articular cartilage, these fibrils have an average diameter of 25 to 40 nm (Fig. 2–2E), but this measurement is highly variable*; scanning electron microscopic studies, for instance, have described fibers with diameters ranging up to 200 nm (e.g., Clarke, 1971). Covalent cross-links form between these tropocollagen molecules, adding to the fibrils' high tensile strength (Torchia et al., 1982).

The collagen in articular cartilage is inhomogeneously distributed, giving the tissue a layered character. Numerous investigations using light, transmission electron, and scanning electron microscopy have identified three separate structural zones. For example, Mow and associates (1974) proposed a zonal arrangement for the collagen network shown schematically in Figure 2–3. In the superficial tangential zone, which represents 10 to 20% of the total thickness, are sheets of fine, densely packed fibers randomly woven in planes parallel to the articular surface (Weiss et al., 1968; Redler and Zimny, 1970).

*Differences in tropocollagen α chains in various body tissues give rise to specific molecular species, or types, of collagen. The collagen type in hyaline cartilage, type II collagen, differs from type I collagen found in bone, ligament, and tendon. Type II collagen forms a thinner fibril than that of type I, permitting maximum dispersion of collagen throughout the cartilage tissue.

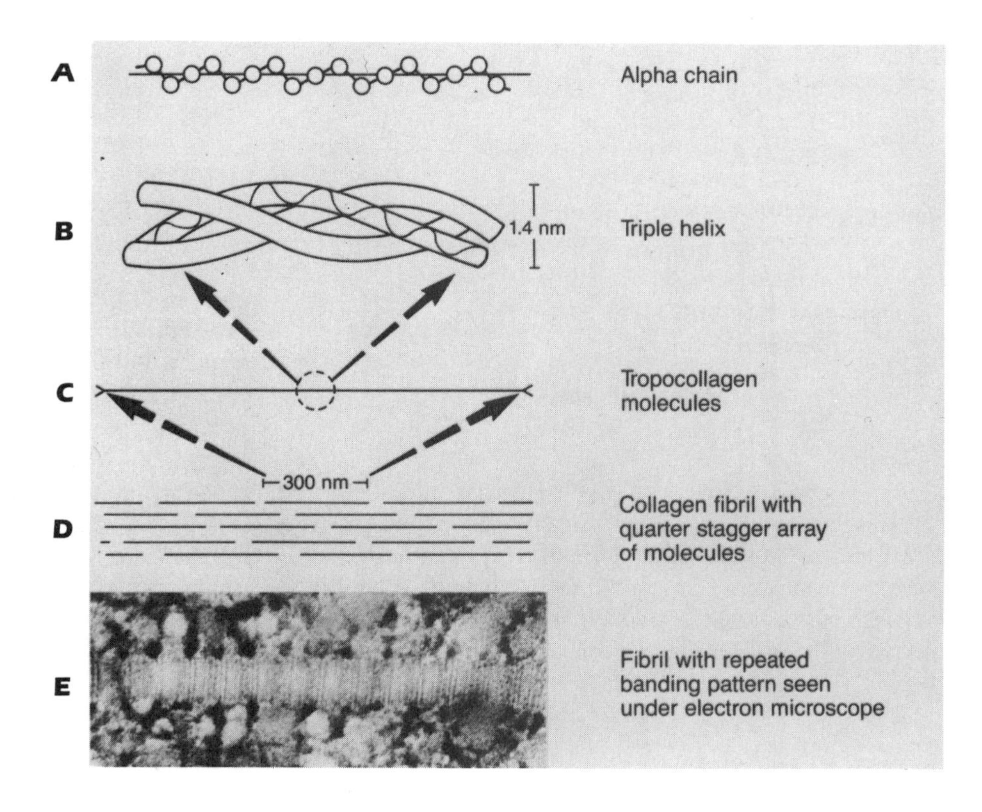

FIG. 2–2
Molecular features of collagen structure from the α chain to the fibril. The flexible amino acid sequence in the α chain **(A)** allows these chains to wind tightly into a right-handed triple helix configuration **(B),** thus forming the tropocollagen molecule **(C).** This tight triple helical arrangement of the chains contributes to the high tensile strength of the collagen fibril. The parallel alignment of the individual tropocollagen molecules, in which each molecule overlaps the other by about one quarter of its length **(D),** results in a repeating banded pattern of the collagen fibril seen by electron microscopy (20,000×) **(E)** (reprinted with permission from Donohue et al., 1983). (Adapted from Eyre, 1980.)

In the middle zone (40 to 60% of the total thickness) the randomly oriented and homogeneously dispersed fibers are farther apart. Below this in the deep zone (about 30% of the total thickness), the fibers come together, forming larger, radially oriented fiber bundles; these bundles then cross the tidemark (Bullough and Jagannath, 1983), the interface between articular cartilage and the calcified cartilage beneath it, to enter the calcified cartilage, forming an interlocking "root" system that anchors the cartilage to the underlying bone. This inhomogeneity of fiber orientation is mirrored by zonal variations in the collagen content, which is highest at the surface and then remains relatively constant throughout the deeper zones (Lipshitz et al., 1975). This layering inhomogeneity appears to serve an important biomechanical function by distributing the stress more uniformly across the loaded regions of the joint tissue (Askew and Mow, 1978).

The most important mechanical properties of collagen fibers are their tensile stiffness and strength (Fig. 2–4A). Although a single collagen fibril has not been tested in tension, the tensile strength of collagen can be inferred from tests on structures with a large collagen content. Tendons, for example, are about 80% collagen (dry weight) and have a tensile stiffness of 1×10^3 megapascals (MPa) and a tensile strength of 50 MPa. Steel, by comparison, has a tensile stiffness of approximately 220×10^3 MPa. (A detailed discussion of the mechanical properties of tissues with a high percentage of collagen is presented in Chapter 3.) Although strong in tension, collagen fibrils offer little resistance to compression because their high slenderness ratio, the ratio of length to thickness, makes it easy for them to buckle under compressive loads (Fig. 2–4B).

Like bone, articular cartilage is anisotropic; its material properties differ with the direction of loading (Woo et al., 1976; Kempson, 1979; Roth and Mow, 1980). It is thought that this anisotropy is related to the varying collagen fiber arrangements within the planes parallel to the articular surface. It is also felt, however, that variations in collagen fiber cross-link density, as well as variations in collagen and PG interactions, contribute to articular cartilage anisotropy. In tension, this anisotropy is usually described with respect to the direction of the articular surface split lines, elongated fissures produced by piercing the articular surface with a small round awl (Fig. 2–5) (Hultkrantz, 1898). The origin of the pattern is related to the directional variation of the tensile stiffness and strength characteristics of articular cartilage described above (Woo et al., 1976; Kempson, 1979; Roth and Mow, 1980). To date, the exact reasons why articular cartilage exhibits such pronounced anisotropies in tension are not known.

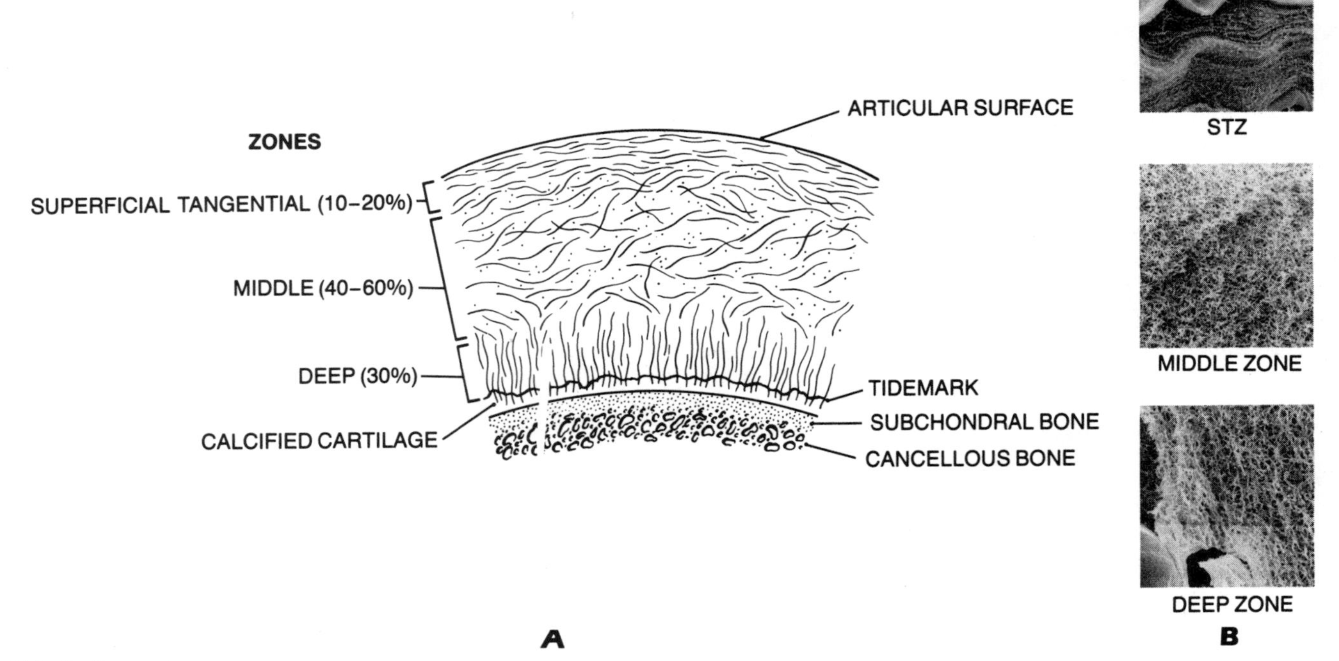

FIG. 2–3
Schematic representation **(A)** (adapted from Mow et al., 1974) and photomicrographs (3000×) **(B)** (courtesy of Dr. T. Takei, Nagano, Japan) of the ultrastructural arrangement of the collagen network throughout the depth of articular cartilage. In the superficial tangential zone (STZ), collagen fibrils are tightly woven into sheets arranged parallel to the articular surface. In the middle zone, randomly arrayed fibrils are less densely packed to accommodate the high concentration of PGs and water. The collagen fibrils of the deep zone form larger, radially oriented fiber bundles that cross the tidemark, enter the calcified zone, and anchor the tissue to the underlying bone. Note the correspondence between this collagen fiber architecture and the spatial arrangement of the chondrocytes shown in Figure 2–1. In the photomicrographs **(B)**, the STZ is shown under compressive loading, while the middle and deep zones are unloaded.

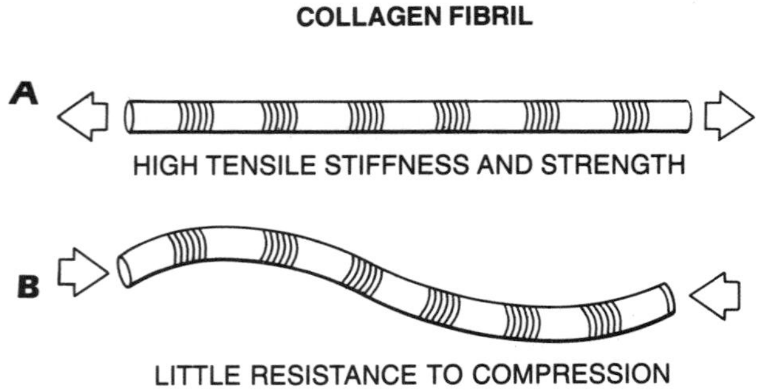

COLLAGEN FIBRIL

A

HIGH TENSILE STIFFNESS AND STRENGTH

B

LITTLE RESISTANCE TO COMPRESSION

FIG. 2–4
Illustration of a collagen fibril's mechanical properties. The fibril is stiff and strong in tension **(A)** but is weak and buckles easily with compression **(B)**. (Adapted from Myers and Mow, 1983.)

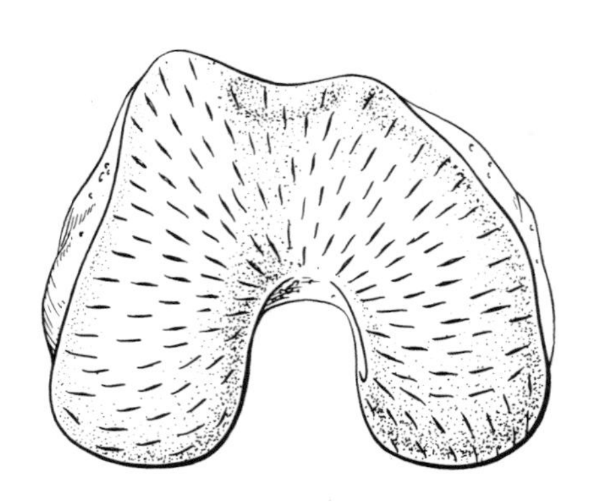

FIG. 2–5
Diagrammatic representation of a split line pattern on the surface of human femoral condyles. (Reprinted from Hultkrantz, 1898.)

PROTEOGLYCAN

Cartilage PGs are large protein-polysaccharide molecules that exist either as monomers or as aggregates (Rosenberg, 1975; Rosenberg et al., 1975; Hascall, 1977; Muir, 1979, 1980). PG monomers (or subunits) consist of an approximately 200-nm-long protein core to which about 150 glycosaminoglycan (GAG) chains and both O-linked and N-linked oligosaccharides are covalently attached (Heinegard and Paulsson, 1984). Keratan sulfate and chondroitin sulfate, the two sulfated GAGs found in articular cartilage, are polymer chains of specific repeating disaccharide units. The chondroitin sulfate chains contain 25 to 30 disaccharide units, whereas the shorter keratan sulfate chains contain only about 13 disaccharide units (Muir, 1979). Furthermore, the distribution of GAGs along the protein core is heterogeneous; there is a region rich in keratan sulfate and O-linked oligosaccharides (Heinegard and Axelsson, 1977) and a region rich in chondroitin sulfate (Hascall, 1977). As shown in Figure 2–6A, the protein core also contains three globular regions (Hardingham et al., 1976, 1987; Perkins et al., 1981; Wiedmann et al., 1984): G_1, the hyaluronic acid–binding region (HABR) located at the N-terminus, which contains a small amount of keratan sulfate and a few N-linked oligosaccharides (Heinegard and Hascall, 1974; Lohmander et al., 1980); G_2, located between the HABR and the keratan sulfate–rich region (Hardingham et al., 1987); and G_3, the core protein C-terminus. The PG monomers may be visualized as having a bottle-brush–like structural arrangement, with the GAGs attached to and radiating perpendicularly from the protein core (Fig. 2–6B, C).

BINDING REGION (G_1)
SECOND GLOBULAR REGION (G_2)
KERATAN SULFATE–RICH REGION
CHONDROITIN SULFATE–RICH REGION
C-TERMINAL (G_3)
LINK PROTEIN
HYALURONIC ACID
PROTEIN CORE
KERATAN SULFATE CHAIN
CHONDROITIN SULFATE CHAIN

A

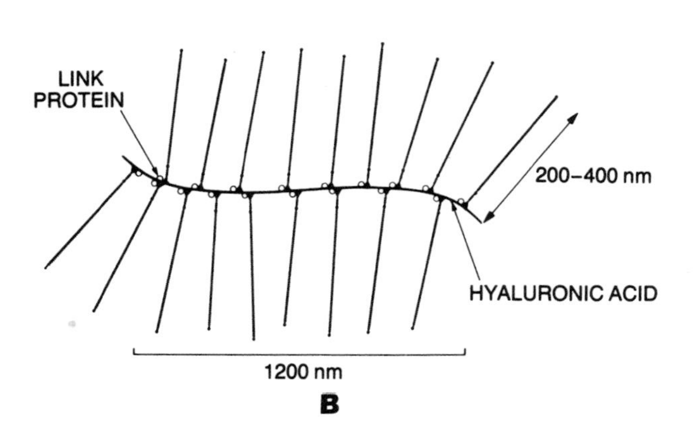

B

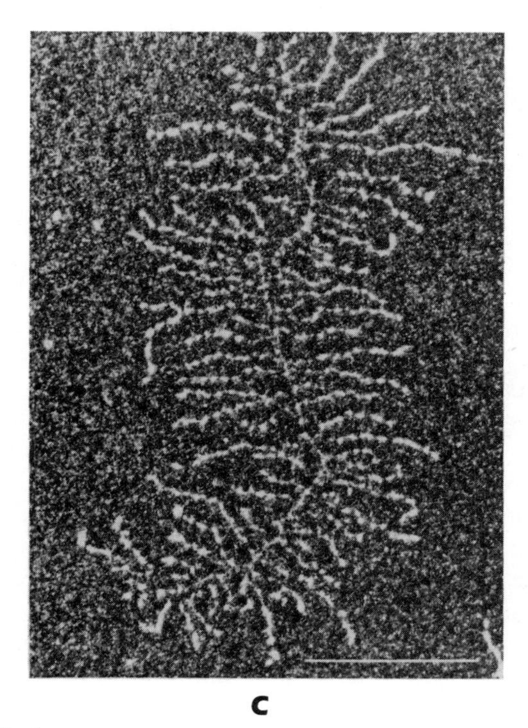

C

FIG. 2–6
A. Schematic depiction of an aggregating PG monomer composed of keratan sulfate and chondroitin sulfate chains bound covalently to a protein core molecule. The PG protein core has three globular regions as well as keratan sulfate– and chondroitin sulfate–rich regions. **B.** Schematic representation of a PG aggregate. In the matrix, monomers noncovalently bind to HA to form a macromolecule with a molecular weight of approximately 200 million daltons. Link protein stabilizes this interaction between the binding region of the monomer and the HA core molecule. **C.** Dark field electron micrograph of a PG aggregate from bovine humeral articular cartilage (120,000×). Horizontal line at lower right represents 0.5 μm. (Reprinted with permission from Rosenberg et al., 1975.)

In native cartilage, most PG monomers associate with hyaluronate to form PG aggregates. These aggregates form when up to 150 PG monomers noncovalently attach to a central hyaluronate core via their HABR (Hardingham and Muir, 1974a, 1974b; Rosenberg, 1975; Rosenberg et al., 1975). The filamentous hyaluronic acid (HA) core molecule is a nonsulfated disaccharide chain, which may be as long as 4 μm. The attachment site between the HABR and the HA is stabilized by small glycoproteins called link proteins (see Fig. 2–6A) (Muir, 1979; Rosenberg, 1980). It is now generally accepted that PG aggregation promotes immobilization of the PGs within the collagen meshwork, adding structural rigidity to the extracellular matrix (Muir, 1980, 1983). Recently, two forms of dermatan sulfate PG have been identified in the extracellular matrix of articular cartilage (Rosenberg et al., 1985). In tendon, dermatan sulfate PGs have been shown to bind noncovalently to the surfaces of collagen fibrils (Scott and Orford, 1981); however, the function of dermatan sulfate in articular cartilage is unknown.

Although PG monomers generally have the basic structure described above, they are not structurally identical. The monomers vary in length, molecular weight, and composition in a variety of ways; in other words, they are polydisperse (Tsiganos et al., 1971; Hardingham et al., 1976; Sweet et al., 1979; Muir, 1983). Recent research has also described two distinct populations of aggregating PG monomers (Inerot and Heinegard, 1983; Thonar et al., 1984; Buckwalter et al., 1985; Heinegard et al., 1985). The first population are present throughout life and contain PGs rich in chondroitin sulfate; the second contain PGs rich in keratan sulfate and are present only in adult cartilage. As articular cartilage matures, other age-related changes in PG composition and structure occur. It is well documented that, with cartilage maturation, the water content (Venn, 1978; Maroudas, 1979; Armstrong and Mow, 1982a) and the carbohydrate-protein ratio progressively decrease (Roughley and White, 1980; Garg and Swann, 1981). These decreases are mirrored by a decrease in the chondroitin sulfate content (Strider et al., 1976; Bayliss and Ali, 1978; Sweet et al., 1979; Roughley and White, 1980). On the other hand, keratan sulfate, which is present in only small amounts at birth, increases throughout development and aging (Inerot et al., 1978; Sweet et al., 1979; Roughley and White, 1980; Garg and Swann, 1981). Thus, the chondroitin sulfate–keratan sulfate ratio, which is about 10 to 1 at birth, is only about 2 to 1 in adult cartilage (Simunek and Muir, 1972; Sweet et al., 1979; Roughley and White, 1980; Thonar et al., 1986). Further, sulfation of the chondroitin sulfate molecules, which can occur at either the 6 or the 4 position, also undergoes age-related changes. In utero, chondroitin-6-sulfate and chondroitin-4-sulfate are present in equal molar amounts; by maturity, the chondroitin-6-sulfate–chondroitin-4-sulfate ratio has increased to about 25 to 1 (Roughley and White, 1980; Roughley et al., 1981). Other studies have also documented an age-related decrease in the hydrodynamic size of the PG monomers (Bayliss and Ali, 1978; Sweet et al., 1979; Roughley and White, 1980; Buckwalter et al., 1985; Thonar et al., 1986). Many of the early changes seen in articular cartilage may reflect cartilage maturation, possibly as a result of increased functional demand with increased weight bearing; however, the functional significance of these changes and those that occur later in life is as yet undetermined.

WATER

Water, the most abundant component of articular cartilage, is most concentrated near the articular surface (approximately 80%) and decreases in a nearly linear fashion with increasing depth to a concentration of approximately 65% in the deep zone (Lipshitz et al., 1976). This fluid contains many free mobile cations (e.g., sodium and calcium) that greatly influence the mechanical behavior of cartilage. The fluid component of articular cartilage is also essential to the health of this avascular tissue in that it permits diffusion of gases, nutrients, and waste products back and forth between chondrocytes and the surrounding nutrient-rich synovial fluid (Mankin and Thrasher, 1975).

Only a small percentage of the water in cartilage is intracellular; about 30% is strongly associated with the collagen fibrils (Torzilli et al., 1982) and is believed to be important in the structural organization of the extracellular matrix. Most of the water thus occupies the intermolecular space and is free to move when a load or pressure gradient is applied to the tissue. When the tissue is loaded, about 70% of the water may be moved. This movement is important in controlling cartilage mechanical behavior and joint lubrication (Mow et al., 1984a; Armstrong and Mow, 1980).

STRUCTURAL INTERACTION AMONG CARTILAGE COMPONENTS

The chemical structure and interactions of the GAGs influence the properties and conformation of the PG aggregates. The closely spaced (0.5 to 1.5 nm) sulfate and carboxyl groups on the GAG disaccharide units dissociate in solution at physiologic pH (Fig. 2–7A), leaving fixed negative charges that create strong intramolecular and intermolecular charge repulsive forces. This charge repulsion extends and stiffens these macromolecules in the interfibrillar space formed by the collagen network. Mobile cations in the solution, such as sodium and calcium, are attracted to the fixed anionic groups on the GAGs (Schubert and Hamerman, 1968; Maroudas, 1970, 1979), creating a substantial osmotic swelling pressure (Donnan osmotic effect) approaching 0.35 MPa (Maroudas, 1979). The magnitude of this swelling pressure is related to the charge group (GAG) density (Maroudas, 1979). This swelling pressure is, in turn, resisted and balanced by tension developed in the collagen network, confining the PGs to only 20% of their free solution domain (Hascall and Hascall, 1981; Muir, 1983). Consequently, this swelling pressure subjects the collagen network to a prestress, even in the absence of external loads. Furthermore, cartilage PGs are inhomogeneously distributed throughout the matrix, generally being most concentrated in the middle zone and least concentrated in the superficial and deep zones (Maroudas et al., 1969; Venn, 1978). The effect on cartilage of inhomogeneous swelling due to varying PG content is as yet unknown, but it is believed to be important in modulating the prestress within the collagen network.

When a stress is applied to the cartilage surface, there is an instantaneous deformation caused primarily by a change in the PG molecular domain (Fig.

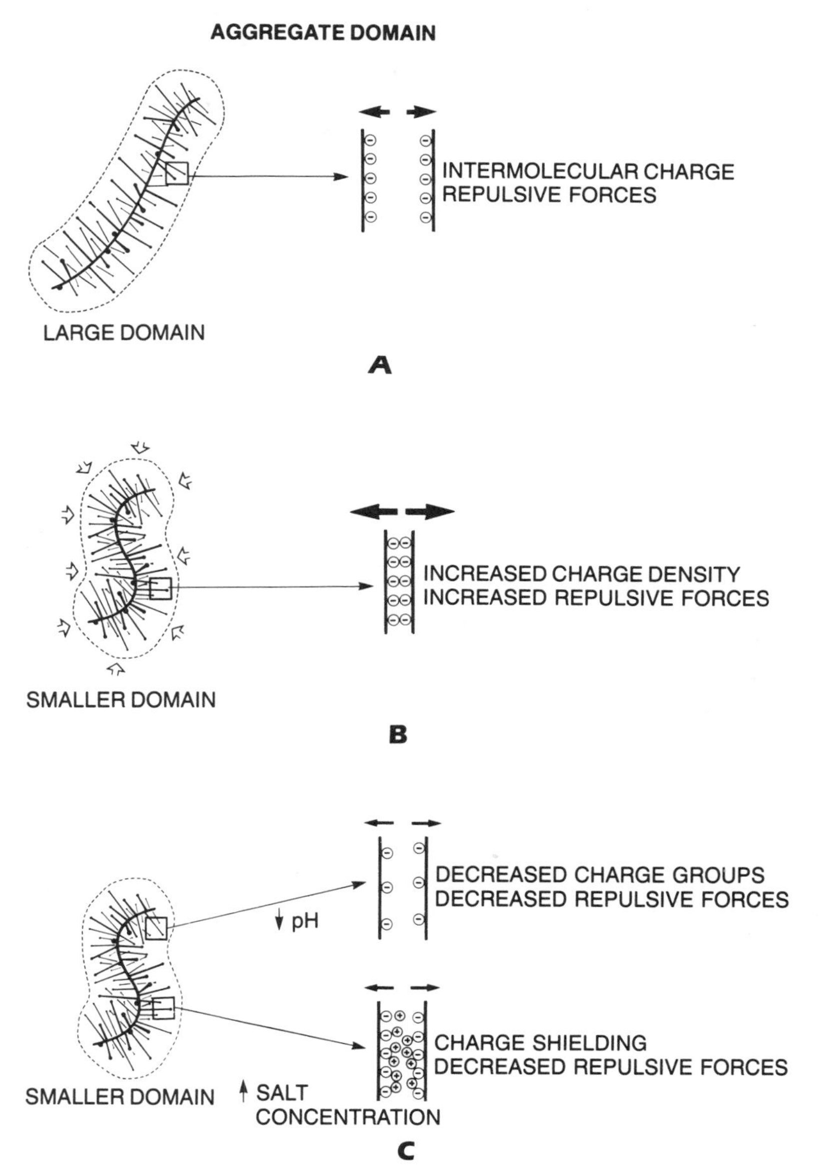

FIG. 2–7
A. Schematic representation of a PG aggregate solution domain (left) and the repelling forces associated with the fixed negative charge groups on the GAGs of a monomer (right). These repulsive forces cause the aggregate to assume a stiffly extended conformation occupying a large solution domain. **B.** Applied compressive stress decreases the aggregate solution domain (left), which in turn increases the charge density and thus the intermolecular charge repulsive forces (right). **C.** Lowering the solution's pH or increasing its ion concentration reduces the intermolecular charge repulsive forces (right), causing the PGs to assume a smaller aggregate domain (left).

2–7B). This external stress causes the internal pressure in the matrix to exceed the swelling pressure, and liquid begins to flow out of the tissue. As the fluid flows out, the PG concentration increases; this change in turn increases the Donnan osmotic swelling pressure and charge-charge repulsive force and bulk compressive stress until they are in equilibrium with the external stress. Furthermore, a change in the solution's pH and/or ion concentration will alter the PG intermolecular charge repulsive forces and cause a change in the size of the aggregate domain (Fig. 2–7C). In this manner, the physicochemical properties of the PG gel trapped in the collagen network enable it to resist compression. This mechanism complements the role played by collagen, which, as previously described, is strong in tension but weak in compression. The ability of PGs to resist compression thus arises from two sources: the Donnan osmotic swelling pressure associated with the tightly packed fixed anionic groups on the GAGs (Maroudas, 1979) and the bulk compressive stiffness of the PG aggregates entangled in the collagen network (Mow and Schoonbeck, 1984). It has been shown that at physiologic ion strength, the Donnan osmotic pressure and the bulk compressive stiffness of the PGs contribute equally to the overall compressive stiffness of the tissue (Mow and Schoonbeck, 1984). The Donnan osmotic pressure as measured by Maroudas (1979) is approximately 0.35 MPa, while the elastic modulus of the collagen-PG matrix as measured by Mow et al. (1980) is approximately 0.78 MPa (Mow and Schoonbeck, 1984).

It is now apparent that collagen and PGs also interact and that these interactions are functionally very important. A small portion of the PGs have been shown to be closely associated with collagen and may serve as a bonding agent between the collagen fibrils, spanning distances too great for collagen cross-links to develop (Shepard and Mitchell, 1977; Muir, 1979, 1983; Hascall and Hascall, 1981; Poole et al., 1982). PGs are also thought to play an important role in maintaining the ordered structure and mechanical properties of the collagen fibrils (Serafini-Fracassini and Smith, 1974; Shepard and Mitchell, 1977; Poole et al., 1982). Recent investigations also show that, in concentrated solutions, PGs interact with each other to form networks of significant strength (Mak et al., 1982, 1983). Moreover, the density and strength of the interaction sites forming the PG network were shown to differ between PG monomers and aggregates. Evidence suggests that there are fewer aggregates in the superficial zone and that the organization of the PGs associated with the collagen fibrils in this zone differs from that of the PGs associated with the collagen fibrils of the deeper zones (Poole et al., 1982). Thus, the interaction between PG and collagen not only plays a direct role in the organization of the extracellular matrix but also contributes directly to the mechanical properties of the tissue (Mow et al., 1984b; Myers et al., 1984).

The specific characteristics of the physical, chemical, and mechanical interactions between collagen and PG have not yet been fully determined. Nevertheless, as discussed above, we know that these structural macromolecules interact to form a porous, composite, fiber-reinforced matrix possessing all the essential mechanical characteristics of a solid that is swollen with water and able to resist the stresses and strains of joint articulation (Maroudas, 1979; Mow et al., 1984a). It has been postulated that these collagen-PG interactions involve a PG monomer, an HA filament, type II collagen, and an unknown bonding agent, and possibly the smaller cartilage components such as collagen type IX, recently identified glycoproteins, and/or polymeric hyaluronate (Poole et al., 1982; Bruckner et al., 1985; Pita et al., 1986). A schematic diagram depicting the structural arrangement within a small volume of articular cartilage is shown in Figure 2–8.

When articular cartilage is subjected to external loads, the collagen-PG solid matrix and interstitial fluid function together in a unique way to protect against high levels of stress and strain. Furthermore,

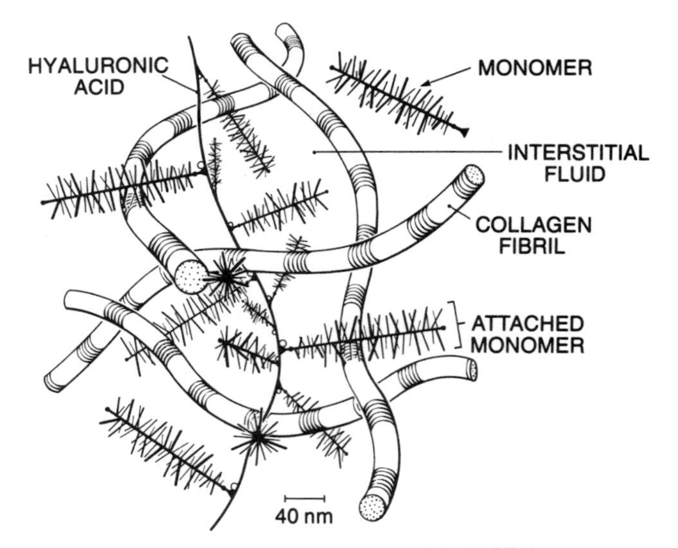

FIG. 2–8
Schematic representation of the molecular organization of cartilage. The structural components of cartilage, collagen and PGs, interact to form a porous composite fiber-reinforced organic solid matrix that is swollen with water.

when the biochemical composition and structural organization of the extracellular matrix change, the biomechanical properties of cartilage also change. In the following section the behavior of articular cartilage under loading and the mechanisms of cartilage fluid flow are discussed in detail.

BIOMECHANICAL BEHAVIOR OF ARTICULAR CARTILAGE

The biomechanical behavior of articular cartilage can best be understood when the tissue is viewed as having two distinct phases: a fluid phase (the interstitial water with the inorganic salts dissolved in it) and a solid phase (the organic solid matrix as described in detail in the previous subsections). In the present context, articular cartilage may be considered as a fluid-filled, porous-permeable medium (the analogy of a water-saturated sponge may be helpful) with both the solid and fluid phases and each distinct constituent of both phases playing roles in the functional behavior of cartilage.

During synovial joint articulation, forces at the joint surface may vary from almost zero to several times body weight (Paul, 1980). The contact areas also vary in a complex manner, and typically they are only of the order of several square centimeters (Ahmed and Burke, 1983). Thus, articular cartilage, under physiologic loading conditions, is a highly stressed material. In order to understand how this tissue responds under these conditions, it is necessary to determine its intrinsic mechanical properties in compression, tension, and shear.

NATURE OF ARTICULAR CARTILAGE VISCOELASTICITY

If a material is subjected to the action of a constant (time-independent) load or a constant deformation and its response varies (is time dependent), then the mechanical behavior of the material is said to be viscoelastic. In general, the response of such a material can be theoretically modeled as a combination of the response of a viscous fluid and an elastic solid, hence viscoelastic.

The two fundamental responses of a viscoelastic material are creep and stress relaxation. Creep occurs when a viscoelastic solid is subjected to the action of a *constant load.* Typically, a viscoelastic solid responds with a rapid initial deformation followed by a slow (time-dependent), progressively increasing deformation known as creep, until an equilibrium is reached. Stress relaxation occurs when a viscoelastic solid is subjected to the action of a *constant deformation.* Typically, a viscoelastic solid responds with a high initial stress followed by a slow (time-dependent), progressively decreasing stress required to maintain the deformation; this phenomenon is known as stress relaxation.

Creep and stress relaxation may be caused by different mechanisms. For solid polymeric materials, these phenomena are due to internal friction caused by the motion of the long polymer chains within the stressed material (Ferry, 1970). The viscoelastic behavior of tendons and ligaments is due primarily to this mechanism (Woo et al., 1981, 1987). The long-term viscoelastic behavior of bone is thought to be due to relative slip of lamellae within the osteons along with the flow of the interstitial fluid (Lakes and Saha, 1979). The compressive viscoelastic behavior of articular cartilage is due primarily to the flow of the interstitial fluid (Mow et al., 1980, 1984a), and in shear it is due primarily to the motion of long polymer chains such as collagen and PGs (Mow et al., 1984a). The component of articular cartilage viscoelasticity due to interstitial fluid flow is known as biphasic viscoelastic behavior (Mow et al., 1980, 1984a), and the component of viscoelasticity due to macromolecular motion is known as flow-independent (Hayes and Bodine, 1978), or intrinsic, viscoelastic behavior of the collagen-PG solid matrix (Roth et al., 1982; Mow et al., 1982; Zhu et al., 1986).

BIPHASIC CREEP RESPONSE OF ARTICULAR CARTILAGE IN COMPRESSION

The biphasic creep response of articular cartilage in a one-dimensional confined compression experiment is depicted in Figure 2–9. In this case, a constant compressive stress is applied to the tissue at a given time (t_o, point A in Fig. 2–9) and the tissue is allowed to creep to its final equilibrium value (ϵ_∞). For articular cartilage, as illustrated in the top diagrams, creep is caused by exudation of the interstitial fluid. Exudation is most rapid initially, as evidenced by the early rapid rate of increased deformation, and it diminishes gradually until flow ceases. During creep, the load applied at the surface is balanced by the compressive stress developed within the collagen-PG solid matrix and frictional drag generated by the flow of the interstitial fluid during exudation. Creep ceases when the compressive stress developed within the solid matrix is sufficient to balance the

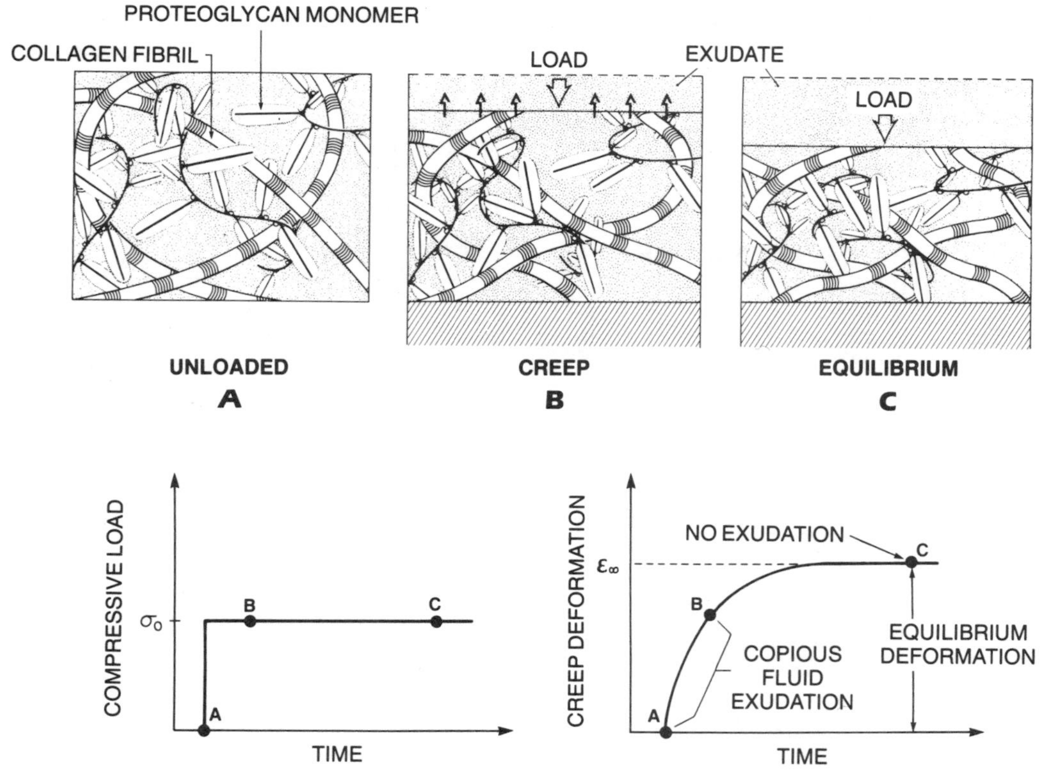

FIG. 2–9
A constant stress (σ_o) applied to a sample of articular cartilage (bottom left) and creep response of the sample under the constant applied stress (bottom right). The drawings of a block of tissue above the curves illustrate that creep is accompanied by copious exudation of fluid from the sample and that the rate of exudation decreases over time from point A to B to C. At equilibrium (ϵ_∞) fluid flow ceases and the load is borne entirely by the solid matrix (point C). (Adapted from Armstrong and Mow, 1980.)

applied stress alone; at this point no fluid flows and the deformation equilibrium is reached.

Typically, relatively thick (2 to 4 mm) human and bovine articular cartilage takes 4 to 16 hours to reach creep equilibrium. Rabbit cartilage, which is generally less than 1.0 mm thick, takes approximately 1 hour to reach creep equilibrium. Theoretically, it can be shown that the time taken to reach creep equilibrium varies inversely with the square of the thickness of the tissue (Mow et al., 1980). Under relatively high loading conditions (greater than 1.0 MPa) 50% of the total fluid content may be squeezed from the tissue (Edwards, 1967). Further, studies in vitro demonstrate that if the tissue is immersed in physiologic saline, the exuded fluid is fully recoverable when the load is removed (Elmore et al., 1963; Sokoloff, 1963).

Since the rate of creep is governed predominantly by the rate of fluid exudation, it can be used to determine the permeability coefficient of the tissue (Mow et al., 1980, 1984a). This is known as the indirect measurement for tissue permeability (k). Average values of human and bovine articular cartilage k obtained in this manner are $4.70 \pm 0.04 \times 10^{-15}$ $m^4/(N \cdot sec)$ and $4.67 \pm 0.04 \times 10^{-15}$ $m^4/(N \cdot sec)$, respectively. At equilibrium, no fluid flow occurs; thus, the equilibrium deformation can be used to measure the *intrinsic* compressive modulus (H_A) of the collagen-PG solid matrix (Mow et al., 1980; Armstrong and Mow, 1982a). Average values of human bovine articular cartilage H_A are 0.79 ± 0.36 and 0.85 ± 0.21 MPa, respectively. Since these coefficients are *intrinsic* material properties of the solid matrix, it is meaningful to determine how they vary with matrix composition. It was determined that k varies directly while H_A varies inversely with water content (Roth et al., 1981; Armstrong and Mow, 1982a). In addition, it was shown that H_A varies directly with tissue uronic acid content (i.e., PG content) (Roth et al., 1981; Armstrong and Mow, 1982b).

BIPHASIC STRESS RELAXATION RESPONSE OF ARTICULAR CARTILAGE IN COMPRESSION

The biphasic viscoelastic stress relaxation response of articular cartilage in a one-dimensional compression experiment is depicted in Figure 2–10. In this case, a constant compression rate (line t_o-A-B of lower left figure) is applied to the tissue until a given deformation (u_o) is reached; beyond point B, the deformation u_o is maintained. For articular cartilage, the typical stress response due to this imposed deformation is shown in the lower right figure (Mow et al., 1984a; Holmes et al., 1986). During the compression phase, the stress rises continuously until a given stress (σ_o) is reached, corresponding to u_o, while during the stress relaxation phase, the stress continuously decays along the curve B-C-D-E until the equilibrium stress (σ_∞) is reached.

The mechanisms responsible for the stress rise and stress relaxation are also depicted in Figure 2–10. The top diagrams illustrate that the stress rise in the compression phase is associated with fluid exudation while stress relaxation is associated with fluid redistribution within the porous solid matrix. During the compressive phase, the high stress is generated by forced exudation of the interstitial fluid and the compaction of the solid matrix near the surface. Stress relaxation is in turn caused by the relief, or rebound, of the high compaction region near the surface of the solid matrix. This stress relaxation process ceases when the compressive stress developed within the solid matrix reaches the stress generated by the intrinsic compressive modulus of the solid matrix corresponding to u_o (Mow et al., 1980, 1984a; Holmes et al., 1986). Analysis of this stress relaxation process leads to the conclusion that, under physiologic loading conditions, excessive

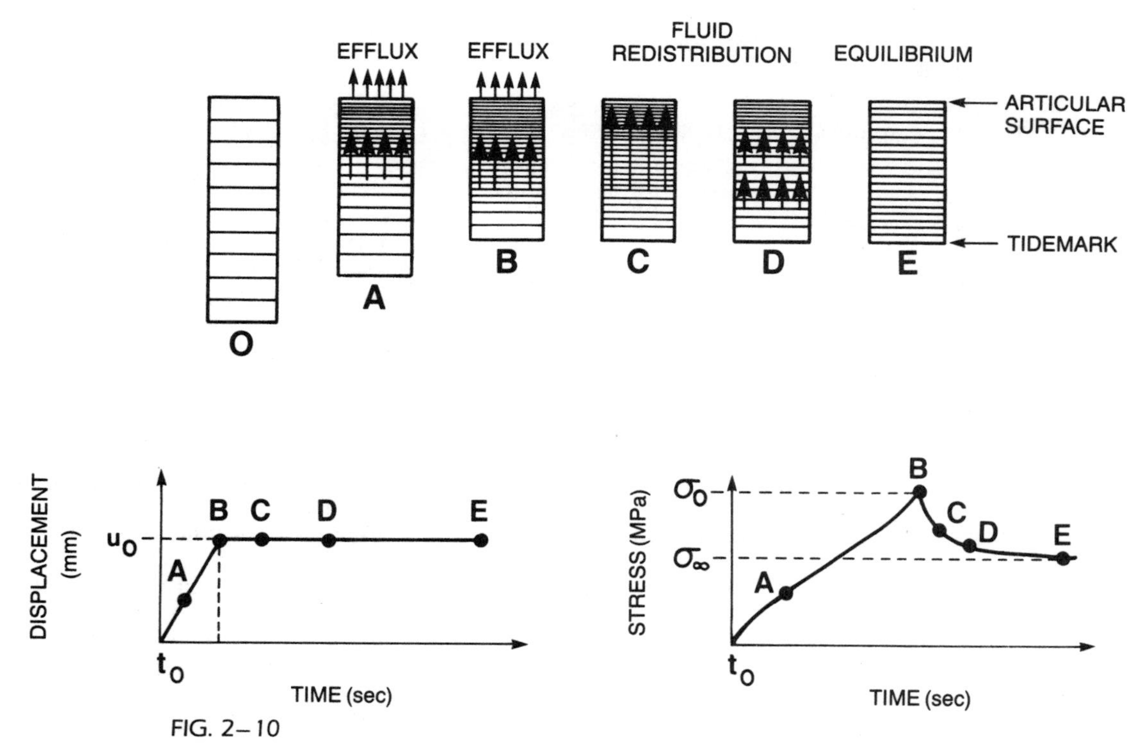

FIG. 2–10
Controlled ramp displacement curve of a cartilage specimen commencing at t_o (bottom left) and the stress response curve of the cartilage in this uniaxial confined compression experiment (bottom right). The sample is compressed until u_o is reached (point B) and maintained over time (points B to E). The history of the stress and response shows a characteristic stress increase during the compressive phase to a peak at σ_o (points t_o to B); the stress then decreases during the relaxation phase (points B to D) until the equilibrium stress σ_∞ is reached (point E). Above these two curves, schematics illustrate interstitial fluid flow (represented by arrows) and solid matrix deformation during this compressive process. Fluid exudation gives rise to the peak stress (point B), and fluid redistribution gives rise to the stress relaxation phenomena.

stress levels are difficult to maintain, since stress relaxation will rapidly attenuate the stress developed within the tissue; this must necessarily lead to rapid spread of the contact area in the joint during articulation (Mow, 1977).

As in the creep experiment, the rate of stress relaxation may be used to determine the permeability coefficient of the tissue, and the equilibrium stress may be used to measure the intrinsic compressive modulus of the solid matrix. The values of k and H_A measured by the biphasic stress relaxation method are consistent with those measured by the biphasic creep method (Stahurski et al., 1981).

PERMEABILITY OF ARTICULAR CARTILAGE

Fluid-filled porous materials may or may not be permeable. The ratio of fluid volume to the total volume of the porous material is known as the porosity (β); thus, porosity is a geometric concept. Articular cartilage is, therefore, a highly porous material. If the pores are interconnected, then the porous material is permeable. Permeability is a measure of the ease with which fluid can flow through a porous permeable material, and it is inversely proportional to the frictional drag (K) exerted by the fluid flowing through this material. Thus, permeability is a physical concept; it is a measure of the resistive force required to cause the fluid to flow at a given speed through the porous permeable material. This resistive force is generated by the interaction of the viscous interstitial fluid and the pore walls of the porous permeable material. The value of k is related to K by the relationship $k = \beta^2/K$ (Lai and Mow, 1980). Articular cartilage has a very low permeability; thus, high frictional resistive forces are generated when fluid is caused to flow through the porous solid matrix.

In the previous sections on cartilage viscoelasticity, we discussed the process of fluid flow through articular cartilage induced by solid matrix compression and how this process influences the viscoelastic behavior of the tissue. This process also provides an indirect method for determining the permeability of the tissue. In this section, we discuss the experimental method for measuring the permeability coefficient directly. Such an experiment is depicted in Figure 2–11A. A specimen of the tissue is held fixed in a chamber subjected to the action of a pressure gradient, the imposed upstream pressure (P_1) being greater than the downstream ambient pressure (P_2). The thickness of the specimen is denoted by h and

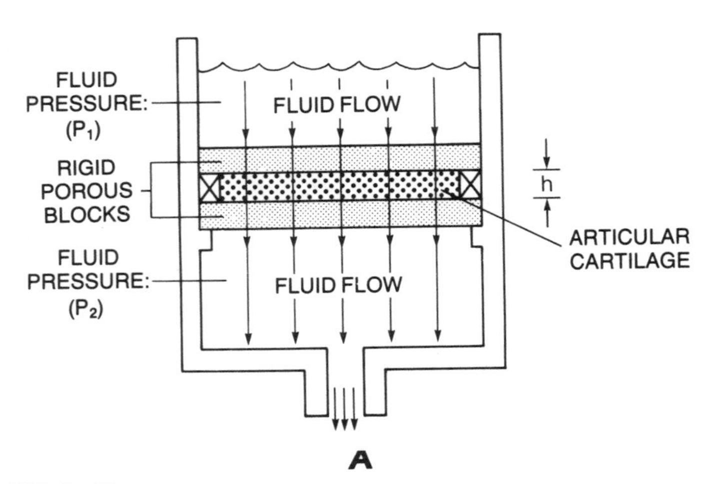

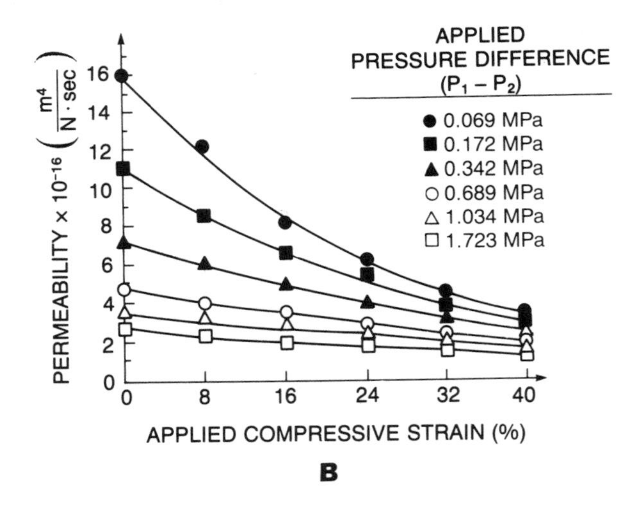

FIG. 2–11
A. Experimental configuration used in measuring the permeability of articular cartilage, involving the application of a pressure gradient $(P_1 - P_2)/h$ across a sample of the tissue (h = tissue thickness). Since the fluid pressure (P_1) above the sample is greater than that beneath it (P_2), fluid will flow through the tissue. The permeability coefficient (k) in this experiment is given by the expression $Qh/A(P_1 - P_2)$, where Q is the volumetric discharge per unit time and A is the area of permeation. (Adapted from Torzilli and Mow, 1976.) **B.** Experimental curves for articular cartilage permeability show its strong dependence on compressive strain and applied pressure. Measurements were taken at applied pressure differential ($P_1 - P_2$) and applied strains. The permeability decreased in an exponential manner as a function of both increasing applied compressive strain and increasing applied pressure. (Adapted from Lai and Mow, 1980.)

the cross-sectional area of permeation by A. Darcy's law, used to determine k in this simple experimental setup, yields $k = Qh/A(P_1 - P_2)$, where Q is the volumetric discharge per unit time through the specimen, whose area of permeation is A. Using low pressures, investigators first used this method to determine the permeability of articular cartilage (McCutchen, 1962; Edwards, 1967; Maroudas, 1970). The value of k obtained in this manner ranged from 1.1×10^{-13} $m^4/(N \cdot sec)$ to 7.6×10^{-13} $m^4/(N \cdot sec)$. In addition, using a uniform straight tube model, these investigators estimated that the average "pore diameter" is 6 nm (McCutchen, 1962; Maroudas, 1979). Thus, the pores in articular cartilage are of molecular size.

The permeability of articular cartilage under compressive strain and at high pressures was first obtained by Mansour and Mow (1976) and Mow and coworkers (1980). The high pressure and compressive strain conditions used in these studies more closely resemble conditions found in diarthrodial joints under physiologic conditions. In these experiments, k was measured as a function of two variables: the pressure gradient across the specimen and the axial compressive strain applied to the sample. The results from these experiments are shown in Figure 2–11B. It was found that the permeability decreased exponentially as a function of both increasing compressive strain and increasing applied fluid pressure. It was later shown, however, that the dependence of k on the applied fluid pressure derives from compaction of the solid matrix, which in turn results from the frictional drag caused by the permeating fluid (Lai and Mow, 1980). From the point of view of pore structure, compaction of the solid matrix reduces the porosity, and hence the average pore diameter, within the solid matrix; thus, solid matrix compaction increases frictional resistance (Mow et al., 1984a).

The nonlinear permeability of articular cartilage demonstrated in Figure 2–11B suggests that the tissue has a mechanical feedback system that may serve important purposes under physiologic conditions. Under high loads, through the mechanism of increased frictional drag against interstitial fluid flow, the tissue appears stiffer and it is more difficult to cause fluid exudation. This mechanism may also be important in joint lubrication (see below).

BEHAVIOR OF ARTICULAR CARTILAGE UNDER UNIAXIAL TENSION

The mechanical behavior of articular cartilage in tension is also highly complex. In tension the tissue is strongly anisotropic (being stiffer and stronger for specimens harvested in the direction parallel to the split line pattern than for those harvested perpendicular to this pattern) and very inhomogeneous (for mature animals, being stiffer and stronger for specimens harvested from the superficial regions than for those harvested deeper in the tissue) (Kempson, 1979; Woo et al., 1976; Roth and Mow, 1980). It is interesting to note that articular cartilage from immature bovine knee joints does not exhibit these layered inhomogeneous variations; however, the superficial zones of both mature and immature bovine cartilage appear to have the same tensile stiffness (Roth and Mow, 1980). These anisotropic and inhomogeneous characteristics in mature joints are believed to be due to the varying collagen and PG structural organizations of the joint surface and the layering structural arrangements found within the tissue. Thus, the collagen-rich superficial zone appears to provide the joint cartilage with a tough, wear-resistant, protective skin (see Fig. 2–3A).

Articular cartilage also exhibits viscoelastic behavior in tension (Simon et al., 1984; Woo et al., 1987), which is attributable to both the internal friction associated with polymer motion and the flow of the interstitial fluid (Li et al., 1984; Woo et al., 1987). To examine the intrinsic mechanical response of the collagen-PG solid matrix in tension, it is necessary to negate the biphasic fluid flow effects. To do this, one must perform slow, low–strain rate experiments (Woo et al., 1976; Roth and Mow, 1980; Li et al., 1983) or perform an incremental strain experiment where stress relaxation is allowed to progress toward equilibration at each increment of strain (Akizuki et al., 1986). Typically, in a low–strain rate tensile experiment, a displacement rate of 0.5 cm per minute is used and the specimens are usually pulled to failure. Unfortunately, using these procedures to negate the effect of interstitial fluid flow also negates the manifestation of the intrinsic viscoelastic behavior of the solid matrix. Thus, only *equilibrium*-intrinsic mechanical properties of the solid matrix may be determined from these tensile tests. The intrinsic viscoelastic properties of the solid matrix must be determined from a pure shear study (see below).

The "equilibrium" stress-strain curve for a specimen of articular cartilage tested under a constant low–strain rate condition is shown in Figure 2–12. A comparison of this curve with that shown for bone (see Fig. 1–6) reveals considerable differences between tensile behaviors of articular cartilage and bone. In contrast to bone, articular cartilage tends to stiffen with increasing strain when the strain be-

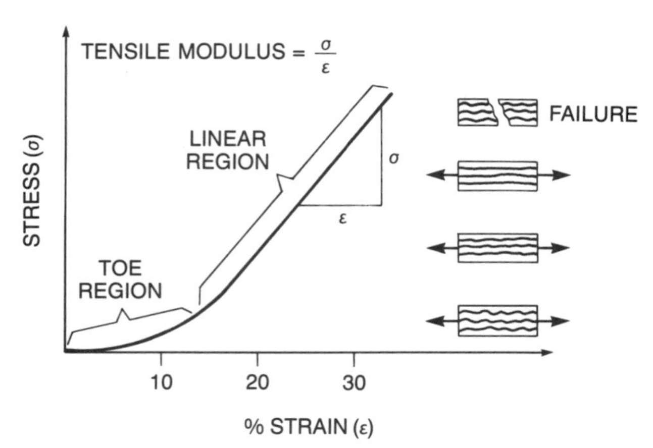

FIG. 2–12
Typical tensile stress-strain curve for articular cartilage. The drawings on the right of the curve show the configuration of the collagen fibrils at various stages of loading. In the toe region, collagen fibril pull-out occurs as the fibrils align themselves in the direction of the tensile load. In the linear region, the aligned collagen fibers are stretched until they fail.

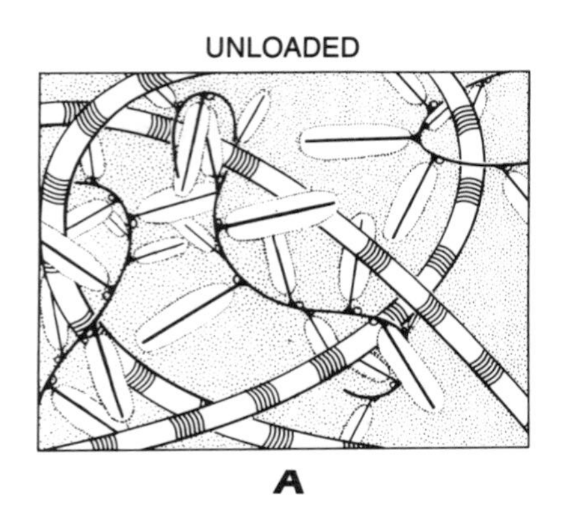

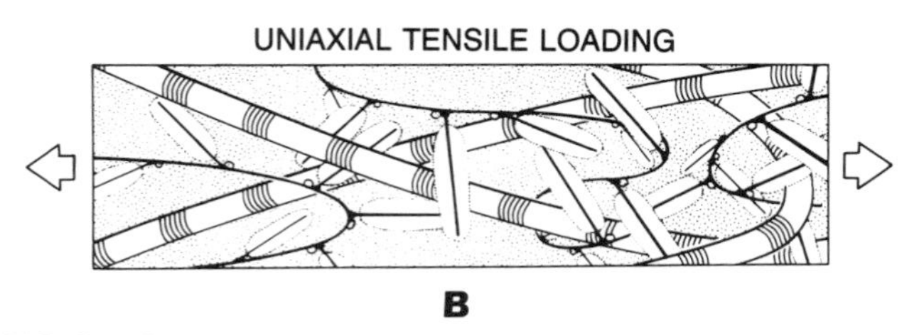

FIG. 2–13
Schematic depiction of the main components of articular cartilage when the tissue is unloaded **(A)** and when a tensile load is applied **(B).** Loading would result in an alignment of collagen fibrils along the axis of tension. (Adapted from Myers and Mow, 1983.)

comes great; thus, over the entire range of strain (up to 60%) in tension, articular cartilage cannot be described by a single Young's modulus. Rather, a tangent modulus, defined by the tangent to the stress-strain curve, must be used to describe the tensile stiffness of the tissue. The use of this tangent modulus results in a widely varying value, ranging from 3 to 100 MPa for the Young's modulus of articular cartilage in tension (Kempson, 1979; Woo et al., 1976; Roth and Mow, 1980). At physiologic strain levels (less than 15%) (Armstrong et al., 1979), the linear Young's modulus of articular cartilage ranges from 5 to 10 MPa (Akizuki et al., 1986).

The morphologic cause for the shape of the tensile stress-strain curve for large strains is depicted in the diagrams on the right of Figure 2–12. The initial toe region is believed to be due to collagen fiber pull-out during the initial portion of the tensile experiment, and the final linear region is believed to be due to stretching of the aligned collagen fibers. Failure occurs when all the collagen fibers contained in the specimen are ruptured. Figure 2–13A depicts an unstretched articular cartilage specimen, while Figure 2–13B depicts a stretched specimen. Figure 2–14A and B present scanning electron micrographs (SEM) of cartilage blocks under zero and 30% stretch (right) and the corresponding histograms of collagen fiber orientation determined from the SEM pictures (left). It can be seen clearly that the collagen network within cartilage responds to tensile stress and strain (Wada and Akizuki, 1987).

Alteration of the molecular structure of collagen, the organization of the collagen fibers within the collagenous network, or the collagen fiber cross-linking (such as that occurring in mild fibrillation or osteoarthritis) changes the tensile properties of the network. Schmidt and coworkers (1987) have shown a definitive relationship between collagen hydroxy-pyridinium cross-linking and tensile stiffness and strength of normal bovine cartilage. Furthermore, Akizuki and associates (1986) showed that progressive degradation of human knee joint cartilage, from mild fibrillation to osteoarthritis, yields a progressive deterioration of the intrinsic tensile properties of the collagen-PG solid matrix. This work supports the belief that disruption of the collagen network is a key factor in the initial events leading to the development of osteoarthritis. Also, loosening of the collagen network is generally believed to be responsible for the increased swelling (hence, water content) of osteoarthritic cartilage (Mankin and Thrasher, 1975; Maroudas, 1979). We have already discussed how increased water content leads to decreased compressive stiffness and increased permeability of articular cartilage.

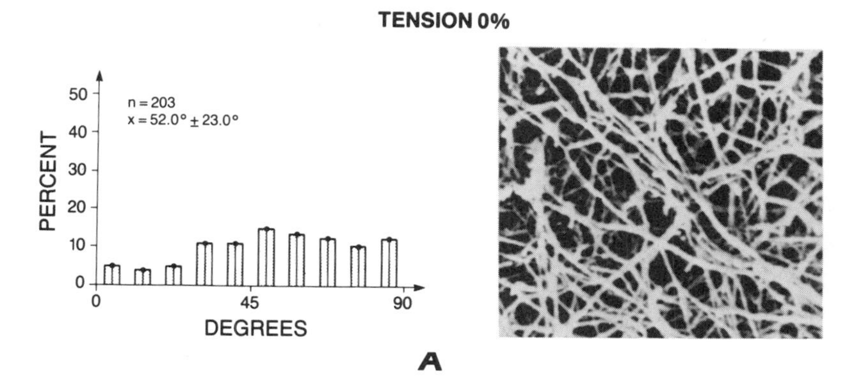

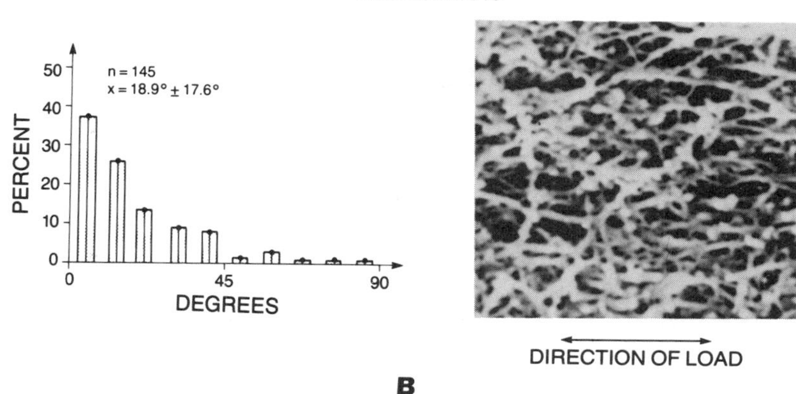

FIG. 2–14
Collagen fibril alignment is clearly demonstrated by the scanning electron micrographs (right, 10,000×) of cartilage blocks under zero stretch **(A)** and 30% stretch **(B)**. The histograms (left), calculated from the micrographs, represent the percentage of collagen fibers oriented in the direction of the applied tension. At zero stretch the fibers have a random orientation; however, at 30% they are aligned in the direction of the applied tension. (Reprinted with permission from Wada and Akizuki, 1987.)

BEHAVIOR OF ARTICULAR CARTILAGE IN PURE SHEAR

In tension and compression, only the *equilibrium*-intrinsic properties of the collagen-PG solid matrix can be determined. This is because a volumetric change always occurs within a material when it is subjected to uniaxial tension or compression. This volumetric change causes interstitial fluid flow and induces biphasic viscoelastic effects within the tissue. If, however, articular cartilage is tested in pure shear under infinitesimal strain conditions, no pressure gradients or volumetric changes will be produced within the material; hence, no interstitial fluid flow will occur (Hayes and Bodine, 1978; Mow et al., 1982) (Fig. 2–15). Thus, a steady dynamic pure shear experiment can be used to assess the intrinsic viscoelastic properties of the collagen-PG solid matrix.

In a steady dynamic shear experiment, the viscoelastic properties of the collagen-PG solid matrix are determined by subjecting a thin circular wafer of tissue to a steady sinusoidal torsional shear (Fig. 2–16). In an experiment of this type, the tissue specimen is held by a precise amount of compression between two rough porous platens. The lower platen is attached to a sensitive torque transducer, and the upper platen is attached to the mechanical spectrometer precision servocontrolled direct current motor. A sinusoidal excitation signal may be provided by the motor in a frequency of excitation ranging from 0.01 to 20 hertz (Hz). For shear strain magnitudes ranging from 0.2 to 2.0%, the storage modulus (G′) and the loss modulus (G″) of the intrinsic viscoelastic response of the collagen-PG solid matrix may be determined as a function of frequency.

Sometimes it is more convenient to determine the magnitude of the dynamic shear modulus ($|G^*|$) given by:

$$|G^*|^2 = (G')^2 + (G'')^2, \qquad (1)$$

and the phase shift angle (δ), given by:

$$\delta = \tan^{-1}(G''/G'). \qquad (2)$$

The magnitude of the dynamic shear modulus is a measure of the total resistance offered by the viscoelastic material. The value of δ, the angle between the steady applied sinusoidal strain and the steady sinusoidal torque response, is a measure of the total

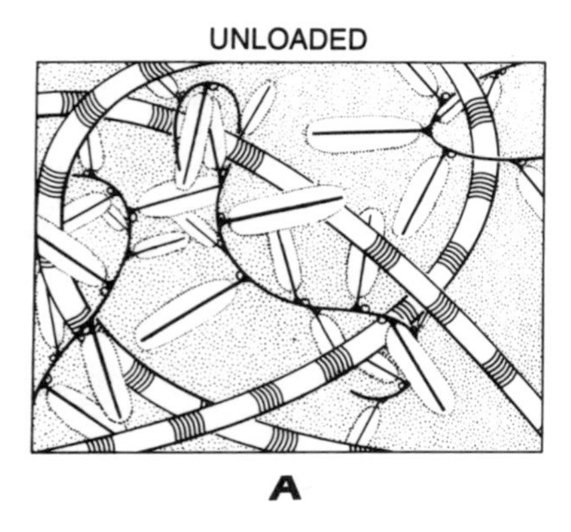

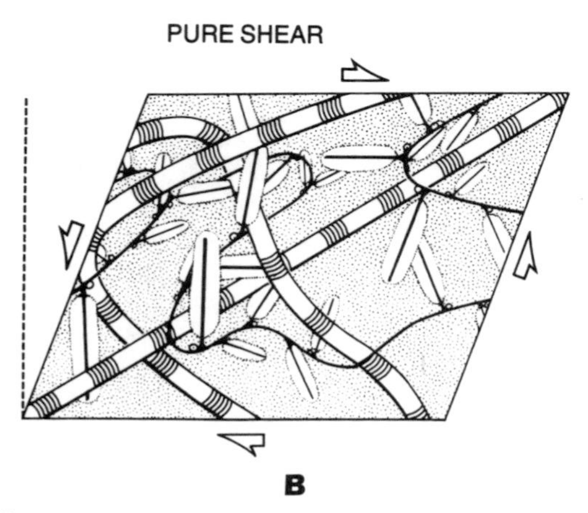

FIG. 2–15
Schematic depiction of unloaded cartilage **(A)**, and cartilage subjected to pure shear **(B)**. When cartilage is tested in pure shear under infinitesimal strain conditions, no volumetric changes or pressure gradients are produced; hence, no interstitial fluid flow occurs. This figure also demonstrates the functional role of collagen fibrils in resisting shear deformation.

frictional energy dissipation within the material. For a pure elastic material with no internal frictional dissipation, δ is zero; for a pure viscous fluid, δ is 90 degrees.

The magnitude of the dynamic shear modulus for normal bovine articular cartilage has been measured to range from 1 to 3 MPa, while δ has been measured to range from 9 to 20 degrees (Hayes and Bodine, 1978; Roth et al., 1982; Mow et al., 1982). Recently Zhu and coworkers (1986) measured the intrinsic transient shear stress relaxation behavior of the collagen-PG solid matrix along with the steady dynamic shear properties. With both the steady and the transient results, the latter investigators showed that the quasilinear viscoelasticity theory proposed by Fung (1981) for biologic materials provides an accurate description of the intrinsic viscoelastic behavior of the collagen-PG solid matrix. Figure 2–17 depicts a comparison of theoretical prediction of the transient stress relaxation phenomenon in shear with the results for the quasilinear viscoelasticity theory.

From these shear studies it is possible to obtain some insight into how the collagen-PG solid matrix functions. First, we note that measurements of PG solutions at concentrations similar to those found in articular cartilage in situ yield a magnitude of shear modulus on the order of 10 Pa and a phase shift angle ranging up to 70 degrees (Mow et al., 1984b). Therefore, it appears that the magnitude of the shear modulus of concentrated PG solution is 100,000 times less and the phase angle is six to seven times greater

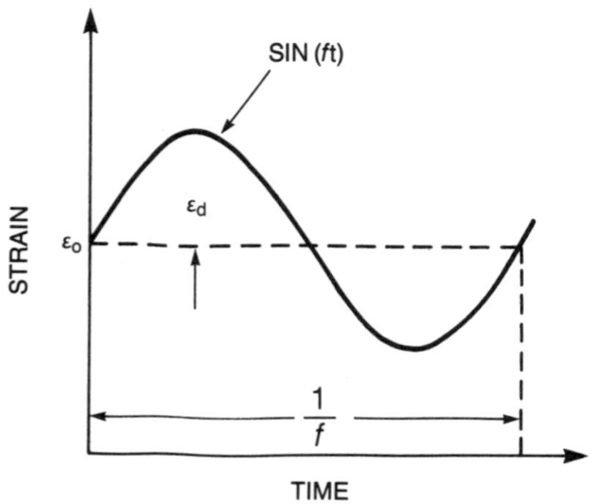

FIG. 2–16
Steady sinusoidal torsional shear imposed on a specimen in pure shear. The fluctuating strain is in the form of a sine wave with a strain amplitude ϵ_d and frequency *f*.

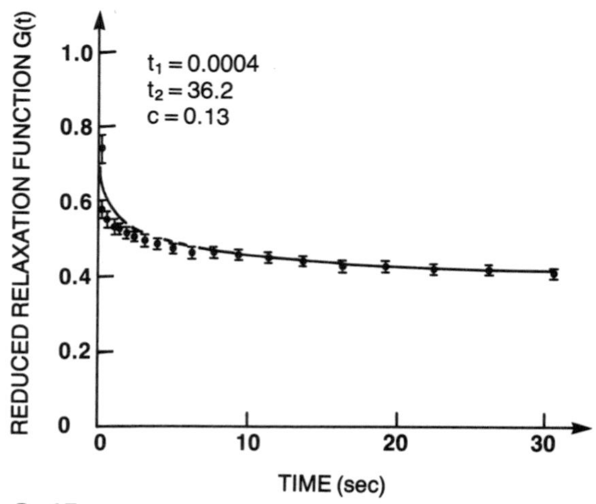

FIG. 2–17
Typical stress-relaxation curve after a step change in shear strain, expressed in terms of the mean of ten cycles of stress relaxation normalized by the initial stress. The solid line represents the theoretical prediction of the quasilinear viscoelasticity theory. (Adapted from Zhu et al., 1986.)

than that of articular cartilage solid matrix. This finding suggests that PGs do not function in situ to provide shear stiffness for articular cartilage. That stiffness must therefore derive from the collagen content or from the collagen-PG interaction (Myers and Mow, 1983; Mow et al., 1984a). From this interpretation, an increase in collagen, which is a much more elastic element than is PG and the predominant load-bearing element of the tissue in shear, would reduce the frictional dissipation and hence the observed phase angle.

LUBRICATION OF ARTICULAR CARTILAGE

Synovial joints are subjected to an enormous range of loading conditions, and under normal circumstances the cartilage surface sustains little wear. The human hip joint, for example, sustains light loads during high-speed motions such as the swing phase of walking or running and impact loads of large magnitude and short duration during such activities as jumping or at the heel strike of walking; furthermore, moderate fixed loads occur during limited activity such as prolonged standing. The minimal wear of normal cartilage associated with such varied loads indicates that sophisticated lubrication processes are at work in the joint. These processes have been attributed to a lubricating fluid film forming *between* the articular cartilage surface and to an adsorbed boundary lubricant *on* the articular cartilage surface during motion and loading. The variety of joint demand also suggests that a number of mechanisms are responsible for diarthrodial joint lubrication. To understand diarthrodial joint lubrication, one should use basic engineering concepts of lubrication.

From an engineering perspective, there are two fundamental types of lubrication. One is boundary lubrication, involving a single monolayer of lubricant molecules adsorbed on each bearing surface. The other is fluid film lubrication, in which a thin fluid film provides greater surface separation (Bowden and Tabor, 1967). Both lubrication types appear to occur in articular cartilage under varying circumstances.

During diarthrodial joint function, relative motion of the articulating surfaces occurs. In boundary lubrication, the surfaces are protected by the adsorbed layer of boundary lubricant; thus, direct, surface-to-surface contact is prevented and most of the surface wear is eliminated. Boundary lubrication is essentially independent of the *physical* properties of either the lubricant (e.g., its viscosity) or the bearing material (e.g., its stiffness), but instead depends almost entirely on the chemical properties of the lubricant (McCutchen, 1966; Dowson, 1966/1967). In synovial joints, a specific glycoprotein, lubricin, appears to be the synovial fluid constituent responsible for boundary lubrication (Swann et al., 1979, 1985). Lubricin is adsorbed as a macromolecule monolayer to each articulating surface (Fig. 2–18). These two layers, ranging in combined thickness from 1 to 100 nm, are able to carry loads and appear to be effective in reducing friction (Swann et al., 1979).

Fluid film lubrication, in contrast to boundary lubrication, utilizes a thin film of lubricant that causes greater bearing surface separation. The load on the bearing is then supported by the pressure in this fluid film. The fluid film thickness associated with engineering bearings is usually less than 20 μm. Two classic modes of fluid film lubrication defined in engineering are hydrodynamic and squeeze film lubrication (Fig. 2–19), which apply to rigid bearings composed of a relatively undeformable material such as stainless steel.

Hydrodynamic lubrication occurs when nonparallel rigid bearing surfaces lubricated by a fluid film move tangentially with respect to each other (i.e., slide on each other), forming a converging wedge of fluid. A lifting pressure is generated in this wedge by the fluid viscosity as the bearing motion drags the fluid into the gap between the surfaces (Fig. 2–19A). Squeeze film lubrication occurs when the rigid bearing surfaces move perpendicularly toward each other. In the gap between the two surfaces, the fluid

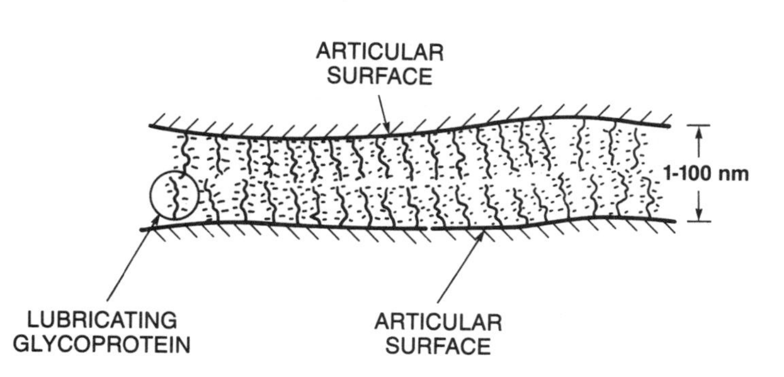

FIG. 2–18
Boundary lubrication of articular cartilage. The load is carried by a monolayer of the lubricating glycoprotein, which is adsorbed onto the articular surfaces. The monolayer effectively serves to reduce friction and helps to prevent cartilage wear. (Adapted from Armstrong and Mow, 1980.)

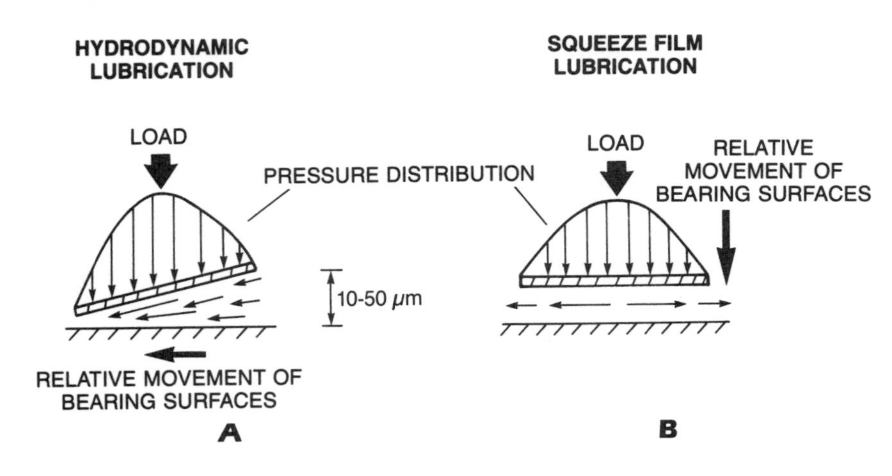

FIG. 2–19
Diagrammatic illustration of the fluid flow pattern and the mechanism by which load support is generated in the two fluid film lubrication modes, hydrodynamic and squeeze film. **A.** In hydrodynamic lubrication, the nonparallel surfaces of the rigid bearings move tangentially with respect to one another, forming a wedge of fluid that is drawn into the gap between the surfaces. The viscosity of the fluid generates a lifting pressure that supports the load. **B.** In squeeze film lubrication the bearings move perpendicularly toward each other, squeezing the fluid film out from between the surfaces. Again, the viscosity of the fluid generates a pressure as the fluid is forced out.

viscosity generates pressure, which is required to force the fluid lubricant out (Fig. 2–19B). The squeeze film mechanism is sufficient to carry high loads for short durations. Eventually, however, the fluid film becomes so thin that the asperities (peaks) on the two bearing surfaces come into contact.

In hydrodynamic and squeeze film lubrication, the thickness and extent of the fluid film, as well as its load-bearing capacity, are characteristics independent of the rigid bearing material properties. These lubrication characteristics are instead determined solely by the lubricant's properties: its rheologic properties (e.g., viscosity), the film geometry (e.g., the shape of the gap between the two bearing surfaces), and the speed of the relative surface motion.

A variation of the hydrodynamic and squeeze film modes of fluid film lubrication occurs when the bearing material is not rigid but relatively soft, such as the articular cartilage covering the joint surface. This type of lubrication, termed elastohydrodynamic, operates when the relatively soft bearing surfaces undergo either a sliding (hydrodynamic) or squeeze film action and the pressure generated in the fluid film substantially deforms the surfaces (Fig. 2–20). These deformations tend to increase the surface area, beneficially altering film geometry. Because the bearing contact area is increased, the lubricant escapes less readily from between the bearing surfaces, a longer lasting lubricant film is generated, and the stress of articulation is lower and more sustainable (Dowson, 1966/1967; Higginson et al., 1976; Armstrong and Mow, 1980; Higginson and Unsworth, 1981). Elastohydrodynamic lubrication enables bearings to greatly increase their load-bearing capacity.

In any bearing the effective mode of lubrication depends on the applied loads and on the velocity (speed and direction of motion) of the bearing surfaces. Adsorption of the synovial fluid glycoprotein lubricin to articular surfaces seems to be most important under severe loading conditions, when contact surfaces sustain high loads, at low relative speeds, and for long periods. Under these conditions, as the surfaces are pressed together, the boundary lubricant monolayers interact, preventing direct contact between the articular surfaces. On the other hand, fluid film lubrication operates under less severe conditions, when loads are low or vary in magnitude and when the contacting surfaces are moving at high relative speeds.

Boundary-lubricated surfaces typically have a coefficient of friction one or two orders of magnitude greater than those of surfaces lubricated by a fluid film. Intact synovial joints have an extremely low coefficient of friction, approximately 0.02 (Dowson, 1966/1967; Linn, 1968; McCutchen, 1962; Unsworth et al., 1975; Armstrong and Mow, 1980), suggesting that synovial joints are lubricated, at least in part, by the fluid film mechanism. It is quite possible that synovial joints use the mechanism that will most effectively provide lubrication under current loading conditions. Unresolved, though, is the manner in which synovial joints generate the fluid lubricant film.

Articular cartilage, like all surfaces, is not perfectly smooth; asperities project from the surface (Gardner and McGillivray, 1971) (see Figs. 2–3B and 2–21). In synovial joints, then, situations may occur wherein the fluid film thickness is of the same order as the mean articular surface asperity. At such times boundary lubrication between the asperities may come into play. If so, a mixed mode of lubrication is operating, with the joint surface load sustained both by the fluid film pressure in areas of noncontact and by the boundary lubricant lubricin in the areas of asperity

RIGID BEARINGS

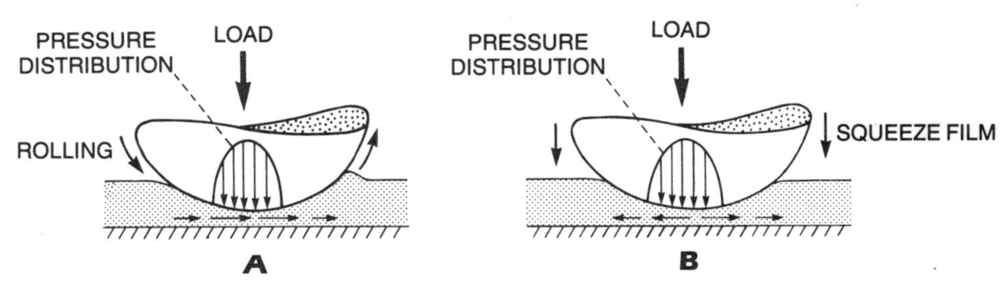

DEFORMABLE BEARINGS

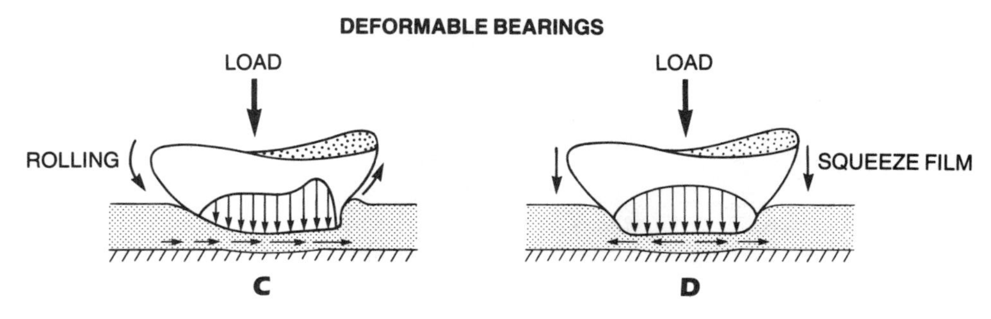

FIG. 2–20
Comparison of hydrodynamic lubrication **(A)** and squeeze film lubrication **(B)** of rigid surfaces, and elastohydrodynamic lubrication of deformable bearing surfaces under a hydrodynamic (sliding) action **(C)** and a squeeze film action **(D)**. As is evident, surface deformation of elastohydrodynamically lubricated bearings increases the contact area, thus increasing the load-carrying capacity of these bearings.

contact (Fig. 2–22). In mixed lubrication, it is probable that most of the friction (which is still extremely low) is generated in the boundary lubricated areas while most of the load is carried by the fluid film (Dowson, 1966/1967).

Synovial joint surfaces are different from typical engineering bearings described above in that they are composed of fluid-filled, porous, permeable cartilage that can exude and imbibe a lubricant fluid. Mansour and Mow (1976) described such a "self-lubrication" mechanism in diarthrodial joints that involves the simultaneous exudation and imbibition of fluid during sliding articulation. As a joint rotates and the articular surfaces slide over each other, they exude fluid in front of and beneath the leading half of the moving load (Fig. 2–23). Once the area of peak stress has passed any given point, the cartilage starts to reabsorb the fluid as it returns to its original dimensions in preparation for the next cycle of movement (Mow and Lai, 1980).

The previously described nonlinear permeability of articular cartilage plays a role in this self-lubrication process. Initially, when the pressures and strains are low, the tissue is most permeable; thus, a large amount of fluid is exuded in front of and at the leading edge of the contact load. As the advancing load moves onto this region of expelled fluid and the pressures and strain increase, the cartilage becomes less permeable, preventing the fluid on the articular surface from flowing back into the cartilage. The reduction in permeability with increased pressure and strain acts to limit additional fluid transport to and from the joint space. The fluid film generated by the self-lubrication mechanism may be no thicker than 10 μm; nevertheless, a film this thick will lubricate the cartilage surfaces extremely well. This mechanism has been verified experimentally through a flow visualization technique focused on the articular surface under a stereomicroscope (Mow and Lai, 1979).

As discussed in an earlier section, this forced fluid circulation into and out of the cartilage matrix may also aid in chondrocyte nutrition, bringing nutrients from the synovial fluid in the joint cavity to the cartilage cells in the matrix. Thus, a natural mecha-

FIG. 2–21
Scanning electron micrograph of the surface of human articular cartilage from a normal young adult showing the typical irregularities characteristic of this tissue (3,000×). (Reprinted with permission from Armstrong and Mow, 1980.)

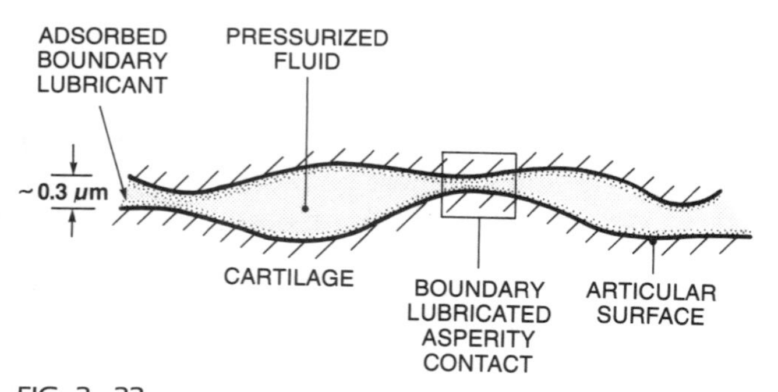

FIG. 2–22
Schematic depiction of mixed lubrication operating in articular cartilage. Boundary lubrication occurs where the thickness of the fluid film is on the same order as the roughness of the bearing surfaces. Fluid film lubrication takes place in areas with more widely separated surfaces. (Adapted from Armstrong and Mow, 1980.)

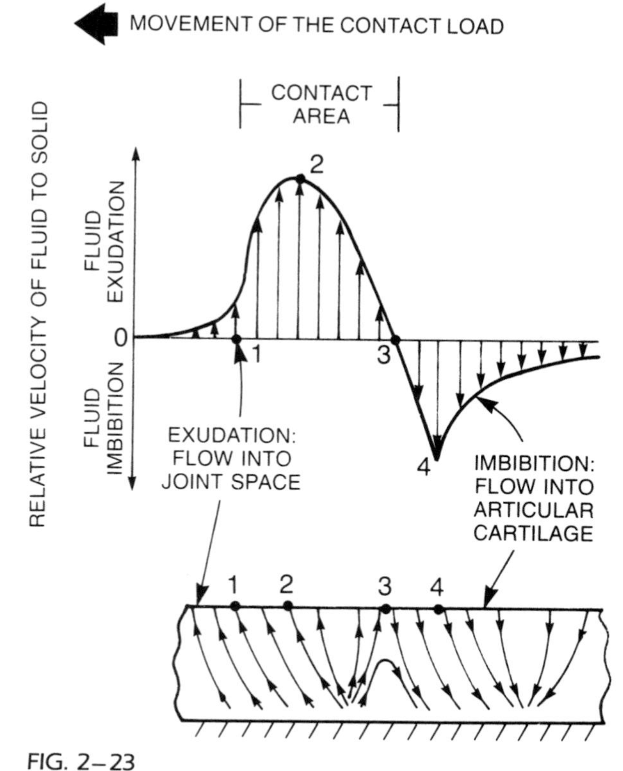

FIG. 2–23
Schematic representation of fluid exudation and imbibition as articular cartilage is subjected to the action of a sliding load. The direction of fluid flow is indicated graphically (top) and within the cartilage (bottom). Fluid is exuded in front of and beneath the leading half of the contact load (points 1 and 2) and imbibed at the trailing edge and behind the moving load (points 3 and 4). The contact area, within which the self-lubrication mechanism occurs, is indicated. (Adapted from Mow and Lai, 1980.)

nism exists to maintain the lubricating fluid film between the opposing articulating surfaces and to maintain a constant nutrient flow to the chondrocytes within the matrix.

Another mechanism by which the articular cartilage surfaces might be protected during joint articulation has been proposed by Walker et al. (1968, 1970) and Maroudas (1967). In their hypothesized mode, termed "boosted lubrication," the solvent component of the synovial fluid actually passes into the articular cartilage during squeeze film action, leaving a concentrated pool of HA protein complex to lubricate the surfaces. According to this theory, it becomes progressively more difficult, as the two articular surfaces approach each other, for the HA macromolecules in the synovial fluid to escape from the gap between the surfaces because they are physically too large (Fig. 2–24). The water and small solute molecules can still escape into the articular cartilage through the cartilage surface and laterally into the joint space at the periphery of the joint.

In summary, it is unlikely that the varied demands on diarthrodial joints during normal function can be satisfied by a single mode of lubrication. As yet, it is impossible to state definitely under which conditions

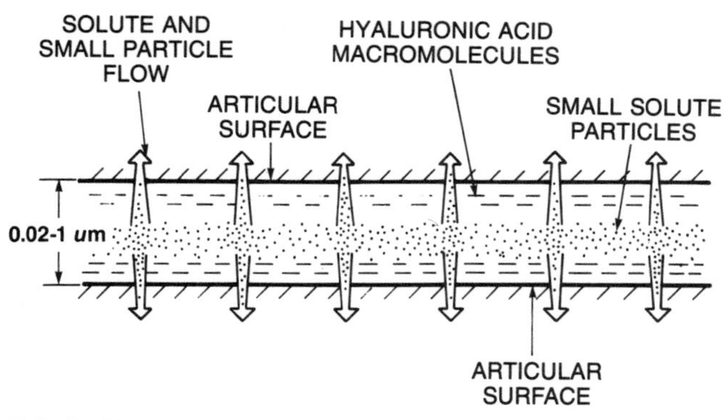

FIG. 2–24
Ultrafiltration of the synovial fluid into a highly viscous gel. As the articular surfaces come together, the small solute molecules escape into the articular cartilage and into the lateral joint space, leaving the large HA macromolecules, which, because of their size, are unable to escape. These macromolecules form a concentrated gel less than 1 μm thick that lubricates the articular surfaces. This hypothesized lubrication mode is called "boosted lubrication" (Walker et al.,1968, 1970).

a particular lubrication mechanism may operate. Nevertheless, using the human hip as an example, some general statements are possible.

1. *Elastohydrodynamic fluid films of both the sliding (hydrodynamic) and the squeeze type probably play an important role in lubricating the joint.* During the swing phase of walking, when loads on the joint are minimal, a substantial layer of synovial fluid film is probably maintained. After the first peak force, at heel strike, a supply of fluid lubricant is generated by articular cartilage. This fluid film thickness, however, will begin to decrease under the high load of stance phase; as a result, squeeze film action occurs. The second peak force during the walking cycle, just before the toe leaves the ground, occurs when the joint is swinging in the opposite direction. Thus, it is possible that a fresh supply of fluid film could be generated at toe-off, thereby providing the lubricant during the next swing phase.
2. *With high loads and low speeds of relative motion, such as during standing, the fluid film will decrease in thickness as the fluid is squeezed out from between the surfaces.* Under these conditions, the fluid exuded from the compressed articular cartilage could become the main contributor to the lubricating film.
3. *Under extreme loading conditions, such as during an extended period of standing following impact, the fluid film may be eliminated, allowing surface-to-surface contact.* The surfaces, however, will probably still be protected, either by a thin layer of ultrafiltered synovial fluid gel or by the adsorbed lubricin monolayer.

WEAR OF ARTICULAR CARTILAGE

Wear is the removal of material from solid surfaces by mechanical action. There are two components of wear: interfacial wear due to the interaction of bearing surfaces and fatigue wear due to bearing deformation under load.

Interfacial wear occurs when bearing surfaces come into direct contact with no lubricant film (boundary or fluid) separating them. This type of wear can take place in either of two ways: adhesion or abrasion. Adhesive wear arises when, as the bearings come into contact, surface fragments adhere to each other and are torn from the surface during sliding. Abrasive wear, on the other hand, occurs when a soft material is scraped by a harder one; the harder material can be either an opposing bearing or loose particles between the bearings. The low rates of interfacial wear observed in articular cartilage tested in vitro (Lipshitz and Glimcher, 1979) suggest that direct surface-to-surface contact between the asperities of the two cartilage surfaces rarely occurs; however, abrasive wear was not ruled out in these experiments. The multiple modes of effective lubrication working in concert are the mechanisms that make interfacial wear of articular cartilage unlikely. Nevertheless, adhesive and abrasive wear may take place in an impaired or degenerated synovial joint. Once the cartilage surface sustains ultrastructural defects and/or decreases in mass, it becomes softer and more permeable (Armstrong and Mow, 1982a; Akizuki et al., 1986); thus, fluid from the lubricant film separating the bearing surfaces may leak away more easily through the cartilage surface. This loss of lubricating fluid from between the surfaces increases the probability of direct contact between the asperities and exacerbates the abrasion process (Armstrong and Mow, 1980).

Fatigue wear of bearing surfaces results not from surface-to-surface contact but from the accumulation of microscopic damage within the bearing material under repetitive stressing. Bearing surface failure may occur with the repeated application of high loads over a relatively short period or with the repetition of low loads over an extended period, even though the magnitude of those loads may be much lower than the material's ultimate strength. This fatigue wear, due to cyclically repeated deformation of the bearing materials, can take place even in well-lubricated bearings.

In synovial joints, the cyclic variation in total joint load during most physiologic activities causes repetitive articular cartilage stress (deformation). In addition, during rotation and sliding, a specific region of the articular surface moves in and out of the loaded contact area and is repeatedly stressed. Loads imposed on articular cartilage are supported by the collagen-PG matrix and by the resistance generated by fluid movement throughout the matrix; thus, repetitive joint movement and loading will cause repetitive stressing of the solid matrix and repeated exudation and imbibition of the tissue's interstitial fluid (Mow and Lai, 1980). These processes give rise to two possible mechanisms by which fatigue damage may accumulate in articular cartilage: disruption of the collagen-PG solid matrix, and PG washout.

First, repetitive collagen-PG matrix stress could disrupt the collagen fibers, the PG macromolecules, or the interface between the two. A popular hypothesis is that cartilage fatigue is due to tensile failure of the collagen fiber network (Freeman, 1975; Weightman, 1976; Weightman and Kempson, 1979). Also, as discussed above, pronounced changes in the articular cartilage PG population have been observed with age and disease (Inerot et al., 1978; McDevitt and Muir, 1976). These PG changes could be considered as part of the accumulated tissue damage. Second, repetitive and massive exudation and imbibition of interstitial fluid may cause a proteoglycan washout from the cartilage matrix near the articular surface, with a resultant decrease in stiffness and increase in permeability of the tissue.

A third mechanism of damage and resultant articular wear is associated with synovial joint impact loading (i.e., rapid application of a high load). With normal physiologic loading, articular cartilage undergoes surface compaction during the compression, the lubricating fluid being exuded through this compacted region (see Fig. 2–10). As described above, however, fluid redistribution within the articular cartilage occurs with time, relieving the stress in this compacted region. This process of stress relaxation takes place quite quickly; the stress may decrease by 63% within 2 to 5 seconds (Mow et al., 1980). If, however, loads are supplied so quickly that there is insufficient time for internal fluid redistribution to relieve the compacted region, the high stresses produced in the collagen-PG matrix may cause damage. This phenomenon could well explain why Radin and Paul (1971, 1976) found dramatic articular cartilage damage with repeated impact loads.

These mechanisms of wear and damage may be the cause of the wide range of structural defects observed in articular cartilage (Meachim and Fergie, 1975; Byers et al., 1977) (Fig. 2–25). One defect commonly noted is the splitting of the cartilage surface. Vertical sections of cartilage that exhibit these lesions, known as fibrillations, show that they eventually extend through the full depth of the articular cartilage. In other specimens, the cartilage layer appears to be eroded rather than split. This erosion is known as smooth-surfaced destructive thinning.

Considering the variety of defects noted in articular cartilage, it is unlikely that a single wear mechanism is responsible for all of them. At any given site, the stress history may be such that fatigue is the initiating failure mechanism. At another, the lubrication conditions may be so unfavorable that interfacial wear dominates the progression of cartilage failure. As yet, there is little experimental information on the type of defect produced by any given wear mechanism.

Once the collagen-PG matrix of cartilage is disrupted, damage resulting from any of the three wear mechanisms mentioned becomes possible: further disruption of the collagen-PG matrix due to repetitive matrix stressing, washout of the PGs due to violent fluid movement and impairment of articular cartilage's self-lubricating ability, or gross alteration of the normal load-bearing mechanism of articular cartilage, resulting in excessively high stress levels in the collagen-PG solid matrix. All these processes may accelerate the rate of interfacial and fatigue wear of the already disrupted cartilage microstructure.

HYPOTHESES ON BIOMECHANICS OF CARTILAGE DEGENERATION

Articular cartilage has only a limited capacity for repair and regeneration, and if subjected to an abnormal range of stresses it can undergo total failure quite quickly. It has been hypothesized that failure progression relates to the following: the magnitude of the imposed stresses, the total number of sustained stress peaks, changes in the intrinsic molecular and microscopic structure of the collagen-PG matrix, and changes in the intrinsic mechanical property of the tissue. The most important failure-initiating factor appears to be loosening of the collagen network, which allows abnormal PG expansion and, thus, tissue swelling (Maroudas, 1979; McDevitt and Muir,

1976). Associated with this change is a decrease in cartilage stiffness and an increase in cartilage permeability (Armstrong and Mow, 1982a), both of which alter cartilage function in a diarthrodial joint during joint motion (Mow and Lai, 1980).

The magnitude of the stresses sustained by articular cartilage is determined by both the total load on the joint and how that load is distributed over the articular surface contact area (Paul, 1967, 1980; Armstrong et al., 1979; Ahmed and Burke, 1983). Any intense stress concentration in the contact area will play a primary role in tissue degeneration. A large number of well-known conditions cause excessive stress concentrations in articular cartilage and result in cartilage failure. Most of these stress concentrations are due to joint surface incongruity, which results in an abnormally small contact area. Examples of conditions that cause such joint incongruities include osteoarthritis subsequent to congenital acetabular dysplasia, slipped capital femoral epiphysis, and intra-articular fractures (Murray, 1965). Two further examples are knee joint meniscectomy, which eliminates the meniscus and its load-distributing function (Lufti, 1975; Shrive et al., 1978; Bourne et al., 1984), and ligament rupture, which allows excessive movement and the generation of abnormal mechanical stresses in the affected joint (Jacobsen, 1977; Arnold et al., 1979). In all the above cases, abnormal joint articulation increases the stress acting on the joint surface, which appears to predispose the cartilage to failure.

Macroscopically, stress localization and concentration at the joint surfaces have a further effect. High contact pressures between the articular surfaces reduce the probability of fluid film lubrication. Subsequent actual surface-to-surface contact of asperities causes microscopic stress concentrations that are responsible for further tissue damage.

The high incidence of specific degenerated joints in persons with certain occupations, such as football players' knees and ballet dancers' ankles, can be explained by the increase in high and abnormal load frequency and magnitude sustained by these joints. It has been suggested that osteoarthritis may in some cases be caused by deficiencies in the mechanisms that minimize peak forces on the joints. Examples of these mechanisms include the active processes of joint flexion and muscle lengthening and the passive absorption of shocks by the subchondral bone (Radin, 1976) and meniscus (Voloshin and Wosk, 1983).

Osteoarthritis may also occur secondary to an insult to the intrinsic molecular and microscopic structure of the collagen-PG matrix. Many conditions

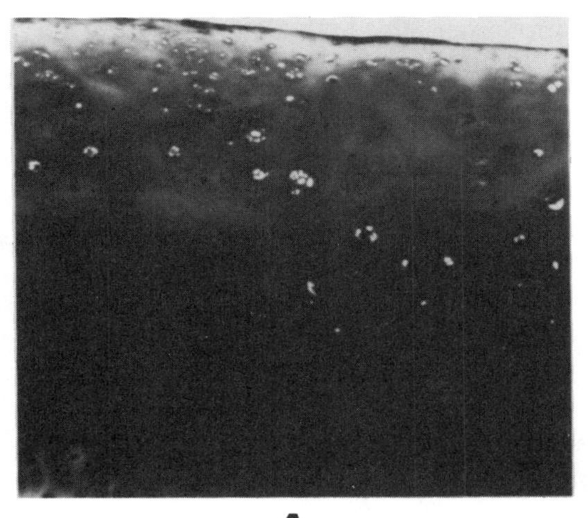

A

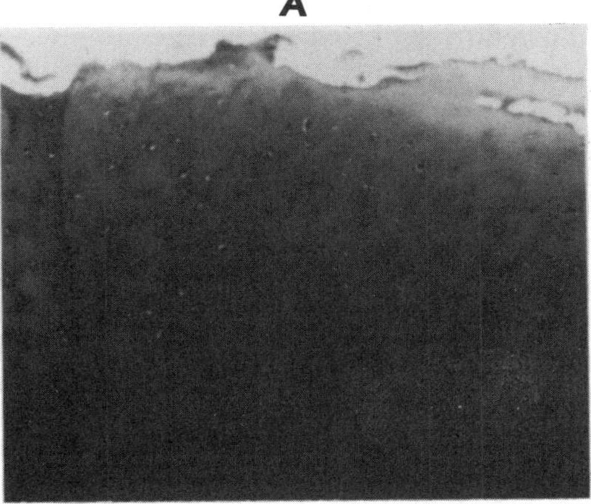

B

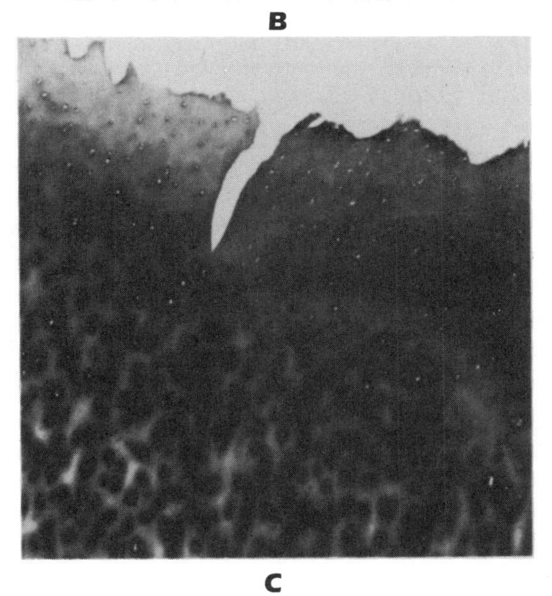

C

FIG. 2–25
Photomicrographs of vertical sections through the surface of articular cartilage showing a normal intact surface **(A)**, an eroded articular surface **(B)**, and a vertical split, or fibrillation, of the articular surface that will eventually extend through the full depth of the cartilage **(C)**. (Courtesy of Dr. S. Akizuki, Nagano, Japan.)

may promote such a breakdown in matrix integrity—degeneration associated with rheumatoid arthritis, joint space hemorrhage associated with hemophilia (Lee et al., 1974), various collagen metabolism disorders, and possibly tissue degradation by proteolytic enzymes (Ali and Evans, 1973). This last process could lead to abnormal tissue swelling and functionally inferior biomechanical properties. In this weakened state, the cartilage ultrastructure is then destroyed by stresses of normal joint articulation.

SUMMARY

1. The function of articular cartilage in diarthrodial joints is to increase the area of load distribution and to provide a smooth, wear-resistant bearing surface.
2. Biomechanically, articular cartilage can be viewed as a biphasic (solid-fluid) material: the collagen-PG solid matrix (approximately 25% by wet weight) surrounded by the freely movable interstitial fluid (approximately 75% by wet weight).
3. Important biomechanical properties of articular cartilage are the intrinsic material properties of the solid matrix and the frictional resistance to the flow of interstitial fluid through the porous permeable solid matrix (a parameter inversely proportional to the tissue permeability).
4. Articular cartilage has the ability to provide for the diarthrodial joint a self-lubrication feature that operates under normal physiologic joint loading conditions.
5. Damage to articular cartilage, from whatever cause, can disrupt the normal load-carrying ability of the tissue and thus the normal lubrication process operating in the joint. Lubrication insufficiency may be a primary factor in the etiology of osteoarthritis.

ACKNOWLEDGMENTS

This work was sponsored by the National Science Foundation grant no. ECE 85-18501 and the National Institutes of Health grants AM19094 and AM26440.

REFERENCES

Ahmed, A. M., and Burke, D. L.: *In vitro* measurement of static pressure distribution in synovial joints. I. Tibial surface of the knee. J. Biomech. Eng., *105*:216, 1983.

Akizuki, S., et al.: Tensile properties of knee joint cartilage. I. Influence of ionic condition, weight bearing, and fibrillation on the tensile modulus. J. Orthop. Res., *4*:379, 1986.

Ali, S. Y., and Evans, L.: Enzymatic degradation of cartilage in osteoarthritis. Fed. Proc., *32*:1494, 1973.

Armstrong, C. G., and Mow, V. C.: Friction, lubrication and wear of synovial joints. *In* Scientific Foundations of Orthopaedics and Traumatology. Edited by R. Owen, J. Goodfellow, and P. Bullough. London, William Heinemann, 1980, pp. 223–232.

Armstrong, C. G., and Mow, V. C.: Variations in the intrinsic mechanical properties of human articular cartilage with age, degeneration, and water content. J. Bone Joint Surg., *64A*:88, 1982a.

Armstrong, C. G., and Mow, V. C.: Biomechanics of normal and osteoarthrotic articular cartilage. *In* Clinical Trends in Orthopaedics. Edited by P. D. Wilson, Jr., and L. R. Straub. New York, Thieme-Stratton, Inc., 1982b, pp. 189–197.

Armstrong, C. G., Bahrani, A. S., and Bardner, D. L.: *In vitro* measurement of articular cartilage deformations in the intact human hip joint under load. J. Bone Joint Surg., *61A*:744, 1979.

Arnold, J. A., et al.: Natural history of anterior cruciate tears. Am. J. Sport Med., *7*:305, 1979.

Askew, M. J., and Mow, V. C.: The biomechanical function of the collagen ultrastructure of articular cartilage. J. Biomech. Eng., *100*:105, 1978.

Bayliss, M., and Ali, Y.: Isolation of proteoglycans from human articular cartilage. Biochem. J., *169*:123, 1978.

Bourne, R. B., Finlay, J. B., Papadopoulos, P., and Andreae, P.: The effect of medial meniscectomy on strain distribution in the proximal part of the tibia. J. Bone Joint Surg., *66A*:1431, 1984.

Bowden, F. P., and Tabor, D.: Friction and Lubrication. London, Methuen, 1967.

Bruckner, P., Vaughan, L., and Winthalter, K. H.: Type IX collagen from sternal cartilage of chicken embryo contains covalently bound glycosaminoglycans. Proc. Natl. Acad. Sci. USA, *82*:2608, 1985.

Buckwalter, J. A., Kuettner, K. E., and Thonar, E. J.-M. A.: Age-related changes in articular cartilage proteoglycans: Electron microscopic studies. J. Orthop. Res., *3*:251, 1985.

Bullough, P. G., and Jagannath, A.: The morphology of the calcification front in articular cartilage. J. Bone Joint Surg., *65B*:72, 1983.

Byers, P. D., et al.: Observations on osteoarthrosis of the hip. Semin. Arthritis Rheum., *6*:277, 1977.

Clarke, I. C.: Articular cartilage: A review and scanning electron microscope study. I. The interterritorial fibrillar architecture. J. Bone Joint Surg., *53B*:732, 1971.

Donohue, J. M., Buss, D., Oegema, T. R., and Thompson, R. C.: The effects of indirect blunt trauma on adult canine articular cartilage. J. Bone Joint Surg., *65A*:948, 1983.

Dowson, D.: Modes of lubrication in human joints. Proc. Inst. Mech. Eng., *1813J*:45, 1966/67.

Edwards, J.: Physical characteristics of articular cartilage. Proc. Inst. Mech. Eng., *181*:16, 1967.

Elmore, S. M., Sokoloff, G. N., and Carmeci, P.: Nature of "imperfect" elasticity of articular cartilage. J. Appl. Physiol., *18*:393, 1963.

Eyre, D. R.: Collagen: Molecular diversity in the body's protein scaffold. Science, *207*:1315, 1980.

Ferry, J. D.: Illustrations of viscoelastic behavior of polymeric systems. *In* Viscoelastic Properties of Polymers. 2nd Ed. New York, John Wiley & Sons, Inc., 1970, pp. 34–57.

Freeman, M. A. R.: The fatigue of cartilage in the pathogenesis of osteoarthrosis. Acta Orthop. Scand., *46*:323, 1975.

Fung, Y. C.: Biomechanics: Mechanical Properties of Living Tissues. New York, Springer-Verlag, 1981, p. 226.

Gardner, S. L., and McGillivray, D. C.: Living articular cartilage is not smooth. The structure of mammalian and avian joint surfaces demonstrated *in vivo* by immersion incident light microscopy. Ann. Rheum. Dis., *30*:3, 1971.

Garg, H. G., and Swann, D. A.: Age-related changes in the chemical composition of bovine articular cartilage. Biochem. J., *193*:459, 1981.

Hardingham, T. E., Beardmore-Garg, M., and Dunham, D. G.: Protein domain structure of the aggregating proteoglycan from cartilage. Trans. Orthop. Res. Soc., *12*:61, 1987.

Hardingham, T. E., Ewins, R. J. F., and Muir, H.: Cartilage proteoglycans: Structure and heterogeneity of the protein core and the effects of specific protein modifications on the binding to hyaluronate. Biochem. J., *157*:127, 1976.

Hardingham, T. E., and Muir, H.: The function of hyaluronic acid in proteoglycan aggregation. *In* Proceedings of the Symposium on Normal and Osteoarthrotic Articular Cartilage. London, Institute of Orthopaedics, 1974a, pp. 51–64.

Hardingham, T. E., and Muir, H.: Hyaluronic acid in cartilage and proteoglycan aggregation. Biochem. J., *139*:565, 1974b.

Hascall, V. C.: Interactions of cartilage proteoglycans with hyaluronic acid. J. Supramol. Structure, *7*:101, 1977.

Hascall, V. C., and Hascall, G. K.: Proteoglycans. *In* Cell Biology of Extracellular Matrix. Edited by E. D. Hay. New York, Plenum Press, 1981, pp. 39–63.

Hayes, W. C., and Bodine, A. J.: Flow-independent viscoelastic properties of articular cartilage matrix. J. Biomech., *11*:407, 1978.

Heinegard, D. K., and Axelsson, I.: Distribution of keratan sulfate in cartilage proteoglycans. J. Biol. Chem., *252*:1971, 1977.

Heinegard, D., and Hascall, V. C.: Aggregation of cartilage proteoglycans. III. Characteristics of the proteins isolated from trypsin digests of aggregates. J. Biol. Chem., *249*:4250, 1974.

Heinegard, D., and Paulsson, M.: Structure and metabolism of proteoglycans. *In* Extracellular Matrix Biochemistry. Edited by K.A. Piez and A. H. Reddi. New York, Elsevier, 1984, pp. 277–328.

Heinegard, D., et al.: Separation and characterization of two populations of aggregating proteoglycans from cartilage. Biochem. J., *225*:95, 1985.

Higginson, G. R., and Unsworth, A.: The lubrication of natural joints. *In* Tribology of Natural and Artificial Joints. Edited by J. H. Dumbleton. Tribology series, 3. Amsterdam, Elsevier, 1981, pp. 47–73.

Higginson, G. R., Litchfield, M. R., and Snaith, J.: Load-displacement-time characteristics of articular cartilage. Int. J. Mech. Sci., *18*:481, 1976.

Holmes, M. H., Lai, W. M., and Mow, V. C.: Compression effects on cartilage permeability. *In* Tissue Nutrition and Viability. Edited by A. R. Hargens. New York, Springer-Verlag, 1986, pp. 73–100.

Hultkrantz, W.: Ueber die Spaltrichtungen der Gelenkknorpel. Verh. Anat. Ges., *12*:248, 1898.

Inerot, S., and Heinegard, D.: Bovine tracheal cartilage proteoglycans. Variations in structure and composition with age. Collagen Relat. Res., *3*:245, 1983.

Inerot, S., Heinegard, D., Audell, L., and Olsson, S.-E.: Articular-cartilage proteoglycans in aging and osteoarthritis. Biochem. J., *169*:143, 1978.

Jacobsen, K.: Osteoarthrosis following insufficiency of the cruciate ligament in man. Acta Orthop. Scand., *48*:520, 1977.

Kempson, G. E.: Mechanical properties of articular cartilage. *In* Adult Articular Cartilage. 2nd Ed. Edited by M. A. R. Freeman. Tunbridge Wells, Pitman Medical, 1979, pp. 333–414.

Kempson, G. E., et al.: The effects of proteolytic enzymes on the mechanical properties of adult human articular cartilage. Biochim. Biophys. Acta, *428*:741, 1976.

Lai, W. M., and Mow, V. C.: Drag-induced compression of articular cartilage during a permeation experiment. J. Biorheol., *17*:111, 1980.

Lakes, R., and Saha, S.: Cement line motion in bone. Science, *204*:501, 1979.

Lee, P., Sturrock, R. D., Kennedy, A. C., and Dick, W. C.: The etiology and pathogenesis of osteoarthrosis: A review. Semin. Arthritis Rheum., *3*:189, 1974.

Li, J. T., Armstrong, C. G., and Mow, V. C.: Effect of strain rate on mechanical properties of articular cartilage in tension. *In* ASME Biomechanics Symposium. Edited by S. L.-Y. Woo and R. Mates. New York, American Society of Mechanical Engineers, 1983, pp. 9–12.

Li, J. T., Mow, V. C., Koob, T. J., and Eyre, D. R.: Effect of chondroitinase ABC treatment on the tensile behavior of bovine articular cartilage. Trans. Orthop. Res. Soc., *9*:35, 1984.

Linn, F. C.: Lubrication of animal joints. I. The mechanism. J. Biomech., *1*:193, 1968.

Linn, F. C., and Sokoloff, L.: Movement and composition of interstitial fluid of cartilage. Arthritis Rheum., *8*:481, 1965.

Lipshitz, H., and Glimcher, M. J.: *In vitro* studies of the wear of articular cartilage. II. Wear, *52*:297, 1979.

Lipshitz, H., Etheredge, R., and Glimcher, M. J.: *In vitro* wear of articular cartilage. I. Hydroxyproline, hexosamine, and amino acid composition of bovine articular cartilage as a function of depth from the surface; hydroxyproline content of the lubricant and the wear debris as a measure of wear. J. Bone Joint Surg., *57A*:527, 1975.

Lipshitz, H., Etheredge, R., and Glimcher, M. J.: Changes in the hexosamine content and swelling ratio of articular cartilage as functions of depth from the surface. J. Bone Joint Surg., *58A*:1149, 1976.

Lohmander, L. S., et al.: Oligosaccharides on proteoglycans from the swarm rat chondrosarcoma. J. Biol. Chem., *255*:6084, 1980.

Lufti, A. M.: Morphological changes in articular cartilage after meniscectomy. J. Bone Joint Surg., *57B*:525, 1975.

Mak, A. F., et al.: Predictions of the number and strength of proteoglycan-proteoglycan interactions from viscometric data. Trans. Orthop. Res. Soc., *8*:3, 1983.

Mak, A. F., et al.: Assessment of proteoglycan-proteoglycan interactions from solution biorheological behaviors. Trans. Orthop. Res. Soc., *7*:169, 1982.

Mankin, H. A., and Thrasher, A. Z.: Water content and binding in normal and osteoarthritic human cartilage. J. Bone Joint Surg., *57A*:76, 1975.

Mansour, J. M., and Mow, V. C.: The permeability of articular cartilage under compressive strain and at high pressures. J. Bone Joint Surg., *58A*:509, 1976.

Maroudas, A.: Hyaluronic acid films. Proc. Inst. Mech. Eng. (London), *181*:122, 1967.

Maroudas, A.: Distribution and diffusion of solutes in articular cartilage. Biophys. J., *10*:365, 1970.

Maroudas, A.: Physicochemical properties of articular cartilage. *In* Adult Articular Cartilage. 2nd Ed. Edited by M. A. R. Freeman. Tunbridge Wells, Pitman Medical, 1979, pp. 215–290.

Maroudas, A., Muir, H., and Wingham, J.: The correlation of fixed negative charge with glycosaminoglycan content of human articular cartilage. Biochim. Biophys. Acta, *177*:492, 1969.

McCutchen, C. W.: The frictional properties of animal joints. Wear, *5*:1, 1962.

McCutchen, C. W.: Boundary lubrication by synovial fluid: Demonstration and possible osmotic explanation. Federation Proc., *25*:1061, 1966.

McDevitt, C. A., and Muir, H.: Biochemical changes in the cartilage of the knee in experimental and natural osteoarthritis in the dog. J. Bone Joint Surg., *58B*:94, 1976.

Meachim, G., and Fergie, I. A.: Morphological patterns of articular cartilage fibrillation. J. Pathol., *115*:231, 1975.

Mow, V. C.: Biphasic rheological properties of cartilage. *In* Proceedings of the 1977 Biomechanics and Biomaterials Symposium. Bull. Hosp. Joint Dis., *38*:121, 1977.

Mow, V. C., and Lai, W. M.: The optical sliding contact analytical rheometer (OSCAR) for flow visualization at the articular surface. *In* Advances in Bioengineering. Edited by M. K. Wells. New York, American Society of Mechanical Engineers, 1979, pp. 97–99.

Mow, V. C., and Lai, W. M.: Recent developments in synovial joint biomechanics. SIAM Rev., *22*:275, 1980.

Mow, V. C., and Schoonbeck, J. M.: Contribution of Donnan osmotic pressure towards the biphasic compressive modulus of articular cartilage. Trans. Orthop. Res. Soc., *8*:262, 1984.

Mow, V. C., Holmes, M. H., and Lai, W. M.: Fluid transport and mechanical properties of articular cartilage: A review. J. Biomech., *17*:377, 1984a.

Mow, V. C., Lai, W. M., and Holmes, M. H.: Advanced theoretical and experimental techniques in cartilage research. *In* Biomechanics: Principles and Applications. Edited by R. Huiskes, D. H. van Campen, and J. R. de Wijin. The Hague, Martinus Niihoff Publishers, 1982, pp. 47–74.

Mow, V. C., Lai, W. M., and Redler, I.: Some surface characteristics of articular cartilage. I. A scanning electron microscopy study and a theoretical model for the dynamic interaction of synovial fluid and articular cartilage. J. Biomech., *7*:449, 1974.

Mow, V. C., Kuei, S. C., Lai, W. M., and Armstrong, C. G.: Biphasic creep and stress relaxation of articular cartilage in compression: Theory and experiments. J. Biomech. Eng., *102*:73, 1980.

Mow, V. C., et al.: Viscoelastic properties of proteoglycan subunits and aggregates in varying solution concentrations. J. Biomech., *17*:325, 1984b.

Muir, H.: Biochemistry. *In* Adult Articular Cartilage. 2nd Ed. Edited by M. A. R. Freeman. Tunbridge Wells, Pitman Medical, 1979, pp. 145–214.

Muir, H.: The chemistry of the ground substance of joint cartilage. *In* The Joints and Synovial Fluid. Vol. II. Edited by L. Sokoloff. New York, Academic Press, 1980, pp. 27–94.

Muir, H.: Proteoglycans as organizers of the extracellular matrix. Biochem. Soc. Trans., *11*:613, 1983.

Murray, R. O.: The aetiology of primary osteoarthritis of the hip. Br. J. Radiol., *38*:810, 1965.

Myers, E. R., and Mow, V. C.: Biomechanics of cartilage and its response to biomechanical stimuli. *In* Cartilage. Vol. I. Structure, Function, and Biochemistry. Edited by B. K. Hall. New York, Academic Press, 1983, pp. 313–341.

Myers, E. R., Armstrong, C. G., and Mow, V. C.: Swelling pressure and collagen tension. *In* Connective Tissue Matrix. Edited by D. W. L. Hukins. London, Macmillan Press, Ltd., 1984, pp. 161–168.

Paul, J. P.: Forces transmitted by joints in the human body. Proc. Inst. Mech. Eng., *181*:8, 1967.

Paul, J. P.: Joint kinetics. *In* The Joints and Synovial Fluid. Vol. II. Edited by L. Sokoloff. New York, Academic Press, 1980, pp. 139–176.

Perkins, S. J., Miller, A., Hardingham, T. E., and Muir, H.: Physical properties of the hyaluronate binding region of proteoglycan from pig laryngeal cartilage. J. Mol. Biol., *150*:69, 1981.

Pita, J. C., Manicourt, D. H., Muller, F. J., and Howell, D. S.: Studies on the potential reversibility of osteoarthritis in some experimental animal models. *In* Articular Cartilage Biochemistry. Edited by K. Kuettner, R. S. Schleyerbach, and V. C. Hascall. New York, Raven Press, 1986, p. 349–363.

Poole, A. R., Pidoux, I., Reiner, A., and Rosenberg, L.: An immunoelectron microscope study of the organization of proteoglycan monomer, link protein, and collagen in the matrix of articular cartilage. J. Cell. Biol., *93*:921, 1982.

Radin, E. L.: Aetiology of osteoarthrosis. Clin. Rheum. Dis., *2*:509, 1976.

Radin, E. L., and Paul, I. L.: Response of joints to impact loading. I. *In vitro* wear. Arthritis Rheum., *14*:356, 1971.

Redler, I., and Zimny, M. L.: Scanning electron microscopy of normal and abnormal articular cartilage and synovium. J. Bone Joint Surg., *53A*:1395, 1970.

Rosenberg, L.: Structure of cartilage proteoglycans. *In* Dynamics of Connective Tissue Macromolecules. Edited by P. M. C. Burleigh and A. R. Poole. Amsterdam, North-Holland Publishing, 1975, pp. 105–128.

Rosenberg, L.: Proteoglycans. *In* Scientific Foundations of Orthopaedics and Traumatology. Edited by R. Owen, J. Goodfellow, and P. Bullough. London, William Heinemann, 1980, pp. 36–42.

Rosenberg, L., et al.: Isolation of dermatan sulfate proteoglycans from mature bovine articular cartilage. J. Biol. Chem., *260*:6304–6313, 1985.

Rosenberg, L., Hellmann, W., and Kleinschmidt, A. K.: Electron microscopic studies of proteoglycan aggregates from bovine articular cartilage. J. Biol. Chem., *250*:1877–1883, 1975.

Roth, V., and Mow, V. C.: The intrinsic tensile behavior of the matrix of bovine articular cartilage and its variation with age. J. Bone Joint Surg., *62A*:1102, 1980.

Roth, V., Schoonbeck, J. M., and Mow, V. C.: Low frequency dynamic behavior of articular cartilage under torsional shear. Trans. Orthop. Res. Soc., *7*:150, 1982.

Roth, V., Mow, V. C., Lai, W. M., and Eyre, D. R.: Correlation of intrinsic compressive properties of bovine articular cartilage with its uronic acid and water content. Trans. Orthop. Res. Soc., *6*:49, 1981.

Roughley, P. J., and White, R. J.: Age-related changes in the structure of the proteoglycan subunits from human articular cartilage. J. Biol. Chem., *255*:217, 1980.

Roughley, P. J., White, R. J., and Santer, V.: Comparison of proteoglycans extracted from high and low-weight bearing human articular cartilage, with particular reference to sialic acid content. J. Biol. Chem., *256*:12699, 1981.

Schmidt, M. B., et al.: The relationship between collagen crosslinking and the tensile properties of articular cartilage. Trans. Orthop. Res. Soc., *12*:134, 1987.

Schubert, M., and Hamerman, D.: A Primer on Connective Tissue Biochemistry. Philadelphia, Lea & Febiger, 1968.

Scott, J. E., and Orford, C. R.: Dermatan sulphate-rich proteoglycan associated with rat tail-tendon collagen at the d band in the gap region. Biochem. J., *197*:213, 1981.

Serafini-Fracassini, A., and Smith, J. W.: The Structure and Biochemistry of Cartilage. Edinburgh, Churchill Livingstone, 1974.

Shepard, N., and Mitchell, N.: The localization of articular cartilage proteoglycan by electron microscopy. Anat. Rec., *187*:463, 1977.

Shrive, N. G., O'Connor, J. J., and Goodfellow, J. W.: Load-bearing in the knee joint. Clin. Orthop., *131*:279, 1978.

Simon, B. R., Coats, R. S., and Woo, S. L.-Y.: Relaxation and creep quasilinear viscoelastic model for normal articular cartilage. J. Biomech. Eng., *106*:159, 1984.

Simunek, A., and Muir, H.: Changes in the protein-polysaccharides of pig articular cartilage during prenatal life, development and old age. Biochim. J., *126*:515, 1972.

Sokoloff, L.: Elasticity of articular cartilage: Effect of ions and viscous solutions. Science, *141*:1055, 1963.

Stahurski, T. M., Armstrong, C. G., and Mow, V. C.: Variation of the intrinsic aggregate modulus and permeability of articular cartilage with trypsin digestion. *In* Biomechanics Symposium. Edited by W. C. VanBuskirk and S. L.-Y. Woo. New York, American Society of Mechanical Engineers, 1981, p. 137.

Stockwell, R. S.: Biology of Cartilage Cells. Cambridge, Cambridge University Press, 1979.

Strider, W., Pal, S., Margolis, R., and Rosenberg, L.: Changes with aging in the size and chemical composition of proteoglycan subunit and aggregate from normal distal femoral human articular cartilages. Trans. Orthop. Res. Soc., 2:33, 1976.

Swann, D. A., Radin, E. L., and Hendren, R. B.: The lubrication of articular cartilage by synovial fluid glycoproteins. Arthritis Rheum., *22*:665, 1979.

Swann, D. A., et al.: The molecular structure and lubricating activity of lubricin from bovine and human synovial fluids. Biochem. J., *225*:195, 1985.

Sweet, M. B. E., Thonar, E. J.-M. A., and Marsh, J.:Age related changes in proteoglycan structure. Arch. Biochem. Biophys., *198*:439, 1979.

Thonar, E. J.-M. A., Bjornsson, S., and Kuettner, K. E.: Age-related changes in cartilage proteoglycans. *In* Articular Cartilage Biochemistry. Edited by K. Kuettner, R. S. Schleyerbach, and V. C. Hascall. New York, Raven Press, 1986, pp. 273–287.

Thonar, E. J.-M. A., et al.: Biochemical basis of the age-related differences in the composition of aggregating proteoglycans in bovine articular cartilage. Trans. Orthop. Res. Soc., *8*:118, 1984.

Torchia, D. A., et al.: Mobility and function in elastin and collagen. *In* Mobility and Function in Proteins and Nucleic Acids (Ciba Foundation Symposium 93). London, Pitman Medical, 1982, pp. 98–115.

Torzilli, P. A., and Mow, V. C.: On the fundamental fluid transport mechanisms through normal and pathological articular cartilage during function. I. The formulation. J. Biomech., *9*:541–552, 1976.

Torzilli, P. A., Rose, D. E., and Dethemers, S. A.: Equilibrium water partition in articular cartilage. Biorheology, *19*:519, 1982.

Tsiganos, C. P., Hardingham, T. E., and Muir, H.: Proteoglycans of cartilage: An assessment of their structure. Biochem. Biophys. Acta, *229*:529, 1971.

Unsworth, A., Dowson, D., and Wright, V.: Some new evidence on human joint lubrication. Ann. Rheum. Dis., *34*:277, 1975.

Venn, M. F.: Variation of chemical composition with age in human femoral head cartilage. Ann. Rheum. Dis., *37*:168, 1978.

Voloshin, A. S., and Wosk, J.: Shock absorption of meniscectomized and painful knees: A comparative *in vivo* study. J. Biomed. Eng., *5*:157, 1983.

Wada, T., and Akizuki, S.: An ultrastructural study of solid matrix in articular cartilage under uniaxial tensile stress. J. Jpn. Orthop. Assoc., *61*:S 344, 1987.

Walker, P. S., Dowson, D., Longfield, M. D., and Wright, V.: "Boosted lubrication" in synovial joints by fluid entrapment and enrichment. Ann. Rheum. Dis., *27*:512, 1968.

Walker, P. S., et al.: Mode of aggregation of hyaluronic acid protein complex on the surface of articular cartilage. Ann. Rheum. Dis., *29*:591, 1970.

Weightman, B.: Tensile fatigue of human articular cartilage. J. Biomech., *9*:193, 1976.

Weightman, B., and Kempson, G.: Load carriage. *In* Adult Articular Cartilage. 2nd Ed. Edited by M. A. R. Freeman. Tunbridge Wells, Pitman Medical, 1979, pp. 291–329.

Weiss, C., Rosenberg, L., and Helfet, A. J.: An ultrastructural study of normal young adult human articular cartilage. J. Bone Joint Surg., *50A*:663, 1968.

Weidmann, H., et al.: Domain structure of cartilage proteoglycans revealed by rotary shadowing of intact and fragmented molecules. Biochem. J., *224*:331, 1984.

Woo, S. L.-Y., Akeson, W. H., and Jemmott, G. F.: Measurements of nonhomogeneous directional mechanical properties of articular cartilage in tension. J. Biomech., *9*:785, 1976.

Woo, S. L.-Y., Gomez, M. A., and Akeson, W. H.: The time and history-dependent viscoelastic properties of the canine medial collateral ligament. J. Biomech. Eng., *103*:293, 1981.

Woo, S. L.-Y., Mow, V. C., and Lai, W. M.: Biomechanical properties of articular cartilage. *In* Handbook of Bioengineering. Edited by R. Skalak and S. Chien. New York, McGraw-Hill, 1987, pp. 4.1–4.44.

Zhu, W. B., Lai, W. M., and Mow, V. C.: Intrinsic quasi-linear viscoelastic behavior of the extracellular matrix of cartilage. Trans. Orthop. Res. Soc., *11*:407, 1986.

SUGGESTED READING

Akizuki, S., et al.: Tensile properties of knee joint cartilage. I. Influence of ionic condition, weight bearing, and fibrillation on the tensile modulus. J. Orthop. Res., *4*:379, 1986.

Armstrong, C. G., and Mow, V. C.: Friction, lubrication and wear of synovial joints. *In* Scientific Foundations of Orthopaedics and Traumatology. Edited by R. Owen, J. Goodfellow, and P. Bullough. London, William Heinemann, 1980, pp. 223–232.

Armstrong, C. G., and Mow, V. C.: Variations in the intrinsic mechanical properties of human articular cartilage with age, degeneration, and water content. J. Bone Joint Surg., *64A*:88–94, 1982.

Buckwalter, J. A., Kuettner, K. E., and Thonar, E. J.-M. A.: Age-related changes in articular cartilage proteoglycans: Electron microscopic studies. J. Orthop. Res., *3*:251, 1985.

Eyre, D. R.: Collagen: Molecular diversity in the body's protein scaffold. Science, *207*:1315–1322, 1980.

Hardingham, T. M., and Muir, H.: Hyaluronic acid in cartilage and proteoglycan aggregation. Biochem. J., *139*:565–581, 1974.

Maroudas, A.: Physicochemical properties of articular cartilage. *In* Adult Articular Cartilage. 2nd Ed. Edited by M. A. R. Freeman. Tunbridge Wells, England, Pitman Medical, 1979, pp. 215–290.

Mow, V. C., and Mak, A. F.: Lubrication of diarthrodial joints. *In* Handbook of Bioengineering. Edited by R. Skalak and S. Chien. New York, McGraw-Hill, 1987, pp. 5.1–5.34.

Mow, V. C., Holmes, M. H., and Lai, W. M.: Fluid transport and mechanical properties of articular cartilage: A review. J. Biomech., *17*:377–394, 1984.

Muir, H.: Biochemistry. *In* Adult Articular Cartilage. 2nd Ed. Edited by M. A. R. Freeman. Tunbridge Wells, Pitman Medical, 1979, pp. 145–214.

Muir, H.: Proteoglycans as organizers of the extracellular matrix. Biochem. Soc. Trans., *11*:613, 1983.

Myers, E. R., and Mow, V. C.: Biomechanics of cartilage and its response to biomechanical stimuli. *In* Cartilage. Vol. I. Structure, Function, and Biochemistry. Edited by B. K. Hall. New York, Academic Press, 1983, pp. 313–341.

Radin, E. L.: Aetiology of osteoarthrosis. Clin. Rheum. Dis., *2*:509, 1976.

Roth, V., and Mow, V. C.: The intrinsic tensile behavior of the matrix of bovine articular cartilage and its variation with age. J. Bone Joint Surg., *62A*:1102–1117, 1980.

Roughley, P. J., and White, R. J.: Age-related changes in the structure of the proteoglycan subunits from human articular cartilage. J. Biol. Chem., *255*:217, 1980.

Woo, S. L.-Y., Mow, V. C., and Lai, W. M.: Biomechanical properties of articular cartilage. *In* Handbook of Bioengineering. Edited by R. Skalak and S. Chien. New York, McGraw-Hill, 1987, pp. 4.1–4.44.

3

BIOMECHANICS OF TENDONS AND LIGAMENTS

Carl A. Carlstedt
Margareta Nordin

The three principal structures that closely surround, connect, and stabilize the joints of the skeletal system are the tendons, the ligaments, and the joint capsules. Although these structures are passive (i.e., they do not actively produce motion as do the muscles), each plays an essential role in joint motion.

The role of the ligaments and joint capsules, which connect bone with bone, is to augment the mechanical stability of the joints, to guide joint motion, and to prevent excessive motion. The function of the tendons is to attach muscle to bone and to transmit tensile loads from muscle to bone, thereby producing joint motion. The tendon also enables the muscle belly to be at an optimal distance from the joint on which it acts without requiring an extended length of muscle between origin and insertion.

Tendon and ligament injuries and derangements are common. Proper management of these disorders requires an understanding of the mechanical properties and function of tendons and ligaments and their capacity for self-repair. This chapter discusses the composition and structure of tendons and ligaments, as well as the biomechanical properties and behavior of normal and injured tendon and ligament tissue. Several factors affecting the biomechanical function of tendons and ligaments—aging, pregnancy, mobilization and immobilization, and nonsteroidal anti-inflammatory drugs (NSAIDs)—are also covered.

COMPOSITION AND STRUCTURE OF TENDONS AND LIGAMENTS

Tendons and ligaments are dense connective tissues known as parallel-fibered collagenous tissues. These sparsely vascularized tissues are composed largely of collagen, a fibrous protein constituting approximately one third of the total protein in the body (White et al., 1964). Collagen constitutes a large portion of the organic matrix of bone and cartilage and has a unique mechanical supportive function in other connective tissues such as vessels, heart, ureters, kidneys, skin, and liver. The great mechanical stability of collagen gives the tendons and ligaments their characteristic strength and flexibility.

Like other connective tissues, tendons and ligaments consist of relatively few cells (fibroblasts) and an abundant extracellular matrix. In general, the cellular material occupies about 20% of the total tissue volume, while the extracellular matrix accounts for the remaining 80%. About 70% of the matrix consists of water, and approximately 30% is solids. These solids are collagen, ground substance, and a small amount of elastin. The collagen content is generally over 75% and is somewhat greater in tendons than in ligaments (Amiel et al., 1984); in extremity tendons the solid material may consist almost entirely of collagen (up to 99% of the dry weight) (Dale, 1974).

The structure and chemical composition of ligaments and tendons are identical in humans and in many animal species such as rats, rabbits, dogs, and monkeys. Hence, extrapolations regarding these structures in humans can be made from the results of studies on these animal species.

COLLAGEN

The collagen molecule is synthesized by the fibroblast (Branwood, 1963; Porter, 1964) within the cell as a larger precursor (procollagen), which is then secreted and cleaved extracellularly to become collagen (Fitton-Jackson, 1965) (Fig. 3–1). Tendons and ligaments, like bone, are composed of the most common collagen molecule, type I collagen. This molecule consists of three polypeptide chains (α chains), each coiled in a left-handed helix (Rich and Crick, 1955) with about 100 amino acids, which give it a total molecular weight of about 340,000 daltons (Ramachandran and Kartha, 1954; Rich and Crick, 1961) (Fig. 3–2). Two of the peptide chains (called α-1 chains) are identical, and one differs slightly (the α-2 chain). The three α chains are combined in a right-handed triple helix, which gives the collagen molecule a rodlike shape. The length of the molecule is about 280 nm, and its diameter is about 1.5 nm (Rich and Crick, 1955; White et al., 1964; Hall, 1965; Diamant et al., 1972; Ham, 1979).

Almost two thirds of the collagen molecule consists of three amino acids: glycine (33%), proline (15%), and hydroxyproline (15%) (Ramachandran, 1963). Every third amino acid in each α chain is glycine, and this repetitive sequence is essential for the proper formation of the triple helix. The small size of this amino acid allows the tight helical packing of the collagen molecule. Moreover, glycine enhances the stability of the molecule by forming hydrogen bonds among the three chains of the superhelix. Hydroxyproline and proline form hydrogen bonds, or hydrogen-bonded water bridges, within each chain. The intra- and interchain bonding, or cross-linking, between specific groups on the chains is essential to the stability of the molecule.

Cross-links are also formed between collagen molecules and are essential to aggregation at the fibril level. It is the cross-linked character of the collagen fibrils that gives strength to the tissues they compose and allows these tissues to function under mechanical stress. Within the fibrils the molecules are apparently cross-linked by "head-to-tail" interactions (Fig. 3–1), but interfibrillar cross-linking of a more complex nature may also occur.

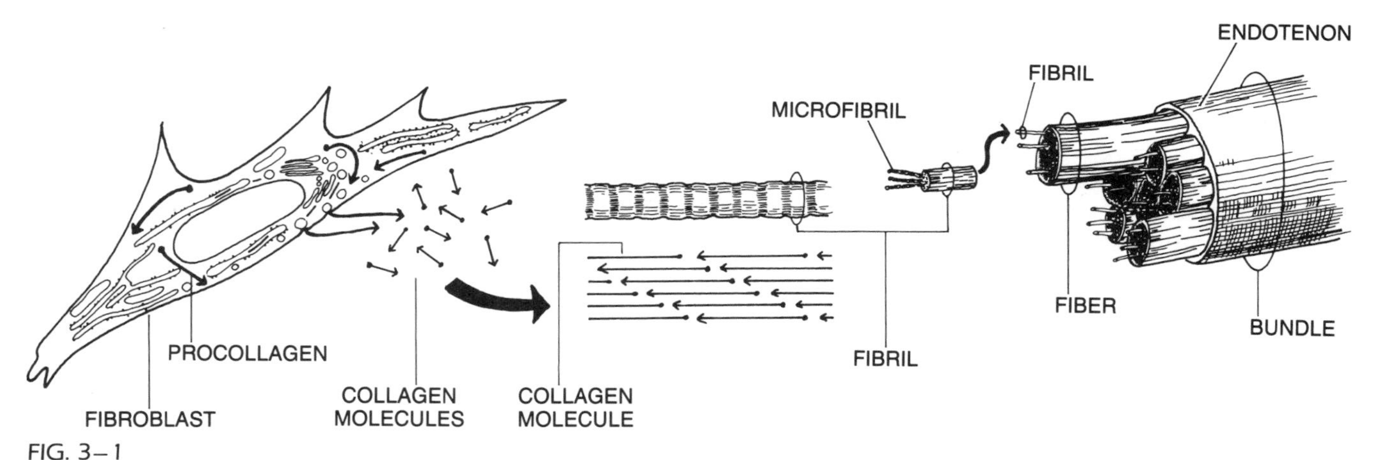

FIG. 3–1
Schematic representation of collagen fibrils, fibers, and bundles in tendons and collagenous ligaments (not drawn to scale). Collagen molecules, triple helices of coiled polypeptide chains, are synthesized and secreted by the fibroblasts. These molecules (depicted with "heads" and "tails" to represent positive and negative polar charges) aggregate in the extracellular matrix in a parallel arrangement to form microfibrils and then fibrils. The staggered array of the molecules, in which each overlaps the other, gives a banded appearance to the collagen fibrils under the electron microscope. The fibrils aggregate further into fibers, which come together into densely packed bundles.

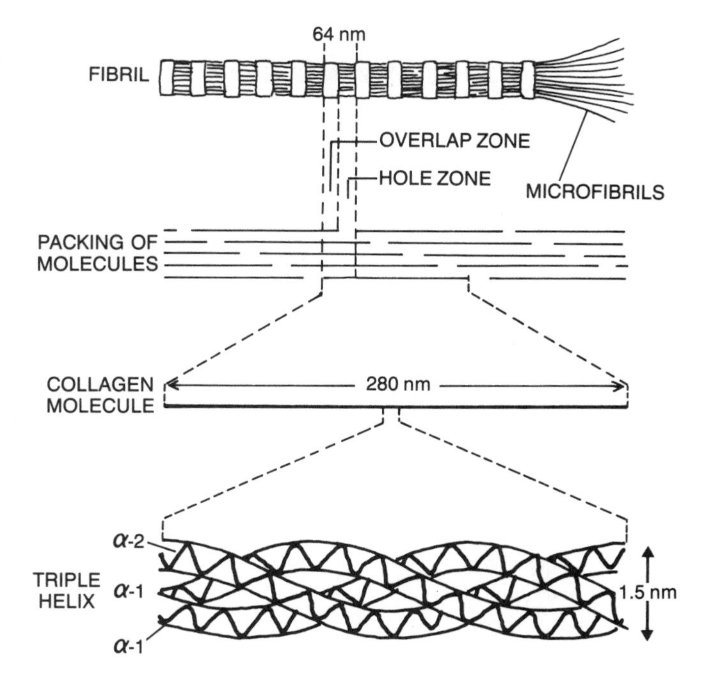

FIG. 3–2
Schematic drawing of collagen microstructure. The collagen molecule consists of three α chains in a triple helix (bottom). Several collagen molecules are aggregated into a staggered parallel array. This staggering, which creates hole zones and overlap zones, causes the cross-striation (banding pattern) visible in the collagen fibril under the electron microscope. (Adapted from Prockop and Guzman, 1977.)

In newly formed collagen the cross-links are relatively few and are reducible; the collagen is soluble in neutral salt solutions and in acid solutions, and the cross-links are fairly easily denatured by heat. As collagen ages, the total number of reducible cross-links decreases to a minimum as a large number of stable, nonreducible cross-links are formed. Mature collagen is not soluble in neutral salt solutions or in acid solutions, and it survives a higher denaturation temperature. (For a review of cross-linkage in collagen, see Viidik et al., 1982.)

A fibril is formed by the aggregation of several collagen molecules in a quaternary structure. This structure, in which each molecule overlaps the other, is responsible for the repeating bands observed on the fibrils under the electron microscope (Fig. 3–2; see Fig. 2–2). The fibrils aggregate further to form collagen fibers, which are visible under the light microscope. These fibers, which range from 1 to 20 μm in diameter, do not branch and may be many centimeters long. They reflect a 64-nm periodicity of the fibrils and have a characteristic undulated form. The fibers aggregate further into bundles. Fibroblasts are aligned in rows between these bundles and are elongated along an axis in the direction of ligament or tendon function (Fig. 3–3).

The arrangement of the collagen fibers differs somewhat in the tendons and ligaments and is suited to the function of each structure. The fibers composing the tendons have an orderly, parallel arrangement, which equips the tendons to handle the high unidirectional (uniaxial) tensile loads to which they are subjected during activity (see Fig. 3–3A). The ligaments generally sustain tensile loads in one predominant direction but may also bear smaller tensile loads in other directions; their fibers may not be completely parallel but are closely interlaced with one another (see Fig. 3–3B). The specific orientation

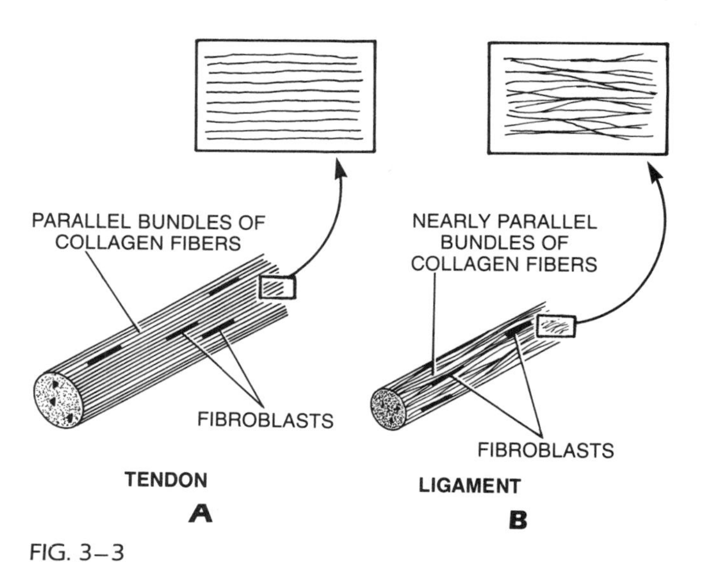

FIG. 3–3
Schematic diagram of the structural orientation of the fibers of tendon **(A)** and ligament **(B)**; insets show longitudinal sections. In both structures the fibroblasts are elongated along an axis in the direction of function. (Adapted from Snell, 1984.)

of the fiber bundles varies to some extent among the ligaments and is dependent on the function of the ligament (Amiel et al., 1984; Kennedy et al., 1976).

The metabolic turnover of collagen may be studied by tritium labeling of hydroxyproline or glycine and by autoradiographic methods. Studies in animals have shown that the half-life of collagen in mature animals is very long: the same collagen molecules may exist throughout the animal's adult life; however, in young animals (Neuberger et al., 1951) and in physically altered (e.g., injured or immobilized) tissue, the turnover is accelerated. Rabbit studies have shown metabolic activity to be somewhat greater in ligaments than in tendons, probably because of different stress patterns (Amiel et al., 1984).

ELASTIN

The mechanical properties of tendons and ligaments are dependent not only on the architecture and properties of the collagen fibers but also on the proportion of elastin that these structures contain. The protein elastin is scarcely present in tendons and extremity ligaments, but in elastic ligaments such as the ligamentum flavum the proportion of elastic fibers is substantial. Nachemson and Evans (1968) found a 2 to 1 ratio of elastic to collagen fibers in the ligamenta flava. These ligaments, which connect the laminae of adjacent vertebrae, appear to have a specialized role, which is to protect the spinal nerve roots from mechanical impingement, to prestress (preload) the motion segment (the functional unit of the spine), and to provide some intrinsic stability to the spine (Nachemson and Evans, 1968).

GROUND SUBSTANCE

The ground substance in ligaments and tendons consists of proteoglycans (PGs) (up to about 20% of the solids) along with structural glycoproteins, plasma proteins, and a variety of small molecules. The PG units, macromolecules composed of various sulfated polysaccharide chains (glycosaminoglycans) bonded to a core protein, bind to a long hyaluronic acid (HA) chain to form an extremely high-molecular-weight PG aggregate like that found in the ground substance of articular cartilage (see Fig. 2–6).

The PG aggregates bind most of the extracellular water of the ligament and tendon, making the matrix a highly structured gel-like material rather than an amorphous solution. Furthermore, by acting as a cementlike substance between the collagen microfibrils they may help stabilize the collagenous skeleton of tendons and ligaments and contribute to the overall strength of these composite structures. Only a small number of these molecules exist in tendon, however, and their importance for its biomechanical properties has been questioned (Viidik, 1973).

OUTER STRUCTURE AND INSERTION INTO BONE

Certain similarities are found in the outer structure of tendons and ligaments, but there are also important differences related to function. Both tendons and ligaments are surrounded by a loose areolar connective tissue. In ligaments this tissue has no specific name, but in tendons it is referred to as the paratenon (Greenlee and Ross, 1967). More structured than the connective tissue surrounding the ligaments, the paratenon forms a sheath that protects the tendon and enhances gliding. In some tendons, such as the flexors tendons of the digits, the sheath runs the length of the tendons, and in others the sheath is found only at the point where the tendon bends in concert with a joint.

In locations where the tendons are subjected to particularly high friction forces (e.g., in the palm, in the digits, and at the level of the wrist joint), a parietal synovial layer is found just beneath the paratenon; this synovium-like membrane, called the epitenon, surrounds several fiber bundles. The synovial fluid produced by the synovial cells of the epitenon facilitates gliding of the tendon. In locations where tendons are subjected to lower friction forces, they are surrounded by the paratenon only.

Each fiber bundle is bound together by the endotenon (see Fig. 3–1), which continues at the musculotendinous junction into the perimysium. At the tendo-osseous junction, the collagen fibers of the endotenon continue into the bone as the perforating fibers of Sharpey and become continuous with the periosteum (Carlstedt, 1987; Woo et al., 1988).

The structure of the insertion into bone is similar in ligaments and tendons and consists of four zones; Figure 3–4 illustrates these zones in a tendon. At the end of the tendon (zone 1), the collagen fibers intermesh with fibrocartilage (zone 2). This fibrocartilage gradually becomes mineralized fibrocartilage (zone 3) and then merges into cortical bone (zone 4). The change from more tendinous to more bony material produces a gradual alteration in the mechanical properties of the tissue (i.e., increased stiffness), which results in a decreased stress concentration

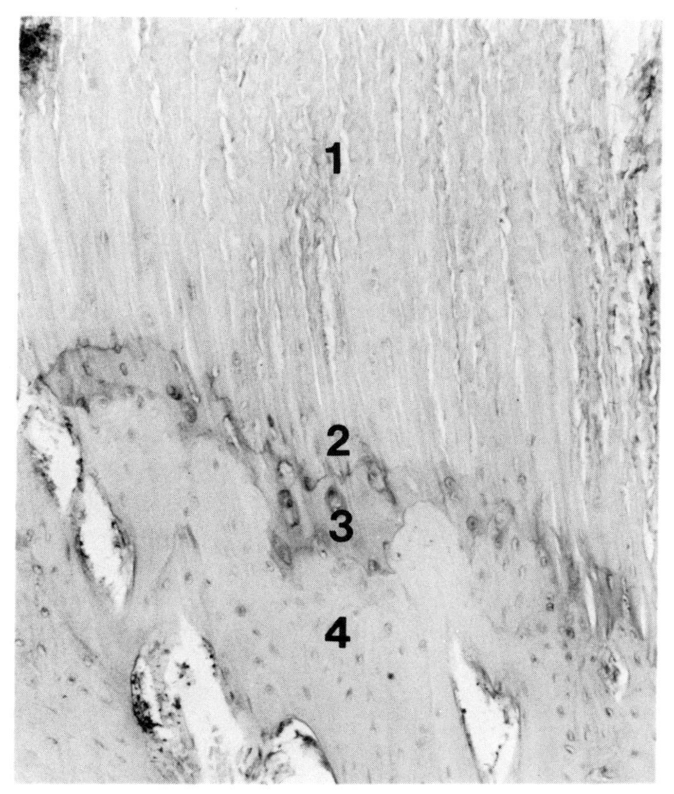

FIG. 3–4
Electron micrograph of a patellar tendon insertion from a dog, showing four zones (25,000×). Zone 1, parallel collagen fibers; zone 2, unmineralized fibrocartilage; zone 3, mineralized fibrocartilage; zone 4, cortical bone. The ligament-bone junction (not pictured) has a similar appearance. (Reprinted with permission from Cooper, R. R., and Misol, S.: Tendon and ligament insertion. A light and electron microscopic study. J. Bone Joint Surg., 52A:1, 1970.)

effect at the insertion of the tendon into the stiffer bone (Cooper and Misol, 1970).

MECHANICAL BEHAVIOR OF TENDONS AND LIGAMENTS

Tendons and ligaments are viscoelastic structures with unique mechanical properties. Tendons are strong enough to sustain the high tensile forces that result from muscle contraction during joint motion yet are sufficiently flexible to angulate around bone surfaces and to deflect beneath retinacula to change the final direction of muscle pull. The ligaments are pliant and flexible, allowing natural movements of the bones to which they attach, but are strong and inextensible so as to offer suitable resistance to applied forces.

Analysis of the mechanical behavior of tendons and ligaments provides important information for the understanding of injury mechanisms. Both structures sustain chiefly tensile loads during normal and excessive loading. When loading leads to injury, the degree of damage is affected by the rate of impact as well as the amount of load.

BIOMECHANICAL PROPERTIES

One means of analyzing the biomechanical properties of tendons and ligaments is to subject specimens to tensile deformation using a constant rate of elongation. The tissue is elongated until it ruptures, and the resulting force, or load (P), is plotted. The resulting load-elongation curve has several regions, which characterize the behavior of the tissue (Fig. 3–5).

The first region of the load-elongation curve is concave and is usually called the toe region. The elongation reflected in this region is believed to be the result of a change in the wavy pattern of the relaxed collagen fibers, which become straighter as loading progresses (Gustavson, 1956; Elliott, 1965; Viidik et

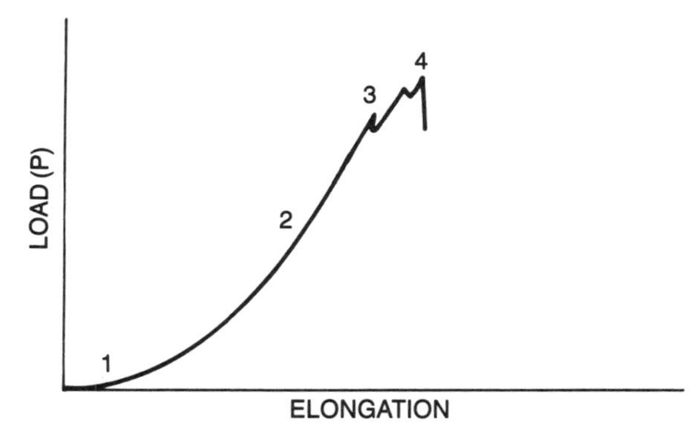

FIG. 3–5
Load-elongation curve for rabbit tendon tested to failure in tension. The numbers indicate the four characteristic regions of the curve. (1) Primary, or "toe," region, in which the tissue elongated with a small increase in load as the wavy collagen fibers straightened out. (2) Secondary, or "linear," region, in which the fibers straightened out and the stiffness of the specimen increased rapidly. Deformation of the tissue began and had a more or less linear relationship with load. (3) End of secondary region. The load value at this point is designated as P_{lin}. Progressive failure of the collagen fibers took place after P_{lin} was reached, and small force reductions (dips) occurred in the curve. (4) Maximum load (P_{max}), reflecting the ultimate tensile strength of the tissue. Complete failure occurred rapidly, and the specimen lost its ability to support loads. (Adapted from Carlstedt, 1987.)

al., 1965; Viidik, 1966, 1967b; Abrahams, 1967; Diamant et al., 1972; Hirsch, 1974) (Fig. 3–6A). Some data suggest, however, that this elongation may be caused mainly by interfibrillar sliding and shear of the interfibrillar gel (ground substance) (for review, see Viidik et al., 1982).

In the toe region, little force is required to elongate the tissue initially. As loading continues, the stiffness of the tissue increases and progressively greater force is required to produce equivalent amounts of elongation. The elongation is often expressed as strain (ϵ), which is the deformation of the tissue calculated as a percentage of the original length of the specimen. The end of the toe region has been reported to have a strain value of between 1.5 and 4.0% (Cronkite, 1936; Rigby et al., 1959; Abrahams, 1967; Diamant et al., 1972; Haut and Littel, 1972; Viidik, 1973).

The second region of the curve represents the response of the tissue to further elongation. Since this region is often more or less linear, it is called the linear region. The collagen fibers become more parallel and lose their wavy appearance (Gratz, 1931; Rigby et al., 1959; Viidik et al., 1965; Viidik 1966, 1973; Abrahams, 1967; Diamant et al., 1972) (Fig. 3–6B). At the end of this region, small force reductions (dips) can sometimes be observed in the loading curves for both tendons and ligaments. These dips are caused by the early sequential failure of a few greatly stretched fiber bundles (Butler et al., 1978).

At the end of the linear region, where the curve levels off toward the strain axis (Viidik, 1968), the load value is designated as P_{lin}. The point at which this value is reached is the yield point for the tissue. The energy to P_{lin} is represented by the area under the curve up to the end of the linear region.

When the linear region is surpassed, major failure of fiber bundles occurs in an unpredictable manner. With the attainment of maximum load (P_{max}), which reflects the ultimate tensile strength of the specimen, complete failure occurs rapidly, and the load-supporting ability of the tendon or ligament is substantially reduced.

The modulus of elasticity for tendons and ligaments has been determined in several investigations (Viidik, 1968; Fung, 1967, 1972). This parameter is based on a linear relationship between load and deformation (elongation), or stress and strain; that is, the stress (force per unit area) is proportional to the strain:

$$E = \frac{\sigma}{\epsilon}$$

where E = modulus of elasticity

σ = stress
ϵ = strain

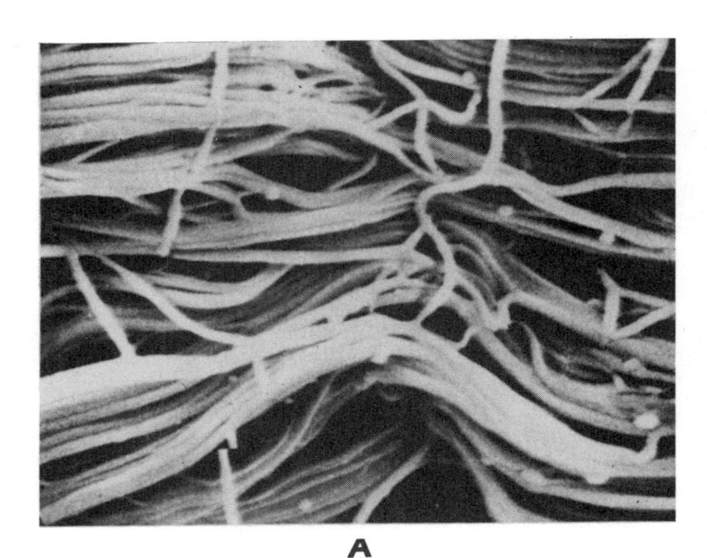

A

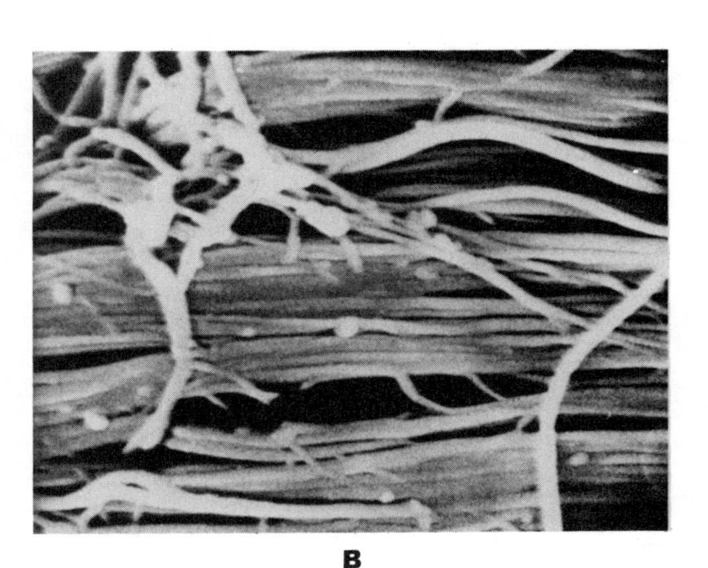

B

FIG. 3–6
Scanning electron micrographs of unloaded (relaxed) and loaded collagen fibers of human knee ligaments (10,000×). **A.** The unloaded collagen fibers have a wavy configuration. **B.** The collagen fibers have straightened out under load. (Reprinted with permission from Kennedy, J. C., et al.: Tension studies of human knee ligaments. Yield point, ultimate failure, and disruption of the cruciate and tibial collateral ligaments. J. Bone Joint Surg., 58A:350–355, 1976.)

In the toe portion of the load-elongation curve (or stress-strain curve), the modulus of elasticity is not constant but increases gradually. The modulus stabilizes in the fairly linear secondary region of the curve.

The load-elongation curve depicted in Figure 3–5 generally applies to tendons and extremity ligaments. The curve for the ligamentum flavum, with its high proportion of elastic fibers, is entirely different (Fig. 3–7). In tensile testing of a human ligamentum flavum, elongation of the specimen reached 50% before the stiffness increased appreciably. Beyond this point the stiffness increased greatly with additional loading and the ligament failed abruptly (reached P_{max}) with little further deformation (Nachemson and Evans, 1968).

PHYSIOLOGIC LOADING OF TENDONS AND LIGAMENTS

The ultimate tensile strength (P_{max}) of ligaments and tendons is of limited interest from a functional standpoint because under normal physiologic conditions in vivo these structures are subjected to a stress magnitude that is only about one third of this value (Viidik, 1980). The upper limit for physiologic strain in tendons and ligaments (when running and jumping, for example) is from 2 to 5% (Fung, 1981).

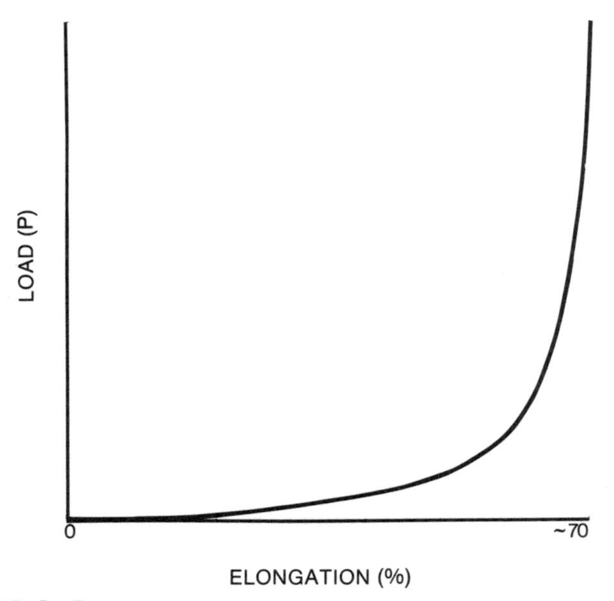

FIG. 3–7
Load-elongation curve for a human ligamentum flavum (60 to 70% elastic fibers) tested in tension to failure. At 70% elongation the ligament exhibited a great increase in stiffness with additional loading and failed abruptly without further deformation. (Adapted from Nachemson and Evans, 1968.)

Few studies of loading of tendons or ligaments in vivo have been performed. Kear and Smith (1975), using the strain gauge method, measured the maximal strain in the lateral digital extensor tendons of sheep. The strain reached 2.6% while the sheep were trotting rapidly and decreased when the trotting speed decreased. This maximal strain occurred for only 0.1 second during each stride. The maximal load imposed on the entire tendon was approximately 45 newtons (N). These results suggest that during normal activity a tendon in vivo is subjected to less than one fourth of its ultimate stress. Findings of this study correlate well with those of other authors (Abrahams, 1967; Elliott, 1967).

LIGAMENT AND TENDON INJURY MECHANISMS

Injury mechanisms are similar for ligaments and tendons, so the following description of ligament injury and failure is generally applicable to tendons. When a ligament in vivo is subjected to loading that exceeds the physiologic range, microfailure takes place even before the yield point (P_{lin}) is reached. When P_{lin} is exceeded, the ligament begins to undergo gross failure, and simultaneously the joint begins to displace abnormally. This displacement can also result in damage to the surrounding structures such as the joint capsule, the adjacent ligaments, and the blood vessels that supply these structures.

Noyes (1977) demonstrated the progressive failure of the anterior cruciate ligament and displacement of the tibiofemoral joint by applying a clinical test, the anterior drawer test, to a cadaver knee up to the point of anterior cruciate ligament failure (Fig. 3–8). At maximum load the joint had displaced several millimeters. The ligament was still in continuity even though it had undergone extensive macro- and microfailure and extensive elongation. In Figure 3–8, the force-elongation curve generated during the experiment, indicating where microfailure of the ligament began, is compared with various stages of joint displacement recorded photographically.

Correlation of the results of this test in vitro with clinical findings sheds light on the microevents that take place in the anterior cruciate ligament during normal daily activity and during injuries of various

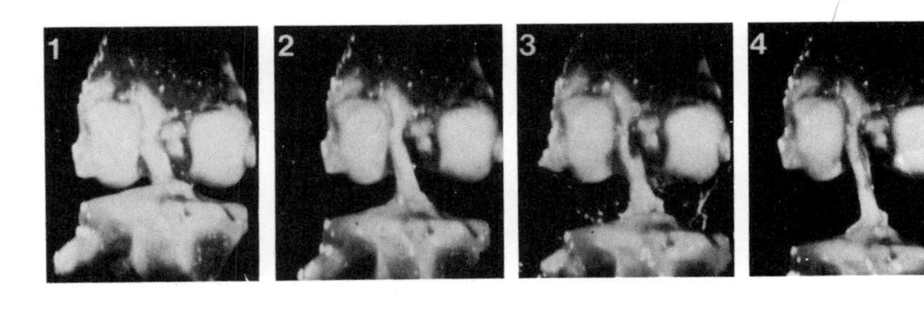

FIG. 3–8
Progressive failure of the anterior cruciate ligament from a cadaver knee tested in tension to failure at a physiologic strain rate (Noyes, 1977). The joint was displaced 7 mm before the ligament failed completely. The force-elongation curve generated during this experiment is correlated with various degrees of joint displacement recorded photographically; photos correspond to similarly numbered points on the curve. (Courtesy of Frank R. Noyes, M.D., and Edward S. Grood, Ph.D.)

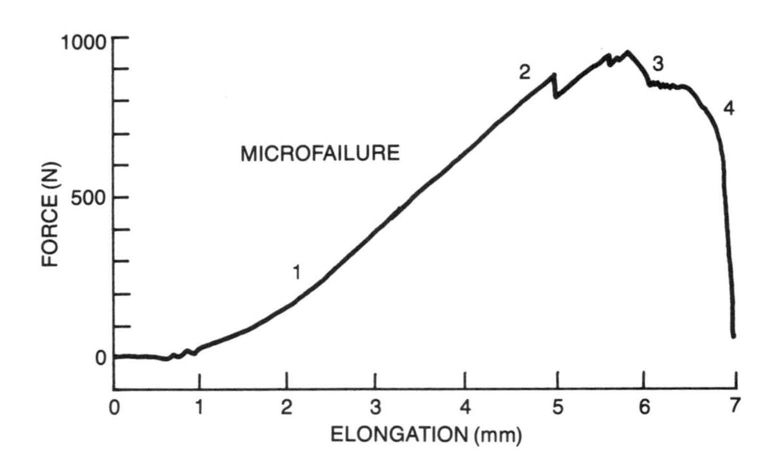

degrees of severity. In Figure 3–9 the curve for the experimental study on cadaver knees presented in Figure 3–8 has been converted into a load-displacement curve and divided into three regions, corresponding, respectively, to (1) the load placed on the anterior cruciate ligament during tests of knee joint stability performed clinically, (2) the load placed on this ligament during physiologic activity, (3) and that imposed on the ligament during injury from the beginning of microfailure to complete rupture. Microfailure begins even before the physiologic loading range is exceeded and indeed can occur throughout the physiologic range in any given ligament.

Ligament injuries are categorized clinically in three ways according to degree of severity. Injuries in the first category produce negligible clinical symptoms. Some pain is felt, but no joint instability can be detected clinically, even though microfailure of the collagen fibers may have occurred.

Injuries in the second category produce severe pain, and some joint instability can be detected clinically. Progressive failure of the collagen fibers has taken place, resulting in partial ligament rupture. The strength and stiffness of the ligament may have decreased by 50% or more, mainly because the

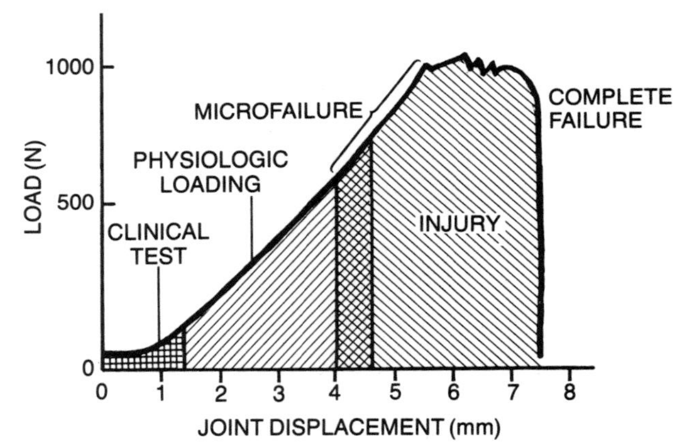

FIG. 3–9
The curve produced during tensile testing of a human anterior cruciate ligament in vitro (Noyes, 1977) (see Fig. 3–8) has been converted to a load-displacement curve and divided into three regions correlating with clinical findings: (1) the load imposed on the anterior cruciate ligament during the anterior drawer test; (2) that placed on the ligament during physiologic activity; and (3) that imposed on the ligament from partial injury to complete rupture. It should be noted that the divisions shown here represent a generalization. Microfailure is shown to begin toward the end of the physiologic loading region, but it may take place well before this point in any given ligament.

amount of undamaged tissue has been reduced. The joint instability produced by a partial rupture of a ligament is often masked by muscle activity, and thus the clinical test for joint stability is usually performed with the patient under anesthesia.

Injuries in the third category produce severe pain during the course of trauma with less pain after injury. Clinically, the joint is found to be completely unstable. Most collagen fibers have ruptured, but a few may still be intact, giving the ligament the appearance of continuity even though it is unable to support any loads.

Loading of a joint that is unstable due to ligament or joint capsule rupture produces abnormally high stresses on the articular cartilage. This abnormal loading of the articular cartilage in the knee has been correlated with early osteoarthritis in humans and in animals (Marshall and Olsson, 1971; Alm et al., 1974).

Although injury mechanisms are generally comparable in ligaments and tendons, two additional factors become important in tendons because of their attachment to muscles: the amount of force produced by contraction of the muscle to which the tendon is attached, and the cross-sectional area of the tendon in relation to that of its muscle. A tendon is subjected to increasing stress as its muscle contracts (see Fig. 5–9). When the muscle is maximally contracted, the tensile stress on the tendon reaches high levels. This stress can be increased further if rapid eccentric contraction of the muscle takes place; for example, rapid dorsiflexion of the ankle, which does not allow for reflex relaxation of the gastrocnemius and soleus muscles, increases the tension on the Achilles tendon. The load imposed on the tendon under these circumstances may exceed the yield point, causing Achilles tendon rupture.

The strength of a muscle depends on its physiologic cross-sectional area. The larger the cross-sectional area of the muscle is, the higher the magnitude of the force produced by the contraction, and thus the greater the tensile loads transmitted through the tendon. Similarly, the larger the cross-sectional area of the tendon is, the greater the loads it can bear. Although the maximal stress to failure for a muscle has been difficult to compute accurately, such measurements have shown that the tensile strength of a healthy tendon may be more than twice that of its muscle (Elliott, 1967). This finding is supported clinically by the fact that muscle ruptures are more common than ruptures through a tendon.

Large muscles usually have tendons with large cross-sectional areas. Examples are the quadriceps muscle with its patellar tendon and the triceps surae muscle with its Achilles tendon. Some small muscles have tendons with large cross-sectional areas, however; the plantaris is a tiny muscle with a large tendon.

VISCOELASTIC BEHAVIOR (RATE DEPENDENCY) IN TENDONS AND LIGAMENTS

Ligaments and tendons exhibit viscoelastic, or rate-dependent (time-dependent), behavior under loading; their mechanical properties change with different rates of loading. When ligament and tendon specimens are subjected to increased strain rates (loading rates), the linear portion of the stress-strain curve becomes steeper, indicating greater stiffness of the tissue at higher strain rates (Frisen et al., 1969a, 1969b; Viidik, 1979). With higher strain rates, ligaments and tendons in isolation store more energy, require more force to rupture, and undergo greater elongation (Kennedy et al., 1976).

During cyclic testing of ligaments and tendons, wherein loads are applied and released at specific intervals, the stress-strain curve is displaced to the right along the deformation (strain) axis with each loading cycle, revealing the presence of a nonelastic (plastic) component; the amount of permanent (nonrecoverable) deformation is progressively greater with every loading cycle. As cyclic loading progresses, the specimen also shows an increase in elastic stiffness due to plastic deformation (molecular displacement) (Viidik, 1968, 1979). Microfailure can occur within the physiologic range if frequent loading is imposed on an already damaged structure the stiffness of which has decreased.

Two standard tests that reveal the viscoelasticity of ligaments and tendons are the stress-relaxation test and the creep test (Fig. 3–10). During a stress-relaxation test, loading is halted safely below the linear region of the stress-strain curve and the strain is kept constant over an extended period. The stress decreases rapidly at first and then gradually more slowly. When the stress-relaxation test is repeated cyclically, the decrease in stress gradually becomes less pronounced.

During a creep test, loading is halted safely below the linear region of the stress-strain curve and the stress is kept constant over an extended period. The strain increases relatively quickly at first and then more and more slowly. When this test is performed

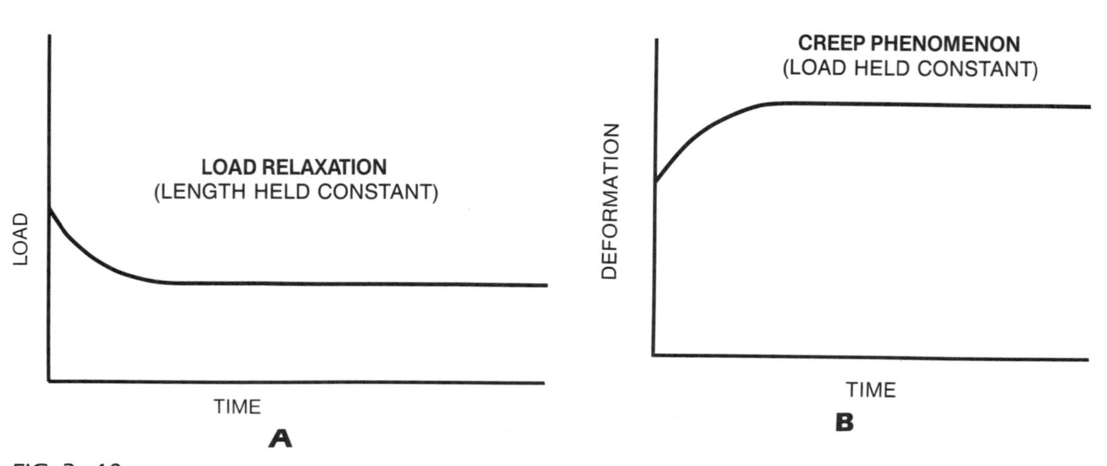

FIG. 3–10
The viscoelasticity (rate dependency, or time dependency) of ligaments and tendons can be demonstrated by two standard tests: the load-relaxation test and the creep test. **A.** Load relaxation is demonstrated when the loading of a specimen is halted safely below the linear region of the load-deformation curve and the specimen is maintained at a constant length over an extended period (i.e., the amount of elongation is constant). The load decreases rapidly at first (i.e., during the first 6 to 8 hours of loading) and then gradually more slowly, but the phenomenon may continue at a low rate for months. **B.** The creep response takes place when loading of a specimen is halted safely below the linear region of the load-deformation curve and the amount of load remains constant over an extended period. The deformation increases relatively quickly at first (within the first 6 to 8 hours of loading) but then progressively more slowly, continuing at a low rate for months.

cyclically, the increase in strain gradually becomes less pronounced.

The clinical application of a constant low load to the soft tissues over a prolonged period, which takes advantage of the creep response, is a useful treatment for several types of deformities. One example is the manipulation of a child's clubfoot by subjecting it to constant loads by means of a plaster cast. Another example is the treatment of idiopathic scoliosis with a brace, whereby constant loads are applied to the spinal area to elongate the soft tissues surrounding the abnormally curved spine.

More complex viscoelastic behavior is observed in the entire bone-ligament-bone complex. Anterior cruciate ligaments in knee specimens taken from 30 primates were tested in tension to failure at a slow and a fast loading rate (Noyes and Grood, 1976). At the slow loading rate (60 seconds), much slower than that of an injury mechanism in vivo, the bony insertion of the ligament was the weakest component of the bone-ligament-bone complex, and a tibial spine avulsion was produced. At the fast loading rate (0.6 seconds), which simulated an injury mechanism in vivo, the ligament was the weakest component in two thirds of the specimens tested. At the slower rate the load to failure decreased by 20%, and 30% less energy was stored to failure, but the stiffness of the bone-ligament-bone complex was nearly the same. These results suggest that as the loading rate is increased bone shows a greater increase in strength than does ligament.

FACTORS THAT AFFECT THE BIOMECHANICAL PROPERTIES OF TENDONS AND LIGAMENTS

Numerous factors affect the biomechanical properties of tendons and ligaments. The most common are aging, pregnancy, mobilization and immobilization, and NSAIDs.

MATURATION AND AGING

The physical properties of collagen and of the tissues it composes are closely associated with the number and quality of the cross-links within and between the collagen molecules. During maturation (up to 20 years of age), the number and quality of cross-links increases, resulting in increased tensile strength of the tendon and ligament (Piez, 1968; Vogel, 1978; Viidik et al., 1982). An increase in collagen fibril diameter is also observed (Parry et al., 1978). After maturation, as aging progresses, collagen reaches a plateau with respect to its mechanical properties, after which the tensile strength and stiffness of the tissue begin to decrease (Rollhäuser, 1950; Yamada, 1970; Vogel, 1978; Viidik, 1980; Viidik et al., 1982). The collagen content of tendons and ligaments also decreases during aging, contributing to the gradual decline in their mechanical properties (strength, stiffness, and ability to withstand deformation).

PREGNANCY AND THE POSTPARTUM PERIOD

A common clinical observation is the increased laxity of the tendons and ligaments in the pubic area during later stages of pregnancy and the postpartum period. This observation has been confirmed in animal studies. Rundgren (1974) found that the tensile strength of the tendons and the pubic symphysis in rats decreased at the end of pregnancy and during the postpartum period. Stiffness of these structures decreased in the early postpartum period but was later restored.

MOBILIZATION AND IMMOBILIZATION

Like bone, ligament and tendon appear to remodel in response to the mechanical demands placed upon them; they become stronger and stiffer when subjected to increased stress and weaker and less stiff when the stress is reduced (Noyes, 1977; Tipton et al., 1970).

Physical training has been found to increase the tensile strength of tendons (Tipton et al., 1967, 1970; Viidik, 1967a, 1979; Woo et al., 1980, 1982) and of the ligament-bone interface (Tipton et al., 1967; Viidik, 1968; Cabaud et al., 1980; Woo et al., 1981). Tipton and coworkers (1970) compared the strength and stiffness of medial collateral ligaments from dogs that were exercised strenuously for 6 weeks with the values for ligaments from a control group of animals. The ligaments of the exercised dogs were stronger and stiffer than those of the control dogs, and the collagen fiber bundles had larger diameters.

Immobilization has been found to decrease the tensile strength of ligaments (Noyes, 1977; Amiel et al., 1982). Noyes (1977) demonstrated a reduction in the mechanical properties of the bone-ligament-bone complex in knees of primates immobilized in body casts for 8 weeks. When tested in tension to failure, the anterior cruciate ligaments from these animals showed a 39% decrease in maximum load to failure and a 32% decrease in energy stored to failure compared with ligaments from a control group of animals (Fig. 3–11A). The immobilized ligaments also displayed more elongation and were significantly less stiff than were the control specimens (Fig. 3–11B).

Amiel and coworkers (1982) showed a similar decrease in the strength and stiffness of lateral collateral ligaments in rabbits immobilized for 9 weeks. As the cross-sectional area of the specimens did not change significantly, the degeneration of mechanical properties was attributed to changes in the ligament substance itself. The tissue metabolism was noted to increase, leading to proportionally more immature collagen with a decrease in the amount and quality of the cross-links between collagen molecules.

In Noyes's 1977 experiment, assessment of the effects of a reconditioning program initiated directly after the 8-week immobilization period demonstrated that considerable time was needed for the immobilized ligaments to regain their former strength and stiffness. After 5 months the reconditioned ligaments still showed considerably less stiffness and 20% less strength than did ligaments from control animals. At 12 months the reconditioned ligaments had strength and stiffness values comparable to those of control group ligaments (see Fig. 3–11A).

NONSTEROIDAL ANTI-INFLAMMATORY DRUGS

NSAIDs (e.g., aspirin, acetaminophen, indomethacin) are frequently used in the treatment of various painful conditions of the musculoskeletal system. NSAIDs are also widely used in the treatment of soft tissue injuries, such as inflammatory disorders and partial ruptures of tendons and ligaments. Vogel (1977) found that treatment with indomethacin resulted in increased tensile strength in rat tail tendons. An increase in the proportion of insoluble collagen and in the total collagen content was also observed. Ohkawa (1982) found increased tensile strength in periodontium of rats after indo-

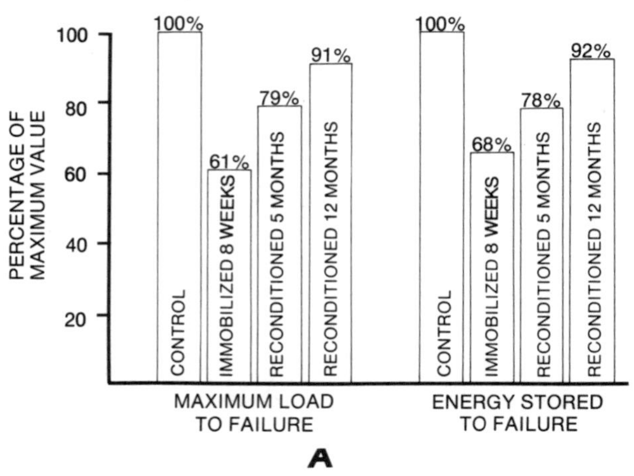

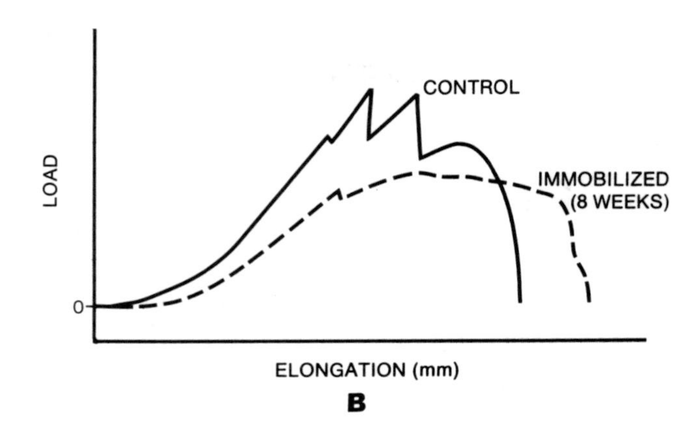

FIG. 3–11
A. Maximal load to failure and energy stored to failure for primate anterior cruciate ligaments tested in tension to failure. Values are shown as a percentage of control values for three groups of experimental animals: (1) those immobilized in body casts for 8 weeks; (2) those immobilized for 8 weeks and given a reconditioning program for 5 months; and (3) those immobilized for 8 weeks and given a reconditioning program for 12 months. **B.** Compared with controls, ligaments immobilized for 8 weeks were significantly less stiff (as indicated by the slope of the curve) and underwent greater elongation. (Adapted from Noyes, 1977.)

methacin treatment. Carlstedt and associates (1986a, 1986b) found that indomethacin treatment increased the tensile strength in developing and healing plantaris longus tendons in the rabbit and noted that the mechanism for this increase was probably an increased cross-linkage of collagen molecules. These animal studies suggest that short-term administration of NSAIDs would not be deleterious for tendon healing but rather would increase the rate of biomechanical restoration of the tissue.

SUMMARY

1. Tendons and extremity ligaments are composed largely of collagen, whose mechanical stability gives these structures their characteristic strength and flexibility. The ligamenta flava of the spine have a substantial proportion of elastin, which lends these structures their great elasticity.
2. The arrangement of the collagen fibers is nearly parallel in tendons, equipping them to withstand high unidirectional loads. The less parallel arrangement of the collagen fibers in ligaments allows these structures to sustain predominant tensile stresses in one direction and smaller stresses in other directions.
3. At the insertion of ligament and tendon into stiffer bone, the gradual change from a more fibrous to a more bony material results in a decreased stress concentration effect.
4. Tendons and ligaments undergo deformation before failure. When the ultimate tensile strength of these structures is surpassed, complete failure occurs rapidly, and their load-bearing ability is substantially decreased.
5. Studies suggest that during normal activity a tendon in vivo is subjected to less than one fourth of its ultimate stress.
6. Injury mechanisms in a tendon are influenced by the amount of force produced by the contraction of the muscle to which the tendon is attached and the cross-sectional area of the tendon in relation to that of its muscle.
7. The biomechanical behavior of ligaments and tendons is viscoelastic, or rate-dependent, so these structures display an increase in strength and stiffness with an increased loading rate.
8. An additional effect of rate dependency is the slow deformation, or creep, that occurs when tendons and ligaments are subjected to a constant low load over an extended period; stress

relaxation takes place when these structures sustain a constant elongation over time.

9. Aging results in a decline in the mechanical properties of tendons and ligaments, i.e., their strength, stiffness, and ability to withstand deformation.
10. Ligaments and tendons remodel in response to the mechanical demands placed on them.

REFERENCES

Abrahams, M.: Mechanical behavior of tendon in vitro. Med. Biol. Eng., *5*:433, 1967.

Alm, A., Ekström, H., Gillquist, J., and Strömberg, B.: The anterior cruciate ligament. A clinical and experimental study on tensile strength, morphology and replacement by patellar ligaments. Acta Chir. Scand., Suppl. *445*:1–49, 1974.

Amiel, D., Woo, S. L.-Y., Harwood, F. L., and Akeson, W. H.: The effect of immobilization on collagen turnover in connective tissue. A biochemical-biomechanical correlation. Acta Orthop. Scand., *53*:325, 1982.

Amiel, D., et al.: Tendons and ligaments: A morphological and biochemical comparison. J. Orthop. Res., *1*:257, 1984.

Branwood, A. W.: The fibroblast. Int. Rev. Connect. Tissue Res., *1*:1, 1963.

Butler, D. L., Grood, E. S., and Noyes, F. R.: Biomechanics of ligaments and tendons. Exerc. Sport Sci. Rev., *6*:125, 1978.

Cabaud, H. E., Chatty, A., Gildengorin, V., and Feltman, R. J.: Exercise effects on the strength of the rat anterior cruciate ligament. Am. J. Sports Med., *8*:79, 1980.

Carlstedt, C. A.: Mechanical and chemical factors in tendon healing. Effects of indomethacin and surgery in the rabbit. Acta Orthop. Scand., Suppl. *224*, 1987.

Carlstedt, C. A., Madsén, K., and Wredmark, T.: The influence of indomethacin on collagen synthesis during tendon healing in the rabbit. Prostaglandins, *32*:353, 1986a.

Carlstedt, C. A., Madsén, K., and Wredmark, T.: The influence of indomethacin on tendon healing. A biomechanical and biochemical study. Arch. Orthop. Trauma Surg., *105*:332, 1986b.

Cooper, R. R., and Misol, S.: Tendon and ligament insertion. A light and electron microscopic study. J. Bone Joint Surg., *52A*:1, 1970.

Cronkite, A. E.: The tensile strength of human tendons. Anat. Rec., *64*:173, 1936.

Dale, W. C.: A composite materials analysis of the structure, mechanical properties, and aging of collagenous tissues. Ph.D. thesis, Case Western Reserve University, Cleveland, Ohio, 1974.

Diamant, J., et al.: Collagen: Ultrastructure and its relations to mechanical properties as a function of ageing. Proc. R. Soc. Lond. [Biol.], *180*:293, 1972.

Elliott, D. H.: Structure and function of mammalian tendon. Biol. Rev., *40*:392, 1965.

Elliott, D. H.: The biomechanical properties of tendon in relation to muscular strength. Ann. Phys. Med., *9*:1, 1967.

Fitton-Jackson, S.: Antecedent phases in matrix formation. *In* Structure and Function of Connective and Skeletal Tissues. London, Butterworth, 1965, p. 277.

Frisen, M., Mägi, M., Sonnerup, L., and Viidik, A.: Rheological analysis of soft collagenous tissue. I. Theoretical consideration. J. Biomech., *2*:13, 1969a.

Frisen, M., Mägi, M., Sonnerup, L., and Viidik, A.: Rheological analysis of soft collagenous tissue. II. Experimental evaluations and verifications. J. Biomech., *2*:21, 1969b.

Fung, Y. C. B.: Elasticity of soft tissues in simple elongation. Am. J. Physiol., *213*:1532, 1967.

Fung, Y. C. B.: Stress-strain-history relations of soft tissues in simple elongation. *In* Biomechanics: Its Foundations and Objectives. Edited by Y. C. Fung, N. Perrone, and M. Anliker. Englewood Cliffs, Prentice-Hall, 1972, pp. 181–208.

Fung, Y. C. B.: Biomechanics: Mechanical Properties of Living Tissues. New York, Springer Verlag, 1981, p. 222.

Gratz, C. M.: Tensile strength and elasticity tests on human fascia lata. J. Bone Joint Surg., *13*:334, 1931.

Greenlee, T. K., and Ross, R.: The development of the rat flexor digital tendon, a fine structure study. J. Ultrastruct. Res., *18*:354, 1967.

Gustavson, K. H.: The Chemistry and Reactivity of Collagen. New York, Academic Press, 1956.

Hall, M. C.: The Locomotor System: Functional Histology. Springfield, Charles C Thomas, 1965.

Ham, A. W., and Cormack, D. H.: Histology. 8th Ed. Philadelphia, J. B. Lippincott, 1979.

Haut, R. C., and Littel, R. W. A.: A constitutive equation for collagen fibers. J. Biomech., *5*:523, 1972.

Hirsch, C.: Tensile properties during tendon healing. Acta Orthop. Scand., Suppl. *153*, 1974.

Kear, M., and Smith, R. N.: A method for recording tendon strain in sheep during locomotion. Acta Orthop. Scand., *46*:896, 1975.

Kennedy, J. C., Hawkins, R. J., Willis, R. B., and Danylchuk, K. D.: Tension studies of human knee ligaments. Yield point, ultimate failure, and disruption of the cruciate and tibial collateral ligaments. J. Bone Joint Surg., *58A*:350, 1976.

Marshall, J. L., and Olsson, S.-E.: Instability of the knee. A long-term study in dogs. J. Bone Joint Surg., *53A*:1561, 1971.

Nachemson, A. L., and Evans, J. H.: Some mechanical properties of the third human lumbar interlaminar ligament (ligamentum flavum). J. Biomech., *1*:211, 1968.

Neuberger, A., Perrone, J. C., and Slack, H. G. B.: The relative metabolic inertia of tendon collagen in the rat. Biochem. J., *49*:199, 1951.

Noyes, F. R.: Functional properties of knee ligaments and alterations induced by immobilization. Clin. Orthop., *123*:210, 1977.

Noyes, F. R., and Grood, E. S.: The strength of the anterior cruciate ligament in humans and Rhesus monkeys. Age-related and species-related changes. J. Bone Joint Surg., *58A*:1074, 1976.

Ohkawa, S.: Effects of orthodontic forces and anti-inflammatory drugs on the mechanical strength of the periodontium in the rat mandibular first molar. Am. J. Orthod., *81*:498, 1982.

Parry, D. A. D., Barnes, G. R. G., and Craig, A. S.: A comparison of the size distribution of collagen fibrils in connective tissues as a function of age and a possible relation between fibril size distribution and mechanical properties. Proc. R. Soc. Lond. [Biol.], *203*:305, 1978.

Piez, K. A.: Cross-linking of collagen and elastin. Ann. Rev. Biochem., *37*:574, 1968.

Porter, K. R.: Cell fine structure and biosynthesis of intercellular macromolecules. Biophys. J., *4*:2, 1964.

Prockop, D. J., and Guzman, N. A.: Collagen diseases and the biosynthesis of collagen. Hosp. Practice, December 1977, pp. 61–68.

Ramachandran, G. N.: Molecular structure of collagen. Int. Rev. Connect. Tissue Res., *1*:127, 1963.

Ramachandran, G. N., and Kartha, G.: Structure of collagen. Nature, *174*:269, 1954.

Rich, A., and Crick, F. H. C.: The structure of collagen. Nature, *176*:915, 1955.

Rich, A., and Crick, F. H. C.: The molecular structure of collagen. J. Mol. Biol., *3*:483, 1961.

Rigby, B. J., Hirai, N., Spikes, J. D., and Eyring, H.: The mechanical properties of rat tail tendon. J. Gen. Physiol., *43*:265, 1959.

Rollhäuser, H.: Die Festigkeit enschlicher Sehnen nach Quellung und Trocknung in Abhängigkeit von Lebensalter. Gegenbauers Morph. Jahrb., *90*:180, 1950.

Rundgren, Å.: Physical properties of connective tissue as influenced by single and repeated pregnancies in the rat. Acta Physiol. Scand., Suppl. *417*, 1974.

Snell, R. S.: Clinical and Functional Histology for Medical Students. Boston, Little, Brown, 1984.

Tipton, C. M., Schild, R. J., and Tomanek, R. J.: Influence of physical activity on the strength of knee ligaments in rats. Am. J. Physiol., *212*:783, 1967.

Tipton, C. M., James, S. L., Mergner, W., and Tcheng, T.: Influence of exercise on strength of medial collateral ligaments of dogs. Am. J. Physiol., *218*:894, 1970.

Viidik, A.: Biomechanics and functional adaptation of tendons and joint ligaments. *In* Studies on the Anatomy and Function of Bones and Joints. Edited by F. G. Evans. Berlin, Springer Verlag, 1966, pp. 17–39.

Viidik, A.: The effect of training on the tensile strength of isolated rabbit tendons. Scand. J. Plast. Reconstr. Surg., *1*:141, 1967a.

Viidik, A.: Experimental evaluation of the tensile strength of isolated rabbit tendons. Biomed. Eng., *2*:64, 1967b.

Viidik, A.: Elasticity and tensile strength of the anterior cruciate ligament in rabbits as influenced by training. Acta Physiol. Scand., *74*:372, 1968.

Viidik, A.: Functional properties of collagenous tissues. Int. Rev. Connect. Tissue Res., *6*:127, 1973.

Viidik, A.: Biomechanical behavior of soft connective tissues. *In* Progress in Biomechanics. Edited by N. Akkas. Alpen Aan den Rijn, Sijthoff and Nordhoff, 1979, pp. 75–113.

Viidik, A.: Mechanical properties of parallel-fibred collagenous tissues. *In* Biology of Collagen. Edited by A. Viidik and J. Vuust. London, Academic Press, 1980, pp. 237–255.

Viidik, A., Danielsen, C. C., and Oxlund, H.: Fourth International Congress of Biorheology Symposium on Mechanical Properties of Living Tissues: On fundamental and phenomenological models, structure and mechanical properties of collagen, elastic and glycosaminoglycan complexes. Biorheology, *19*:437, 1982.

Viidik, A., Sandqvist, L., and Mägi, M. L.: Influence of postmortal storage on tensile strength characteristics and histology of rabbit ligaments. Acta Orthop. Scand., Suppl. *79*, 1965.

Vogel, H. C.: Mechanical and chemical properties of various connective tissue organs in rats as influenced by non-steroidal antirheumatic drugs. Connect. Tissue Res., *5*:91, 1977.

Vogel, H. C.: Influence of maturation and age on mechanical and biochemical parameters of connective tissue of various organs in the rat. Connect. Tissue Res., *6*:161, 1978.

White, A., Handler, P., and Smith, E. L.: Principles of Biochemistry. New York, McGraw-Hill, 1964.

Woo, S. L.-Y., Gomez, M. A., Woo, Y.-K., and Akeson, W. H.: Mechanical properties of tendons and ligaments. Biorheology, *19*:397, 1982.

Woo, S. L.-Y., et al.: The effects of exercise on the biomechanical and biochemical properties of swine digital flexor tendons. J. Biomech. Eng., *103*:51, 1981.

Woo, S. L.-Y., et al.: The biomechanical and biochemical properties of swine tendons. Long-term effects of exercise on the digital extensors. Connect. Tissue Res., *7*:177, 1980.

Woo, S., et al.: Ligament, tendon, and joint capsule insertions to bone. *In* Injury and Repair of the Musculoskeletal Soft Tissues. Edited by S. L.-Y. Woo and J. Buckwalter. Park Ridge, IL, American Academy of Orthopaedic Surgeons, 1988, pp. 133–166.

Yamada, H.: Strength of Biological Materials. Edited by F. G. Evans. Baltimore, Williams & Wilkins, 1970.

SUGGESTED READING

Abrahams, M.: Mechanical behavior of tendon in vitro. Med. Biol. Eng., *5*:433–443, 1967.

Alm, A., Ekström, H., Gillquist, J., and Strömberg, B.: The anterior cruciate ligament. A clinical and experimental study on tensile strength, morphology and replacement by patellar ligaments. Acta Chir. Scand., Suppl. *445*:1–49, 1974.

Amiel, D., Woo, S. L.-Y., Harwood, F. L., and Akeson, W. H.: The effect of immobilization on collagen turnover in connective tissue. A biochemical-biomechanical correlation. Acta Orthop. Scand., *53*:325–332, 1982.

Amiel, D., et al.: Tendons and ligaments: A morphological and biochemical comparison. J. Orthop. Res., *1*:257–265, 1984.

Branwood, A. W.: The fibroblast. Int. Rev. Connect. Tissue Res., *1*:1, 1963.

Butler, D. L., Grood, E. S., and Noyes, F. R.: Biomechanics of ligaments and tendons. Exerc. Sport Sci. Rev., *6*:125–181, 1978.

Cabaud, H. E., Chatty, A., Gildengorin, V., and Feltman, R. J.: Exercise effects on the strength of the rat anterior cruciate ligament. Am. J. Sports Med., *8*:79–86, 1980.

Carlstedt, C. A.: Mechanical and chemical factors in tendon healing. Effects of indomethacin and surgery in the rabbit. Acta Orthop. Scand., Suppl. *224*:1–75, 1987.

Carlstedt, C. A., Madsén, K., and Wredmark, T.: The influence of indomethacin on collagen synthesis during tendon healing in the rabbit. Prostaglandins, *32*:353–358, 1986a.

Carlstedt, C. A., Madsén, K., and Wredmark, T.: The influence of indomethacin on tendon healing. A biomechanical and biochemical study. Arch. Orthop. Trauma Surg., *105*:332–336, 1986b.

Cooper, R. R., and Misol, S.: Tendon and ligament insertion. A light and electron microscopic study. J. Bone Joint Surg., *52A*:1–20, 1970.

Cronkite, A. E.: The tensile strength of human tendons. Anat. Rec., *64*:173–186, 1936.

Dale, W. C.: A composite materials analysis of the structure, mechanical properties, and aging of collagenous tissues. Ph.D. thesis, Case Western Reserve University, Cleveland, Ohio, 1974.

Diamant, J., et al.: Collagen: Ultrastructure and its relations to mechanical properties as a function of ageing. Proc. R. Soc. Lond. [Biol.], *180*:293–315, 1972.

Elliott, D. H.: Structure and function of mammalian tendon. Biol. Rev., *40*:392–421, 1965.

Elliott, D. H.: The biomechanical properties of tendon in relation to muscular strength. Ann. Phys. Med., *9*:1, 1967.

Ellis, P. G.: Cross-sectional area measurements for tendon specimens: A comparison of several methods. J. Biomech., *2*:175, 1969.

Eyre, D. R.: Collagen: Molecular diversity in the body's protein scaffold. Science, *207*:1315–1322, 1980.

Fitton-Jackson, S.: Antecedent phases in matrix formation. *In* Structure and Function of Connective and Skeletal Tissues. London, Butterworth, 1965, p. 277.

Frank, C., et al.: Normal ligament: Structure, function, and composition. *In* Injury and Repair of the Musculoskeletal Soft Tissues. Edited by S. L.-Y. Woo and J. Buckwalter. Park Ridge, IL, American Academy of Orthopaedic Surgeons, 1988, pp. 45–101.

Frisen, M., Mägi, M., Sonnerup, L., and Viidik, A.: Rheological analysis of soft collagenous tissue. I. Theoretical consideration. J. Biomech., *2*:13–20, 1969.

Frisen, M., Mägi, M., Sonnerup, L., and Viidik, A.: Rheological analysis of soft collagenous tissue. II. Experimental evaluations and verifications. J. Biomech., *2*:21–28, 1969.

Fung, Y. C. B.: Elasticity of soft tissues in simple elongation. Am. J. Physiol., *213*:1532–1544, 1967.

Fung, Y. C. B.: Stress-strain-history relations of soft tissues in simple elongation. *In* Biomechanics: Its Foundations and Objectives. Edited by Y. C. Fung, N. Perrone, and M. Anliker. Englewood Cliffs, NJ, Prentice-Hall, 1972, pp. 181–208.

Garrett, W., and Tidball, J.: Myotendinous junction: Structure, function, and failure. *In* Injury and Repair of the Musculoskeletal Soft Tissues. Edited by S. L.-Y. Woo and J. Buckwalter. Park Ridge, IL, American Academy of Orthopaedic Surgeons, 1988, pp. 171–207.

Gelberman, R., Goldberg, V., An, K.-N., and Banes, A.: Tendon. *In* Injury and Repair of the Musculoskeletal Soft Tissues. Edited by S. L.-Y. Woo and J. Buckwalter. Park Ridge, IL, American Academy of Orthopaedic Surgeons, 1988, pp. 5–44.

Gratz, C. M.: Tensile strength and elasticity tests on human fascia lata. J. Bone Joint Surg., *13*:334–340, 1931.

Greenlee, T. K., and Ross, R.: The development of the rat flexor digital tendon, a fine structure study. J. Ultrastruct. Res., *18*:354–376, 1967.

Gustavson, K. H.: The Chemistry and Reactivity of Collagen. New York, Academic Press, 1956.

Hall, M. C.: The Locomotor System: Functional Histology. Springfield, Charles C Thomas, 1965.

Harkness, R. D.: Mechanical properties of collagenous tissues. *In* Treatise on Collagen. Edited by B. S. Gould. New York, Academic Press, 1968, pp. 247–310.

Haut, R. C., and Littel, R. W. A.: A constitutive equation for collagen fibers. J. Biomech., *5*:523–530, 1972.

Hirsch, C.: Tensile properties during tendon healing. Acta Orthop. Scand., Suppl. *153*, 1974.

Jonsson, U., Ranta, H., and Stromberg, L.: Growth changes of collagen cross-linking, calcium and water content in bone. Arch. Orthop. Trauma Surg., *104*:89–93, 1985.

Kear, M., and Smith, R. N.: A method for recording tendon strain in sheep during locomotion. Acta Orthop. Scand., *46*:896–905, 1975.

Kennedy, J. C., Hawkins, R. J., Willis, R. B., and Danylchuk, K. D.: Tension studies of human knee ligaments. Yield point, ultimate failure, and disruption of the cruciate and tibial collateral ligaments. J. Bone Joint Surg., *58A*:350–355, 1976.

Marshall, J. L., and Olsson, S.-E.: Instability of the knee. A long-term study in dogs. J. Bone Joint Surg., *53A*:1561–1570, 1971.

Nachemson, A. L., and Evans, J. H.: Some mechanical properties of the third human lumbar interlaminar ligament (ligamentum flavum). J. Biomech., *1*:211–220, 1968.

Neuberger, A., Perrone, J. C., and Slack, H. G. B.: The relative metabolic inertia of tendon collagen in the rat. Biochem. J., *49*:199, 1951.

Noyes, F. R.: Functional properties of knee ligaments and alterations induced by immobilization. Clin. Orthop., *123*:210–242, 1977.

Noyes, F. R., and Grood, E. S.: The strength of the anterior cruciate ligament in humans and Rhesus monkeys. Age-related and species-related changes. J. Bone Joint Surg., *58A*:1074–1082, 1976.

Ohkawa, S.: Effects of orthodontic forces and anti-inflammatory drugs on the mechanical strength of the periodontium in the rat mandibular first molar. Am. J. Orthod., *81*:498–502, 1982.

Parry, D. A. D., Barnes, G. R. G., and Craig, A. S.: A comparison of the size distribution of collagen fibrils in connective tissues as a function of age and a possible relation between fibril size distribution and mechanical properties. Proc. R. Soc. Lond. [Biol.], *203*:305–321, 1978.

Piez, K. A.: Cross-linking of collagen and elastin. Ann. Rev. Biochem., *37*:574, 1968.

Porter, K. R.: Cell fine structure and biosynthesis of intercellular macromolecules. Biophys. J., *4*:2–167, 1964.

Prockop, D. J., and Guzman, N. A.: Collagen diseases and the biosynthesis of collagen. Hosp. Practice, December 1977, pp. 61–68.

Ramachandran, G. N.: Molecular structure of collagen. Int. Rev. Connect. Tissue Res., *1*:127–182, 1963.

Ramachandran, G. N., and Kartha, G.: Structure of collagen. Nature, *174*:269, 1954.

Rich, A., and Crick, F. H. C.: The structure of collagen. Nature, *176*:915, 1955.

Rich, A., and Crick, F. H. C.: The molecular structure of collagen. J. Molec. Biol., *3*:483, 1961.

Rigby, B. J., Hirai, N., Spikes, J. D., and Eyring, H.: The mechanical properties of rat tail tendon. J. Gen. Physiol., *43*:265–283, 1959.

Rollhäuser, H.: Die Festigkeit enschlicher Sehnen nach Quellung und Trocknung in Abhängigkeit von Lebensalter. Gegenbauers Morph. Jahrb., *90*:180–181, 1950.

Rundgren, Å.: Physical properties of connective tissue as influenced by single and repeated pregnancies in the rat. Acta Physiol. Scand., Suppl. *417*, 1974.

Steiner, M.: Biomechanics of tendon healing. J. Biomech., *15*:951–958, 1982.

Tipton, C. M., Schild, R. J., and Tomanek, R. J.: Influence of physical activity on the strength of knee ligaments in rats. Am. J. Physiol., *212*:783–787, 1967.

Tipton, C. M., James, S. L., Mergner, W., and Tcheng, T.: Influence of exercise on strength of medial collateral ligaments of dogs. Am. J. Physiol., *218*:894–902, 1970.

Viidik, A.: Biomechanics and functional adaptation of tendons and joint ligaments. *In* Studies on the Anatomy and Function of Bones and Joints. Edited by F. G. Evans. Berlin, Springer Verlag, 1966, pp. 17–39.

Viidik, A.: The effect of training on the tensile strength of isolated rabbit tendons. Scand. J. Plast. Reconstr. Surg., *1*:141–147, 1967.

Viidik, A.: Experimental evaluation of the tensile strength of isolated rabbit tendons. Biomed. Eng., *2*:64–67, 1967.

Viidik, A.: A rheological model for uncalcified parallel-fibered collagenous tissue. J. Biomech., *1*:3–11, 1968.

Viidik, A.: Elasticity and tensile strength of the anterior cruciate ligament in rabbits as influenced by training. Acta Physiol. Scand., *74*:372–380, 1968.

Viidik, A.: Functional properties of collagenous tissues. Int. Rev. Connect. Tissue Res., *6*:127, 1973.

Viidik, A.: Biomechanical behavior of soft connective tissues. *In* Progress in Biomechanics. Edited by N. Akkas. Alpen Aan den Rijn, Sijthoff and Nordhoff, 1979, pp. 75–113.

Viidik, A.: Interdependence between structure and function in collagenous tissue. *In* Biology of Collagen. Edited by A. Viidik and J. Vuust. London, Academic Press, 1980, pp. 257–280.

Viidik, A.: Mechanical properties of parallel-fibred collagenous tissues. *In* Biology of Collagen. Edited by A. Viidik and J. Vuust. London, Academic Press, 1980, pp. 237–255.

Viidik, A., Danielsen, C. C., and Oxlund, H.: Fourth International Congress of Biorheology Symposium on Mechanical Properties of Living Tissues: On fundamental and phenomenological models, structure and mechanical properties of collagen, elastic and glycosaminoglycan complexes. Biorheology, *19*:437–451, 1982.

Viidik, A., Sandqvist, L., and Mägi, M. L.: Influence of postmortal storage on tensile strength characteristics and histology of rabbit ligaments. Acta Orthop. Scand., Suppl. *79*, 1965.

Vogel, H. C.: Mechanical and chemical properties of various connective tissue organs in rats as influenced by non-steroidal antirheumatic drugs. Connect. Tissue Res., *5*:91–95, 1977.

Vogel, H. C.: Influence of maturation and age on mechanical and biochemical parameters of connective tissue of various organs in the rat. Connect. Tissue Res., *6*:161–166, 1978.

Vogel, H. C.: Age dependence of mechanical properties of rat tail tendons (hysteresis experiments). Akt. Gerontol., *13*:22–27, 1983.

Woo, S. L.-Y., Gomez, M. A., Woo, Y.-K., and Akeson, W. H.: Mechanical properties of tendons and ligaments. Biorheology, *19*:397–408, 1982.

Woo, S. L.-Y., et al.: The effects of exercise on the biomechanical and biochemical properties of swine digital flexor tendons. J. Biomech. Eng., *103*:51–56, 1981.

Woo, S. L.-Y., et al.: The biomechanical and biochemical properties of swine tendons. Long-term effects of exercise on the digital extensors. Connect. Tissue Res., *7*:177–183, 1980.

Woo, S., et al.: Ligament, tendon, and joint capsule insertions to bone. *In* Injury and Repair of the Musculoskeletal Soft Tissues. Edited by S. L.-Y. Woo and J. Buckwalter. Park Ridge, IL, American Academy of Orthopaedic Surgeons, 1988, pp. 133–166.

Yamada, H.: Strength of Biological Materials. Edited by F. G. Evans. Baltimore, Williams & Wilkins, 1970.

4

BIOMECHANICS OF PERIPHERAL NERVES

Björn Rydevik
Göran Lundborg
Richard Skalak

The nervous system serves as the body's control center and communications network. As such, it has three broad roles: it senses changes in the body and in the external environment, it interprets these changes, and it responds to this interpretation by initiating action in the form of muscle contraction or gland secretion.

For descriptive purposes, the nervous system can be divided into two parts: the central nervous system, consisting of the brain and spinal cord, and the peripheral nervous system, composed of the various nerve processes that extend from the brain and spinal cord. These peripheral nerve processes provide input to the central nervous system from sensory receptors in skin, joints, muscles, tendons, viscera, and sense organs and output from it to effectors (muscles and glands).

The peripheral nervous system includes 12 pairs of cranial nerves and their branches and 31 pairs of spinal nerves and their branches (Fig. 4–1A). These branches are called peripheral nerves.

Each spinal nerve is connected to the spinal cord through a posterior (dorsal) root and an anterior (ventral) root, which unite to form the spinal nerve at the intervertebral foramen (Fig. 4–1B–D). The posterior roots contain fibers of sensory neurons (those conducting sensory information from receptors in the skin, muscles, tendons, and joints to the central nervous system), and the anterior roots contain mainly fibers of motor neurons (those that conduct impulses from the central nervous system to distal targets such as muscle fibers).

Shortly after the spinal nerves leave their intervertebral foramina, they divide into two main branches: the dorsal rami, which innervate the muscles and skin of the head, neck, and back, and the generally larger and more important ventral rami, which innervate the ventral and lateral parts of these structures as well as the upper and lower extremities. Except in the thoracic region, the ventral rami do not run directly to the structures that they innervate but first form interlacing networks, or plexuses, with adjacent nerves (see Fig. 4–1A).

This chapter focuses on the spinal peripheral nerves, which contain not only nerve fibers but also connective tissue elements and vascular structures that encompass the nerve fibers. The peripheral nerves possess some special anatomic properties that may serve to protect the nerve fibers from mechanical damage, for instance, stretching (tension) and compression. In this chapter the basic microanatomy of the peripheral nerve is reviewed with special reference to these built-in mechanisms of protection. The mechanical behavior of peripheral nerves subjected to tension and compression is also described in some detail. Further, regeneration of injured nerve tissue and aging of normal nerve tissue are discussed.

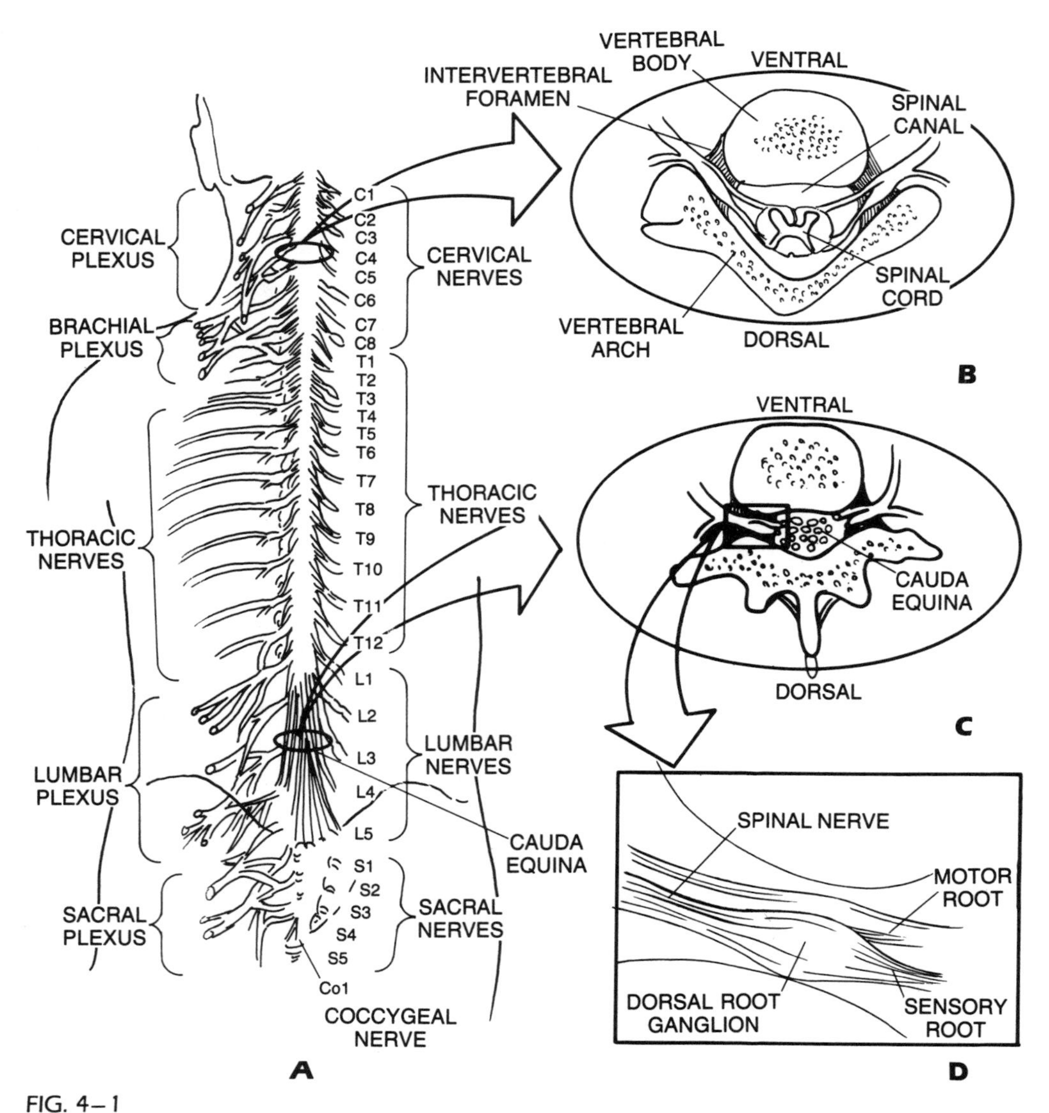

FIG. 4–1

A. Schematic drawing of the spinal cord and the spinal nerves (posterior view). The spinal nerves emerge from the spinal canal through the intervertebral foramina. There are eight pairs of cervical nerves, 12 pairs of thoracic nerves, five pairs of lumbar nerves, five pairs of sacral nerves, and one pair of coccygeal nerves. Except in the region of the second to the 11th thoracic vertebrae (T2–T11), the nerves form complex networks called plexuses after exiting the intervertebral foramina. Only the main branch of each nerve, the ventral ramus, is depicted. (Adapted from Tortora and Anagnostakos, 1984.) **B.** Cross section of the cervical spine showing the spinal cord in the spinal canal and the nerve roots exiting through the intervertebral foramina. **C.** Cross section of the lumbar spine showing the nerve roots of the cauda equina in the spinal canal. **D.** Each exiting nerve root complex in the intervertebral foramen consists of a motor root, a sensory root, and a dorsal root ganglion.

COMPOSITION AND STRUCTURE OF PERIPHERAL NERVES

The peripheral nerves are complex composite structures consisting of nerve fibers, connective tissue, and blood vessels. Because the three tissue elements that make up these nerves react to trauma in different ways and may each play distinct roles in the functional deterioration of the nerve after injury, each element is described separately.

THE NERVE FIBERS: STRUCTURE AND FUNCTION

The term "nerve fiber" refers to the elongated process (axon) extending from the nerve cell body along with its myelin sheath and Schwann cells (Figs. 4–2 and 4–3). The nerve fibers of sensory neurons conduct impulses from the skin, skeletal muscles, and joints to the central nervous system. The nerve fibers of the motor neurons convey impulses from the central nervous system to the skeletal muscles, causing muscle contraction. (A detailed description of the mechanics of muscle contraction is given in Chapter 5.)

The nerve fibers not only transmit impulses, but also serve as an anatomic connection between the nerve cell body and its end organs. This connection is maintained by so-called axonal transport systems, through which various substances synthesized within the cell body (e.g., proteins) are transported from the cell body to the periphery and in the opposite direction. The axonal transport takes place at speeds that vary from about 1 mm per day to about 400 mm per day (Weiss and Gorio, 1982).

Most axons of the peripheral nervous system are surrounded by multilayered, segmented coverings known as myelin sheaths (see Fig. 4–3). Fibers with this covering are said to be myelinated, whereas those without it (mainly small sensory fibers conducting impulses for pain from the skin) are unmyelinated. The myelin sheath of the axons of the peripheral nerves is produced by flattened cells called Schwann cells arranged along the axon (see Fig. 4–3). A sheath is formed as the Schwann cell encircles the axon and winds around it many times, pushing its cytoplasm and nucleus to the outside layer. Between the segments of the myelin sheath about 1 to 2 mm apart are unmyelinated gaps called nodes of Ranvier.

The myelin sheath increases the speed of conduction of nerve impulses and insulates and maintains the axon. Impulses are propagated along the unmyelinated nerve fibers in a slow, continuous way, whereas in the myelinated nerve fibers the impulses "jump" at a higher speed from one node of Ranvier to the next, a process called saltatory conduction. The conduction velocity of a myelinated nerve is directly proportional to the diameter of the fiber (Gasser, 1935; Sunderland, 1978), which usually ranges from 2 to 20 μm. Motor fibers that innervate skeletal muscle have large diameters, as do sensory fibers that relay impulses associated with touch, pressure, heat, cold,

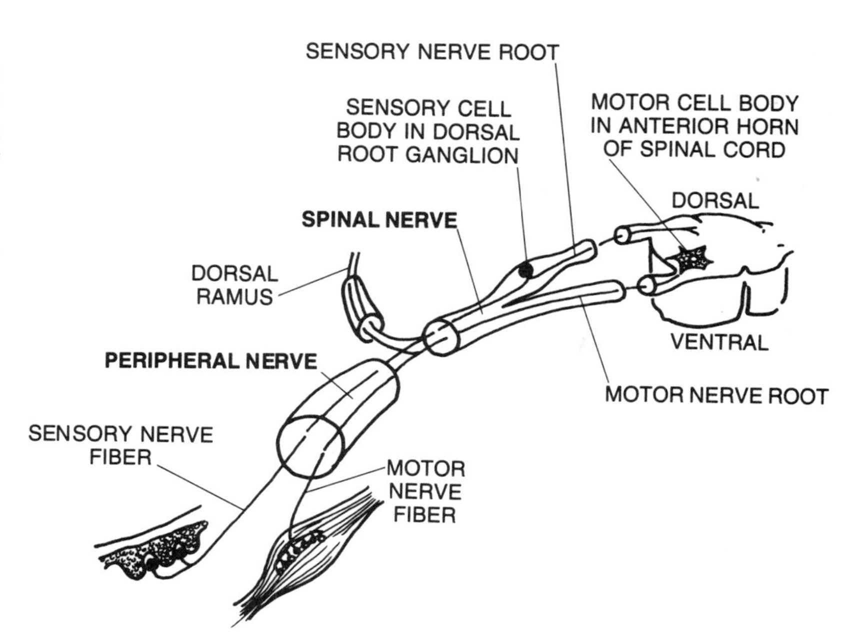

FIG. 4–2
Schematic representation of the arrangement of a typical spinal nerve as it emerges from its dorsal and ventral nerve roots. The peripheral nerve begins after the dorsal ramus branches off. (For the sake of simplicity, the nerve is not shown entering a plexus.) Spinal nerves and most peripheral nerves are mixed nerves: they contain both sensory (afferent) and motor (efferent) nerve fibers. The cell body and its nerve fibers make up the neuron. The cell bodies of the motor neurons are located in the anterior horn of the spinal cord and those of the sensory neurons are found in the dorsal root ganglia. Here a motor nerve fiber is shown innervating muscle and a sensory nerve fiber is depicted innervating skin. (Adapted from Rydevik et al., 1984.)

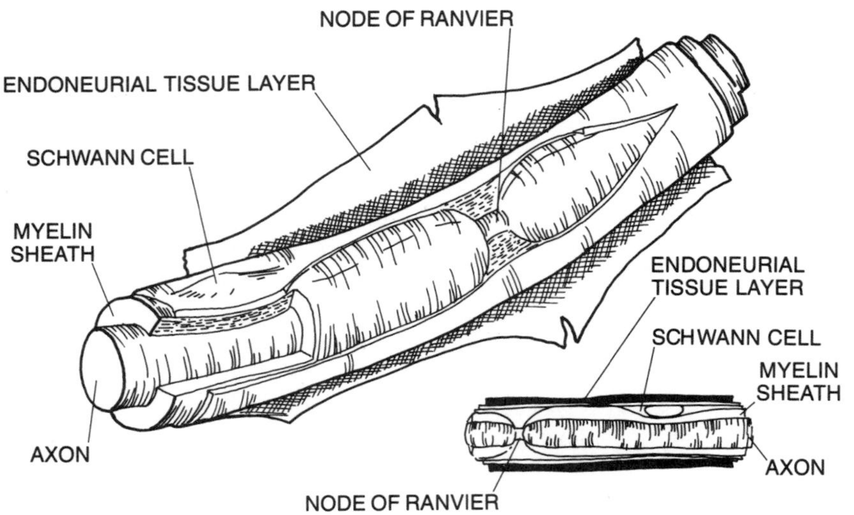

FIG. 4–3
Schematic drawings of the structural features of a myelinated nerve fiber. (Adapted from Sunderland, 1978.)

and kinesthetic sense such as skeletal muscle tension and joint position; sensory fibers that conduct impulses for dull, diffuse pain (as opposed to sharp, immediate pain) have the smallest diameters. Nerve fibers are packed closely in fascicles, which are further arranged into bundles that make up the nerve itself. The fascicles are the functional subunits of the nerve.

INTRANEURAL CONNECTIVE TISSUE

Surrounding the nerve fibers are successive layers of connective tissue—generally called the endoneurium, perineurium, and epineurium—which protect the fibers' continuity (Figs. 4–4 and 4–5). The protective function of these connective tissue layers is essential, since the nerve fibers are extremely susceptible to stretching and compression.

The outermost layer, the epineurium, is located between the fascicles and superficially in the nerve. This rather loose connective tissue layer serves as a cushion during movements of the nerve, protecting the fascicles from external trauma and maintaining the oxygen supply system via the epineural blood vessels. The amount of epineural connective tissue varies among nerves and at different levels within the same nerve (Sunderland and Bradley, 1949). Where the nerves lie close to bone or pass joints, the epineurium is often more abundant than elsewhere, as the need for protection may be greater in these locations (Sunderland, 1978). The spinal nerve roots are devoid of both epineurium and perineurium, and the nerve fibers in the nerve root may therefore be more susceptible to trauma (Sunderland and Bradley, 1961b; Rydevik et al., 1984).

The perineurium is a lamellar sheath that encompasses each fascicle. This sheath has great mechanical strength as well as specific biochemical barrier properties (Shanta and Bourne, 1968; Sunderland, 1978). Its strength is demonstrated by the fact that the fascicles can be inflated by fluid to a pressure of about 1000 mm of mercury (Hg) before the perineurium ruptures (Selander, Lundborg, and Rydevik, unpublished data).

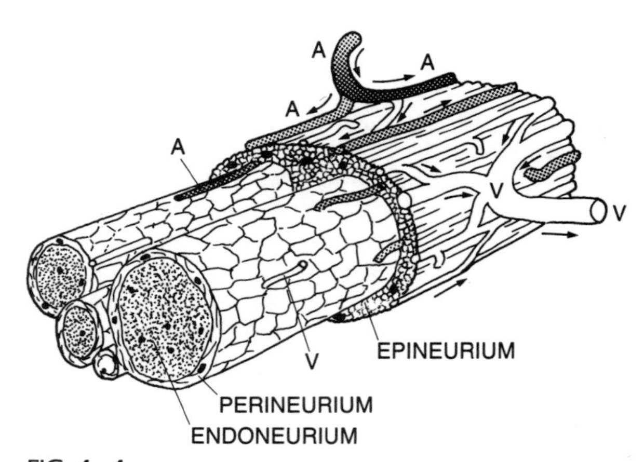

FIG. 4–4
Schematic drawing of a segment of a peripheral nerve. The individual nerve fibers are located within the endoneurium. They are closely packed in fascicles, each of which is surrounded by a strong sheath, the perineurium. A bundle of fascicles is embedded in a loose connective tissue, the epineurium. Blood vessels are present in all layers of the nerve. A, arterioles (shaded); V, venules (unshaded). The arrows indicate the direction of blood flow. (Adapted from Dahlin et al., 1986.)

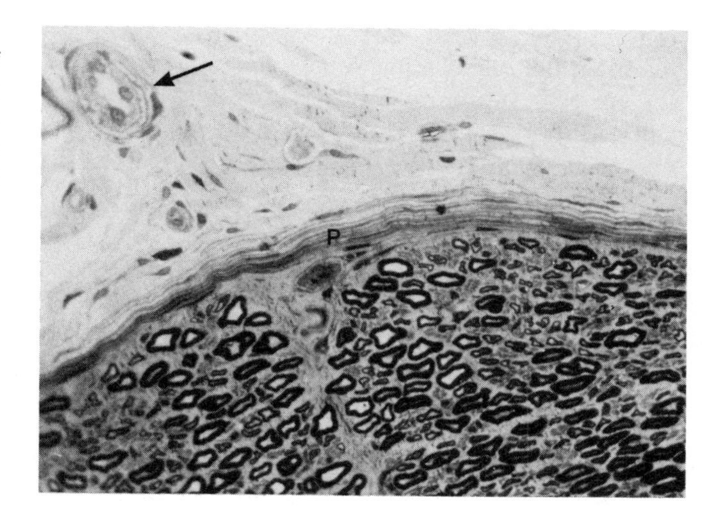

FIG. 4–5
Microscopic appearance of a cross section of part of one fascicle of a human peripheral nerve. The myelin sheaths of the nerve fibers are seen as either oval or round black forms in the endoneurium. Note the perineurium (P), which surrounds the fascicle, and the vessel (arrow) in the epineurium. Staining was done with osmium tetroxide and toluidine blue (300×).

The barrier function of the perineurium chemically isolates the nerve fibers from their surroundings, thus preserving an ionic environment of the interior of the fascicles, a special *milieu interieur*. The endoneurium, the connective tissue inside the fascicles, is composed principally of fibroblasts and collagen, which form tubules that surround each nerve fiber.

The interstitial tissue pressure in the fascicles, the endoneurial fluid pressure, is normally slightly elevated (+1.5±0.7 mm Hg [Low and Dyck, 1977; Myers and Powell, 1981]) compared with the pressure in surrounding tissues such as subcutaneous tissue (−4.7±0.8 mm Hg [Chen et al., 1976]) and muscle tissue (−2±2 mm Hg [Hargens et al., 1978]). The elevated endoneurial fluid pressure is illustrated by the phenomenon whereby incision of the perineurium results in herniation of nerve fibers (Spencer et al., 1975). The endoneurial fluid pressure may increase further as a result of trauma to the nerve with subsequent edema. Such a pressure increase may affect the microcirculation and the function of the nerve.

THE MICROVASCULAR SYSTEM

The peripheral nerve is a well-vascularized structure containing vascular networks in the epineurium, the perineurium, and the endoneurium. Because both impulse propagation and axonal transport depend on a local oxygen supply, it is natural that the microvascular system has a large reserve capacity (Lundborg, 1970).

The blood supply to the peripheral nerve as a whole is provided by large vessels that approach the nerve segmentally along its course. When these local nutrient vessels reach the nerve, they divide into ascending and descending branches. These vessels run longitudinally and frequently anastomose with the vessels in the perineurium and endoneurium. Within the epineurium large arterioles and venules, 50 to 100 μm in diameter, constitute a longitudinal vascular system (see Figs. 4–4 and 4–5).

Within each fascicle lies a longitudinally oriented capillary plexus with loop formations at various levels. The capillary system is fed by arterioles 25 to 150 μm in diameter that penetrate the perineurial membrane. These vessels run an oblique course through the perineurium, and it is believed that because of this structural peculiarity they are easily closed like valves in the event that tissue pressure inside the fascicles increases (Lundborg, 1975). The phenomenon may explain why even a limited increase in endoneurial fluid pressure is associated with a reduction in intrafascicular blood flow (Lundborg et al., 1983).

The built-in safety system of longitudinal anastomoses provides a wide margin of safety if the regional segmental vessels are transected. In an experimental animal it is extremely difficult to induce complete ischemia to a nerve by local surgical procedures. For example, if the whole sciatic-tibial nerve complex of a rabbit (15 cm long) is surgically separated from its surrounding structures and the regional nutrient vessels are cut, there is no detectable reduction in the intrafascicular blood flow as studied by intravital microscopic techniques (Lundborg, 1970). Even if such a mobilized nerve is cut distally or proximally, the intraneural longitudinal vascular systems can maintain the microcirculation at least 7 to 8 cm from the cut end. If a nonmobilized nerve is cut, there is still perfect microcirculation even at the very tip of the nerve; this phenomenon demonstrates the sufficiency of the intraneural vascular collaterals.

BIOMECHANICAL BEHAVIOR OF PERIPHERAL NERVES

External trauma to the extremities and nerve entrapment may produce mechanical deformation of peripheral nerves that results in deterioration of nerve function. If the mechanical trauma exceeds a certain degree, the nerves' built-in mechanisms of protection may not be sufficient and changes in nerve structure and function result. Common modes of nerve injury are stretching and compression, which may be inflicted, respectively, by rapid extension and crushing.

STRETCHING (TENSILE) INJURIES OF PERIPHERAL NERVES

Nerves are strong structures with considerable tensile strength. The maximal load that can be sustained by the median and ulnar nerves is in the range of 70 to 220 newtons (N) and 60 to 150 N, respectively (Sunderland, 1978). These figures are of academic interest only, because severe intraneural tissue damage is produced by tension long before a nerve breaks.

A discussion of the elasticity and biomechanical properties of nerves is complicated by the fact that nerves are not homogeneous isotropic materials but are composite structures each tissue component of which has its own biomechanical properties. The connective tissues of the epineurium and perineurium are primarily longitudinal structures. There is some controversy regarding the contributions of the various tissue components to the mechanical properties of nerves. Sunderland (1978) believes that the tensile strength of the peripheral nerve depends essentially on the perineurium and that the elastic properties of nerves are maintained as long as the perineurium remains intact. Haftek (1970) claims, however, that the elasticity of the nerves depends largely on the epineurium, much less on the perineurium, and only a little on the intrafascicular tissue.

When tension is applied to a nerve, initial elongation of the nerve under a very small load is followed by an interval in which stress and elongation show a linear relationship characteristic of an elastic material (Fig. 4–6). As the limit of the linear region is approached, the nerve fibers start to rupture inside the endoneurial tubes and inside the intact perineurium (Haftek, 1970). Finally, the epineurium and perineurium rupture, there is a disintegration of the elastic properties, and the nerve behaves more like a plastic material (i.e., its response to the release of loads is incomplete recovery).

Although variations exist in the tensile strength of various nerves, the maximal elongation at the elastic limit is about 20%, and complete structural failure seems to occur at a maximum elongation of approximately 30% (Sunderland and Bradley, 1961a; Sunder land, 1978). These values are for normal nerves; injury to a nerve may induce changes in its mechan-

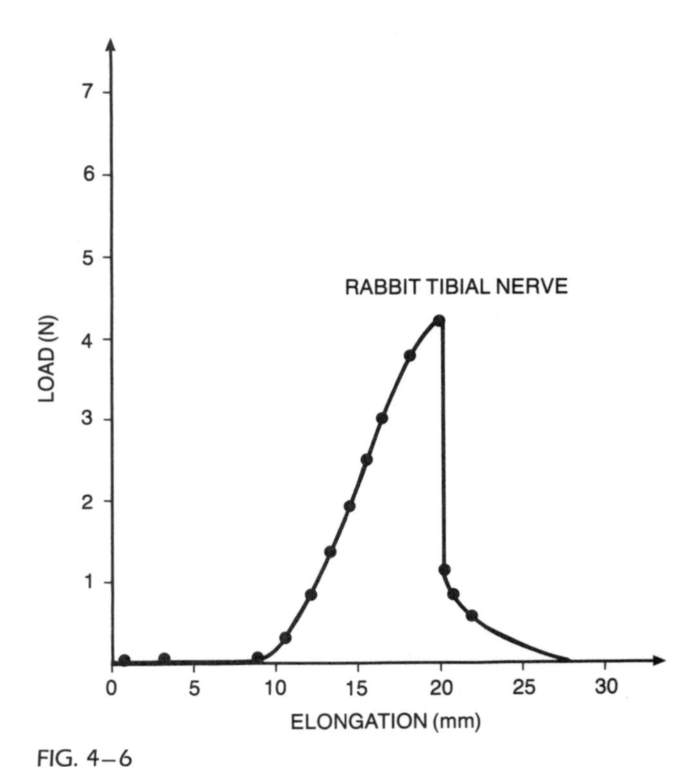

FIG. 4–6
Load-elongation curve obtained by testing a freshly excised rabbit tibial nerve with an Instron testing machine at a loading rate of 1.0 mm per minute. The initial portion of the curve indicates considerable elongation without any measurable load owing to the slackness of the nerve at the beginning of the test. The initial length of the test segment was 36.6 mm. (It is known that the nerve is under a small amount of tension in vivo.) After the load begins to increase, the curve shows a region in which the nerve behaves like a linearly elastic material. This linear portion extends nearly to the point of peak load, in this case 4.3 N. Beyond the peak load, the mechanical capacity of the nerve deteriorates sharply.

ical properties, namely increased stiffness and decreased elasticity (Beel et al., 1984).

Stretching, or tensile, injuries of peripheral nerves are usually associated with severe accidents, such as when high-energy tension is applied to the brachial plexus, for instance, in association with a high-speed vehicular collision or a fall from a height. Such plexus injuries may result in partial or total functional loss of some or all of the nerves in the upper extremity, and the consequent functional deficits represent a considerable disability in terms of sensory and motor loss. The outcome depends on which tissue components of the nerves are damaged as well as on the extent of the tissue injury.

Of clinical importance is the observation that the nerve fibers in a nerve under tension seem to rupture before their endoneurial tubes and perineurium (Sunderland, 1978). This finding indicates that after a moderate stretching injury that involves rupture of axons only, regenerating axons may have intact pathways to follow during their growth toward the periphery.

High-energy plexus injuries represent an extreme type of stretching lesion caused by sudden violent trauma. A different stretching situation of considerable clinical interest is the suturing of the two ends of a cut nerve under moderate tension. This situation occurs when a substantial gap exists in the continuity of a nerve trunk and restoration of continuity requires the application of tension to bring the nerve ends back together. The moderate, gradual tension applied to the nerve in these cases may stretch and angulate local feeding vessels. It may also be sufficient to reduce the transverse fascicular cross-sectional area and impair the intraneural nutritive capillary flow (Fig. 4–7).

As the sutured nerve is stretched, the perineurium tightens; as a result the endoneurial fluid pressure is increased and the intrafascicular capillaries may be obliterated. Also, the flow is impaired in the segmental, feeding, and draining vessels, as it is in larger vessels in the epineurium, and at a certain stage the intraneural microcirculation ceases. Intravital observations of intraneural blood flow in rabbit tibial nerves (Lundborg and Rydevik, 1973) showed that an elongation of 8% induced impaired venular flow and that even greater tension produced continuous impairment of capillary and arteriolar flow until, at 15% elongation, all intraneural microcirculation ceased completely. In studies using microangiographic techniques to evaluate microvascular flow in dog nerves sutured under tension, 5% elongation was found to induce injury to intrafascicular vessels (Miyamoto et al., 1979; Miyamoto, 1979), and nerves sutured under

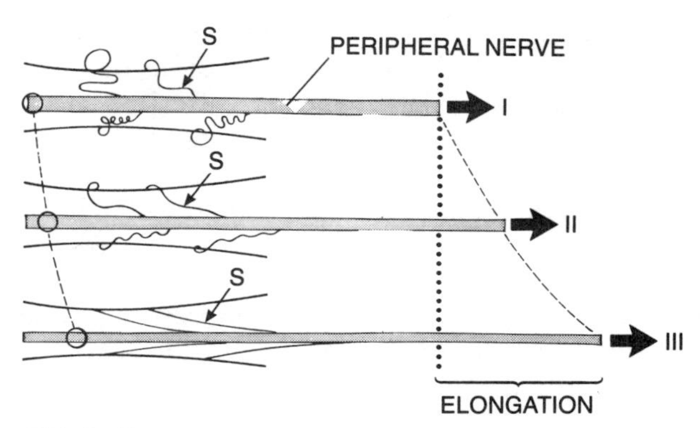

FIG. 4–7
Schematic representation of a peripheral nerve and its blood supply at three stages during stretching. Stage I: The segmental blood vessels (S) are normally coiled to allow for the physiologic movements of the nerve. Stage II: Under gradually increasing elongation these regional vessels become stretched, and the blood flow in them is impaired. Stage III: The cross-sectional area of the nerve (represented within the circle) is reduced during stretching and the intraneural blood flow is further impaired. Complete cessation of all blood flow in the nerve usually occurs at about 15% elongation (Lundborg and Rydevik, 1973). (Adapted from Lundborg, 1970.)

such tension did not regenerate as well as nerves sutured under no tension at all.

A situation of even more gradual stretching, applied over a long time, is the growth of intraneural tumors such as schwannomas. In this case the nerve fibers are forced into a circumferential course around the gradually expanding tumor. Functional changes in cases of such very gradual stretching are often minimal or nonexistent.

COMPRESSION INJURIES OF PERIPHERAL NERVES

It has long been known that compression of a nerve can induce symptoms such as numbness, pain, and muscle weakness (Sunderland, 1978). The biologic basis for the functional changes has been extensively investigated (Meek and Leaper, 1911; Grundfest, 1936; Bentley and Schlapp, 1943; Denny-Brown and Brenner, 1944; Fowler et al., 1972; Rydevik and Nordborg, 1980).

Grundfest (1936) showed that nerves enclosed in a compression chamber could maintain normal function, even when the pressure in the chamber was very high, as long as an adequate concentration of oxygen was present. If oxygen was removed from the chamber in such a situation, nerve function deteriorated rapidly, even when the pressure around the nerve was normal. On the basis of these observations, it was assumed that the pressure was much less important than ischemia for changes in nerve function during compression (Denny-Brown and Brenner, 1944). Grundfest's (1936) experiments, however, were performed on whole nerves enclosed in a compression chamber. Different findings were obtained when nerves were locally compressed in one defined segment (Bentley and Schlapp, 1943; Fowler et al., 1972; Rydevik and Nordborg, 1980). In these investigations (Fig. 4–8), even mild compression was observed to induce structural and functional changes, and the significance of mechanical factors such as pressure level and mode of compression became apparent.

Critical Pressure Levels

Experimental and clinical observations have revealed some data on the critical pressure levels at which disturbances occur in intraneural blood flow, axonal transport, and nerve function. Certain pressure levels seem to be well defined with respect to structural and functional changes induced in the nerve. The duration of the compression also influences the development of these changes.

At 30 mm Hg of local compression, functional changes may occur in the nerve, and its viability may be jeopardized during prolonged compression (4 to 6 hours) at this pressure level (Lundborg et al., 1982). Such changes seem to be due to impairment of the blood flow in the compressed part of the nerve (Rydevik et al., 1981). Corresponding pressure levels (about 32 mm Hg) were recorded close to the median nerve in the carpal tunnel in patients with carpal tunnel syndrome, while in a group of control subjects the pressure in the carpal tunnel averaged only 2 mm Hg (Gelberman et al., 1981). Longstanding or intermittent compression at low pressure levels (about 30 to 80 mm Hg) may induce intraneural edema, which in turn may become organized into a fibrotic scar in the nerve (Rydevik and Lundborg, 1977).

Compression at about 30 mm Hg also brings about changes in the axonal transport systems (Dahlin et al., 1984a), and longstanding compression may thus lead to depletion of axonally transported proteins distal to the compression site. Such blockage of axonal transport induced by local compression (pinching) may cause the axons to be more susceptible to additional compression distally, the so-called double-crush syndrome (Upton and McComas, 1973).

Slightly higher pressure, 80 mm Hg, for example, causes complete cessation of intraneural blood flow; the nerve in the locally compressed segment becomes completely ischemic. Yet, even after 2 hours or more of compression, blood flow is rapidly restored when the pressure is released (Rydevik et al., 1981). Even higher levels of pressure—for example, 200 to 400 mm Hg—applied directly to a nerve can induce structural nerve fiber damage and rapid deterioration of nerve function, with incomplete recovery after even shorter periods of compression (Rydevik et al., 1980; Rydevik and Nordborg, 1980). Hence, the magnitude of the applied pressure and the severity of the induced compression lesion appear to be correlated (Dahlin et al., 1986).

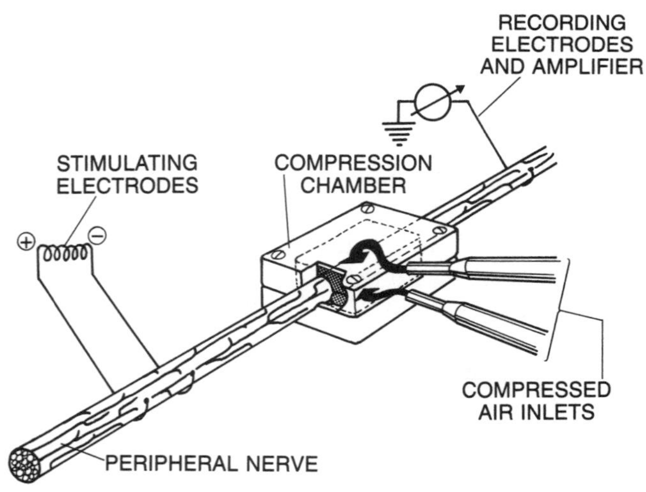

FIG. 4–8
Schematic drawing of an experimental setup for studying deterioration of nerve function during compression. (Adapted from Dahlin et al., 1986.)

Mode of Pressure Application

The pressure level is not the only factor that influences the severity of nerve injury brought about by compression. Experimental and clinical evidence indicates that the mode of pressure application is also of major significance. Its importance is illustrated by the fact that direct compression of a nerve at 400 mm Hg by means of a small inflatable cuff around the nerve induces a more severe nerve injury than does indirect compression of the nerve at 1000 mm Hg via a tourniquet applied around the extremity (Fowler et al., 1972; Rydevik and Nordborg, 1980). Even though the hydrostatic pressure acting on the nerve in the former situation is less than half that in the latter, the nerve lesion is more severe, probably because direct compression causes a more pronounced deformation of the nerve (especially at its edges) than does indirect compression, in which the tissue layers between the compression device and the nerve "bolster" the nerve. It may also be concluded that the nerve injury caused by compression is not directly related to the high hydrostatic pressure in the center of the compressed nerve segment but instead is more dependent on the specific mechanical deformation induced by the applied pressure (Dahlin et al., 1984b).

Mechanical Aspects of Nerve Compression

Electron microscopic analysis of the deformation of the nerve fibers in the peroneal nerve of the baboon hind limb induced by tourniquet compression (Ochoa et al., 1972) demonstrated the so-called edge effect; that is, a specific lesion was induced in the nerve fibers at both edges of the compressed nerve segment: the nodes of Ranvier were displaced toward the noncompressed parts of the nerve. The nerve fibers in the center of the compressed segment, where the hydrostatic pressure is highest, generally were not affected acutely. The large-diameter nerve fibers were usually affected, but the thinner fibers were spared. This finding is in accord with theoretical calculations indicating that larger nerve fibers undergo a relatively greater deformation than do thinner fibers at a given pressure (McGregor et al., 1975). It is also known clinically that a compression lesion of a nerve first affects the large fibers (e.g., those that carry motor function), while the thin fibers (e.g., those that mediate pain sensation) are often preserved (Sunderland, 1978). The intraneural blood vessels have also been shown to be injured at the edges of the compressed segment (Rydevik and Lundborg, 1977). Basically, the lesions of nerve fibers and blood vessels seem to be consequences of the pressure gradient, which is maximal just at the edges of the compressed segment.

In considering the mechanical effects on nerve compression, one should keep in mind that the effect of a given pressure depends on the way in which it is applied and on its magnitude and duration. Although pressure may be applied with a variety of spatial distributions, two basic types of pressure applications are generally encountered in experimental settings and in pathologic conditions. One type is uniform pressure applied around the entire circumference of a longitudinal segment of a nerve or extremity. This is the kind of purely radial pressure that is applied by the common pneumatic tourniquet. It has also been used in miniature apparatus to produce controlled compression of individual nerves (Rydevik and Lundborg, 1977) (see Fig. 4–8). Clinically, this type of loading on a nerve probably occurs when the pressure on the median nerve is elevated in the carpal tunnel, producing a characteristic syndrome.

Another type of mechanical action takes place when the nerve is compressed laterally. This is the kind of deformation that occurs if a nerve or extremity is placed between two parallel flat rigid surfaces that are then moved toward each other, squeezing the nerve or extremity. This type of deformation occurs if a sudden blow by a rigid object squeezes a nerve against the surface of an underlying bone. It may also occur when a spinal nerve is compressed by a herniated disc.

The details of the deformation of a nerve may be quite different in these two cases of loading. In uniform circumferential compression like that applied by a pneumatic tourniquet (Fig. 4–9), the cross

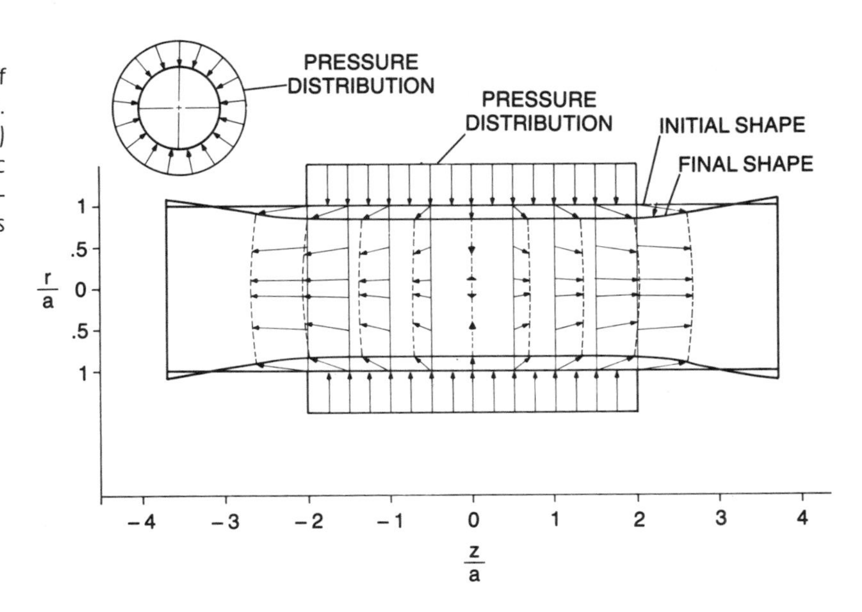

FIG. 4–9
Theoretical displacement field under a pressure cuff applied to a model cylindrical nerve of radius a. Displacements in the radial (r) and longitudinal (z) directions are computed on the basis of isotropic material properties and elastic theory, so the deformation is proportional to the pressure. The arrows represent displacement vectors.

section of the nerve or extremity tends to remain circular but decreases in diameter in the loaded region. Since the material of the tissues is relatively incompressible, this radial compression requires a squeezing out of the tissue under the tourniquet, moving it outward from the center line toward each of the free edges. It can be seen readily that the displacement of the tissue builds up from zero at the center line to a maximum at the edge of the tourniquet. It is this large displacement, along with accompanying shear stresses, that is believed to cause the edge effect mentioned above that is observed in experiments in vivo. This region sustains both the maximum pressure gradient and the maximum displacement.

Lateral compression does not necessarily produce any axial motion of material, but it may simply deform the cross section from nearly circular to more elliptical as shown in Figure 4–10. In this kind of compression it is clear that in the direction perpendicular to the direction of compression (x) the nerve must be extended. This extension is illustrated by the movement of point G to G′ during compression. At the same time point A moves to A′, indicating shortening, or compression, in the direction of loading. The degree of compression can be measured by the maximum extension ratio (λ), which is defined as the maximum diameter divided by the initial diameter of the nerve. The theoretically computed shapes are shown for λ values of 1.1, 1.3, and 1.5. The theoretical results shown in the figure are based on the theory of elasticity (Green and Zerna, 1968). Point B moves to B′, C moves to C′, and so on, during the deformation.

The effects of a deformation such as that shown in Figure 4–10 on the functioning of the axoplasm and the neural membrane are not known. It seems likely that the initial degeneration of function would be associated with damage to the membrane. It can be shown that if the cross-sectional area of the nerve shown in Figure 4–10 remains constant during the deformation the perimeter must increase in moving from the initial circular shape to the final elliptical shape. This increase indicates that there must be stretching of the membrane, which is likely to affect its permeability and electrical properties. This deformation is similar to that of a Pacinian corpuscle, which senses pressure applied to the skin (Loewenstein and Skalak, 1966). It may be that this kind of deformation can trigger firing of nerves, resulting in a sensation of pain when the nerve fibers are laterally compressed. The details of such deformation of nerves and their functional consequences have not been studied extensively and require further research for their elucidation.

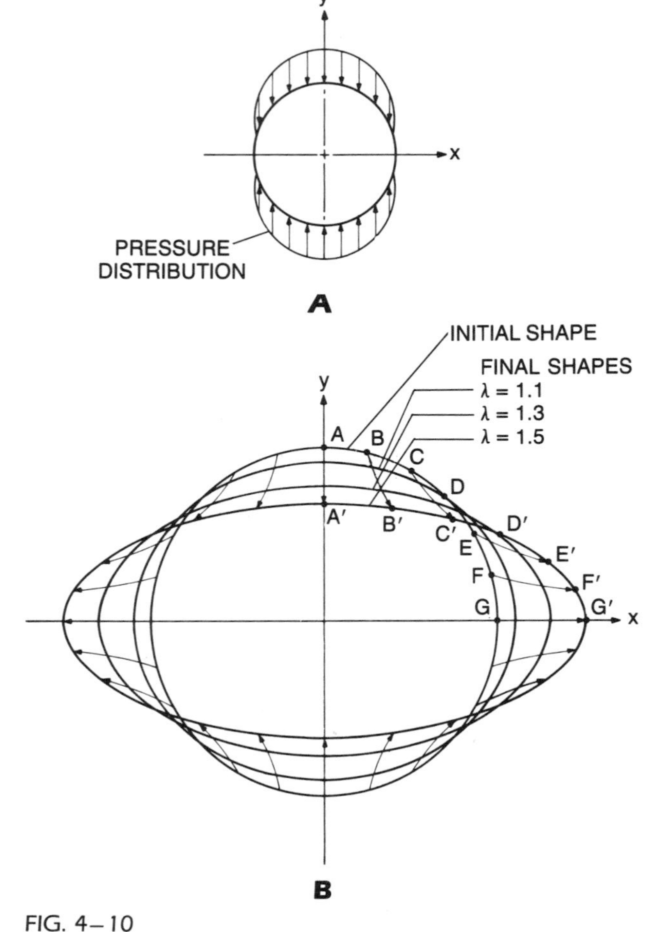

FIG. 4–10
A. Theoretical displacement field under lateral compression due to uniform clamping pressure. **B.** The original and deformed cross sections are shown for maximum elongation in the x direction of 10, 30, and 50%. The vectors shown from A to A′, B to B′, and so forth, indicate the paths followed by the particular points A, B, and so forth, during the deformation.

Duration of Pressure versus Pressure Level

Knowledge is limited regarding the relative importance of pressure and time, respectively, in the production of nerve compression lesions. Mechanical factors seem to be relatively more important at higher than at lower pressures. Time, however, is a significant factor at both high and low pressures. This phenomenon is illustrated by the fact that direct nerve compression at 30 mm Hg for 2 to 4 hours produces reversible changes, whereas prolonged compression at this pressure level may cause irreversible damage to the nerve (Rydevik et al., 1981; Lundborg et al., 1982). Compression at 400 mm Hg causes a much more severe nerve injury after 2 hours than after 15 minutes (Dahlin et al., 1986). Such information indicates that even high pressure has to "act" for a certain period of time for injury to occur. These data also give some information about the viscoelastic (time-dependent) properties of peripheral nerve tissue. Sufficient time must elapse for permanent deformation to develop.

REGENERATION OF PERIPHERAL NERVES

The regeneration of peripheral nerves following injury has been extensively described in several reviews (Sunderland, 1978; Lundborg, 1987a, 1987b). Following a crush lesion or a transection, the continuity of the axons is broken and the distal parts of the axons undergo so-called wallerian degeneration. The Schwann cells of the distal nerve segment proliferate during the first weeks after injury, lining up longitudinally to form so-called Bungner bands. The myelin and axoplasm disintegrate and are resorbed by macrophage activity. At the site of the lesion, the proximal parts of the severed axons start sending out a great number of sprouts, which grow toward the distal segment and then advance along the Schwann cell columns (endoneurial tubes), ultimately to reinnervate the target organs. In humans the maximum rate of axonal outgrowth is around 1 mm per day.

Following a crush lesion, the continuity of the Schwann cell columns at the site of injury is usually preserved, and these columns help to guide the growing axons toward their proper distal end organs. When the nerve is severed, however, no such guiding structures are preserved at the injury site and a great deal of axonal misdirection usually takes place. As a result, incorrect target organs may be reinnervated, and peripheral territories are thereby projected in new and inappropriate cortical areas in the brain. Clinically, this situation is often reflected in unsatisfactory recovery of nerve function.

The regulation of axon growth and orientation is complex and is based on a variety of biochemical and biomechanical mechanisms. Polarized structures, such as the strands in a fibrin clot, may provide guidance for the advancing sprouts. Locally occurring proteins and protein-related substances in the microenvironment at the injury zone and in the distal nerve segment seem to play a crucial role in the regulation of axonal outgrowth and orientation.

AGING OF PERIPHERAL NERVE TISSUE

Several structural and functional changes occur in the peripheral nervous system with aging (for reviews, see Schaumburg et al., 1983, and Kimura, 1983). Most persons show evidence of dysfunction in the peripheral nerves by the seventh decade; examples of such dysfunction are alterations in vibratory sensation, two-point discrimination, ankle jerks, and peripheral nerve conduction velocity.

There is usually little impairment of vibratory sense in the upper extremities of the normal elderly person. By contrast, loss of vibratory perception in the lower extremities is common with aging. Other changes in peripheral nerve function in elderly persons are diminished touch sensitivity and a slightly elevated pain threshold. Motor dysfunction also may be observed and can be brought about by changes in either the innervating peripheral nerve in question or the muscle itself.

Peripheral nerve conduction velocities, which range from 50 to 70 m per second in young adults, are slightly decreased in elderly individuals. The reduction starts to occur in the thirties or forties, but the change is normally less than 10 m per second at age 60 to 80 years (Kimura, 1983).

The structural changes underlying the functional disorders in elderly persons are not completely understood. Degenerative changes in both unmyelinated and myelinated fibers, as well as thickening of the perineurium and an increase in endoneurial connective tissue, have been reported (Schaumburg et al., 1983). It is not known whether biologic aging per se is the cause of these neural alterations or whether factors such as trauma, disuse, and vascular compromise may account for them.

Clinically, it has been solidly established that functional recovery of an elderly person after a peripheral nerve injury is much less satisfactory than that of a very young person. The difference in recovery may be based not only on a more efficient axonal outgrowth in the younger person, but also on a greater capacity of the young person's brain to adapt to the peripheral nerve injury.

SUMMARY

1. The peripheral nerves are composed of nerve fibers, layers of connective tissue, and blood vessels.
2. The nerve fibers are extremely susceptible to trauma, but because they are surrounded by successive layers of connective tissue (the epineurium and perineurium) they are mechanically protected.
3. Stretching induces changes in intraneural blood flow before the nerve fibers rupture.
4. Compression of a nerve can cause injury to both nerve fibers and blood vessels in the nerve, mainly at the edges of the compressed nerve segment.
5. Pressure level, duration of compression, and mode of pressure application are significant variables in the development of nerve injury.
6. Regeneration following transection of a peripheral nerve is a complex issue in which biologic and biomechanical factors are important for the outcome of the injury. Nerve fibers grow at a maximum speed of about 1 mm per day.
7. Aging of peripheral nerves can manifest itself as alterations in vibratory perception, ankle jerks, and nerve conduction velocity.

ACKNOWLEDGMENTS

This review was based on work supported by the Swedish Medical Research Council (projects no. 5188, 7092, and 7651), the Gothenburg Medical Society, and the University of Gothenburg, Sweden.

REFERENCES

Beel, J. A., Groswald, D. E., and Luttges, M. W.: Alterations in the mechanical properties of peripheral nerve following crush injury. J. Biomech., *17*:185, 1984.

Bentley, F. H., and Schlapp, W.: The effects of pressure on conduction in peripheral nerve. J. Physiol., *102*:72, 1943.

Chen, H. I., Granger, H. J., and Taylor, A. E.: Interaction of capillary interstitial and lymphatic forces in the canine hindpaw. Circ. Res., *39*:245, 1976.

Dahlin, L. B., Rydevik, B., and Lundborg, G.: The pathophysiology of nerve entrapments and nerve compression injuries. *In* Effects of Mechanical Stress on Tissue Viability. Edited by A. R. Hargens. New York, Springer Verlag, 1986.

Dahlin, L. B., Rydevik, B., McLean, W. G., and Sjöstrand, J.: Changes in fast axonal transport during experimental nerve compression at low pressures. Exper. Neurol., *84*:29, 1984a.

Dahlin, L. B., et al.: Distribution of tissue fluid pressure beneath a pneumatic tourniquet. Trans. Orthop. Res. Soc., *9*:362, 1984b.

Denny-Brown, D., and Brenner, C.: Paralysis of nerve induced by direct pressure and by tourniquet. Arch. Neurol. Psychiatry, *51*:1026, 1944.

Fowler, R. J., Danta, G., and Gilliatt, R. W.: Recovery of nerve conduction after a pneumatic tourniquet: Observations on the hind-limb of the baboon. J. Neurol. Neurosurg. Psychiatry, *35*:638, 1972.

Gasser, H. S.: Conduction in nerves in relation to fiber types. Proc. Assoc. Res. Mental Nerv. Dis., *15*:35, 1935.

Gelberman, R. H., et al.: The carpal tunnel syndrome. A study of carpal canal pressures. J. Bone Joint Surg., *61A*:380, 1981.

Grundfest, H.: Effects of hydrostatic pressure upon the excitability, the recovery and the potential sequence of frog nerve. Cold Spring Harbor Symp. Quant. Biol., *4*:1979, 1936.

Haftek, J.: Stretch injury of peripheral nerves. Acute effects of stretching on rabbit nerve. J. Bone Joint Surg., *52B*:354, 1970.

Hargens, A. R., et al.: Fluid balance within the canine anterolateral compartment and its relationship to compartment syndromes. J. Bone Joint Surg., *60A*:499, 1978.

Kimura, J.: Electrodiagnosis in Diseases of Nerve and Muscle: Principles and Practice. Philadelphia, F. A. Davis, 1983.

Loewenstein, W. R., and Skalak, R.: Mechanical transmission in a Pacinian corpuscle (an analysis and a theory). J. Physiol., *182*:346, 1966.

Low, P. A., and Dyck, P. J.: Increased endoneurial fluid pressure in experimental lead neuropathy. Science, *269*:427, 1977.

Lundborg, G.: Ischemic nerve injury. Experimental studies on intraneural microvascular pathophysiology and nerve function in a limb subjected to temporary circulatory arrest. Scand. J. Plast. Reconstr. Surg., Suppl. *6*:1–113, 1970.

Lundborg, G.: Structure and function of the intraneural microvessels as related to trauma, edema formation and nerve function. J. Bone Joint Surg., *57A*:938, 1975.

Lundborg, G.: Nerve Injury and Repair. Edinburgh, Churchill Livingstone, 1987a.

Lundborg, G.: Nerve repair and regeneration. Acta Orthop. Scand., *58*:145, 1987b.

Lundborg, G., and Rydevik, B.: Effects of stretching the tibial nerve of the rabbit: A preliminary study of the intraneural circulation and the barrier function of the perineurium. J. Bone Joint Surg., *55B*:390, 1973.

Lundborg, G., Myers, R. R., and Powell, H. C.: Nerve compression injury and increase in endoneurial fluid pressure—A "miniature compartment syndrome." J. Neurol. Neurosurg. Psychiatry, *46*:1119, 1983.

Lundborg, G., et al.: Median nerve compression in the carpal tunnel: The functional response to experimentally induced controlled pressure. J. Hand Surg., *7*:252, 1982.

McGregor, R. J., Sharpless, S. K., and Luttges, M. N.: A pressure vessel model for nerve compression. J. Neurol. Sci., *24*:299, 1975.

Meek, W. J., and Leaper, W. E.: The effect of pressure on conductivity of nerve and muscle. Am. J. Physiol., *27*:308, 1911.

Miyamoto, Y.: Experimental study of results of nerve suture under tension versus nerve grafting. Plast. Reconstr. Surg., *64*:540, 1979.

Miyamoto, Y., Watary, S., and Tsuge, K.: Experimental studies on the effects of tension on intraneurial microcirculation in sutured peripheral nerves. Plast. Reconstr. Surg., *63*:398, 1979.

Myers, R. R., and Powell, H. C.: Endoneurial fluid pressure in peripheral neuropathies. *In* Tissue Fluid Pressure and Composition. Edited by A. R. Hargens. Baltimore, Williams & Wilkins, 1981, p. 193.

Ochoa, J., Fowler, T. J., and Gilliatt, R. W.: Anatomical changes in peripheral nerves compressed by a pneumatic tourniquet. J. Anat., *113*:433, 1972.

Rydevik, B., and Lundborg, G.: Permeability of intraneural microvessels and perineurium following acute, graded experimental nerve compression. Scand. J. Plast. Reconstr. Surg., *11*:179, 1977.

Rydevik, B., and Nordborg, C.: Changes in nerve function and nerve fiber structure induced by acute, graded compression. J. Neurol. Neurosurg. Psychiatry, *43*:1070, 1980.

Rydevik, B., Brown, M. D., and Lundborg, G.: Pathoanatomy and pathophysiology of nerve root compression. Spine, *9*:7, 1984.

Rydevik, B., Lundborg, G., and Bagge, U.: Effects of graded compression on intraneural blood flow. An in vivo study on rabbit tibial nerve. J. Hand Surg., *6*:3, 1981.

Rydevik, B., McLean, W. G., Sjöstrand, J., and Lundborg, G.: Blockage of axonal transport induced by acute, graded compression of the rabbit vagus nerve. J. Neurol. Neurosurg. Psychiatry, *43*:690, 1980.

Schaumburg, H. H., Spencer, P. S., and Ochoa, J.: The aging of human peripheral nervous system. *In* The Neurology of Aging. Edited by R. Katzman and R. Terry. Philadelphia, F. A. Davis, 1983.

Selander, D., Lundborg, G., and Rydevik, B.: Unpublished data.

Shanta, T. R., and Bourne, G. H.: The perineurial epithelium. A new concept. *In* Structure and Function of Nervous Tissue. Vol. 1. Structure. London, Academic Press, 1968.

Spencer, P. S., Weinberg, H. J., Raine, C. S., and Prineas, J. W.: The perineurial window—A new model of focal demyelination and remyelination. Brain Res., *96*:323, 1975.

Sunderland, S.: Nerves and Nerve Injuries. 2nd Ed. Edinburgh, Churchill Livingstone, 1978.

Sunderland, S., and Bradley, K. C.: The cross-sectional area of peripheral nerve trunks devoted to nerve fibers. Brain, *72*:428, 1949.

Sunderland, S., and Bradley, K. C.: Stress-strain phenomena in human peripheral nerve trunks. Brain, *84*:102, 1961a.

Sunderland, S., and Bradley, K. C.: Stress-strain phenomena in human nerve roots. Brain, *84*:120, 1961b.

Tortora, G. J., and Anagnostakos, N. P.: Principles of Anatomy and Physiology. 4th Ed. New York, Harper & Row, 1984.

Upton, R. M., and McComas, A. J.: The double-crush in nerve entrapment syndromes. Lancet, *2*:359, 1973.

Weiss, D. G., and Gorio, A. (eds.): Axoplasmic Transport in Physiology and Pathology. Berlin, Springer Verlag, 1982.

SUGGESTED READING

Dahlin, L. B., Rydevik, B., McLean, W. G., and Sjöstrand, J.: Changes in fast axonal transport during experimental nerve compression at low pressures. Exper. Neurol., *84*:29, 1984.

Fowler, R. J., Danta, G., and Gilliatt, R. W.: Recovery of nerve conduction after a pneumatic tourniquet: Observations on the hind-limb of the baboon. J. Neurol. Neurosurg. Psychiatry, *35*:638, 1972.

Gelberman, R. H., et al.: The carpal tunnel syndrome. A study of carpal canal pressures. J. Bone Joint Surg., *61A*:380, 1981.

Haftek, J.: Stretch injury of peripheral nerves. Acute effects of stretching on rabbit nerve. J. Bone Joint Surg., *52B*:354, 1970.

Loewenstein, W. R., and Skalak, R.: Mechanical transmission in a Pacinian corpuscle (an analysis and a theory). J. Physiol., *182*:346, 1966.

Low, P. A., and Dyck, P. J.: Increased endoneurial fluid pressure in experimental lead neuropathy. Science, *269*:427, 1977.

Lundborg, G.: Structure and function of the intraneural microvessels as related to trauma, edema formation and nerve function. J. Bone Joint Surg., *57A*:938, 1975.

Lundborg, G., and Rydevik, B.: Effects of stretching the tibial nerve of the rabbit: A preliminary study of the intraneural circulation and the barrier function of the perineurium. J. Bone Joint Surg., *55B*:390, 1973.

Lundborg, G., et al.: Peripheral nerve: The physiology of injury and repair. *In* Injury and Repair of the Musculoskeletal Soft Tissues. Edited by S. L.-Y. Woo and J. A. Buckwalter. Park Ridge, IL, American Academy of Orthopaedic Surgeons, 1987, pp. 297–352.

McGregor, R. J., Sharpless, S. K., and Luttges, M. N.: A pressure vessel model for nerve compression. J. Neurol. Sci., *24*:299, 1975.

Myers, R. R., and Powell, H. C.: Endoneurial fluid pressure in peripheral neuropathies. *In* Tissue Fluid Pressure and Composition. Edited by A. R. Hargens. Baltimore, Williams & Wilkins, 1981, p. 193.

Ochoa, J., Fowler, T. J., and Gilliatt, R. W.: Anatomical changes in peripheral nerves compressed by a pneumatic tourniquet. J. Anat., *113*:433, 1972.

Rydevik, B., Lundborg, G., and Bagge, U.: Effects of graded compression on intraneural blood flow. An in vivo study on rabbit tibial nerve. J. Hand Surg., *6*:3, 1981.

Sumner, A. J. (ed.): The Physiology of Peripheral Nerve Disease. Philadelphia, W. B. Saunders, 1980.

Sunderland, S.: Nerves and Nerve Injuries. 2nd Ed. Edinburgh, Churchill Livingstone, 1978.

Weiss, D. G., and Gorio, A. (eds.): Axoplasmic Transport in Physiology and Pathology. Berlin, Springer Verlag, 1982.

5

BIOMECHANICS OF SKELETAL MUSCLE

Mark I. Pitman
Lars Peterson

The muscular system consists of three muscle types: cardiac muscle, which composes the heart; smooth (nonstriated or involuntary) muscle, which lines the hollow internal organs; and skeletal (striated or voluntary) muscle, which attaches to the skeleton via the tendons and causes it to move. The focus of this chapter is the role and function of skeletal muscle.

Skeletal muscle is the most abundant tissue in the human body, accounting for 40 to 45% of the total body weight. There are more than 430 skeletal muscles, found in pairs on the right and left sides of the body. The most vigorous movements are produced by fewer than 80 pairs. The muscles provide strength and protection to the skeleton by distributing loads and absorbing shock, and they enable the bones to move at the joints. Such movement usually represents the action of muscle groups rather than of individual muscles.

The skeletal muscles perform both dynamic and static work. Dynamic work permits locomotion and the positioning of the body segments in space. Static work maintains body posture.

In this chapter the macroscopic and microscopic composition and structure of skeletal muscle are described. The process of muscle contraction is presented in some detail, and the various types of muscle contraction involved in static and dynamic muscle work are discussed. The mechanical properties of muscle are examined, as are the differences in muscle fiber makeup. Finally, the effects of disuse and immobilization and of physical training on muscle are covered.

COMPOSITION AND STRUCTURE OF SKELETAL MUSCLE

An understanding of the biomechanics of muscle function requires knowledge of the gross anatomic structure and function of the musculotendinous unit and the basic microscopic structure and chemical composition of the muscle fiber.

STRUCTURE AND ORGANIZATION OF MUSCLE

The structural unit of skeletal muscle is the fiber, a long cylindrical cell with many hundreds of nuclei. Muscle fibers range in thickness from about 10 to 100 μm and in length from about 1 to 30 cm, the longest fibers being found in the sartorius muscle.

Each fiber is encompassed by a loose connective tissue called the endomysium, and the fibers are organized into various-sized bundles, or fascicles (Fig. 5–1A and B), which are in turn encased in a dense connective tissue sheath known as the perimy-

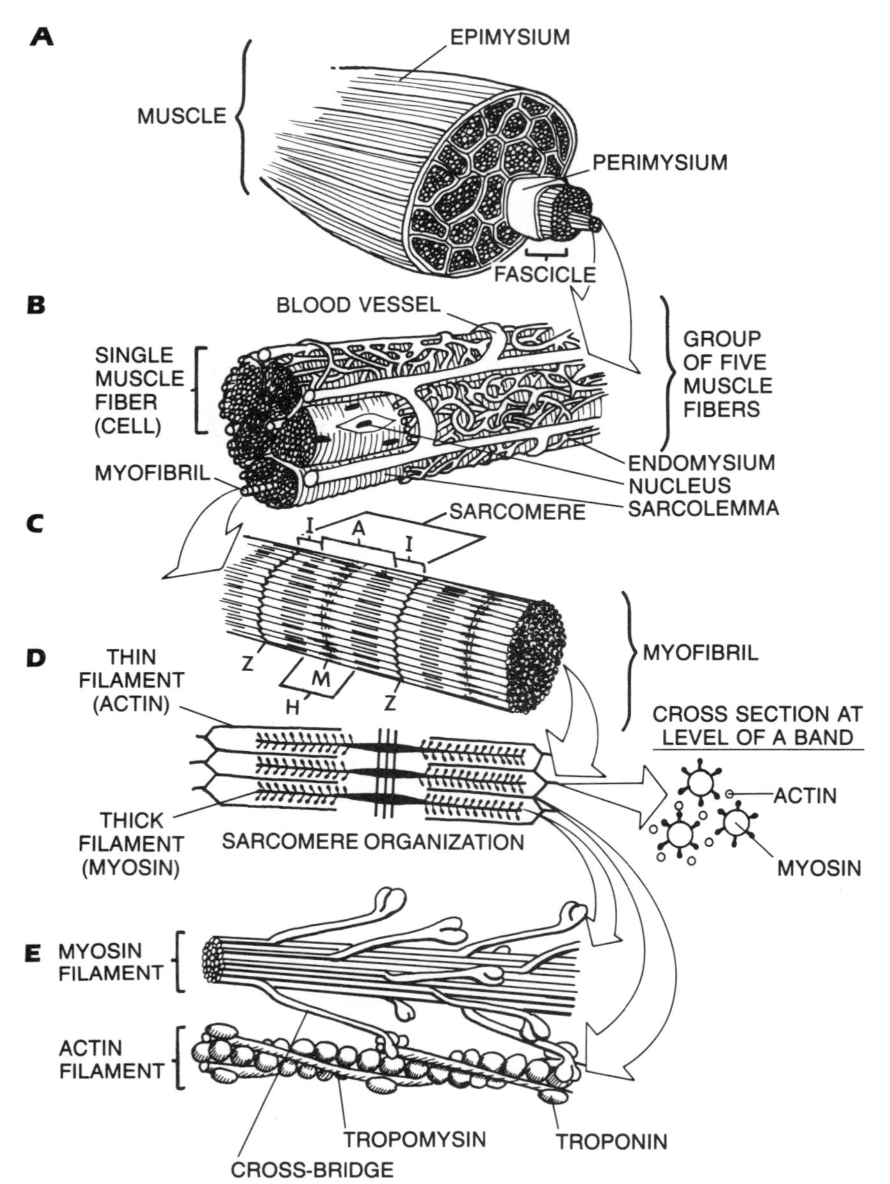

FIG. 5–1
Schematic drawings of the structural organization of muscle. **A.** A fibrous connective tissue fascia, the epimysium, surrounds the muscle, which is composed of many bundles, or fascicles. The fascicles are encased in a dense connective tissue sheath, the perimysium. **B.** The fascicles are composed of muscle fibers, which are long, cylindrical, multinucleated cells. Between the individual muscle fibers are capillary blood vessels. Each muscle fiber is surrounded by a loose connective tissue called the endomysium. Just beneath the endomysium lies the sarcolemma, a thin elastic sheath with infoldings that invaginate the fiber interior. Each muscle fiber is composed of numerous delicate strands, myofibrils, the contractile elements of muscle. **C.** Myofibrils consist of smaller filaments, which form a repeating banding pattern along the length of the myofibril. One unit of this serially repeating pattern is called a sarcomere. The sarcomere is the functional unit of the contractile system of muscle. **D.** The banding pattern of the sarcomere is formed by the organization of thick and thin filaments, composed of the proteins myosin and actin, respectively. The actin filaments are attached at one end but are free along their length to interdigitate with the myosin filaments. The thick filaments are arranged in a hexagonal fashion. A cross section through the area of overlap shows the thick filaments surrounded by six equally spaced thin filaments. **E.** The lollipop-shaped molecules of each myosin filament are arranged so that the long tails form a sheaf with the heads, or cross-bridges, projecting from it. The cross-bridges point in one direction along half of the filament and in the other direction along the other half. Only a portion of one half of a filament is shown here. The cross-bridges are an essential element in the mechanism of muscle contraction, extending outward to interdigitate with receptor sites on the actin filaments. Each actin filament is a double helix, appearing as two strands of beads spiraling around each other. Two additional proteins, tropomyosin and troponin, are associated with the actin helix and play an important role in regulating the interdigitation of the actin and myosin filaments. Tropomyosin is a long polypeptide chain that lies in the grooves between the helices of actin. Troponin is a globular molecule attached at regular intervals to the tropomysin. (Adapted from Williams and Warwick, 1980.)

sium. Surrounding the entire muscle is a fascia of fibrous connective tissue called the epimysium.

In general, each end of a muscle is attached to bone by tendons, which have no active contractile properties. The collagen fibers in the perimysium and epimysium are continuous with those in the tendons, and together these fibers act as a structural framework for the attachment of bones and muscle fibers. The forces produced by the contracting muscles are transmitted to bone through these connective tissues and tendons.

Each muscle fiber is composed of a large number of delicate strands, the myofibrils. These are the contractile elements of muscle. Their structure and function have been studied exhaustively by light and electron microscopy, and their histochemistry and biochemistry have been explained (Arvidson et al., 1984; Åstrand and Rodahl, 1977; Fuchs, 1974; Guyton, 1986; Henriksson, 1976; Huxley, 1965, 1969). About 1 μm in diameter, the myofibrils lie parallel to each other within the cytoplasm (sarcoplasm) of the muscle fiber and extend throughout its length. They vary in number from a few to several thousand, depending on the diameter of the fiber.

The individual myofibrils are aligned so that areas with the same internal organization lie at the same level. When the surface of a muscle fiber is viewed with the light microscope, this alignment appears as transverse light and dark striations (hence the term "striated muscle"). These striations, or banding patterns, are continuous throughout the muscle fiber.

The transverse banding pattern repeats itself along the length of the muscle fiber, each repeat being known as a sarcomere (Fig. 5–1C). The sarcomere is the functional unit of the contractile system in muscle, and the events that take place in one sarcomere are duplicated in the others.

Examination of the myofibrils with the electron microscope reveals the structures responsible for the repeating pattern. Each myofibril is composed of fibrous filaments of two types: thin filaments (approximately 5 nm in diameter) composed of the protein actin and thick filaments (about 15 nm in diameter) composed of the protein myosin (Fig. 5–1D and E). The thick filaments are located in the central region of the sarcomere, where their orderly, parallel arrangement gives rise to dark bands, known as A bands because they are strongly anisotropic. The thin filaments are attached at either end of the sarcomere to a structure known as the Z line, which consists of short elements that link the thin filaments of adjacent sarcomeres, defining the limits of each sarcomere. The thin filaments extend from the Z line toward the center of the sarcomere, where they overlap with the thick filaments.

Bisected by the Z lines is the I band, so called because the region is nearly isotropic. This band contains the portion of the thin filaments that does not overlap with the thick filaments. Because the I band contains only thin filaments, it usually appears as a light band separating the dark A bands. In the center of the A band, in the gap between the ends of the thin filaments, is the H zone, a light band containing only thick filaments. A narrow dark area in the center of the H zone is the M line, produced by transversely and longitudinally oriented proteins that link adjacent thick filaments, maintaining their parallel arrangement. The various areas of the banding pattern are apparent in the photomicrograph of human skeletal muscle shown in Figure 5–2.

Closely correlated with the repeating pattern of the sarcomeres is an organized network of tubules and sacs known as the sarcoplasmic reticulum. This network is distributed throughout the sarcoplasm of the muscle fiber, forming a structure resembling a lacy sleeve around each myofibril (Fig. 5–3). The tubules of the sarcoplasmic reticulum lie parallel to the myofibrils and tend to enlarge and fuse at the level of the junctions between the A and I bands, forming transverse sacs that surround the individual myofibril completely. These transverse outpouchings of the tubules are called terminal cisternae.

The terminal cisternae enclose a smaller tubule that is separated from them by its own membrane. The smaller tubule and the terminal cisterna above and below it are known as a triad. The enclosed tubule is part of the transverse tubule system, or T system, which constitutes a deep infolding into the interior of the fiber by the sarcolemma, a tough, exceedingly thin elastic membrane that surrounds each muscle fiber just beneath the endomysium. This extracellular T system serves as a duct system for the movement of fluids and other substances into and out of the muscle cell and also allows for propagation of the electric stimulus for contraction (the action potential) from the sarcolemma inward into the deep regions of the cell (Arvidson et al., 1984; Brobeck, 1979; Williams and Warwick, 1980; Ham and Cormack, 1979). Although not in direct contact with the inner regions of the muscle fiber, the sarcoplasmic reticulum, by having a great capacity for storing calcium, also plays an important role in the mechanics of muscle contraction, as described in the following section.

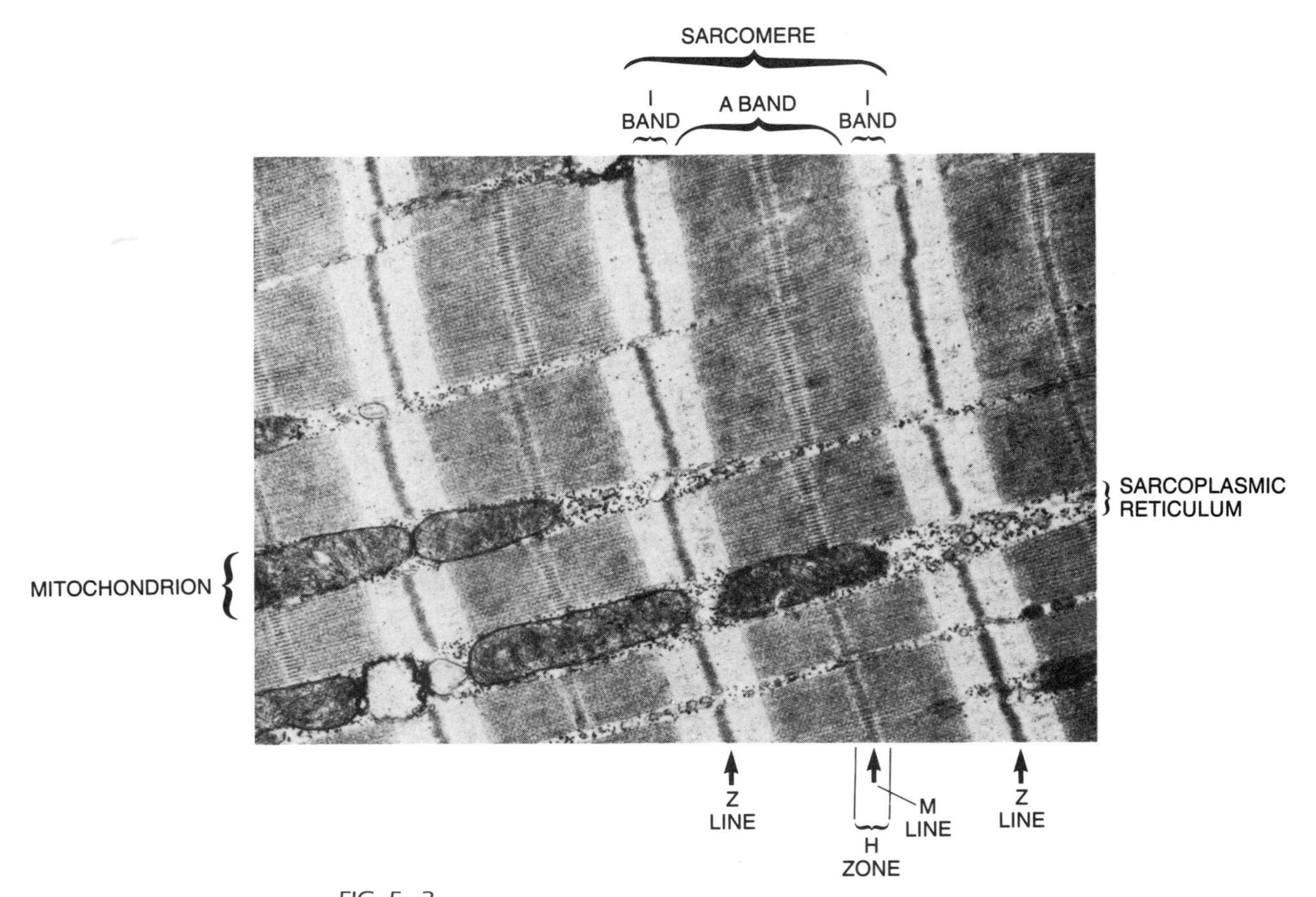

FIG. 5–2
Electron photomicrograph of a cross section of human skeletal muscle. The sarcomeres are apparent along the myofibrils. Characteristic regions of the sarcomere are indicated.

MOLECULAR COMPOSITION OF THE MYOFIBRIL

At a very high magnification the structure and spatial orientation of the various proteins that compose the contractile filaments of a myofibril can be observed (see Fig. 5–1D, and E). Myosin, the thicker filament, is composed of individual molecules each of which resembles a lollipop with a globular "head" projecting from a long shaft, or "tail." Several hundred such molecules are packed tail to tail in a sheaf with their heads pointed in one direction along half of the filament and in the opposite direction along the other half, leaving a head-free region (the H zone) in between. The globular heads spiral about the myosin filament in the region of actin and myosin overlap (the A band) and extend as cross-bridges to interdigitate with sites on the actin filaments, thus forming the structural and functional link between the two filament types.

Actin, the chief component of the thin filament, has the shape of a double helix and appears as two strands of beads spiraling around each other. Two additional proteins, troponin and tropomyosin, are important constituents of the actin helix, as they appear to regulate the making and breaking of contacts between the actin and myosin filaments during contraction. Tropomyosin is a long polypeptide chain that lies in the grooves between the helices of actin. Troponin is a globular molecule attached at regular intervals to the tropomyosin.

MOLECULAR BASIS OF MUSCLE CONTRACTION

The most widely held theory of muscle contraction is the sliding filament theory, proposed simultaneously by A. F. Huxley and H. E. Huxley in 1964 and subsequently refined (Huxley, H. E., 1965, 1969,

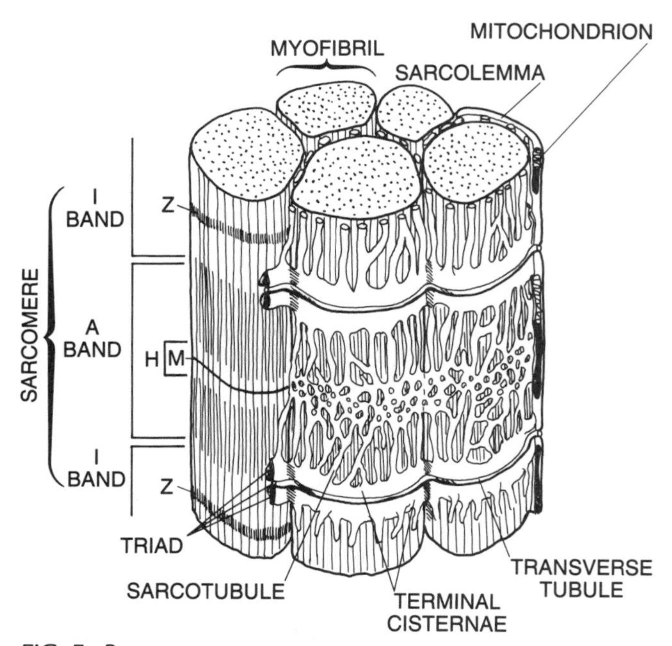

FIG. 5–3
Diagram of a portion of a skeletal muscle fiber illustrating the sarcoplasmic reticulum that surrounds each myofibril. The various regions of the sarcomere are indicated on the left myofibril to show the correlation of these regions with the sarcoplasmic reticulum, shown surrounding the middle and right myofibrils. The transverse tubules represent an infolding of the sarcolemma, the plasma membrane that encompasses the entire muscle fiber. Two transverse tubules supply each sarcomere at the level of the junctions of the A band and I bands. Terminal cisternae are located on each side of the transverse tubule, and together these structures constitute a triad. The terminal cisternae connect with a longitudinal network of sarcotubules spanning the region of the A band. (Adapted from Ham and Cormack, 1979.)

1972; Ebashi et al., 1969; Needham, 1971; Weber and Murray, 1973; Huxley, A. F., 1974). According to this theory, active shortening of the sarcomere, and hence of the muscle, results from the relative movement of the actin and myosin filaments past one another while each retains its original length. The force of contraction is developed by the myosin heads, or cross-bridges, in the region of overlap between actin and myosin (the A band). These cross-bridges swivel in an arc around their fixed positions on the surface of the myosin filament, much like the oars of a boat. This movement of the cross-bridges in contact with the actin filaments produces the sliding of the actin filaments toward the center of the sarcomere.

Since a single movement of a cross-bridge produces only a small displacement of the actin filament relative to the myosin filament, each individual cross-bridge detaches itself from one receptor site on the actin filament and reattaches itself to another site farther along, repeating the process five or six times, "with an action similar to a man pulling on a rope hand over hand" (Wilkie, 1968). The cross-bridges do not act in a synchronized manner; each acts independently. Thus, at any given moment only about half of the cross-bridges actively generate force and displacement, and when these detach others take up the task so that shortening is maintained. The shortening is reflected in the sarcomere as a decrease in the I band and a decrease in the H zone as the Z lines move closer together; the width of the A band remains constant.

A key to the sliding mechanism is the calcium ion (Ca^{2+}), which turns the contractile activity on and off. Muscle contraction is initiated when calcium is made available to the contractile elements and ceases when calcium is removed. The mechanisms that regulate the availability of calcium ions to the contractile machinery are coupled to electric events occurring in the muscle membrane (sarcolemma). An action potential in the sarcolemma provides the electric signal for the initiation of contractile activity. The mechanism by which the electric signal triggers the chemical events of contraction is known as excitation-contraction coupling.

When the motor neuron stimulates the muscle at the neuromuscular junction (Fig. 5–4), and the propagated action potential depolarizes the muscle cell membrane (sarcolemma), there is an inward spread of the action potential along the T system. (Details of this process are given in Table 5–1, which summarizes the events during excitation, contraction, and relaxation of muscle.) Depolarization of the membrane of the terminal cisternae of the sarcoplasmic reticulum makes this membrane permeable for Ca^{2+} (Ebashi, 1976).

In the muscle at rest, the sarcoplasm surrounding the myofibrils is almost free of Ca^{2+}. Thus, calcium ions are quickly released into the sarcoplasm from the vesicles of the sarcoplasmic reticulum, in which they are stored, and are bound to troponin, which has a strong affinity for Ca^{2+}. This action affects tropomysin, which in resting muscle is believed to inhibit actin-myosin interaction and to prevent their permanent bonding by blocking the myosin receptors on the actin fibers. Apparently, the binding of Ca^{2+} to troponin allows movement of the tropomyosin molecule away from the myosin receptor sites on the actin that it had been blocking (Huxley, 1972). The troponin molecules appear to undergo a conformational change that somehow "tugs" on the tropomy

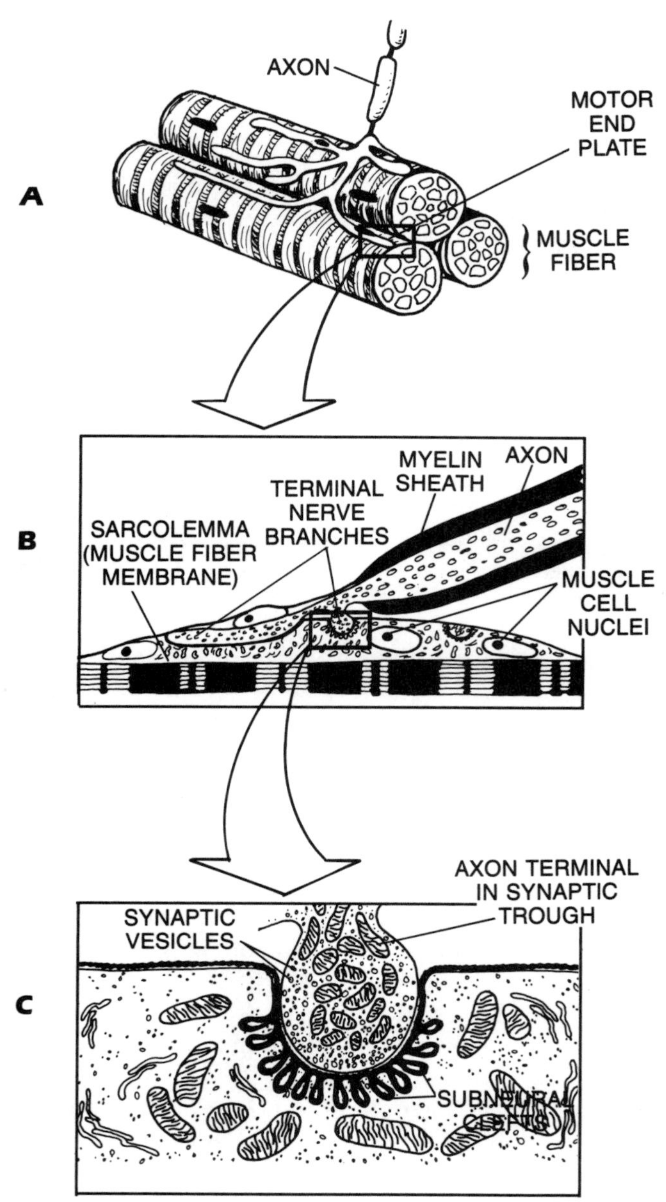

FIG. 5–4
Schematic representation of the innervation of muscle fibers. **A.** An axon of a motor neuron (originating from the cell body in the anterior horn of the spinal cord) branches near its end to innervate several skeletal muscle fibers, forming a neuromuscular junction with each fiber. The region of the muscle membrane (sarcolemma) lying directly under the terminal branches of the axon has special properties and is known as the motor end plate, or motor end-plate membrane. The rectangular area is shown in detail in **B.** (Adapted from Prichard, 1979.) **B.** The fine terminal branches of the nerve (axon terminals), devoid of myelin sheaths, lie in grooves on the sarcolemma. The rectangular area in this section is shown in detail in **C.** Ultrastructure of the junction of an axon terminal and the sarcolemma. The invagination of the sarcolemma forms the synaptic trough into which the axon terminal protrudes. The invaginated sarcolemma has many folds, or subneural clefts, which greatly increase its surface area. Acetylcholine is stored in synaptic vesicles in the axon terminal. (**B** and **C**, adapted from Brobeck, 1979.)

TABLE 5–1

EVENTS DURING EXCITATION, CONTRACTION, AND RELAXATION OF MUSCLE FIBER

1. An action potential is initiated and propagated in a motor axon.

2. This action potential causes the release of acetylcholine from the axon terminals at the neuromuscular junction.

3. Acetylcholine is bound to receptor sites on the motor end-plate membrane.

4. Acetylcholine increases the permeability of the motor end plate to sodium and potassium ions, producing an end-plate potential.

5. The end-plate potential depolarizes the muscle membrane (sarcolemma), generating a muscle action potential which is propagated over the membrane surface.

6. Acetylcholine is rapidly destroyed by acetylcholinesterase on the end-plate membrane.

7. The muscle action potential depolarizes the transverse tubules.

8. Depolarization of the transverse tubules leads to the release of calcium ions from the terminal cisternae of the sarcoplasmic reticulum surrounding the myofibrils. These ions are released into the sarcoplasm in the direct vicinity of the regulatory proteins, tropomyosin and troponin.

9. Calcium ions bind to troponin, allowing movement of the tropomyosin molecule away from the myosin receptor sites on the actin filament that it had been blocking, and releasing the inhibition that had prevented actin from combining with myosin.

10. Actin combines with myosin ATP (receptor sites on the myosin cross-bridges bind to receptor sites on the actin chain):

$$A + M \cdot ATP \rightarrow A \cdot M \cdot ATP$$

11. Actin activates the myosin ATPase found on the myosin cross-bridge, enabling ATP to be split (hydrolyzed). This process releases energy used to produce movement of the myosin cross-bridges:

$$A \cdot M \cdot ATP \rightarrow A \cdot M + ADP + P_i$$

12. Oarlike movements of the cross-bridges produce relative sliding of the thick and thin filaments past each other.

13. Fresh ATP binds to the myosin cross-bridge, breaking the actin-myosin bond and allowing the cross-bridge to dissociate from actin:

$$A \cdot M + ATP \rightarrow A + M \cdot ATP$$

14. Cycles of binding and unbinding of actin with the myosin cross-bridges at successive sites along the actin filament (steps 11, 12, and 13) continue as long as the concentration of calcium remains high enough to inhibit the action of the troponin-tropomyosin system.

15. Concentration of calcium ions falls as they are pumped into the terminal cisternae of the sarcoplasmic reticulum by an energy-requiring process that splits ATP.

16. Calcium dissociates from troponin, restoring the inhibitory action of troponin-tropomyosin. The actin filament slides back and the muscle lengthens. In the presence of ATP, actin and myosin remain in the dissociated, relaxed state.

(Modified from Luciano et al., 1978.)

osin protein strand, moving the tropomyosin deeper into the groove between the two actin helices.

Contraction begins now that the actin monomers, released from the inhibitory influence of the troponin-tropomyosin complex, are able to react with the myosin cross-bridges, which move out perpendicularly from the filament core and attach to the actin molecules within reach. The energy for cross-bridge movement is provided by the breakdown (hydrolysis) of adenosine triphosphate (ATP) to diphosphate (ADP + P_i) by an ATPase on the myosin cross-bridge that is activated by actin. This hydrolytic splitting of ATP releases a large amount of energy used for mechanical work. The cross-bridges undergo an energy-yielding conformational change that causes them to alter their angular relationship to the axis of the myosin core, and the actin filaments are pulled along the myosin filaments toward the H zone. When the ATP becomes hydrolyzed, the ADP and free phosphate leave the binding sites on the myosin head.

The energy required for the release of contraction is also provided by ATP. The muscle relaxes when fresh ATP is taken up by (bound to) the myosin head, which promptly dissociates actin from myosin, breaking the bridge between them. Ca^{2+} is dissociated from the troponin as calcium pumps in the sarcoplasmic reticulum pump the released calcium ions back into the terminal cisternae. The energy for this process comes from the splitting of ATP. This process of reaccumulation of the released calcium takes much longer than the initial release (which lasts only a few milliseconds), and the contractile activity of the cross-bridges persists for several hundred milliseconds, until the concentration of free calcium becomes so low that calcium ions dissociate from troponin and the blocking action of troponin-tropomyosin is no longer repressed. The tropomyosin again changes its position on the actin filament and once more inhibits actin from interacting with the cross-bridges by blocking the active receptor sites. The cross-bridges switch back to their original conformation, the actin filaments slide back, and the muscle lengthens.

The presence of ATP is essential for the dissociation of actin and myosin. This fact is demonstrated by the phenomenon of rigor mortis, in which the muscles of the body become rigid shortly after death. The absence of ATP in the dead muscle cells results in permanent bonding of actin and myosin.

THE MOTOR UNIT

The functional unit of skeletal muscle is the motor unit, which includes a single motor neuron and all of the muscle fibers innervated by it. This unit is the smallest part of the muscle that can be made to contract independently. When stimulated, all muscle fibers in the motor unit respond as one. The fibers of a motor unit are said to show an all-or-none response to stimulation: they either contract maximally or not at all.

The number of muscle fibers forming a motor unit is closely related to the degree of control required of the muscle. In small muscles that perform very fine movements, such as the extraocular muscles, each motor unit may contain less than a dozen muscle fibers, whereas in large muscles that perform coarse movements, such as the gastrocnemius, the motor unit may contain 1,000 to 2,000 muscle fibers.

The fibers of each motor unit are not contiguous but are dispersed throughout the muscle with fibers of other units. Thus, if a single motor unit is stimulated, a large portion of the muscle appears to contract. If additional motor units of the nerve innervating the muscle are stimulated, the muscle contracts with greater force. The calling in of additional motor units in response to greater stimulation of the motor nerve is called recruitment.

THE MUSCULOTENDINOUS UNIT

The tendons and the connective tissues in and around the muscle belly are viscoelastic structures that help determine the mechanical characteristics of whole muscle during contraction and passive extension. Hill (1938, 1950, 1970) showed that the tendons represent a springlike elastic component located in series with the contractile component (the contractile proteins of the myofibril, actin and myosin), while the epimysium, perimysium, endomysium, and sarcolemma represent a second elastic component located in parallel with the contractile component (Fig. 5–5).

When the parallel and series elastic components stretch during active contraction or passive extension of a muscle, tension is produced and energy is stored; when they recoil with muscle relaxation, this energy is released. The series elastic fibers are more important in the production of tension than are the parallel elastic fibers (Wilkie, 1956). Several investigators have suggested that the cross-bridges of the myosin filaments have a springlike property and also contribute to the elastic properties of muscle (Hill, 1968; Huxley, 1969; Ottoson, 1983).

The distensibility and elasticity of the elastic components are valuable to the muscle in several ways (Wilkie, 1956; Gowitzke and Milner, 1987): they tend

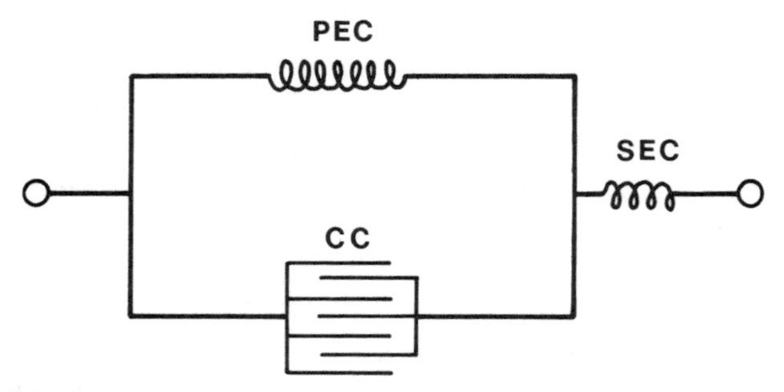

FIG. 5–5
The musculotendinous unit may be depicted as consisting of a contractile component (CC) in parallel with an elastic component (PEC) and in series with another elastic component (SEC) (Hill, 1938, 1950). The contractile component is represented by the contractile proteins of the myofibril, actin and myosin. (The myosin cross-bridges may also exhibit some elasticity.) The parallel elastic component comprises the connective tissue surrounding the muscle fibers (the epimysium, perimysium, and endomysium) and the sarcolemma. The series elastic component is represented by the tendons. (Adapted from Keele et al., 1982.)

to keep the muscle in readiness for contraction and assure that muscle tension is produced and transmitted smoothly during contraction; they assure that the contractile elements return to their original (resting) positions when contraction is terminated; and they may help prevent the passive overstretch of the contractile elements when these elements are relaxed, thereby lessening the danger of muscle injury.

The viscous property of the series and parallel elastic components allows them to absorb energy proportional to the rate of force application and to dissipate energy in a time-dependent manner (for a discussion of viscoelasticity, see Chapter 3). This viscous property, combined with the elastic properties of the musculotendinous unit, is demonstrated in everyday activities. For example, when a person attempts to stretch and touch the toes, the stretch is initially elastic. As the stretch is held, however, further elongation of the muscle results from the viscosity of the muscle-tendon structure, and the fingers slowly reach closer to the floor.

MECHANICS OF MUSCLE CONTRACTION

Electromyography provides a mechanism for evaluating and comparing neural effects on muscle and the contractile activity of the muscle itself in vivo and in vitro. Much has been learned by using electromyography to study various aspects of the contractile process, particularly the time relationship between the onset of electrical activity in the muscle and of actual contraction of the muscle or muscle fiber. The following sections discuss the mechanical response of a muscle to electrical (neural) stimulation and the various ways in which the muscle contracts in order to move a joint, control its motion, or maintain its position.

SUMMATION AND TETANIC CONTRACTION

The mechanical response of a muscle to a single stimulus of its motor nerve is known as a twitch, which is the fundamental unit of recordable muscle activity. Following stimulation there is an interval of a few milliseconds known as the latency period before the tension in the muscle fibers begins to rise. This period represents the time required for the "slack" in the elastic components to be taken up. The time from the start of tension development to peak tension is the contraction time, and the time from peak tension until the tension drops to zero is the relaxation time. The contraction time and relaxation time vary among muscles, as they depend largely on the muscle fiber makeup (described below). Some muscle fibers contract with a speed of only 10 msec, while others may take 100 msec or longer.

An action potential lasts only about 1 to 2 msec. This is a small fraction of the time taken for the subsequent mechanical response, or twitch, even in muscles that contract quickly; so it is possible for a series of action potentials to be initiated before the first twitch is completed if the activity of the motor axon is maintained. When mechanical responses to successive stimuli are added to an initial response, the result is known as summation (Fig. 5–6). If a second stimulus occurs during the latency period of the first muscle twitch, it produces no additional response, and the muscle is said to be completely refractory.

The frequency of stimulation is variable and is modulated by individual motor units. The greater the frequency of stimulation of the muscle fibers, the greater will be the tension produced in the muscle as a whole. However, a maximal frequency will be reached beyond which the tension of the muscle no longer increases. When this maximal tension is sustained as a result of summation, the muscle is said to contract tetanically. In this case the rapidity of stimulation outstrips the contraction-relaxation time of the muscle so that little or no relaxation can

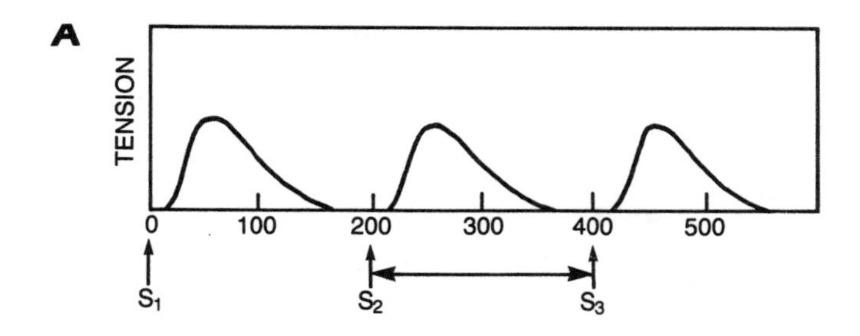

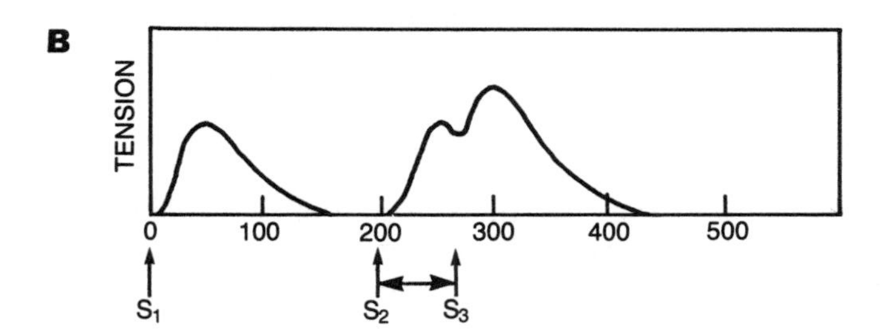

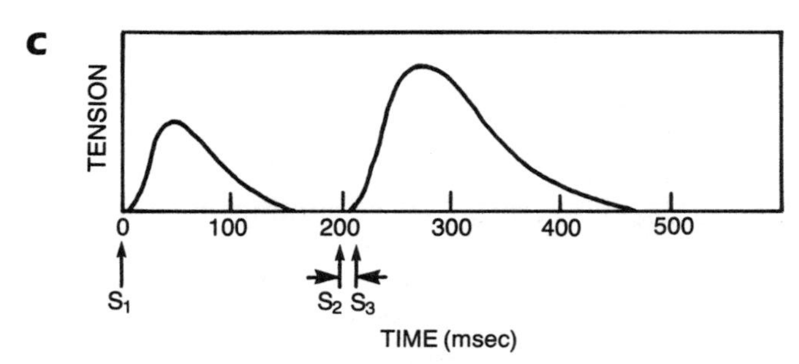

FIG. 5–6
Summation of contractions in a muscle held at a constant length. **A.** An initial stimulus (S_1) is applied to the muscle, and the resulting twitch lasts 150 msec. The second (S_2) and third (S_3) stimuli are applied to the muscle after 200-msec intervals when the muscle has relaxed completely; thus no summation occurs. **B.** S_3 is applied 60 msec after S_2, when the mechanical response from S_2 is beginning to decrease. The resulting peak tension is greater than that of the single twitch. **C.** The interval between S_2 and S_3 is further reduced to 10 msec. The resulting peak tension is even greater than in **B,** and the increase in tension produces a smooth curve. The mechanical response evoked by S_3 appears as a continuation of that evoked by S_2. (Adapted from Luciano et al., 1978.)

occur before the next contraction is initiated (Fig. 5–7).

The considerable gradation of contraction exhibited by whole muscles is achieved by the differential activity of their motor units, in both stimulation frequency and the number of units activated. The repetitive twitching of all recruited motor units of a muscle in an asynchronous manner results in brief summations or more prolonged subtetanic or tetanic contractions of the muscle as a whole and is a principal factor responsible for the smooth movements produced by the skeletal muscles.

TYPES OF MUSCLE CONTRACTION

During contraction, the force exerted by a contracting muscle on the bony lever(s) to which it is attached is known as the muscle tension, and the external force exerted on the muscle is known as the resistance, or load. As the muscle exerts its force, it generates a turning effect, or moment (torque), on the involved joint, as the line of application of the muscle force usually lies at a distance from the center of motion of the joint. The moment is calculated as the product of the muscle force and the perpendicular distance between its point of application and the center of motion (this distance is known as the lever arm, or moment arm, of the force). Muscle contractions can be classified according to the relationship between either the muscle tension and the resistance to be overcome or the muscle moment generated and the resistance to be overcome (Table 5–2).

When muscles develop sufficient tension to overcome the resistance of the body segment, the muscles shorten and cause joint movement. Such a contraction is called a concentric contraction (Norkin, 1983; Rodgers and Cavanagh, 1984; Komi, 1986). The net moment generated by the muscle is in the same direction as the change in joint angle (Rodgers and Cavanagh, 1984). An example of a concentric contraction is the action of the quadriceps in extending the knee when one ascends stairs.

When a muscle cannot develop sufficient tension and is overcome by the external load, it progressively lengthens instead of shortening; in this case the muscle is said to contract eccentrically (Norkin, 1983; Rodgers and Cavanagh, 1984; Komi, 1986). The net muscle moment is in the opposite direction from the change in joint angle (Rodgers and Cavanagh, 1984). One purpose of eccentric contraction is to decelerate the motion of a joint. For example, when one descends stairs, the quadriceps works eccentrically to decelerate flexion of the knee, thus decelerating the limb. The tension that it applies is less than the force of gravity pulling the body downward, but it is sufficient to allow controlled lowering of the body.

Both concentric and eccentric contractions involve dynamic muscle work; i.e., the muscle moves a joint or controls its movement. Muscles are not always directly involved in the production of joint movements, however. They may exercise either a restraining or a holding action, such as that needed to maintain the body in an upright position in opposing the force of gravity. In this case the muscle attempts to shorten (i.e., the myofibrils shorten and in doing so stretch the series elastic component,

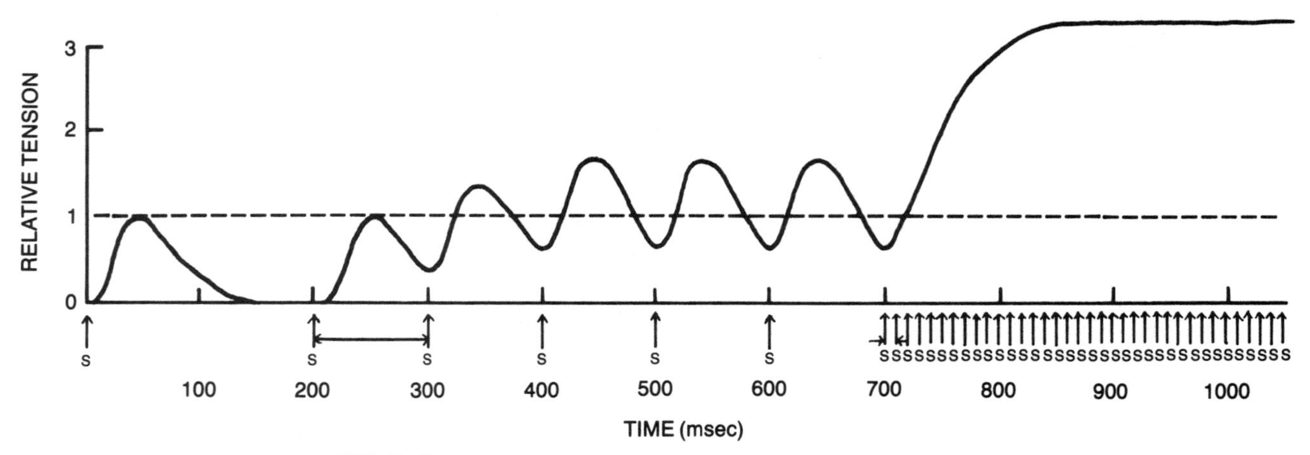

FIG. 5–7
Generation of muscle tetanus. As the frequency of stimulation (S) increases (i.e., the intervals shorten from 200 to 100 msec), the muscle tension rises as a result of summation. When the frequency is increased to 100 per second, summation becomes maximal and the muscle contracts tetanically, exerting sustained peak tension. (Adapted from Luciano et al., 1978.)

thereby producing tension), but it does not overcome the load and cause movement; instead, it produces a moment that supports the load in a fixed position (e.g., maintains posture). Such a contraction is termed isometric (*iso*, constant; *metric*, length), as no change takes place in the distance between the muscle's points of attachment (Rodgers and Cavanagh, 1984).

Although no motion is accomplished and no mechanical work is performed during an isometric contraction (Komi, 1986), muscle work (physiologic work) is performed: energy is expended and is mostly dissipated as heat. This type of muscle work is called static work. All dynamic contractions involve what may be considered an initial static (isometric) phase as the muscle first develops tension equal to the load it is expected to overcome.

The tension in a muscle varies with the type of contraction. Isometric contractions produce greater tension than do concentric contractions. Studies suggest that the tension developed in an eccentric contraction may even exceed that developed during an isometric contraction (Norkin, 1983). These differences are thought to be due in large part to the varying amounts of supplemental tension produced in the series elastic component of the muscle and to differences in contraction time (Komi, 1979). The longer contraction time of the isometric and eccentric contractions allows greater cross-bridge formation by the contractile components, thus permitting greater tension to be generated (Kroll, 1987). More time is also available for this tension to be transmitted to the series elastic component as the muscle-tendon unit is stretched (Cavanagh and Komi, 1979). Further, the longer contraction time allows the recruitment of additional motor units.

MAX CONT ONLY

Komi (1986) has pointed out that concentric, isometric, and eccentric muscle contractions seldom occur alone in normal human movement. Rather, one type of contraction or load is preceded by a different type. An example is the eccentric loading prior to the concentric contraction that occurs at the ankle from midstance to toe-off during gait.

Isokinetic (*iso*, constant; *kinetic*, motion) muscle contraction is a type of dynamic muscle work in which movement of the joint is kept at a constant velocity, and hence the velocity of shortening or lengthening of the muscle is constant (Rodgers and Cavanagh, 1984). Since velocity is held constant, muscle energy cannot be dissipated through acceleration of the body part and is entirely converted to a resisting moment. The muscle force varies with changes in its lever arm throughout the range of joint motion (Hislop and Perrine, 1967). The muscle contracts concentrically and eccentrically with different directions of joint motion. For example, the flexor muscles of a joint contract concentrically during flexion and eccentrically during extension, acting as decelerators during the latter.

Since muscles normally shorten or lengthen at varying velocities and with varying amounts of tension, performance and measurement of isokinetic work require the use of an isokinetic dynamometer. This device provides constant velocity of joint motion and maximum external resistance throughout the range of motion of the involved joint, thereby requiring maximal muscle torque. The use of the isokinetic dynamometer provides a method of selec-

TABLE 5–2

TYPES OF MUSCLE WORK AND CONTRACTION

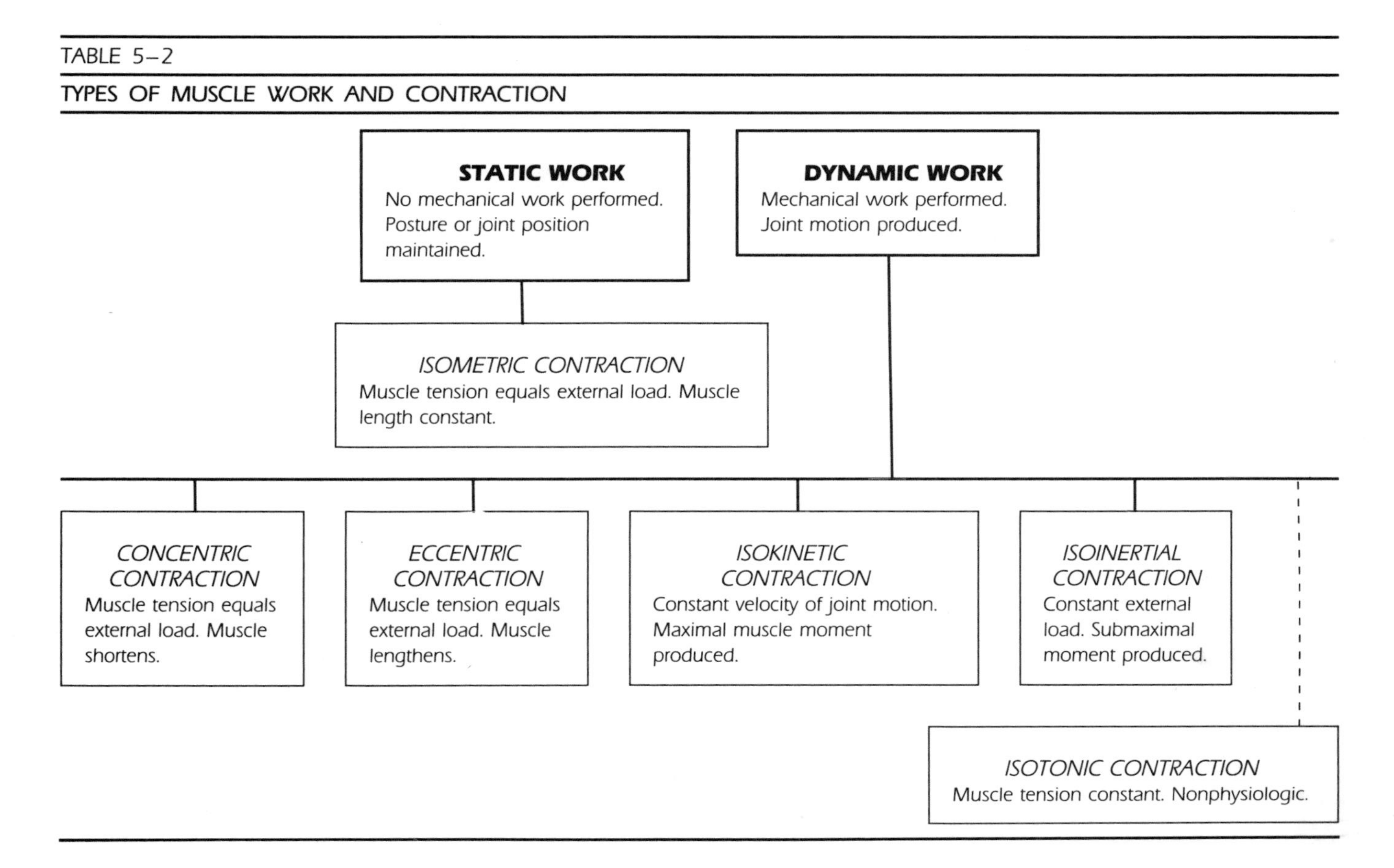

tive training and measurement, but physiologic movement is not simulated.

Isoinertial (*iso*, constant; *inertial*, resistance) contraction is a type of dynamic muscle work wherein the resistance against which the muscle must contract remains constant (Kroemer, 1983). If the moment (torque) produced by the muscle is equal to or less than the resistance to be overcome, the muscle length remains unchanged and the muscle contracts isometrically. If the moment is greater than the resistance, the muscle shortens (contracts concentrically) and causes acceleration of the body part.

Isoinertial contraction occurs, for example, when a constant external load is lifted. At the extremes of motion the inertia of the load must be overcome; the involved muscles contract isometrically and muscle torque is maximal. In the midrange of the motion, with the inertia overcome, the muscles contract concentrically and the torque is submaximal.

The term "isotonic" (*iso*, constant; *tonic*, force) is commonly used to define muscle contraction in which the tension is constant throughout a range of joint motion (Rodgers and Cavanagh, 1984). This term does not take into account the leverage effects at the joint, however. Because the muscle force moment arm changes throughout the range of joint motion, the muscle tension must also change. Thus, isotonic muscle contraction in the truest sense does not exist in the production of joint motion (Kroll, 1987).

FORCE PRODUCTION IN MUSCLE

The total force that a muscle can produce is influenced by its mechanical properties, which can be described by examining the length-tension, load-velocity, and force-time relationships of the muscle. Other principal factors in force production are muscle temperature, muscle fatigue, and prestretching.

LENGTH-TENSION RELATIONSHIP

The force, or tension, that a muscle exerts varies with the length at which it is held when stimulated. This relationship can be observed in a single fiber contracting isometrically and tetanically, as illustrated by the length-tension curve in Figure 5–8. Maximal tension is produced when the muscle fiber is

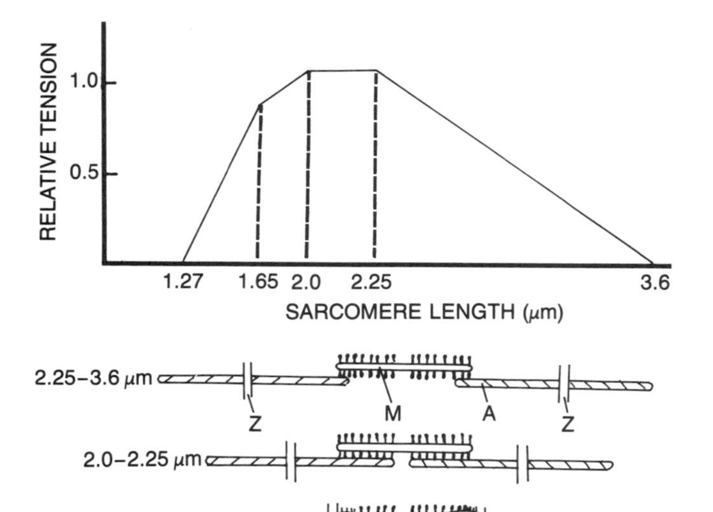

FIG. 5–8
Tension-length curve from part of an isolated muscle fiber stimulated at different lengths. The isometric tetanic tension is closely related to the number of cross-bridges on the myosin filament overlapped by the actin filament. The tension is maximal at the slack length, or resting length, of the sarcomere (2 μm), where overlap is greatest, and falls to zero at the length where overlap no longer occurs (3.6 μm). The tension also decreases when the sarcomere length is reduced below the resting length, falling sharply at 1.65 μm and reaching zero at 1.27 μm, as the extensive overlap interferes with cross-bridge formation. The structural relationship of the actin and myosin filaments at various stages of sarcomere shortening and lengthening is portrayed below the curve. A, actin filaments; M, myosin filaments; Z, Z lines. (Adapted from Crawford and James, 1980, as modified from Gordon et al., 1966b.)

approximately at its "slack," or resting, length. If the fiber is held at shorter lengths, the tension falls off slowly at first and then rapidly. If the fiber is lengthened beyond the resting length, tension progressively decreases.

The changes in tension when the fiber is stretched or shortened are due primarily to structural alterations in the sarcomere. Maximal isometric tension can be exerted when the sarcomeres are at their resting length (2.0 to 2.25 μm), because the actin and myosin filaments overlap along their entire length and the number of cross-bridges is maximal. If the sarcomeres are lengthened, there are fewer junctions between the filaments, and the active tension decreases. At a sarcomere length of about 3.6 μm, there is no overlap and hence no active tension. Sarcomere shortening to less than its resting length decreases the active tension because it allows overlapping of the thin filaments at opposite ends of the sarcomere, which are functionally polarized in opposite directions. At a sarcomere length of less than 1.65 μm, the thick filaments abut on the Z line and the tension diminishes sharply.

The length-tension relationship illustrated in Figure 5–8 is for an individual muscle fiber. If this relationship is measured in a whole muscle contracting isometrically and tetanically, the tension produced by both active components and passive components must be taken into account (Fig. 5–9).

The curve labeled *Active tension* in Figure 5–9 represents the tension developed by the contractile elements of the muscle, and it resembles the curve for the individual fiber. The curve labeled *Passive tension* reflects the tension developed when the muscle surpasses its resting length and the noncontractile muscle belly is stretched. This passive tension is mainly developed in the parallel and series elastic components (see Fig. 5–5). When the belly contracts, the combined active and passive tensions produce the total tension exerted. The curve demonstrates that as a muscle is progressively stretched beyond its resting length the passive tension rises and the active tension decreases.

Most muscles that cross only one joint normally are not stretched enough for the passive tension to play an important role, but the case is different for

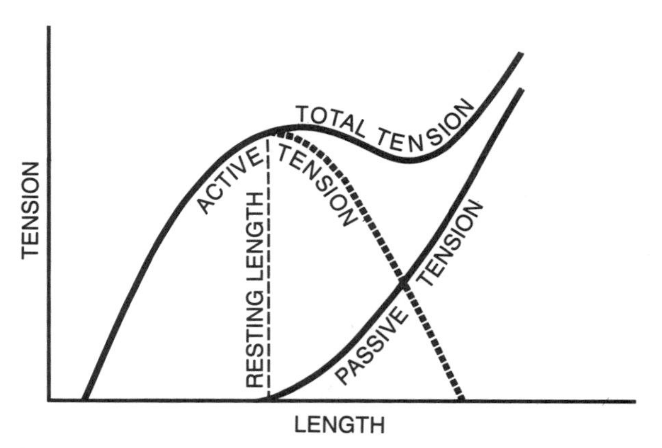

FIG. 5–9
The active and passive tension exerted by a whole muscle contracting isometrically and tetanically is plotted against the muscle's length. The active tension is produced by the contractile muscle components and the passive tension by the series and parallel elastic components, which develop stress when the muscle is stretched beyond its resting length. The greater the amount of stretching, the larger is the contribution of the elastic component to the total tension. The shape of the active curve is generally the same in different muscles, but the passive curve, and hence the total curve, varies depending on how much connective tissue (elastic component) the muscle contains. (Adapted from Crawford and James, 1980.)

two-joint muscles, in which the extremes of the length-tension relationship may be functioning (Crawford and James, 1980). For example, the hamstrings shorten so much when the knee is fully flexed that the tension that they can exert decreases considerably. Conversely, when the hip is flexed and the knee extended, the muscles are so stretched that it is the magnitude of their passive tension that prevents further elongation and causes the knee to flex if hip flexion is increased.

LOAD-VELOCITY RELATIONSHIP

The relationship between the velocity of shortening or eccentric lengthening of a muscle and different constant loads can be determined by plotting the velocity of motion of the muscle lever arm at various external loads, thereby generating a load-velocity curve (Fig. 5–10). The velocity of shortening of a muscle contracting concentrically is inversely related to the external load applied (Brobeck, 1979; Gordon et al., 1966a, 1966b; Guyton, 1986; Phillips and Petrofsky, 1983; Ottoson, 1983). The velocity of shortening is greatest when the external load is zero, but as the load increases the muscle shortens more and more slowly. When the external load equals the maximal force that the muscle can exert, the velocity of shortening becomes zero and the muscle contracts isometrically. When the load is increased still further, the muscle contracts eccentrically: it elongates during contraction. The load-velocity relationship is reversed from that of the concentrically contracting muscle; the muscle eccentrically lengthens more quickly with increasing load (Kroll, 1987).

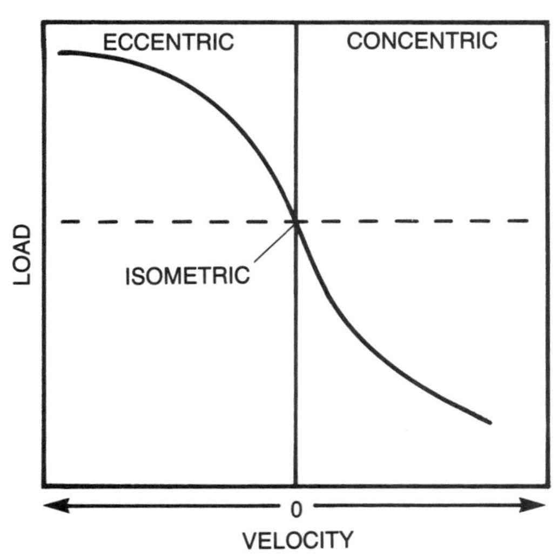

FIG. 5–10
Load-velocity curve generated by plotting the velocity of motion of the muscle lever arm against the external load. When the external load imposed on the muscle is negligible, the muscle contracts concentrically with maximal speed. With increasing loads the muscle shortens more slowly. When the external load equals the maximum force that the muscle can exert, the muscle fails to shorten (i.e., has zero velocity) and contracts isometrically. When the load is increased further, the muscle lengthens eccentrically; this lengthening is more rapid with greater load. (Adapted from Kroll, 1987.)

FORCE-TIME RELATIONSHIP

The force, or tension, generated by a muscle is proportional to the contraction time: the longer the contraction time, the greater is the force developed, up to the point of maximum tension. In Figure 5–11, this relationship is illustrated by a force-time curve for a whole muscle contracting isometrically. Slower contraction leads to greater force production because time is allowed for the tension produced by the contractile elements to be transmitted through the parallel elastic components to the tendon. Although tension production in the contractile component can

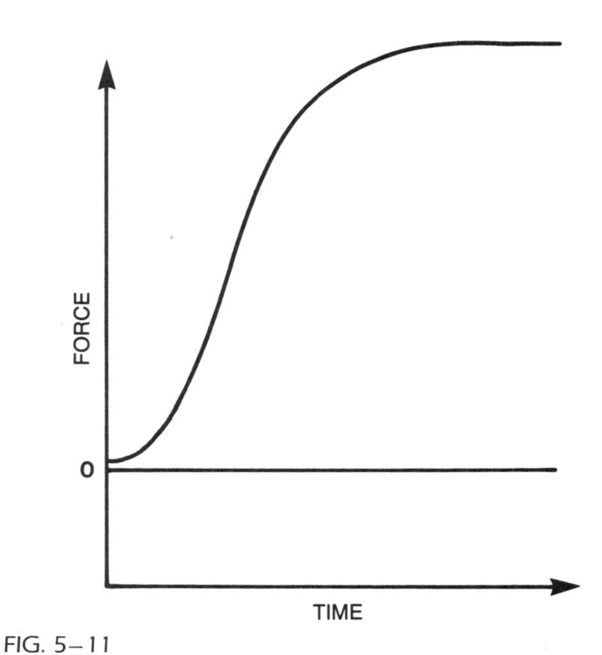

FIG. 5–11
Force-time curve for a whole muscle contracting isometrically. The force exerted by the muscle is greater when the contraction time is longer, since time is required for the tension created by the contractile components to be transferred to the parallel elastic component and then to the series elastic component as the musculotendinous unit is stretched. (Adapted from Kroll, 1987.)

reach a maximum in as little as 10 msec (Cavanagh and Komi, 1973), up to 300 msec may be needed for that tension to be transferred to the elastic components (Sukop and Nelson, 1974). The tension in the tendon will reach the maximum tension developed by the contractile element only if the active contraction process is of sufficient duration (Sukop and Nelson, 1974; Ottoson, 1983).

EFFECT OF PRESTRETCHING

It has been demonstrated in amphibians and in humans (Cavagna et al., 1968; Ciullo and Zarins, 1983) that a muscle performs more work when it shortens immediately after being stretched in the concentrically contracted state than when it shortens from a state of isometric contraction. This phenomenon is not entirely accounted for by the elastic energy stored in the series elastic component during stretching but must also be due to energy stored in the contractile component.

EFFECT OF TEMPERATURE

A rise in muscle temperature causes an increase in conduction velocity across the sarcolemma (Phillips and Petrofsky, 1983), increasing the frequency of stimulation and hence the production of muscle force. A rise in temperature also causes greater enzymatic activity of muscle metabolism, thus increasing the efficiency of muscle contraction. A further effect of a rise in temperature is increased elasticity of the collagen in the series and parallel elastic components, which enhances the extensibility of the muscle-tendon unit. This increased prestretch increases the force production of the muscle.

Muscle temperature increases by means of two mechanisms: an increase in blood flow, which occurs when an athelete "warms up" his or her muscles; and production of the heat of reaction generated by metabolism, by the release of the energy of contraction, and by friction as the contractile components slide over each other.

EFFECT OF FATIGUE

The ability of a muscle to contract and relax is dependent on the availability of adenosine triphosphate (ATP) (see Table 5–1). If a muscle has an adequate supply of oxygen and nutrients that can be broken down to provide ATP, it can sustain a series of low-frequency twitch responses for a long time. The frequency must be low enough to allow the muscle to synthesize ATP at a rate sufficient to keep up with the rate of ATP breakdown during contraction. If the frequency of stimulation increases and outstrips the rate of replacement of ATP, the twitch responses soon grow progressively weaker and eventually fall to zero (Fig. 5–12). This drop in tension following prolonged stimulation is muscle fatigue. If the frequency is high enough to produce tetanic contractions, fatigue occurs even sooner. If a period of rest is allowed before stimulation is continued, the ATP concentration rises and the muscle briefly recovers its contractile ability before again undergoing fatigue.

Three sources supply ATP in muscle: creatine phosphate, oxidative phosphorylation in the mitochondria, and substrate phosphorylation during anaerobic glycolysis. When contraction begins, the myosin ATPase breaks down ATP very rapidly. The increase in ADP and P_i concentrations resulting from this breakdown ultimately leads to increased rates of oxidative phosphorylation and glycolysis. After a short lapse these metabolic pathways begin to deliver ATP at a high rate, however. During this interval the energy for ATP formation is provided by creatine phosphate, which offers the most rapid means of forming ATP in the muscle cell.

At moderate rates of muscle activity, most of the required ATP can be formed by the process of oxidative phosphorylation. During very intense exercise, when ATP is being broken down very rapidly, the cell's ability to replace ATP by oxidative phosphorylation may be limited, primarily by inadequate delivery of oxygen to the muscle by the circulatory system.

Even when oxygen delivery is adequate, the rate at which oxidative phosphorylation can produce ATP may be insufficient to sustain very intense exercise, since the enzymatic machinery of this pathway is relatively slow. Anaerobic glycolysis then begins to

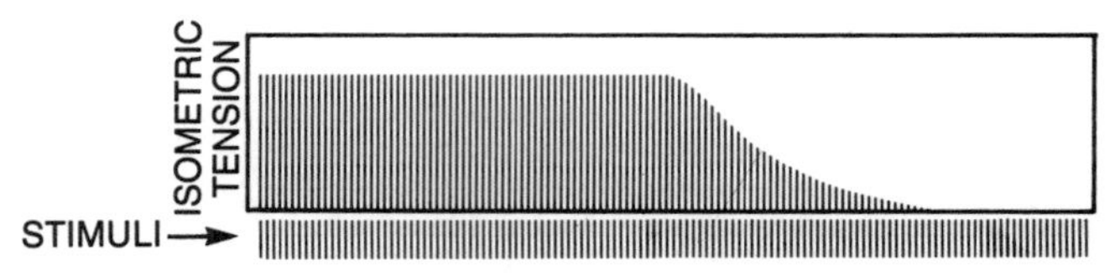

FIG. 5–12
Fatigue in a muscle contracting isometrically. Prolonged stimulation occurs at a frequency that outstrips the muscle's ability to produce sufficient ATP for contraction. As a result, tension production declines and eventually ceases. (Adapted from Luciano et al., 1978.)

contribute an increasing portion of the ATP. The glycolytic pathway, although it produces much smaller amounts of ATP from the breakdown of glucose, operates at a much faster rate. It can also proceed in the absence of oxygen, with the formation of lactic acid as its end product. Thus, during intense exercise anaerobic glycolysis becomes an additional source for rapidly supplying the muscle with ATP.

The glycolytic pathway has the disadvantage of requiring large amounts of glucose for the production of small amounts of ATP. Thus, even though muscle stores glucose in the form of glycogen, existing glycogen supplies may be depleted quickly when muscle activity is intense. Finally, myosin ATPase may break down ATP faster than even glycolysis can replace it, and fatigue occurs rapidly as ATP concentrations drop.

After a period of intense exercise, creatine phosphate levels have become low and much of the muscle glycogen may have been converted to lactic acid. For the muscle to be returned to its original state, creatine phosphate must be resynthesized and the glycogen stores must be replaced. Since both processes require energy, the muscle will continue to consume oxygen at a rapid rate even though it has stopped contracting. This sustained high oxygen uptake is demonstrated by the fact that a person continues to breath heavily and rapidly after a period of strenuous exercise.

When the energy necessary to return glycogen and creatine phosphate to their original levels is taken into account, the efficiency with which muscle converts chemical energy to work (movement) is usually no more than 20 to 25%, the majority of the energy being dissipated as heat. Even when muscle is operating in its most efficient state, a maximum of only about 45% of the energy is used for contraction (Arvidson et al., 1984; Brobeck, 1979; Guyton, 1986).

MUSCLE FIBER DIFFERENTIATION

In the preceding section, the major factors that determine the total tension developed by the whole muscle when it contracts were described. In addition, individual muscle fibers display distinct differences in their rates of contraction, development of tension, and susceptibility to fatigue.

Many methods of classifying muscle fibers have been devised. As early as 1678 Lorenzini observed anatomically the gross difference between red and white muscle, and in 1873 Ranvier typed muscle on the basis of speed of contractility and fatigability. Although considerable confusion has existed concerning the method and terminology for classifying skeletal muscle (Close, 1972; Engel, 1974), recent histologic and histochemical observations have led to the identification of three distinct types of muscle fibers on the basis of differing contractile and metabolic properties (Engel, 1962; Brandstater and Lambert, 1969; Buchtal and Sohmalburch, 1980; Burke et al., 1971; Dubowitz, 1973) (Table 5–3).

The fiber types are distinguished mainly by the metabolic pathways by which they can generate ATP and the rate at which its energy is made available to the contractile system of the sarcomere, which determines the speed of contraction. The three fiber types are termed type I, slow-twitch oxidative (SO) fibers; type IIA, fast-twitch oxidative-gycolytic (FOG) fibers; and type IIB, fast-twitch glycolytic (FG) fibers.

Type I (SO) fibers are characterized by a low activity of myosin ATPase in the muscle fiber and, therefore, a relatively slow contraction time. The glycolytic (anaerobic) activity is low in this fiber type, but a high content of mitochondria produces a high potential for oxidative (aerobic) activity (Essen et al., 1977; Gollnick, 1973; Jansson and Kaijser, 1977). Type I fibers are very difficult to fatigue, as the high rate of blood flow to these fibers delivers oxygen and nutrients at a sufficient rate to keep up with the relatively slow rate of ATP breakdown by myosin ATPase. Thus, the fibers are well-suited for prolonged, low-intensity work (Gollnick and Hermansen, 1973; Gollnick and Sembrowich, 1977). These fibers are relatively small in diameter and thus produce relatively little tension. The high myoglobin content of type I fibers gives the muscle a red color.

Type II muscle fibers have been divided into two main subgroups, IIA and IIB, on the basis of differing susceptibility to treatment with different buffers prior to incubation (Brooke and Kaiser, 1970). A third subgroup, the type IIC fibers, are rare, undifferentiated fibers that may be transitory (Jansson et al., 1978). Type IIA and IIB fibers are characterized by a high activity of myosin ATPase, which results in relatively fast contraction.

Type IIA (FOG) fibers are considered intermediate between type I and type IIB, because their fast contraction time is combined with a moderately well developed capacity for both aerobic (oxidative) and anaerobic (glycolytic) activity. These fibers also have a well-developed blood supply. They can maintain their contractile activity for relatively long periods; however, at high rates of activity, the high rate of ATP splitting exceeds the capacity of both oxidative

TABLE 5–3

PROPERTIES OF THREE TYPES OF SKELETAL MUSCLE FIBERS

	TYPE I SLOW-TWITCH OXIDATIVE (SO)	*TYPE IIA* FAST-TWITCH OXIDATIVE-GLYCOLYTIC (FOG)	*TYPE IIB* FAST-TWITCH GLYCOLYTIC (FG)
Speed of contraction	Slow	Fast	Fast
Myosin-ATPase activity	Low	High	High
Primary source of ATP production	Oxidative phosphorylation	Oxidative phosphorylation	Anaerobic glycolysis
Glycolytic enzyme activity	Low	Intermediate	High
Number of mitochondria	Many	Many	Few
Capillaries	Many	Many	Few
Myoglobin content	High	High	Low
Muscle color	Red	Red	White
Glycogen content	Low	Intermediate	High
Fiber diameter	Small	Intermediate	Large
Rate of fatigue	Slow	Intermediate	Fast

(Modified from Luciano et al., 1978.)

phosphorylation and glycolysis to supply ATP and these fibers eventually fatigue. Because the myoglobin content of this muscle type is quite high, the muscle is often categorized as red muscle.

Type IIB (FG) fibers rely primarily on glycolytic (anaerobic) activity for ATP production. Very few capillaries are found in the vicinity of these fibers, and because they contain little myoglobin they are often referred to as white muscle. Although type IIB fibers are able to produce ATP rapidly, they fatigue very easily, as their high rate of ATP splitting quickly depletes the glycogen needed for glycolysis. These fibers generally are of large diameter and are thus able to produce great tension but for only short periods before they fatigue.

It has been well demonstrated that the nerve innervating the muscle fiber determines its type (Burke et al., 1971; Buller et al., 1960; Dubowitz, 1967; Estrom and Kugelberg, 1969); thus, the muscle fibers of each motor unit are of a single type. In humans and other species, electrical stimulation was found to change the fiber type (Munsat et al., 1976). In animal studies, transecting the nerves that innervate slow-twitch and fast-twitch muscle fibers and then crossing these nerves was noted to reverse the fiber types (Buller et al., 1960; Dubowitz, 1967). After recovery from the cross-innervation, the slow-twitch fibers became fast in their contractile and histochemical properties and the fast-twitch fibers became slow.

The fiber composition of a given muscle depends on the function of that muscle. Some muscles perform predominantly one form of contractile activity and are often composed mostly of one muscle fiber type. An example is the soleus muscle in the calf, which primarily maintains posture and is composed of a high percentage of type I fibers. More commonly, however, a muscle is required to perform endurance-type activity under some circumstances and high-intensity strength activity under others. These muscles generally contain a mixture of the three muscle fiber types.

In a typical mixed muscle exerting low tension, some of the small motor units, composed of type I fibers, contract. As the muscle force increases, more motor units are recruited and their frequency of stimulation increases. As the frequency becomes maximal, greater muscle force is achieved by recruitment of larger motor units composed of type IIA (FOG) fibers and eventually type IIB (FG) fibers. As the peak muscle force decreases, the larger units are the first to cease activity (Crawford and James, 1980; Guyton, 1986; Luciano et al., 1978).

It is generally, but not universally, accepted that fiber types are genetically determined (Jansson et al.,

1978; Costill et al., 1976; Gollnick et al., 1973; Gollnick, 1982). In the average population about 50 to 55% of muscle fibers are type I, about 30 to 35% are type IIA, and about 15% are type IIB, but these percentages vary greatly among individuals.

In elite athletes the relative percentage of fiber types differs from that in the general population and appears to depend on whether the athlete's principal activity requires a short, explosive, maximal effort or involves submaximal endurance. Sprinters and shot putters, for example, have a high percentage of type II fibers, whereas distance runners and cross-country skiers have a higher percentage of type I fibers. Endurance athletes may have as many as 80% type I fibers, and those engaged in short, explosive efforts as few as 30% of these fibers (Essen, 1977; Saltin et al., 1977).

The genetically determined fiber typing may be responsible for the natural selective process by which athletes are drawn to the type of sport for which they are most suited. Since fiber types are determined by the nerve that innervates the muscle fiber, there may be some cortical control of this innervation that influences an athlete to choose the sport in which he or she is genetically able to excel.

MUSCLE REMODELING

The remodeling of muscle tissue is similar to that of other skeletal tissues such as bone, articular cartilage, and ligaments (Costill et al., 1977; Eriksson, 1976; Gollnick et al., 1974; Häggmark and Eriksson, 1979a; Häggmark and Eriksson, 1979b; Lippman and Selig, 1928; Patel et al., 1969; Sargeant et al., 1977). As in these other tissues, muscle atrophies in response to disuse and immobilization and hypertrophies when subjected to greater use than usual.

EFFECTS OF DISUSE AND IMMOBILIZATION

Clinical and laboratory studies of human and animal muscle tissue suggest that a program of immediate or early motion may prevent muscle atrophy after injury or surgery. In a study of crush injuries to rat muscle, the effect of immobilization of the crushed limb was compared with that of immediate motion (Jarvinen, 1976). The muscle fibers were found to regenerate in a more parallel orientation in the mobilized animal than in the immobilized animal, capillarization occurred more rapidly, and tensile strength returned more quickly.

It has been found clinically that atrophy of the quadriceps muscle that develops while the limb is immobilized in a rigid plaster cast cannot be reversed through the use of isometric exercises. Atrophy may be limited by allowing early motion such as that permitted by a partly mobile cast brace (Häggmark and Eriksson, 1979a). In this case dynamic exercises can be performed.

Human muscle biopsy studies have shown that it is mainly the type I fibers that atrophy with immobilization; their cross-sectional area decreases, and their potential for oxidative enzyme activity is reduced (Davies and Sargeant, 1975; Eriksson, 1976; Eriksson, 1981; Häggmark, 1978; Häggmark et al., 1981; Sargeant et al., 1977). Early motion may prevent this atrophy. It appears that if the muscle is placed under tension when the body segment moves, afferent (sensory) impulses from the intrafusal muscle spindles will increase, leading to increased stimulation of the type I fiber (Häggmark and Eriksson, 1979a, 1979b; Vrbova, 1963). While intermittent isometric exercise may be sufficient to maintain the metabolic capacity of the type II fiber, the type I fiber (the postural fiber) requires a more continuous impulse. There is also evidence that electric stimulation may prevent the decrease in type I fiber size and the decline in its oxidative enzyme activity caused by immobilization (Eriksson and Häggmark, 1981; Eriksson et al., 1981; Hymes et al., 1974).

In elite athletes, inactivity following injury, surgery, or immobilization rapidly decreases the size and aerobic capability of muscle fibers, particularly in the fiber type affected by the chosen sport (Häggmark, 1978; Häggmark et al., 1981; Henriksson and Reitman, 1977; Saltin et al., 1977). In endurance athletes, type I fibers are affected, while in athletes engaged in an explosive activity such as sprinting type II fibers are affected (Arvidson et al., 1984; Henriksson and Reitman, 1976; Saltin et al., 1977).

EFFECTS OF PHYSICAL TRAINING

Physical training has been found to increase the cross-sectional area of all muscle fibers, accounting for the increase in muscle bulk and strength (Arvidson et al., 1984; Eriksson and Häggmark, 1979; Häggmark, 1978; Häggmark et al., 1981; Thorstensson et al., 1977). Some evidence suggests that the relative percentage of fiber types composing a person's muscles may also change with physical training (Arvidson et al., 1984; Eriksson, 1980; Jansson et al., 1978; Saltin et al., 1977). The cross-sectional area of the fibers affected by the athlete's principal activity increases. For example, in endurance athletes, the area of muscle taken up by type I and type IIA fibers increases at the expense of the total area of type IIB fibers.

A common aspect of physical training is the performance of specific exercises to stretch out the muscle-tendon complex. The efficacy of such a muscle stretching regimen in preventing injury and increasing efficiency of performance has been well demonstrated in athletes (Ekstrand, 1982; Holt, 1973; Jacobs, 1976). Stretching increases muscle flexibility, maintains and augments the range of joint motion, and increases the elasticity and length of the musculotendinous unit (Brobeck, 1979; Ciullo and Zarins, 1983; Frankel et al., 1971; Frost, 1973; Hill, 1938, 1950, 1968; Huxley, 1974). It also permits the musculotendinous unit to store more energy in its viscoelastic and contractile components.

The events that take place during muscle stretching are complex and incompletely understood (Guyton, 1986; Brobeck, 1979; Costill et al., 1977; Gollnick et al., 1973; Gollnick et al., 1974; Gollnick, 1982; Kabat, 1952; Kabat, 1958; Stanish, 1982). It appears that these events are controlled or modified by both the intrafusal muscle spindles, located in parallel with the extrafusal fibers of the muscle belly, and the Golgi tendon organs, located in series with these fibers. The spindles respond to an increase in muscle length and the Golgi apparatus to an increase in muscle tension. The resulting spindle reflex increases muscle contraction, while the Golgi reflex inhibits contraction and enhances muscle relaxation.

The intrafusal muscle spindles are of two types: primary and secondary. The primary spindles respond to changes in the rate of muscle lengthening (dynamic response) and the actual amount of lengthening. The secondary spindles respond only to the actual length change (static response). The static response is weak and the dynamic response is strong; therefore, keeping the rate of stretch low may allow the dynamic response to be bypassed, essentially negating the effect of the spindles. Conversely, the increase in muscle tension during stretching may activate the relaxing effect of the Golgi apparatus and thus enhance further stretching. The various methods and theories of stretching all have as a common goal inhibition of the spindle effect and enhancement of the Golgi effect to relax the muscle and promote further lengthening.

SUMMARY

1. The structural unit of skeletal muscle is the fiber, which is encompassed by the endomysium and organized into fascicles encased in the perimysium. The epimysium surrounds the entire muscle.
2. The fibers are composed of myofibrils, aligned so as to create a band pattern. Each repeat of this pattern is a sarcomere, the functional unit of the contractile system.
3. The myofibrils are composed of thin filaments of the protein actin and thick filaments of the protein myosin.
4. According to the sliding filament theory, active shortening of the muscle results from the relative movement of the actin and myosin filaments past one another. The force of contraction is developed by movement of the myosin heads, or cross-bridges, in contact with the actin filaments. Troponin and tropomyosin, two proteins in the actin helix, regulate the making and breaking of the contacts between filaments.
5. A key to the sliding mechanism is the calcium ion, which turns the contractile activity on and off.
6. The motor unit, a single motor neuron and all muscle fibers innervated by it, is the smallest part of the muscle that can contract independently. The calling in of additional motor units in response to greater stimulation of the motor nerve is known as recruitment.
7. The tendons and the endomysium, perimysium, and epimysium represent parallel and series elastic components that stretch with active contraction or passive muscle extension and recoil with muscle relaxation.
8. Summation occurs when mechanical responses of the muscle to successive stimuli are added to an initial response. When maximal tension is sustained as a result of summation, the muscle contracts tetanically.
9. Muscles may contract concentrically, isometrically, or eccentrically depending on the relationship between the muscle tension and the resistance to be overcome. Concentric and eccentric contractions involve dynamic work, in which the muscle moves a joint or controls its movement. Isometric contractions involve static work, in which the joint position is maintained.

10. Force production in muscle is influenced by the length-tension, load-velocity, and force-time relationships of the muscle. The length-tension relationship in a whole muscle is influenced by both active (contractile) and passive (series and parallel elastic) components.
11. Two other factors that increase force production are prestretching of the muscle and a rise in muscle temperature.
12. The energy for muscle contraction and its release is provided by the hydrolytic splitting of ATP. Muscle fatigue occurs when the ability of the muscle to synthesize ATP is insufficient to keep up with the rate of ATP breakdown during contraction.
13. Three main fiber types have been identified: type I, slow-twitch oxidative; type IIA, fast-twitch oxidative-glycolytic; and type IIB, fast-twitch glycolytic fibers. Most muscles contain a mixture of these types.
14. Muscle atrophies in response to disuse and immobilization and hypertrophies when subjected to greater use than usual.

REFERENCES

Arvidson, I., Eriksson, E., and Pitman, M.: Neuromuscular basis of rehabilitation. *In* Rehabilitation of the Injured Knee. Edited by E. Hunter and J. Funk. St. Louis, C. V. Mosby, 1984, pp. 210–234.

Åstrand, P.-O., and Rodahl, K.: Textbook of Work Physiology: Physiological Bases of Exercise. 2nd Ed. New York, McGraw Hill, 1977, pp. 37–52.

Brandstater, M. E., and Lambert, E. H.: A histologic study of the spatial arrangements of muscle fibers in single motor units within rat tibialis anterior muscle. Bull. Am. Assoc. Electromyog. Electro. Diag., *82*:15, 1969.

Brobeck, J. R. (ed.): Best & Taylor's Physiological Basis of Medical Practice. 10th Ed. Baltimore, Williams & Wilkins, 1979, pp. 59–113.

Brooke, M. H., and Kaiser, K. K.: Three myosin adenosine triphosphatase systems: The nature of their pH liability and sulfhydryl dependence. J. Histochem. Cytochem., *18*:670, 1970.

Buchtahl, F., and Sohmalburch, H.: Motor units of mammalian muscle. Physiol. Rev., *60*:90, 1980.

Buchtahl, P., Kaiser, E., and Rosenfalch, P.: The rheology of the cross striated muscle fibre. Biol. Med. Dansk. vid Selsk., *21*:318, 1951.

Buller, A. J., Eccles, J. C., and Eccles, R. M.: Differentiation of fast and slow muscles in the cat hind limb. J. Physiol., *150*:399, 1960.

Burke, R. E., et al.: Mammalian motor units: Physiological histochemical correlation in three types of motor units in cat gastrocnemius. Science, *174*:709, 1971.

Buxton, P. H.: Skeletal muscle structure and development. *In* Scientific Foundations of Orthopaedics and Traumatology. Edited by R. Owen, J. Goodfellow, and P. Bullough. London, William Heinemann, 1980, pp. 22–30.

Cavagna, G. A., Dusman, B., and Margaria, R.: Positive work done by a previously stretched muscle. J. Appl. Physiol., *24*:21, 1968.

Cavanagh, P. R., and Komi, P. V.: Electromechanical delay in human muscle under concentric and eccentric contractions. Eur. J. Appl. Physiol., *42*:159, 1973.

Ciullo, J. V., and Zarins, B.: Biomechanics of the musculotendinous unit: Relation to athletic performance and injury. Clin. Sports Med., *2*:71, 1983.

Close, R. I.: Dynamic properties of mammalian skeletal muscles. Physiol. Rev., *52*:97, 1972.

Costill, P. L., Fink, W. J., and Habansky, A. J.: Muscle rehabilitation after knee surgery. Physician Sports Med., *5*:71, 1977.

Costill, P. L., et al.: Adaptations in skeletal muscles following strength training. J. Appl. Physiol., *46*:96, 1976.

Crawford, G. N. C., and James, N. T.: The design of muscles. *In* Scientific Foundations of Orthopaedics and Traumatology. Edited by R. Owen, J. Goodfellow, and P. Bullough. London, William Heinemann, 1980, pp. 67–74.

Davies, C. T. M., and Sargeant, A. J.: Effects of exercise therapy on total and component tissue leg volumes of patients undergoing rehabilitation from lower limb injury. Ann. Hum. Biol., *2*:327, 1975.

Dubowitz, V.: Cross innervated mammalian skeletal muscle. J. Physiol., *193*:481, 1967.

Dubowitz, V., and Brooke, M. H.: Muscle Biopsy: A Modern Approach. Philadelphia, W. B. Saunders, 1973.

Ebashi, S.: Excitation-contraction coupling. Ann. Rev. Physiol., *38*:293, 1976.

Edstrom, L.: Selective atrophy of red muscle fibers in the quadriceps in long standing knee joint dysfunction: Injuries to the anterior cruciate ligament. J. Neurol. Sci., *11*:55, 1970.

Edstrom, L., and Kugelberg, E.: Histochemical mapping of motor units in experimentally reinnervated skeletal muscles. Experientia, *25*:1044, 1969.

Ekstrand, J.: Soccer injuries and their prevention. Linkoping University Medical Dissertations, No. 130, Linkoping, Sweden, 1982.

Engel, W. K.: The essentiality of histo- and cystochemical studies of skeletal muscle in investigation of neuromuscular disease. Neurology, *12*:778, 1962.

Engel, W. K.: Fiber-type nomenclature of human skeletal muscle for histochemical purposes. Neurology, *25*:344, 1974.

Eriksson, E.: Sport injuries of the knee ligaments: Their diagnosis, treatment, rehabilitation and prevention. Med. Sci. Sports, *8*:133, 1976.

Eriksson, E.: Muscle physiology: Adaptation of the musculoskeletal system to exercise. Contemp. Orthop., *2*:228, 1980.

Eriksson, E.: Rehabilitation of muscle function after sport injury. Major problems in sports medicine. Int. J. Sports Med., *2*:1, 1981.

Eriksson, E., and Häggmark, T.: Comparison of isometric muscle training and electrical stimulation supplementing isometric muscle training in the recovery after major knee ligament surgery. Am. J. Sports Med., *7*:169, 1979.

Eriksson, E., Häggmark, T., Kiessling, K. H., and Karlsson, J.: Effect of electrical stimulation on human skeletal muscle. Int. J. Sports Med., *2*:18, 1981.

Essen, B.: Intramuscular substrate utilization during prolonged exercise. Ann. N. Y. Acad. Sci., *301*:30, 1977.

Frankel, V. H., Burstein, A. H., and Brooks, D. B.: Biomechanics of internal derangement of the knee in pathomechanics as determined by analysis of the instant centers of motion. J. Bone Joint Surg., *53A*:945, 1971.

Frost, H. M.: Orthopaedic Biomechanics. Springfield, Charles C Thomas, 1973.

Fuchs, F.: Striated muscle. Ann. Rev. Physiol., *36*:461, 1974.

Gollnick, P. D.: Relationship of strength and endurance with skeletal muscle structure and metabolic potential. Int. J. Sports Med. [Suppl.], *3*:26, 1982.

Gollnick, P. D., Eriksson, E., Häggmark, T., and Saltin, B.: Recovery with a movable or standard cast following intra-articular reconstruction of the anterior cruciate ligament. AMA Med. Aspects Sport, *15*:56, 1974.

Gollnick, P. D., et al.: Effect of training on enzyme activity and fiber composition of human skeletal muscle. J. Appl. Physiol., *34*:107, 1973.

Gordon, A. M., Huxley, A. F., and Julian, F. J.: Tension development in highly stretched vertebrate muscle fibers. J. Physiol., *184*:143, 1966a.

Gordon, A. M., Huxley, A. F., and Julian, F. J.: The variation in isometric tension with sarcomere length in vertebrate muscle fibers. J. Physiol., *184*:170, 1966b.

Gowitzke, B. A., and Milner, M.: Scientific Bases of Human Movement. 3rd Ed. Baltimore, Williams & Wilkins, 1987.

Guyton, A. C.: Textbook of Medical Physiology. 7th Ed. Philadelphia, W. B. Saunders, 1986.

Häggmark, T.: A study of morphological and enzymatic properties of the skeletal muscles after injuries and immobilization in man. Thesis, Karolinska Institute, University of Stockholm, 1978.

Häggmark, T., and Eriksson, E.: Cylinder or mobile cast brace after knee ligament surgery: A clinical analysis and morphologic and enzymatic study of changes in the quadriceps muscle. Am. J. Sports Med., *7*:48, 1979a.

Häggmark, T., and Eriksson, E.: Hypotrophy of the soleus muscle in man after achilles tendon rupture. Am. J. Sports Med., *7*:121, 1979b.

Häggmark, T., Jansson, E., and Eriksson, E.: Fibre type area and metabolic potential of the thigh muscle in man after knee surgery and immobilization. Int. J. Sports Med., *2*:12, 1981.

Ham, A. W., and Cormack, D. H.: Histology, 8th Ed. Philadelphia, J. B. Lippincott, 1979.

Henriksson, J.: Human skeletal muscle adaptation to physical activity. Thesis, University of Stockholm, 1976.

Henriksson, J., and Reitman, J. S.: Quantitative measures of enzyme activities in type I and type II muscle fibers of man after training. Acta Physiol. Scand., *97*:392, 1976.

Henriksson, J., and Reitman, J. S.: Time course of activity changes in human muscle succinate dehydrogenase and cytochrome oxidase activities and maximal oxygen uptake with physical activity and inactivity. Acta Physiol. Scand., *99*:91, 1977.

Hill, A. V.: The heat of shortening and the dynamic constants of muscle. Proc. R. Soc. Biol., *126*:136, 1938.

Hill, A. V.: The series elastic component of muscle. Proc. R. Soc. Biol., *137*:273, 1950.

Hill, A. V.: First and Last Experiments in Muscle Mechanics. Cambridge, Cambridge University Press, 1970.

Hill, D. K.: Tension due to interaction between the sliding filaments of resting striated muscle. The effect of stimulation. J. Physiol. [Lond.], *199*:637, 1968.

Hill, D. K.: The effect of temperature in the range 0–35°C on the resting tension of frog's muscle. J. Physiol., *208*:725, 1970.

Hislop, H. J., and Perrine, J. J.: The isokinetic concept of exercise. Phys. Ther., *47*:114, 1967.

Holt, L. E.: Scientific Stretches for Sport. Dalhousie, Dalhousie University Sport Research Ltd., 1973.

Huxley, A. F.: Muscular contraction. J. Physiol., *243*:1, 1974.

Huxley, A. F., and Huxley, H. E.: Organizers of a discussion of the physical and chemical basis of muscular contraction. Proc. R. Soc., *B160*:433, 1964.

Huxley, A. F., and Simmons, R. M.: Mechanical properties of the crossbridges of frog striated muscle. J. Physiol., *218*:59, 1971.

Huxley, H. E.: The mechanism of muscular contraction. Sci. Am., *213*:18, 1965.

Huxley, H. E.: The mechanism of muscular contraction. Science, *164*:1356, 1969.

Huxley, H. E.: Molecular basis of contraction in cross-striated muscles. *In* The Structure and Function of Muscle. Vol. 1, Part 1. 2nd Ed. Edited by G. H. Bourne. New York, Academic Press, 1972a, pp. 301–387.

Huxley, H. E.: Structural changes in the actin- and myosin-containing filaments during contraction. Cold Spring Harbor Symp. Quant. Biol., *37*:361, 1972b.

Hymes, A. C., Raab, D. E., and Yonehird, E. G.: Acute pain control by electricostimulation: A preliminary report. Adv. Neurol., *4*:761, 1974.

Jacobs, M.: Neurophysiological implications of slow, active stretching. Am. Corr. Ther. J., *30*:151, 1976.

Jansson, E., and Kaijser, L.: Muscle adaptation to extreme endurance training in man. Acta Physiol. Scand., *100*:315, 1977.

Jansson, E., Sjodin, B., and Tesch, P.: Changes in muscle fibre type distribution in man after physical training. Acta Physiol. Scand., *104*:235, 1978.

Jarvinen, M.: Healing of a crush injury in rat striated muscle—with special reference to treatment by early mobilization or immobilization. Thesis, University of Turku, Finland, 1976.

Kabat, H.: Studies of neuromuscular dysfunction. The role of central facilitation in restoration of motor function in paralysis. Arch. Phys. Med., *33*:523, 1952.

Kabat, H.: Proprioceptive facilitation in therapeutic exercise. *In* Therapeutic Exercise. Edited by S. Lecht. New Haven, Elizabeth Lecht, 1958.

Keele, C. A., Neil, E., and Joels, N.: Muscle and the nervous system. *In* Samson Wright's Applied Physiology. 13th Ed. Oxford, Oxford University Press, 1982, pp. 248–259.

Komi, P. V.: The stretch-shortening cycle and human power output. *In* Human Muscle Power. Edited by N. L. Jones, N. McCartney, and A. J. McConas. Champaign, IL, Human Kinetics Publishers, 1986, pp. 27–39.

Komi, P. V., and Bosco, C.: Utilization of stored elastic energy in leg extensor muscles by men and women. Med. Sci. Sports, *10*:261, 1978.

Kroemer, K. H.: An isoinertial technique to assess individual lifting capability. Human Factors, *24*:493, 1983.

Kroll, P. G.: The effect of previous contraction condition on subsequent eccentric power production in elbow flexor muscles. Ph.D. dissertation, New York University, 1987.

Lippman, R. K., and Selig, S.: An experimental study of muscle atrophy. Surg. Gynecol. Obstet., *47*:512, 1928.

Luciano, D. S., Vander, A. J., and Sherman, J. H.: Human Function and Structure. New York, McGraw-Hill, 1978, pp. 113–136.

Munsat, T. L., NcNeal, D., and Waters, R.: Effects of nerve stimulation on human muscle. Arch. Neurol., *33*:608, 1976.

Needham, D. M.: Machinia Carnis. Cambridge, Cambridge University Press, 1971.

Norkin, C., and Levange, P.: Joint Structure and Function: A Comprehensive Analysis. Philadelphia, F. A. Davis, 1983.

Ottoson, D.: Physiology of the Nervous System. New York, Oxford University Press, 1983, pp. 78–116.

Patel, A. N., Razzak, Z. A., and Pastur, P. K.: Disuse atrophy of human skeletal muscles. Arch. Neurol., *20*:413, 1969.

Perrine, J. J.: Isokinetic exercises and the mechanical energy potentials of muscle. J. Health Phys. Educ. Rec., *4*:40, 1968.

Phillips, C. A., and Petrofsky, J. S.: Mechanics of Skeletal and Cardiac Muscle. Springfield, Charles C Thomas, 1983.

Prichard, J. W.: Nerve. *In* The Scientific Basis of Orthopaedics. Edited by J. A. Albright and R. A. Brand. New York, Appleton-Century-Crofts, 1979, pp. 385–414.

Rodgers, M. M., and Cavanagh, P. R.: Glossary of biomechanical terms, concepts, and units. Phys. Ther., *64*:1886, 1984.

Saltin, B., et al.: Fiber types and metabolic potentials of skeletal muscles in sedentary man and endurance runners. Ann. NY Acad. Sci., *301*:3, 1977.

Sargeant, A. J., et al.: Functional and structural changes after disuse of human muscle. Clin. Sci., *52*:337, 1977.

Stanish, W. P.: Neurophysiology of stretching. *In* Prevention and Treatment of Running Injuries. Edited by I. D'Ambrosio and O. Drey. Thorofare, NJ, Chas. B. Leach, 1982.

Sukop, J., and Nelson, R. C.: Effects of isometric training in the force-time characteristics of muscle contraction. *In* Biomechanics IV. Edited by R. C. Nelson and C. A. Morehouse. Baltimore, University Press, 1974, pp. 440–447.

Thorstensson, A., Larsson, L., Tesch, P., and Karlsson, J.: Muscle strength and fiber composition in athletes and sedentary men. Med. Sci. Sports, *9*:26, 1977.

Vrbova, G.: Change in the motor reflex produced by tenotomy. J. Physiol., *166*:241, 1963.

Weber, A., and Murray, J. M.: Molecular control mechanisms in muscle contraction. Physiol. Rev., *53*:612, 1973.

Wilkie, D. R.: The mechanical properties of muscle. Br. Med. Bull., *12*:177, 1956.

Wilkie, D. R.: Muscle. London, Edward Arnold, 1968.

Williams, P., and Warwick, R.: Gray's Anatomy. 36th Ed. Edinburgh, Churchill Livingstone, 1980, pp. 506–515.

SUGGESTED READING

Arvidson, I., Eriksson, E., and Pitman, M.: Neuromuscular basis of rehabilitation. *In* Rehabilitation of the Injured Knee. Edited by E. Hunter and J. Funk. St. Louis, C. V. Mosby, 1984, pp. 210–234.

Åstrand, P.-O., and Rodahl, K.: Textbook of Work Physiology: Physiological Bases of Exercise. 2nd Ed. New York, McGraw Hill, 1977, pp. 37–52.

Basmajian, J. V.: Primary Anatomy. Baltimore, Williams & Wilkins, 1982, pp. 111–120.

Basmajian, J. V.: Muscles Alive. 5th Ed. Baltimore, Williams & Wilkins, 1985.

Brandstater, M. E., and Lambert, E. H.: A histologic study of the spatial arrangements of muscle fibers in single motor units within rat tibialis anterior muscle. Bull. Am. Assoc. Electromyog. Electro. Diag., *82*:15–16, 1969.

Brobeck, J. R. (ed.): Best & Taylor's Physiological Basis of Medical Practice. 10th Ed. Baltimore, Williams & Wilkins, 1979, pp. 59–113.

Brooke, M. H., and Kaiser, K. K.: Three myosin adenosine triphosphatase systems: The nature of their pH liability and sulfhydryl dependence. J. Histochem. Cytochem., *18*:670–672, 1970.

Buchtahl, F., and Sohmalburch, H.: Motor units of mammalian muscle. Physiol. Rev., *60*:90–142, 1980.

Buchtahl, P., Kaiser, E., and Rosenfalch, P.: The rheology of the cross-striated muscle fibre. Biol. Med. Dansk. vid Selsk., *21*:1–318, 1951.

Buller, A. J., Eccles, J. C., and Eccles, R. M.: Differentiation of fast and slow muscles in the cat hind limb. J. Physiol., *150*:399–416, 1960.

Burke, R. E., et al.: Mammalian motor units: Physiological histochemical correlation in three types of motor units in cat gastrocnemius. Science, *174*:709–712, 1971.

Buxton, P. H.: Skeletal muscle structure and development. *In* Scientific Foundations of Orthopaedics and Traumatology. Edited by R. Owen, J. Goodfellow, and P. Bullough. London, William Heinemann, 1980, pp. 22–30.

Caplan, A., et al.: Skeletal muscle. *In* Injury and Repair of the Musculoskeletal Soft Tissues. Edited by S. L.-Y. Woo and J. A. Buckwalter. Park Ridge, IL, American Academy of Orthopaedic Surgeons, 1987, pp. 213–291.

Cavagna, G. A., Dusman, B., and Margaria, R.: Positive work done by a previously stretched muscle. J. Appl. Physiol., *24*:21–32, 1968.

Cavanagh, P. R., and Komi, P. V.: Electromechanical delay in human skeletal muscle under concentric and eccentric contractions. Eur. J. Appl. Physiol., *42*:159–163, 1973.

Ciullo, J. V., and Zarins, B.: Biomechanics of the musculotendinous unit: Relation to athletic performance and injury. Clin. Sports Med., *2*:71–86, 1983.

Close, R. I.: Dynamic properties of mammalian skeletal muscles. Physiol. Rev., *52*:97–129, 1972.

Costill, P. L., Fink, W. J., and Habansky, A. J.: Muscle rehabilitation after knee surgery. Physician Sports Med., *5*:71–74, 1977.

Costill, P. L., et al.: Adaptations in skeletal muscles following strength training. J. Appl. Physiol., *46*:96–99, 1976.

Crawford, G. N. C., and James, N. T.: The design of muscles. *In* Scientific Foundations of Orthopaedics and Traumatology. Edited by R. Owen, J. Goodfellow, and P. Bullough. London, William Heinemann, 1980, pp. 67–74.

Davies, C. T. M., and Sargeant, A. J.: Effects of exercise therapy on total and component tissue leg volumes of patients undergoing rehabilitation from lower limb injury. Ann. Hum. Biol., *2*:327–337, 1975.

Dubowitz, V.: Cross innervated mammalian skeletal muscle. J. Physiol., *193*:481, 1967.

Dubowitz, V., and Brooke, M. H.: Muscle Biopsy: A Modern Approach. Philadelphia, W. B. Saunders, 1973.

Ebashi, S.: Excitation-contraction coupling. Ann. Rev. Physiol., *38*:293, 1976.

Edstrom, L.: Selective atrophy of red muscle fibers in the quadriceps in long standing knee joint dysfunction: Injuries to the anterior cruciate ligament. J. Neurol. Sci., *11*:55–59, 1970.

Edstrom, L., and Kugelberg, E.: Histochemical mapping of motor units in experimentally reinnervated skeletal muscles. Experientia, *25*:1044–1045, 1969.

Elftman, H.: Biomechanics of muscle. With particular application to studies of gait. J. Bone Joint Surg., *48A*:363–377, 1966.

Engel, W. K.: The essentiality of histo- and cystochemical studies of skeletal muscle in investigation of neuromuscular disease. Neurology, *12*:778, 1962.

Engel, W. K.: Fiber-type nomenclature of human skeletal muscle for histochemical purposes. Neurology, *25*:344, 1974.

Eriksson, E.: Sport injuries of the knee ligaments: Their diagnosis, treatment, rehabilitation and prevention. Med. Sci. Sports, *8*:133–144, 1976.

Eriksson, E.: Muscle physiology: Adaptation of the musculoskeletal system to exercise. Contemp. Orthop., *2*:228–232, 1980.

Eriksson, E.: Rehabilitation of muscle function after sport injury. Major problems in sports medicine. Int. J. Sports Med., *2*:1–5, 1981.

Eriksson, E., and Häggmark, T.: Comparison of isometric muscle training and electrical stimulation supplementing isometric muscle training in the recovery after major knee ligament surgery. Am. J. Sports Med., *7*:169–171, 1979.

Eriksson, E., Häggmark, T., Kiessling, K. H., and Karlsson, J.: Effect of electrical stimulation on human skeletal muscle. Int. J. Sports Med., *2*:18–22, 1981.

Essen, B.: Intramuscular substrate utilization during prolonged exercise. Ann. N. Y. Acad. Sci., *301*:30, 1977.

Frankel, V. H., Burstein, A. H., and Brooks, D. B.: Biomechanics of internal derangement of the knee in pathomechanics as determined by analysis of the instant centers of motion. J. Bone Joint Surg., *53A*:945, 1971.

Frost, H. M.: Orthopaedic Biomechanics. Springfield, Charles C Thomas, 1973.

Fuchs, F.: Striated muscle. Ann. Rev. Physiol., *36*:461–502, 1974.

Fujiwara, M., and Basmajian, J. V.: Electromyographic study of two-joint muscle. Am. J. Med., *54*:234–242, 1975.

Gollnick, P. D.: Relationship of strength and endurance with skeletal muscle structure and metabolic potential. Int. J. Sports Med. [Suppl.], *3*:26–32, 1982.

Gollnick, P. D., and Matoba, H.: The muscle fiber composition of skeletal muscle as a predictor of athletic success. An overview. Am. J. Sports Med., *12*:212–217, 1984.

Gollnick, P. D., Eriksson, E., Häggmark, T., and Saltin, B.: Recovery with a movable or standard cast following intra-articular reconstruction of the anterior cruciate ligament. AMA Med. Aspects Sport, *15*:56–60, 1974.

Gollnick, P. D., et al.: Effect of training on enzyme activity and fiber composition of human skeletal muscle. J. Appl. Physiol., *34*:107–111, 1973.

Gordon, A. M., Huxley, A. F., and Julian, F. J.: Tension development in highly stretched vertebrate muscle fibers. J. Physiol., *184*:143–169, 1966.

Gordon, A. M., Huxley, A. F., and Julian, F. J.: The variation in isometric tension with sarcomere length in vertebrate muscle fibers. J. Physiol., *184*:170–192, 1966.

Gowitzke, B. A., and Milner, M.: Scientific Bases of Human Movement. 3rd Ed. Baltimore, Williams & Wilkins, 1987.

Grimby, G., Gustafsson, E., Petterson, L., and Renström, P.: Quadriceps function and training after knee ligament surgery. Med. Sci. Sports Exer., *12*:70–75, 1980.

Guyton, A. C.: Textbook of Medical Physiology. 7th Ed. Philadelphia, W. B. Saunders, 1986.

Häggmark, T., and Eriksson, E.: Cylinder or mobile cast brace after knee ligament surgery: A clinical analysis and morphologic and enzymatic study of changes in the quadriceps muscle. Am. J. Sports Med., *7*:48–56, 1979.

Häggmark, T., and Eriksson, E.: Hypotrophy of the soleus muscle in man after Achilles tendon rupture. Am. J. Sports Med., *7*:121–126, 1979.

Häggmark, T., Jansson, E., and Eriksson, E.: Fibre type area and metabolic potential of the thigh muscle in man after knee surgery and immobilization. Int. J. Sports Med., *2*:12–17, 1981.

Henriksson, J., and Reitman, J. S.: Quantitative measures of enzyme activities in type I and type II muscle fibers of man after training. Acta Physiol. Scand., *97*:392–397, 1976.

Henriksson, J., and Reitman, J. S.: Time course of activity changes in human muscle succinate dehydrogenase and cytochrome oxidase activities and maximal oxygen uptake with physical activity and inactivity. Acta Physiol. Scand., *99*:91–97, 1977.

Hill, A. V.: The heat of shortening and the dynamic constants of muscle. Proc. R. Soc. Biol., *126*:136–195, 1938.

Hill, A. V.: The series elastic component of muscle. Proc. R. Soc. Biol., *137*:273–180, 1950.

Hill, A. V.: First and Last Experiments in Muscle Mechanics. Cambridge, Cambridge University Press, 1970.

Hill, D. K.: Tension due to interaction between the sliding filaments of resting striated muscle. The effect of stimulation. J. Physiol., *199*:637–684, 1968.

Hill, D. K.: The effect of temperature in the range 0–35°C on the resting tension of frog's muscle. J. Physiol., *208*:725–739, 1970.

Hislop, H. J., and Perrine, J. J.: The isokinetic concept of exercise. Phys. Ther., *47*:114–117, 1967.

Holt, L. E.: Scientific Stretches for Sport. Dalhousie, Dalhousie University Sport Research Ltd., 1973.

Huxley, A. F.: Muscular contraction. J. Physiol., *243*:1–43, 1974.

Huxley, A. F., and Huxley, H. E.: Organizers of a discussion of the physical and chemical basis of muscular contraction. Proc. R. Soc., *B160*:433–542, 1964.

Huxley, A. F., and Simmons, R. M.: Mechanical properties of the crossbridges of frog striated muscle. J. Physiol., *218*:59–60, 1971.

Huxley, H. E.: The mechanism of muscular contraction. Sci. Am., *213*:18–27, 1965.

Huxley, H. E.: The mechanism of muscular contraction. Science, *164*:1356, 1969.

Huxley, H. E.: Molecular basis of contraction in cross-striated muscles. *In* The Structure and Function of Muscle. Vol. 1, Part 1. 2nd Ed. Edited by G. H. Bourne. New York, Academic Press, 1972, pp. 301–387.

Huxley, H. E.: Structural changes in the actin- and myosin-containing filaments during contraction. Cold Spring Harbor Symp. Quant. Biol., *37*:361–376, 1972.

Hymes, A. C., Raab, D. E., and Yonehird, E. G.: Acute pain control by electricostimulation: A preliminary report. Adv. Neurol., *4*:761–767, 1974.

Jacobs, M.: Neurophysiological implications of slow, active stretching. Am. Corr. Ther. J., *30*:151–154, 1976.

Jansson, E., and Kaijser, L.: Muscle adaptation to extreme endurance training in man. Acta Physiol. Scand., *100*:315–324, 1977.

Jansson, E., Sjodin, B., and Tesch, P.: Changes in muscle fibre type distribution in man after physical training. A sign of fiber type transformation. Acta Physiol. Scand., *104*:235–237, 1978.

Kabat, H.: Studies of neuromuscular dysfunction. The role of central facilitation in restoration of motor function in paralysis. Arch. Phys. Med., *33*:523, 1952.

Kabat, H.: Proprioceptive facilitation in therapeutic exercise. *In* Therapeutic Exercise. Edited by S. Lecht. New Haven, Elizabeth Lecht, 1958.

Keele, C. A., Neil, E., and Joels, N.: Muscle and the nervous system. *In* Samson Wright's Applied Physiology. 13th Ed. Oxford, Oxford University Press, 1982, pp. 248–259.

Komi, P. V.: The stretch-shortening cycle and human power output. *In* Human Muscle Power. Edited by N. L. Jones, N. McCartney, and A. J. McConas. Champaign, IL, Human Kinetics Publishers, 1986, pp. 27–39.

Komi, P. V., and Bosco, C.: Utilization of stored elastic energy in leg extensor muscles by men and women. Med. Sci. Sports, *10*:261–265, 1978.

Kroemer, K. H.: An isoinertial technique to assess individual lifting capability. Human Factors, *24*:493–506, 1983.

Lippman, R. K., and Selig, S.: An experimental study of muscle atrophy. Surg. Gynecol. Obstet., *47*:512–522, 1928.

Luciano, D. S., Vander, A. J., and Sherman, J. H.: Human Function and Structure. New York, McGraw-Hill, 1978, pp. 113–136.

McArdle, W. D., Katch, F. I., and Katch, V. L.: Exercise Physiology. Energy, Nutrition, and Human Performance. Philadelphia, Lea & Febiger, 1981, pp. 234–248.

Munsat, T. L., NcNeal, D., and Waters, R.: Effects of nerve stimulation on human muscle. Arch. Neurol., *33*:608–617, 1976.

Needham, D. M.: Machinia Carnis. Cambridge, Cambridge University Press, 1971.

Norkin, C., and Levange, P.: Joint Structure and Function: A Comprehensive Analysis. Philadelphia, F. A. Davis, 1983.

Ottoson, D.: Physiology of the Nervous System. New York, Oxford University Press, 1983, pp. 78–116.

Patel, A. N., Razzak, Z. A., and Pastur, P. K.: Disuse atrophy of human skeletal muscles. Arch. Neurol., *20*:413–421, 1969.

Perrine, J. J.: Isokinetic exercises and the mechanical energy potentials of muscle. J. Health Phys. Educ. Rec., *4*:40–44, 1968.

Perry, J.: Anatomy and biomechanics of the shoulder in throwing, swimming, gymnastics and tennis. Clin. Sports Med., *2*:247–270, 1983.

Phillips, C. A., and Petrofsky, J. S.: Mechanics of Skeletal and Cardiac Muscle. Springfield, IL, Charles C Thomas, 1983.

Rodgers, M. M., and Cavanagh, P. R.: Glossary of biomechanical terms, concepts, and units. Phys. Ther., *64*:1886–1902, 1984.

Saltin, B., et al.: Fiber types and metabolic potentials of skeletal muscles in sedentary man and endurance runners. Ann. NY Acad. Sci., *301*:3–29, 1977.

Sargeant, A. J., et al.: Functional and structural changes after disuse of human muscle. Clin. Sci., *52*:337–342, 1977.

Stanish, W. P.: Neurophysiology of stretching. *In* Prevention and Treatment of Running Injuries. Edited by I. D'Ambrosio and O. Drey. Thorofare, NJ, Chas. B. Leach, 1982.

Sukop, J., and Nelson, R. C.: Effects of isometric training in the force-time characteristics of muscle contraction. *In* Biomechanics IV. Edited by R. C. Nelson and C. A. Morehouse. Baltimore, University Press, 1974, pp. 440–447.

Thorstensson, A.: Muscle strength, fiber types and enzyme activities in man. Acta Physiol. Scand., Suppl. *443*, 1976.

Thorstensson, A., Larsson, L., Tesch, P., and Karlsson, J.: Muscle strength and fiber composition in athletes and sedentary men. Med. Sci. Sports, *9*:26–30, 1977.

Vrbova, G.: Change in the motor reflex produced by tenotomy. J. Physiol., *166*:241–250, 1963.

Weber, A., and Murray, J. M.: Molecular control mechanisms in muscle contraction. Physiol. Rev., *53*:612, 1973.

Wilkie, D. R.: The mechanical properties of muscle. Br. Med. Bull., *12*:177–182, 1956.

Wilkie, D. R.: Muscle. London, Edward Arnold, 1968.

Williams, P., and Warwick, R.: Gray's Anatomy. 36th Ed. Edinburgh, Churchill Livingstone, 1980, pp. 506–515.

part two

BIOMECHANICS OF JOINTS

6

BIOMECHANICS OF THE KNEE

Margareta Nordin
Victor H. Frankel

The knee transmits loads, participates in motion, aids in conservation of momentum, and provides a force couple for activities involving the leg. The human knee, the largest and perhaps the most complex joint in the body, is a two-joint structure composed of the tibiofemoral joint and the patellofemoral joint (Fig. 6–1). The fact that the knee sustains high forces and is situated between the body's two longest lever arms makes it particularly susceptible to injury. This chapter utilizes the knee to introduce the basic terms, explain the methods, and demonstrate the calculations necessary for analyzing joint motion and the forces and moments acting on a joint. This information is applied to other joints in subsequent chapters.

The knee is particularly well suited for demonstrating biomechanical analyses of joints because these analyses can be simplified in the knee and will still yield useful data. Although knee motion occurs simultaneously in three planes, the motion in one plane is so great that it accounts for nearly all of the motion. Also, although many

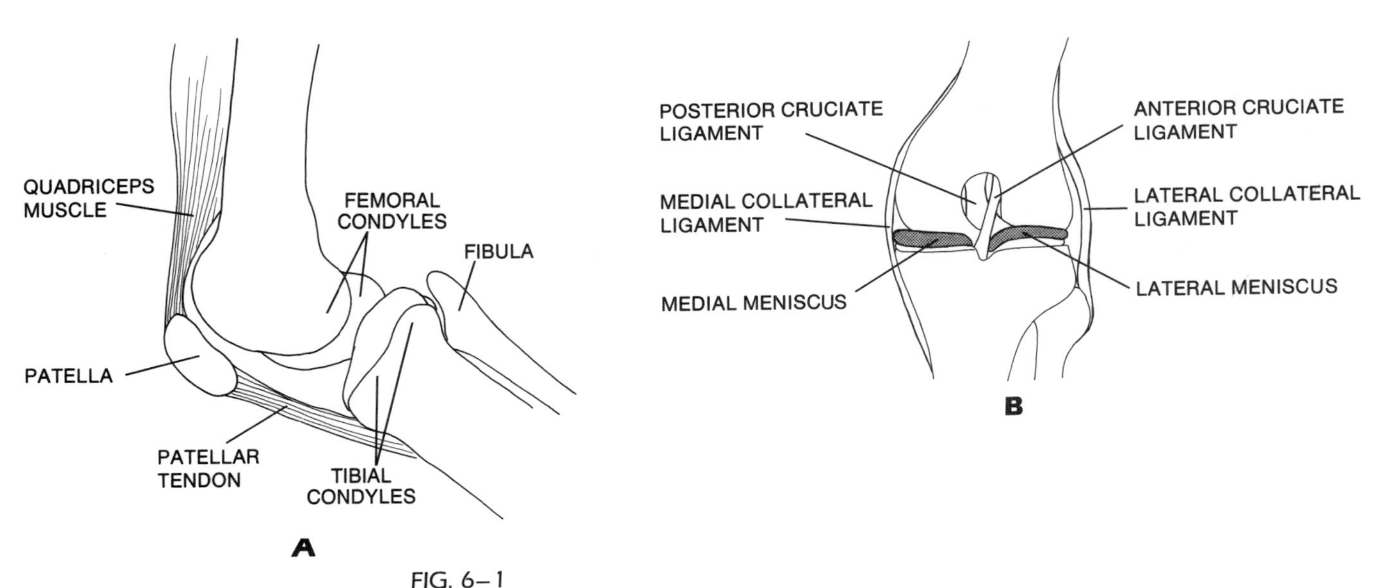

FIG. 6–1
Two-joint structure of the knee. **A.** Lateral view. **B.** Anterior view without patella.

muscles produce forces on the knee, at any particular instant one muscle group predominates, generating a force so great that it accounts for most of the muscle force acting on the knee. Thus, basic biomechanical analyses can be limited to motion in one plane and to the force produced by a single muscle group and can still give an understanding of knee motion and an estimation of the magnitude of the principal forces and moments on the knee.

Analysis of motion in any joint requires the use of kinematic data. Kinematics is the branch of mechanics that deals with motion of a body without reference to force or mass. Analysis of the forces and moments acting on a joint necessitates the use of both kinematic and kinetic data. Kinetics is the branch of mechanics that deals with the motion of a body under the action of given forces and/or moments.

KINEMATICS

Kinematics defines the range of motion and describes the surface motion of a joint in three planes: frontal (coronal or longitudinal), sagittal, and transverse (horizontal) (Fig. 6–2). Of the two joints composing the knee, the tibiofemoral joint lends itself particularly well to an analysis of range of joint motion. Analysis of surface joint motion can be performed easily for both the tibiofemoral and the patellofemoral joint.

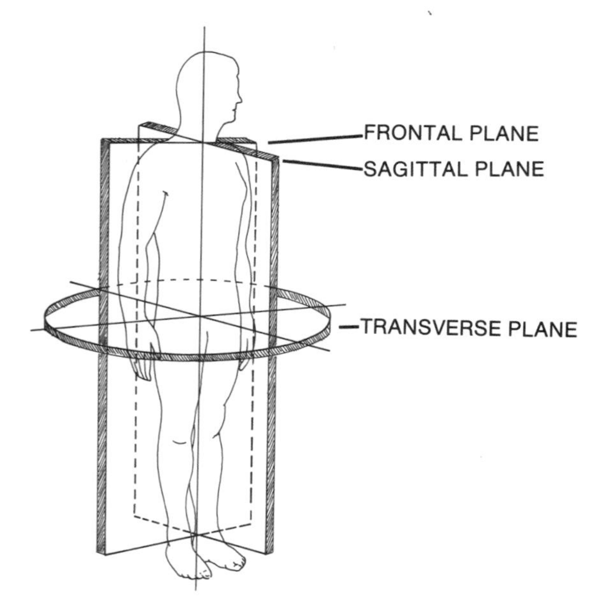

FIG. 6–2
Frontal (coronal or longitudinal), sagittal, and transverse (horizontal) planes in the human body.

RANGE OF MOTION

The range of motion of any joint can be measured in any plane. Gross measurements can be made with a goniometer, but more specific measurements require the use of more precise methods such as electrogoniometry, roentgenography, stereophotogrammetry, or photographic and video techniques using skeletal pins.

In the tibiofemoral joint, motion takes place in all three planes, but the range of motion is greatest by far in the sagittal plane. Motion in this plane, from full extension to full flexion of the knee, is from zero to approximately 140 degrees.

Motion in the transverse plane, internal and external rotation, is influenced by the position of the joint in the sagittal plane. With the knee in full extension, rotation is almost completely restricted by the interlocking of the femoral and tibial condyles, which occurs mainly because the medial femoral condyle is longer than the lateral condyle. The range of rotation increases as the knee is flexed, reaching a maximum at 90 degrees of flexion: with the knee in this position external rotation ranges from zero to approximately 45 degrees and internal rotation ranges from zero to approximately 30 degrees. Beyond 90 degrees of flexion the range of internal and external rotation decreases, primarily because the soft tissues restrict rotation.

Motion in the frontal plane, abduction and adduction, is similarly affected by the amount of joint flexion. Full extension of the knee precludes almost all motion in the frontal plane. Passive abduction and adduction increase with knee flexion up to 30 degrees, but each reaches a maximum of only a few degrees. With the knee flexed beyond 30 degrees, motion in the frontal plane again decreases because of the limiting function of the soft tissues.

The range of tibiofemoral joint motion required for the performance of various physical activities can be determined from kinematic analysis. Motion in this joint during walking has been measured in all planes. The range of motion in the sagittal plane during level walking was measured with an electrogoniometer by Murray and coworkers (1964). During the entire gait cycle the knee was never fully extended. Nearly full extension (5 degrees of flexion) was noted at the beginning of the stance phase at heel strike and at the end of the stance phase just before toe-off (Fig. 6–3). Maximum flexion (75 degrees) was observed during the middle of the swing phase.

Motion in the transverse plane during walking has been measured by several investigators. Using a photographic technique involving placement of skeletal pins through the femur and tibia, Levens and associates (1948) found that total rotation of the tibia with respect to the femur ranged from about 4 to 13 degrees in 12 subjects (mean 8.6 degrees). Greater rotation (mean 13.3 degrees) was noted by Kettelkamp and coworkers (1970), who used electrogoniometry on 22 subjects. In both studies external rotation began during knee extension in the stance phase and reached a peak value at the end of the swing phase just before heel strike. Internal rotation was noted during flexion in the swing phase.

Motion in the frontal plane during walking was also measured with an electrogoniometer by Kettelkamp's group (1970). In nearly all of the 22 subjects, maximal abduction of the tibia was observed during extension at heel strike and at the beginning of the stance phase, and maximal adduction occurred as the knee was flexed during the swing phase. The total amount of abduction and adduction averaged 11 degrees.

Values for the range of motion of the tibiofemoral joint in the sagittal plane during several common activities are presented in Table 6–1. Maximal knee flexion occurs during lifting. A range of motion from full extension to at least 117 degrees of flexion appears to be required in order for an individual to carry out the activities of daily living in a normal manner. Any restriction of knee motion is compensated for by increased motion in other joints. In studying the range of tibiofemoral joint motion during various activities, Kettelkamp and coworkers (1970) noted a significant relationship between the length of the lower leg and the range of motion in the sagittal plane. The longer the leg was, the greater the range of motion. An increased speed of movement requires a greater range of motion in the tibiofemoral joint (Perry et al., 1977). As the pace accelerates from walking slowly to running, progressively more knee flexion is needed during the stance phase (Table 6–2).

TABLE 6–1

RANGE OF TIBIOFEMORAL JOINT MOTION IN THE SAGITTAL PLANE DURING COMMON ACTIVITIES

ACTIVITY	RANGE OF MOTION FROM KNEE EXTENSION TO KNEE FLEXION (DEGREES)
Walking	0– 67*
Climbing stairs	0– 83†
Descending stairs	0– 90
Sitting down	0– 93
Tying a shoe	0–106
Lifting an object	0–117

*Data from Kettelkamp et al., 1970. Mean for 22 subjects. A slight difference was found between right and left knees (mean for right knee 68.1 degrees; mean for left knee 66.7 degrees).

†These and subsequent data from Laubenthal et al., 1972. Mean for 30 subjects.

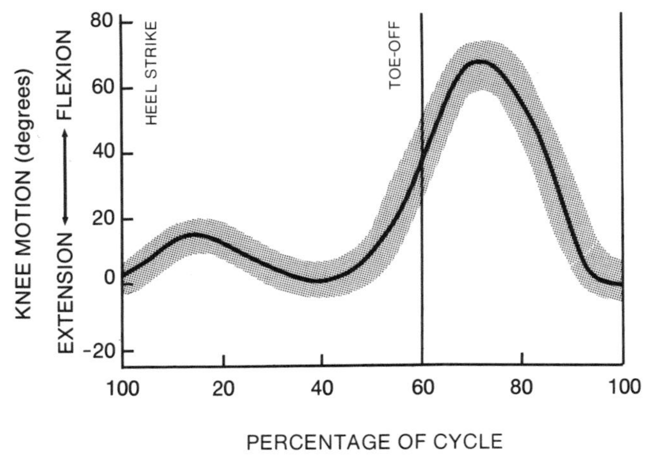

FIG. 6–3
Range of motion of the tibiofemoral joint in the sagittal plane during level walking, one gait cycle. Shaded area indicates variation among 60 subjects (age range 20 to 65 years). (Adapted from Murray et al., 1964.)

SURFACE JOINT MOTION

Surface joint motion, the motion between the articulating surfaces of a joint, can be described for any joint in any plane with the use of stereophotogrammetric methods (Selvik, 1978, 1983). Because these methods are highly technical and complex, a simpler method, evolved in the nineteenth century, is still used (Reuleaux, 1876). This method, called the instant center technique, allows surface joint motion to be analyzed in the sagittal and frontal planes but not in the transverse plane. The instant center

TABLE 6–2

AMOUNT OF KNEE FLEXION DURING STANCE PHASE OF WALKING AND RUNNING

ACTIVITY	RANGE IN AMOUNT OF KNEE FLEXION DURING STANCE PHASE (DEGREES)
Walking	
Slow	0– 6
Free	6–12
Fast	12–18
Running	18–30

(Data from Perry et al., 1977. Range for seven subjects.)

technique provides a description of the relative uniplanar motion of two adjacent segments of a body and the direction of displacement of the contact points between these segments.

The skeletal portion of a body segment is called a link. As one link rotates about the other, at any instant there is a point that does not move, that is, a point that has zero velocity. This point constitutes an instantaneous center of motion, or instant center. The instant center is found by identifying the displacement of two points on a link as the link moves from one position to another in relation to an adjacent link, which is considered to be stationary. The points on the moving link in its original position and in its displaced position are designated on a graph, and lines are drawn connecting the two sets of points. The perpendicular bisectors of these two lines are then drawn. The intersection of the perpendicular bisectors is the instant center.

Clinically, a pathway of the instant center for a joint can be determined by taking successive roentgenograms of the joint in different positions (usually 10 degrees apart) throughout the range of motion in one plane and applying the Reuleaux method for locating the instant center for each interval of motion.

When the instant center pathway has been determined for joint motion in one plane, the surface joint motion can be described. For each interval of motion the point at which the joint surfaces make contact is located on the roentgenograms used for the instant center analysis, and a line is drawn from the instant center to the contact point. A second line drawn at right angles to this line indicates the direction of displacement of the contact points. The direction of displacement of these points throughout the range of motion describes the surface motion in the joint. In most joints the instant centers lie at a distance from the joint surface and the line indicating the direction of displacement of the contact points is tangential to the load-bearing surface, demonstrating that one joint surface is gliding on the other (load-bearing) surface. In the rare instance in which the instant center is found on the surface, the joint has a rolling motion and there is no gliding function. (Three types of surface joint motion—rotation, rolling, and gliding [translation]—are depicted in Figure 12–7). Since the instant center technique allows a description of motion in one plane only, it is not useful for describing the surface joint motion if more than 15 degrees of motion takes place in any plane other than the one being measured.

In the knee, surface joint motion occurs between the tibial and femoral condyles and between the femoral condyles and the patella. In the tibiofemoral joint, surface motion takes place in all three planes simultaneously but is minimal in the transverse and frontal planes. Surface motion in the patellofemoral joint takes place in two planes simultaneously, the frontal and transverse, but is far greater in the frontal plane.

Tibiofemoral Joint

An example will illustrate the use of the instant center technique in describing the surface motion of the tibiofemoral joint in the sagittal plane. To determine the pathway of the instant center of this joint during flexion, a lateral roentgenogram is taken of the knee in full extension, and successive films are taken at 10-degree intervals of increased flexion. Care is taken to keep the tibia parallel to the x-ray table and to prevent rotation about the femur. When a patient has limited knee motion, the knee is flexed or extended only as far as the patient can tolerate.

Two points on the femur that are easily identified on all roentgenograms are selected and designated on each roentgenogram (Fig. 6–4A). The films are then compared in pairs, with the images of the tibiae superimposed on each other. Roentgenograms with marked differences in tibial alignment are not used. Lines are drawn between the points on the femur in the two positions, and the perpendicular bisectors of these lines are then drawn. The point at which these perpendicular bisectors intersect is the instant center of the tibiofemoral joint for each 10-degree interval of

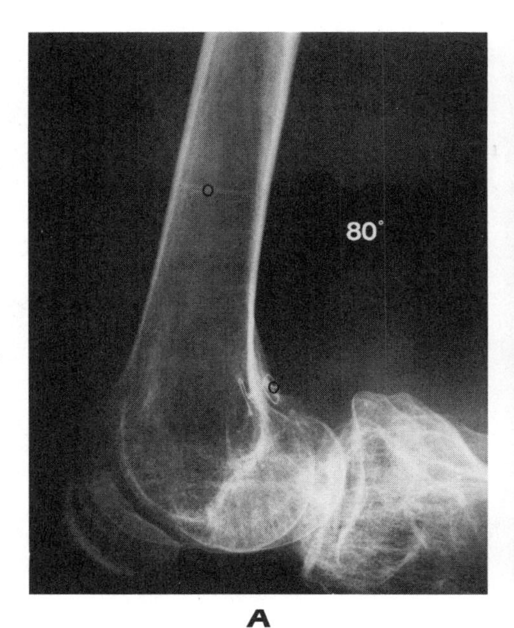

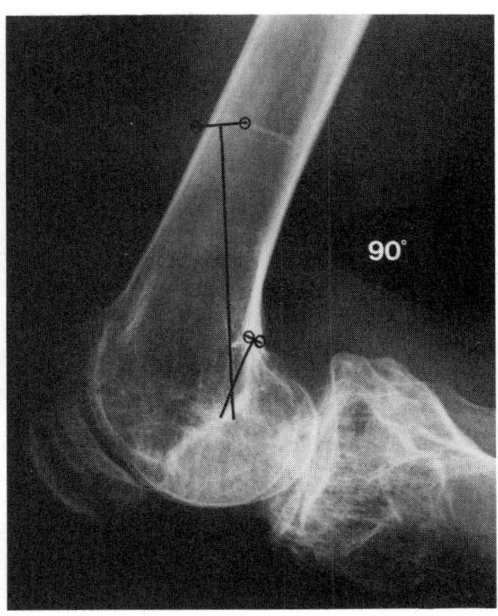

FIG. 6–4
Locating the instant center. **A.** Two easily identifiable points on the femur are designated on a roentgenogram of a knee flexed 80 degrees. **B.** This roentgenogram is compared with a roentgenogram of the knee flexed 90 degrees, on which the same two points have been indicated. The images of the tibiae are superimposed, and lines are drawn connecting each set of points. The perpendicular bisectors of these two lines are then drawn. The point at which these perpendicular bisectors intersect is the instant center of the tibiofemoral joint for the motion between 80 and 90 degrees of flexion. (Courtesy of Ian Goldie, M.D.)

motion (Fig. 6–4B). The instant center pathway throughout the entire range of knee flexion and extension can then be plotted. In a normal knee the instant center pathway for the tibiofemoral joint is semicircular (Fig. 6–5).

After the instant center pathway has been determined for the tibiofemoral joint, the surface motion can be described. On each set of superimposed roentgenograms the point of contact of the tibiofemoral joint surfaces (the narrowest point in the joint space) is determined and a line is drawn connecting this point with the instant center. A second line drawn at right angles to this line indicates the direction of displacement of the contact points. In a normal knee this line is tangential to the surface of the tibia for each interval of motion from full extension to full flexion, demonstrating that the femur is gliding on the tibial condyles (Fig. 6–6).

Frankel and associates (1971) determined the instant center pathway and analyzed the surface motion of the tibiofemoral joint from 90 degrees of flexion to full extension in 25 normal knees. Tangential gliding was noted in all cases. They also determined the instant center pathway for the tibiofemoral joint in 30 knees with internal derangements and found that in all cases the instant center was displaced from the normal position during some portion of the motion examined. The abnormal instant center pathway for one subject, a 35-year-old

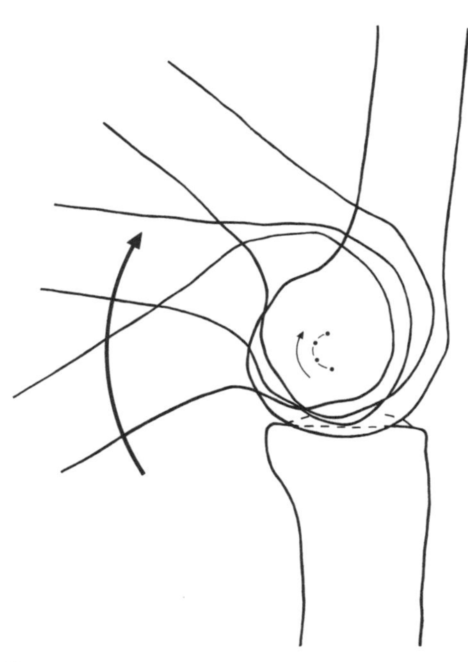

FIG. 6–5
Semicircular instant center pathway for the tibiofemoral joint in a 19-year-old man with a normal knee.

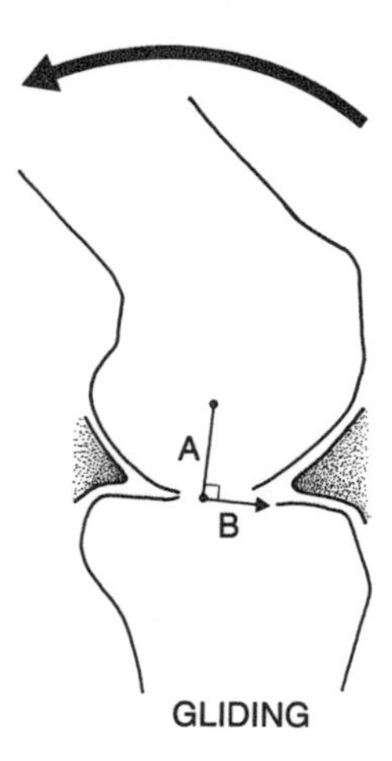

FIG. 6–6
In a normal knee a line drawn from the instant center of the tibiofemoral joint to the tibiofemoral contact point (line A) forms a right angle with a line tangential to the tibial surface (line B). The arrow indicates the direction of displacement of the contact points. Line B is tangential to the tibial surface, indicating that the femur glides on the tibial condyles during the measured interval of motion.

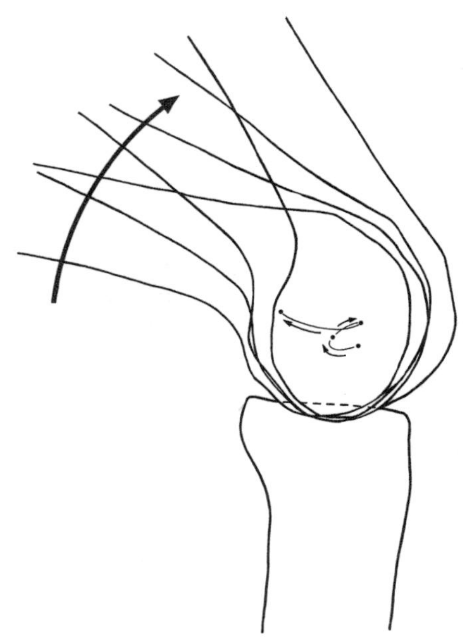

FIG. 6–7
Abnormal instant center pathway for a 35-year-old man with a bucket-handle derangement. The instant center jumps at full extension of the knee. (Adapted from Frankel et al., 1971.)

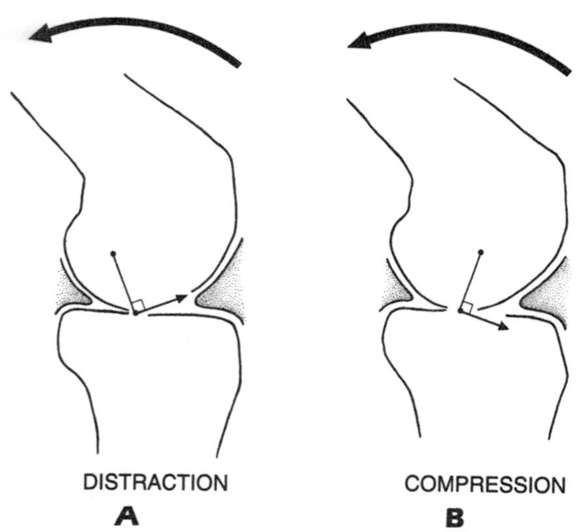

FIG. 6–8
Surface motion in two tibiofemoral joints with displaced instant centers. In both joints the arrowed line at right angles to the line between the instant center and the tibiofemoral contact point indicates the direction of displacement of the contact points. **A.** The small arrow indicates that with further flexion the tibiofemoral joint will be distracted. **B.** With increased flexion this joint will be compressed.

man with a bucket-handle derangement, is shown in Figure 6–7.

If the knee is extended and flexed about a displaced instant center, the tibiofemoral joint surfaces do not glide tangentially throughout the range of motion but become either distracted or compressed (Fig. 6–8). Such a knee is analogous to a door with a bent hinge, which no longer fits into the door jamb. If the knee is continually forced to move about a displaced instant center, a gradual adjustment to the situation will be reflected either by stretching of the ligaments and other supporting soft tissues or by the imposition of abnormally high pressure on the articular surfaces.

Internal derangements of the tibiofemoral joint may interfere with the so-called screw-home mechanism, which is a combination of knee extension and external rotation of the tibia (Fig. 6–9). The tibiofemoral joint is not a simple hinge joint but has a spiral, or helicoid, motion. The spiral motion of the tibia about the femur during flexion and extension results from the anatomical configuration of the medial femoral condyle; in a normal knee this condyle is approximately 1.7 cm longer than the lateral condyle. As the tibia glides on the femur from the fully flexed to the fully extended position, it

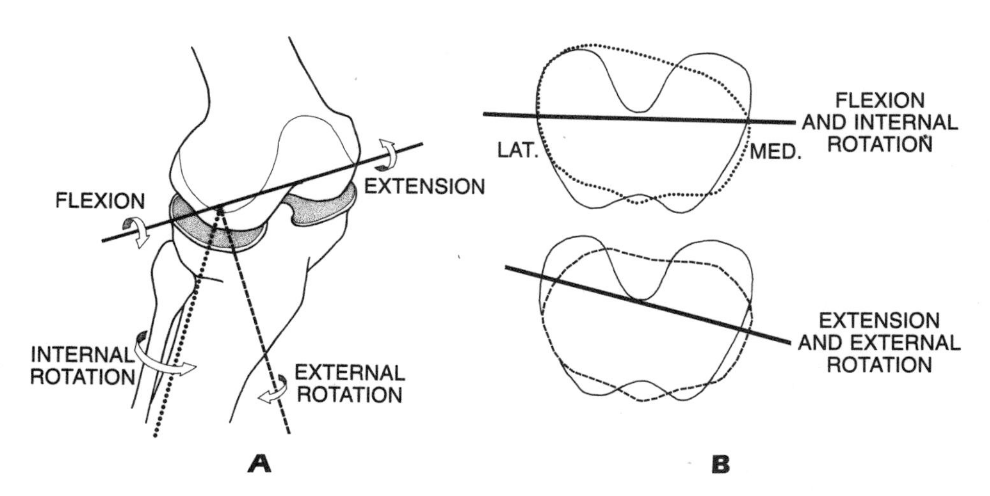

FIG. 6–9
Screw-home mechanism of the tibiofemoral joint. During knee extension the tibia rotates externally. This motion is reversed as the knee is flexed. **A.** Oblique view of the femur and tibia. Shaded area indicates the tibial plateau. **B.** Top view showing the position of the tibial plateau on the femoral condyles in knee flexion (top) and extension (bottom). The solid outlines represent the femoral condyles; the broken lines represent the tibial plateau. (Adapted from Helfet, 1974.)

descends and then ascends the curves of the medial femoral condyle and simultaneously rotates externally. This motion is reversed as the tibia moves back into the fully flexed position. This screw-home mechanism provides more stability to the knee in any position than would a simple hinge configuration of the tibiofemoral joint.

A clinical test, the Helfet test, is often used to determine whether external rotation of the tibiofemoral joint takes place during knee extension, thereby indicating whether the screw-home mechanism is intact. This clinical test is performed with the patient sitting with the knee and hip flexed 90 degrees and the leg hanging free. The medial and lateral borders of the patella are marked on the skin. The tibial tuberosity and the midline of the patella are then designated, and the alignment of the tibial tuberosity with the patella is checked. In a normal knee flexed 90 degrees the tibial tuberosity aligns with the medial half of the patella (Fig. 6–10A). The knee is then extended fully and the movement of the tibial tuberosity is observed. In a normal knee the tibial tuberosity moves laterally during extension and aligns with the lateral half of the patella at full extension (Fig. 6–10B). Rotatory motion in a normal knee may be as great

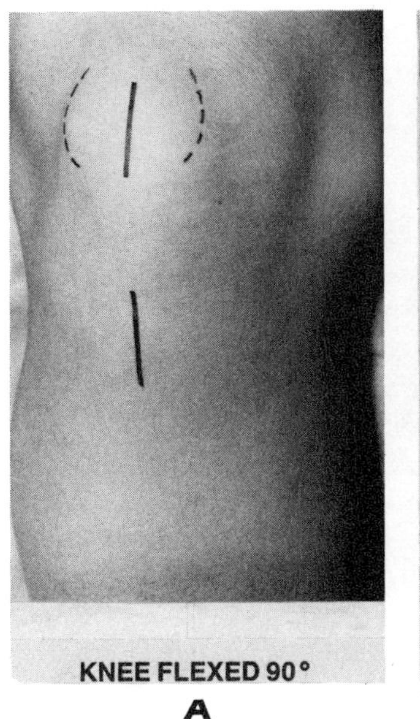

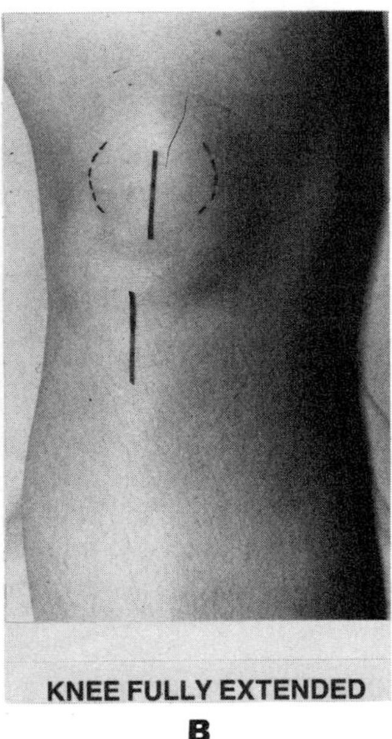

FIG. 6–10
Helfet test. **A.** In a normal knee flexed 90 degrees the tibial tuberosity aligns with the medial half of the patella. **B.** When the knee is fully extended the tibial tuberosity aligns with the lateral half of the patella.

as half the width of the patella. In a deranged knee the tibia may not rotate externally during extension. Because of the altered surface motion in such a knee, the tibiofemoral joint will be abnormally compressed if the knee is forced into extension and the joint surfaces may be damaged.

Patellofemoral Joint

The surface motion of the patellofemoral joint in the frontal plane may also be described by means of the instant center technique. This joint is shown to have a gliding motion (Fig. 6–11). From full extension to full flexion of the knee the patella glides caudally approximately 7 cm on the femoral condyles. Both the medial and lateral facets of the femur articulate with the patella from full extension to 90 degrees of flexion (Fig. 6–12). Beyond 90 degrees of flexion the patella rotates externally, and only the medial femoral facet articulates with the patella (Fig. 6–12B). At full flexion the patella sinks into the intercondylar groove (Goodfellow et al., 1976).

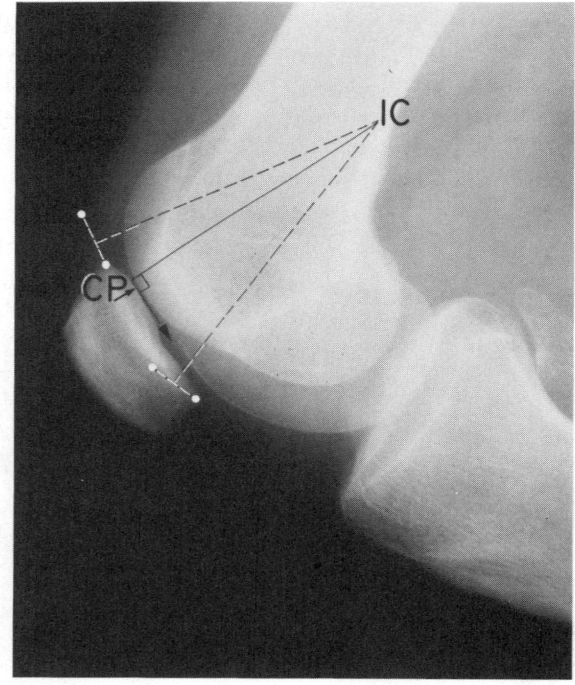

FIG. 6–11
After the instant center (IC) is determined for the patellofemoral joint for the motion from 75 to 90 degrees of knee flexion, a line is drawn from the instant center to the contact point (CP) between the patella and the femoral condyle. A line drawn at right angles to this line is tangential to the surface of the patella, indicating gliding.

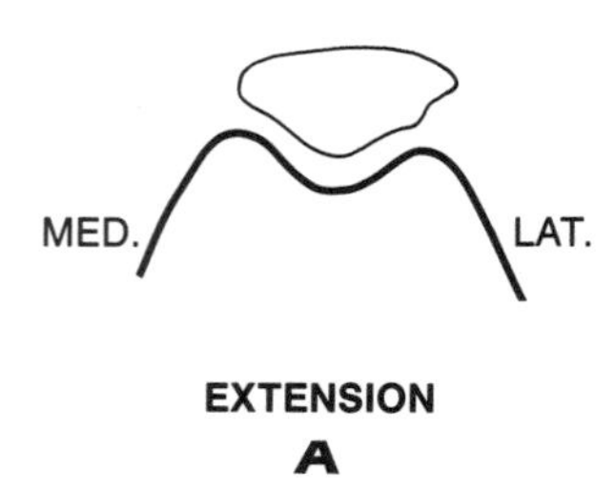

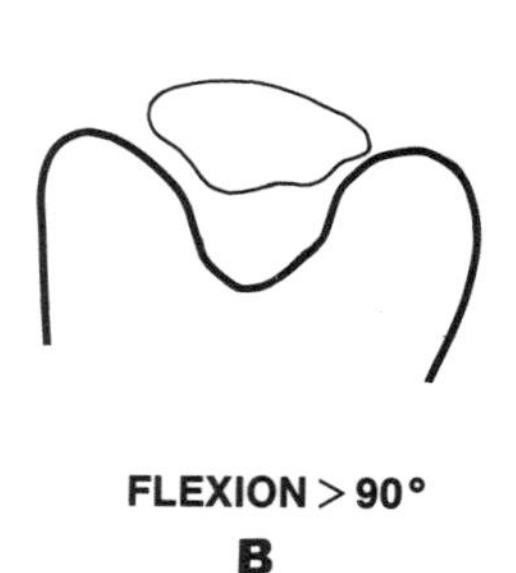

FIG. 6–12
A. At full extension both the medial and lateral femoral facets articulate with the patella. **B.** Beyond 90 degrees of flexion the patella rotates externally and only the medial femoral facet articulates with the patella. (Adapted from Goodfellow et al., 1976.)

KINETICS

Kinetics involves both static and dynamic analysis of the forces and moments acting on a joint. Statics is the study of the forces and moments acting on a body in equilibrium (a body at rest or moving at a constant speed). For a body to be in equilibrium, two equilibrium conditions must be met: force (translatory) equilibrium, in which the sum of the forces is zero, and moment (rotatory) equilibrium, in which the sum of the moments is zero. Dynamics is the study of the moments and forces acting on a body in motion (an accelerating or decelerating body). In this case the forces do not add up to zero and the body displaces and/or the moments do not add up to zero and the body rotates around an axis perpendicular to the plane of the forces producing the moments. Kinetic analysis allows one to determine the magnitude of the moments and forces on a joint produced by body weight, muscle action, soft tissue resistance, and externally applied loads in any situation, either static or dynamic, and to identify those situations that produce excessively high moments or forces.

In this and subsequent chapters, the discussion of statics and dynamics of the joints of the skeletal system concerns the magnitude of the forces and moments acting to move a joint about an axis or to maintain its position. It does not take into account the deforming effect of these forces and moments on the joint structures. This effect is indeed present, but for the purposes of our analysis it may be disregarded.

STATICS OF THE TIBIOFEMORAL JOINT

Static analysis may be used to determine the forces and moments acting on a joint when no motion takes place or at one instant in time during a dynamic activity such as walking, running, or lifting an object. It can be performed for any joint in any position and under any loading configuration. In such analysis either graphic or mathematical methods may be used to solve for the unknown forces or moments.

A complete static analysis involving all moments and all forces imposed on a joint in three dimensions is complicated. For this reason a simplified technique is often used. The technique utilizes a free body diagram and limits the analysis to one plane, to the three main coplanar forces acting on the free body, and to the main moments acting about the joint under consideration. The minimum magnitudes of the forces and moments are obtained.

When the simplified free body technique is used to analyze coplanar forces, one portion of the body is considered as distinct from the entire body, and all forces acting on this free body are identified. A diagram is drawn of the free body in the loading situation to be analyzed. The three principal coplanar forces acting on the free body are identified and designated on the free body diagram.

These forces are designated as vectors if four characteristics are known: magnitude, sense, line of application, and point of application. (The term "direction" includes line of application and sense.) If the points of application for all three forces and the directions for two forces are known, all remaining characteristics can be obtained for a force equilibrium situation. When the free body is in equilibrium, the three principal coplanar forces are concurrent; that is, they intersect at a common point. In other words, these forces form a closed system with no resultant (i.e., their vector sum is zero). For this reason the line of application for one force can be determined if the lines of application for the other two forces are known. Once the lines of application for all three forces are known, a triangle of forces can be constructed and the magnitudes of all three forces can be scaled from this triangle.

An example will illustrate the application of the simplified free body technique for coplanar forces to the knee. In this case, the technique is used to estimate the minimum magnitude of the joint reaction force acting on the tibiofemoral joint of the weight-bearing leg when the other leg is lifted during stair climbing. The lower leg is considered as a free body, distinct from the rest of the body, and a diagram of this free body in the stair-climbing situation is drawn. From all forces acting on the free body the three main coplanar forces are identified as the ground reaction force (equal to body weight), the tensile force through the patellar tendon exerted by the quadriceps muscle, and the joint reaction force on the tibial plateau. The *ground reaction force* (W) has a known magnitude (equal to body weight*), sense, line of application, and point of application (point of contact between the foot and the ground). The *patellar tendon force* (P) has a known sense (away from the knee joint), line of application (along the patellar tendon), and point of application (point of insertion of the patellar tendon on the tibial tuberosity), but an unknown magnitude. The *joint reaction force* (J) has a known point of application on the surface of the tibia (the contact point of the joint surfaces between the tibial and femoral condyles, estimated from a roentgenogram of the joint in the proper loading configuration), but an unknown magnitude, sense, and line of application.

These three forces are designated on the free body diagram (Fig. 6–13). Because the lower leg is in equilibrium, the lines of application for all three forces intersect at one point. Since the lines of application for two forces (W and P) are known, the line of application for the third force (J) can be determined. The lines of application for forces W and P are extended until they intersect. The line of

*In this case the ground reaction force is actually equal to body weight minus the weight of the lower leg. Since the weight of the lower leg is minimal (less than one tenth of the body weight), it can be disregarded, and the figure for total body weight can be utilized in the calculation.

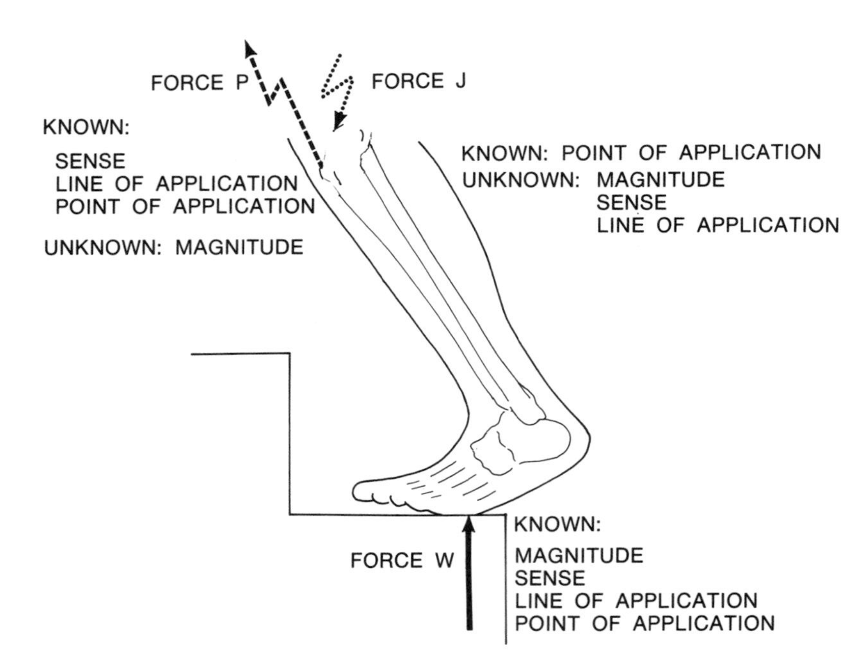

FIG. 6–13
The three main coplanar forces acting on the lower leg are designated on the free body diagram. Force W is the ground reaction force, force P is the patellar tendon force, and force J is the tibiofemoral joint reaction force.

application for J can then be drawn from its point of application on the tibial surface through the intersection point (Fig. 6–14).

Now that the line of application for J has been determined, it is possible to construct a triangle of forces. First, a vector representing W is drawn. Next, P is drawn from the head of vector W. The line of application and sense of P can be indicated, but its length cannot be determined because the magnitude is unknown. Since the lower leg is in force equilibrium, however, it is known that when J is added the triangle must close (that is, the head of P must touch the origin of J). The line of application of J is then drawn from the origin of vector W. The point at which J intersects P is the head of vector P and the origin of vector J. The magnitudes of P and J can now be scaled from the drawing (Fig. 6–15). In this case the patellar tendon force (P) is 3.2 times body weight and the joint reaction force (J) is 4.1 times body weight.

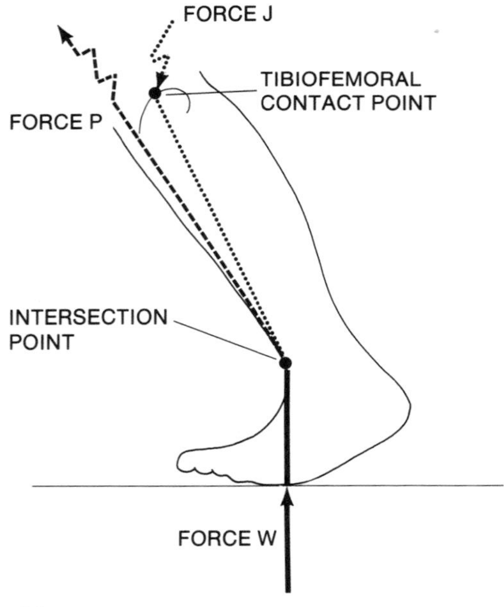

FIG. 6–14
On the free body diagram of the lower leg the lines of application for forces W and P are extended until they intersect (intersection point). The line of application for force J is then determined by connecting its point of application (tibiofemoral contact point) with the intersection point for forces W and P.

It can be seen that the main muscle force has a much greater influence on the magnitude of the joint reaction force than does the ground reaction force produced by body weight. It should be noted that in this example only the minimum magnitude of the joint reaction force has been calculated. If other muscle forces are considered, such as the force produced by contraction of the hamstring muscles in stabilizing the knee, the joint reaction force increases.

The next step in the static analysis is analysis of the moments acting around the center of motion of the tibiofemoral joint with the knee in the same position and loading configuration shown in Figure 6–13 (Fig. 6–16). The moment analysis is used to estimate the minimum magnitude of the moment produced through the patellar tendon, which counterbalances

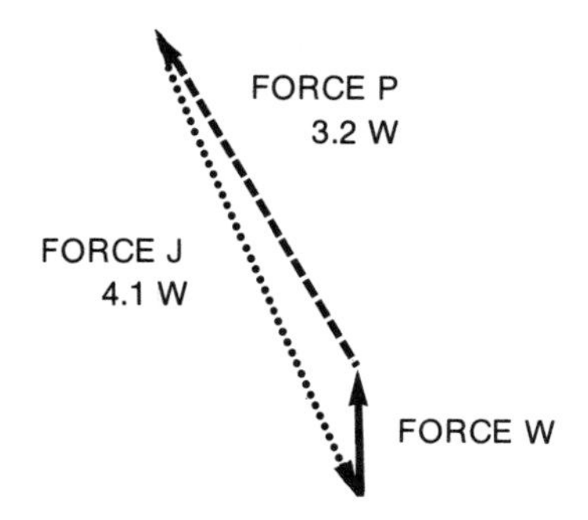

FIG. 6–15
A triangle of forces is constructed. First vector W is drawn. Next force P is drawn from the head of vector W. Then, to close the triangle, force J is drawn from the origin of vector W. The point at which forces P and J intersect defines the length of these vectors. Now that the length of all three vectors is known, the magnitudes of forces P and J can be scaled from force W, which is equal to body weight. Force P is 3.2 times body weight, and force J is 4.1 times body weight.

the moment on the lower leg produced by the weight of the body as the subject ascends stairs. The flexing moment on the lower leg is the product of the weight of the body* (W, the ground reaction force) and its lever arm (a) (also called moment arm), which is the perpendicular distance of the force from the center of motion of the tibiofemoral joint. The counterbalancing extending moment is the product of the quadriceps muscle force through the patellar tendon (P) and its lever arm (b). Because the lower leg is in equilibrium, the sum of these two moments must equal zero.

$$\Sigma M = 0.$$

In this example, the counterclockwise moment is arbitrarily designated as positive:

$$W \times a - P \times b = 0$$

$$W \times a = P \times b.$$

Values for lever arms a and b can be measured from anatomic specimens or on soft tissue roentgenograms, and the magnitude of W can be determined from the body weight of the individual. The magnitude of P can then be found from the moment equilibrium equation:

$$P = \frac{W \times a}{b}.$$

*Again the weight of the lower leg is disregarded because it is less than one tenth of body weight.

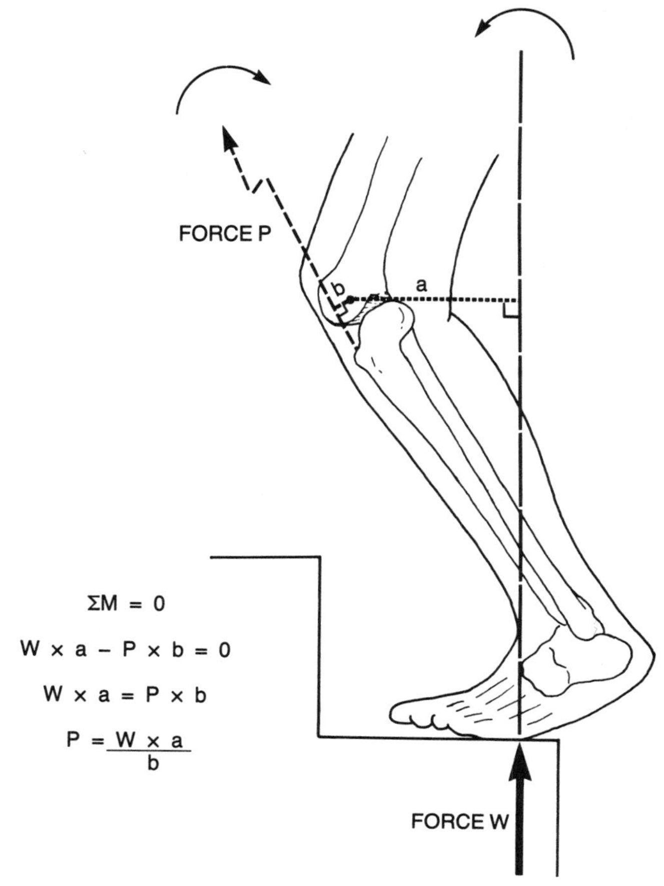

FIG. 6–16
The two main moments acting around the center of motion of the tibiofemoral joint (solid dot) are designated on the free body diagram of the lower leg during stair climbing. Since the lower leg is in equilibrium, the extending moment produced by the patellar tendon force P times its lever arm (b) counterbalances the flexing moment produced by the ground reaction force W times its lever arm (a). The weight of the lower leg is disregarded.

DYNAMICS OF THE TIBIOFEMORAL JOINT

Although estimations of the magnitude of the forces and moments imposed on a joint in static situations are useful, most of our activities are of a dynamic rather than a static nature. Analysis of the forces and moments acting on a joint during motion requires the use of a technique for solving dynamic problems.

As in static analysis, the main forces considered in dynamic analysis are those produced by body weight, muscles, other soft tissues, and externally applied loads. Friction forces, which are negligible in a normal joint, are not considered. In dynamic analysis, two factors in addition to those in static analysis must be taken into account: the acceleration of the body part under consideration, and the mass moment of inertia of the body part. (The mass moment of inertia is the unit used to express the amount of torque needed to accelerate a body and depends on the shape of the body.)

The steps for calculating the minimum magnitudes of the forces acting on a joint at a particular instant in time during a dynamic activity are as follows:

1. The anatomic structures involved in the production of forces are identified.
2. The angular acceleration of the moving body part is determined.
3. The mass moment of inertia of the moving body part is determined.
4. The torque (moment) acting about the joint is calculated.
5. The magnitude of the main muscle force accelerating the body part is calculated.
6. The magnitude of the joint reaction force at a particular instant in time is calculated by static analysis.

In the first step, the structures of the body involved in producing forces on the joint are identified. These are the moving body part and the main muscles in that body part that are involved in the production of the motion.

In joints of the extremities, acceleration of the body part involves a change in joint angle. To determine this angular acceleration of the moving body part, the entire movement of the body part is recorded photographically. Recording can be done with a stroboscopic light and movie camera, with video photogrammetry, with Selspot systems, or with other methods. The maximal angular acceleration for a particular motion is calculated (Frankel and Burstein, 1970).

Next, the mass moment of inertia for the moving body part is determined. Anthropometric data on the body part can be used for this determination. Since calculating these data is a complicated procedure, tables are commonly used (Drillis et al., 1964).

The torque about the joint can now be calculated using Newton's second law of motion, which states that when motion is angular, the torque is a product of the mass moment of inertia of the body part and the angular acceleration of that part:

$$T = I\alpha,$$

where T is the torque expressed in newton meters (Nm)
I is the mass moment of inertia expressed in newton meters times seconds squared (Nm sec^2)
α is the angular acceleration expressed in radians per second squared (r/sec^2).

Not only is the torque a product of the mass moment of inertia and the angular acceleration of the body part, but it is also a product of the main muscle force accelerating the body part and the perpendicular distance of the force from the center of motion of the joint (lever arm). Thus,

$$T = Fd,$$

where F is the force expressed in newtons (N)
d is the perpendicular distance expressed in meters (m).

Since T is known and d can be measured on the body part from the line of application of the force to the center of motion of the joint, the equation can be solved for F. When F has been calculated, the remaining problem can be solved like a static problem using the simplified free body technique to determine the minimum magnitude of the joint reaction force acting on the joint at a certain instant in time.

An example will illustrate the use of dynamic analysis in calculating the joint reaction force on the tibiofemoral joint at a particular instant during a dynamic activity, that of kicking a football (Frankel and Burstein, 1970). A stroboscopic film of the knee and lower leg was taken, and the angular acceleration was found to be maximal at the instant the foot struck the ball; the lower leg was almost vertical at this instant. From the film the maximal angular acceleration was computed to be 453 r/sec^2. From anthropometric data tables (Drillis et al., 1964) the mass moment of inertia for the lower leg was determined to be 0.35 Nm sec^2. The torque about the tibiofemoral joint was calculated according to the equation torque equals mass moment of inertia times angular acceleration ($T = I\alpha$):

$$0.35 \text{ Nm sec}^2 \times 453 \text{ r/sec}^2 = 158.5 \text{ Nm}.$$

After the torque had been determined to be 158.5 Nm and the perpendicular distance from the subject's patellar tendon to the instant center for the tibiofemoral joint had been found to be 0.05 m, the muscle force acting on the joint through the patellar tendon was calculated using the equation torque equals force times distance (T = Fd):

$$158.5\ \text{Nm} = F \times 0.05\ \text{m}$$

$$F = \frac{158.5\ \text{Nm}}{0.05\ \text{m}}$$

$$F = 3170\ \text{N}.$$

Thus, 3,170 N was the maximal force exerted by the quadriceps muscle during the kicking motion.

Static analysis can now be performed to determine the minimum magnitude of the joint reaction force on the tibiofemoral joint. The main forces on this joint are identified as the patellar tendon force (P), the gravitational force of the lower leg (T), and the joint reaction force (J). P and T are known vectors. J has an unknown magnitude, sense, and line of application. The free body technique for three coplanar forces is used to solve for J, which is found to be only slightly lower than P.

As is evident from the calculations, the two main factors that influence the magnitude of the forces on a joint in dynamic situations are the acceleration of the body part and its mass moment of inertia. An increase in angular acceleration of the body part will produce a proportional increase in the torque about the joint. Although in the body the mass moment of inertia is anatomically set, it can be manipulated externally. For example, it is increased when a weight boot is applied to the foot during rehabilitative exercises of the extensor muscles of the knee. Normally a joint reaction force of approximately 50% of body weight results when the knee is extended from 90 degrees of flexion to full extension. In a person weighing 70 kg this force is approximately 350 N. If a 10-kg weight boot is placed on the foot, it will exert a gravitational force of 100 N. This will increase the joint reaction force by 1,000 N, making this force almost four times greater than without the boot.

Dynamic analysis has been used to investigate the peak magnitudes of the joint reaction forces, muscle forces, and ligament forces on the tibiofemoral joint during walking. Morrison (1970) calculated the magnitude of the joint reaction force transmitted through the tibial plateau in men and women subjects during level walking. He simultaneously recorded muscle activity electromyographically to determine which muscles produced the peak magnitudes of this force on the tibial plateau during various stages of the gait cycle (Fig. 6–17).

Just after heel strike the joint reaction force ranged from two to three times body weight and was associated with contraction of the hamstring muscles, which have a decelerating and stabilizing effect on the knee. During knee flexion in the beginning of the stance phase the joint reaction force was approximately two times body weight and was associated with contraction of the quadriceps muscle, which acts to prevent buckling of the knee. The peak joint reaction force occurred during the late stance phase just before toe-off. This force ranged from two to four times body weight, varying among the subjects tested, and was associated with contraction of the gastrocnemius muscle. In the late swing phase contraction of the hamstring muscles resulted in a joint reaction force approximately equal to body weight. No significant difference was found between the joint reaction force magnitudes for men and women when

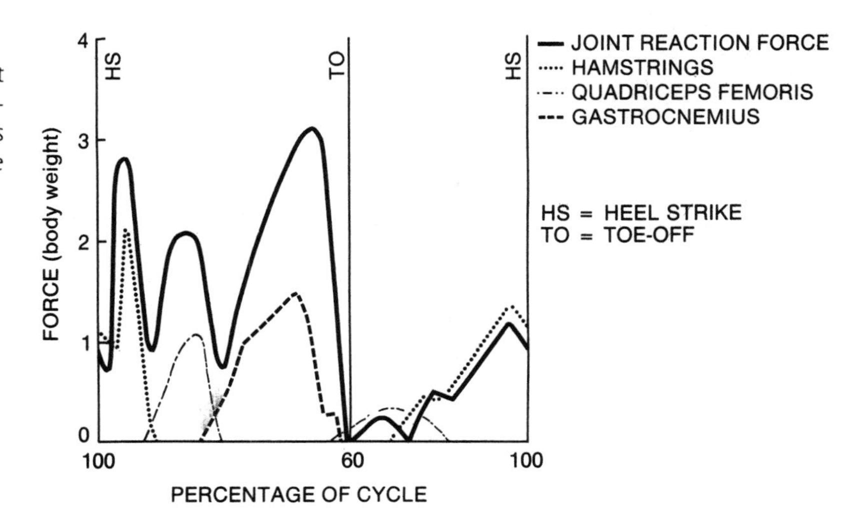

FIG. 6–17
Joint reaction force in terms of body weight transmitted through the tibial plateau during walking, one gait cycle (12 subjects). The muscle forces producing the peak magnitudes of this force are also designated. (Adapted from Morrison, 1970.)

the values were normalized by dividing them by body weight.

During the gait cycle the joint reaction force shifted from the medial to the lateral tibial plateau. In the stance phase, when the force reached its peak value, it was sustained mainly by the medial plateau; in the swing phase, when the force was minimal, it was sustained primarily by the lateral plateau. The contact area of the medial tibial plateau is approximately 50% larger than that of the lateral tibial plateau (Kettelkamp and Jacobs, 1972). Also, the cartilage on this plateau is approximately three times thicker than that on the lateral plateau. The larger surface area and the greater thickness of the medial plateau allow it to sustain more easily the higher forces imposed upon it.

In a normal knee, joint reaction forces are sustained by the menisci as well as by the articular cartilage. The function of the menisci was investigated by Seedhom and coworkers (1974), who examined the distribution of stresses in knees of human autopsy subjects with and without menisci. Their results suggest that in load-bearing situations the magnitude of the stresses on the tibiofemoral joint when the menisci have been removed may be as much as three times greater than when these structures are intact.

In a normal knee stresses are distributed over a wide area of the tibial plateau. If the menisci are removed, the stresses are no longer distributed over such a wide area but are limited to a contact area in the center of the plateau (Fig. 6–18). Thus, not only does removal of the menisci increase the magnitude of the stresses on the cartilage at the center of the tibial plateau, but it also diminishes the size and changes the location of the contact area. Over the long term the high stresses placed on this smaller contact area may be harmful to the exposed cartilage, which is usually soft and fibrillated in that area.

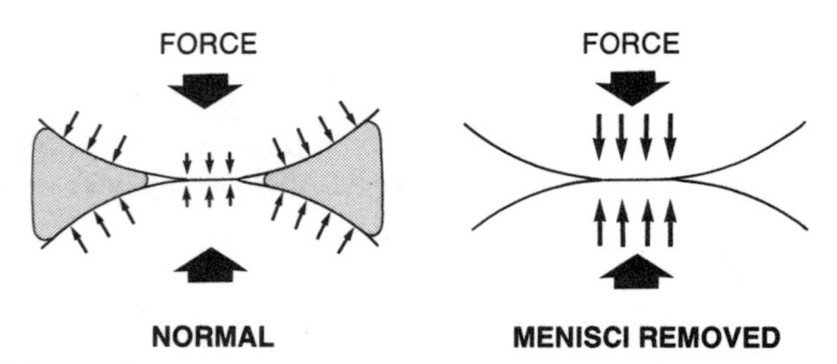

FIG. 6–18
Stress distribution in a normal knee and in a knee with the menisci removed. Removal of the menisci increases the magnitude of stresses on the cartilage of the tibial plateau and changes the size and location of the tibiofemoral contact area. With the menisci intact the contact area encompasses nearly the entire surface of the tibial plateau. With the menisci removed the contact area is limited to the center of the tibial plateau.

The forces sustained by the ligaments in the tibiofemoral joint are lower than those acting on the tibial plateau and are mainly tensile. Morrison (1970) calculated the forces on the knee ligaments during walking. The posterior cruciate ligament sustained the highest forces, about one half body weight; peak forces occurred just after heel strike and in the later part of the stance phase.

FUNCTION OF THE PATELLA

The patella serves two important biomechanical functions in the knee. First, it aids knee extension by producing anterior displacement of the quadriceps tendon throughout the entire range of motion, thereby lengthening the lever arm of the quadriceps muscle force. Second, it allows a wider distribution of compressive stress on the femur by increasing the area of contact between the patellar tendon and the femur. The contribution of the patella to the length of the quadriceps muscle force lever arm varies from full flexion to full extension of the knee (Smidt, 1973; Lindahl and Movin, 1967). At full flexion, when the patella is in the intercondylar groove, it produces little anterior displacement of the quadriceps tendon, and it contributes the least to the length of the quadriceps muscle force lever arm (about 10% of the total length). As the knee is extended, the patella rises from the intercondylar groove, producing significant anterior displacement of the tendon. The length of the quadriceps force lever arm rapidly increases with extension up to 45 degrees, at which point the patella lengthens the lever arm by about 30%.

With knee extension beyond 45 degrees, the length of the lever arm is diminished slightly. With this decrease in its lever arm, the quadriceps muscle force must increase for the torque about the knee to remain the same. In a study in vitro of normal knees Lieb and Perry (1968) showed that the quadriceps muscle force required to extend the knee the last 15 degrees increased by approximately 60% (Fig. 6–19).

If the patella is removed from a knee, the patellar tendon lies closer to the center of motion of the

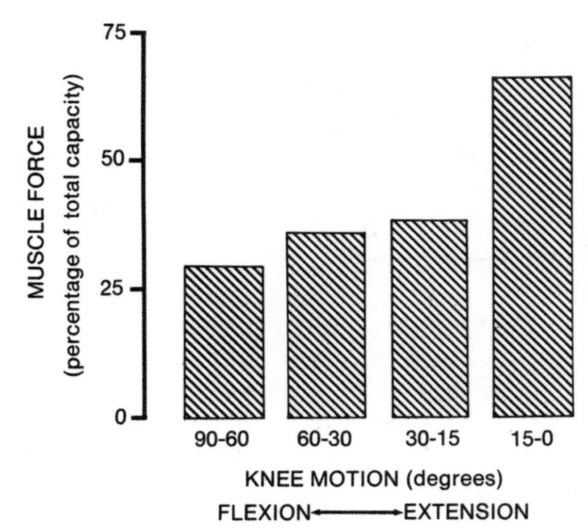

FIG. 6–19
Quadriceps muscle force required during knee motion from 90 degrees of flexion to full extension. (Adapted from Lieb and Perry, 1968.)

tibiofemoral joint than in an intact knee (Fig. 6–20). Acting with a shorter lever arm, the quadriceps muscle must produce even more force than is normally required in order for a certain torque about the knee to be maintained during the last 45 degrees of extension. Full active extension of such a knee may require as much as 30% more quadriceps force than is normally required (Kaufer, 1971). This increase in force may be beyond the capacity of the quadriceps muscle in some patients, particularly those who have intra-articular disease or are advanced in age.

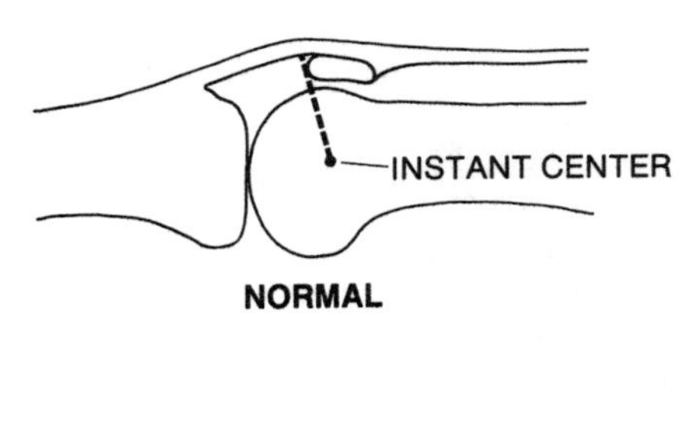

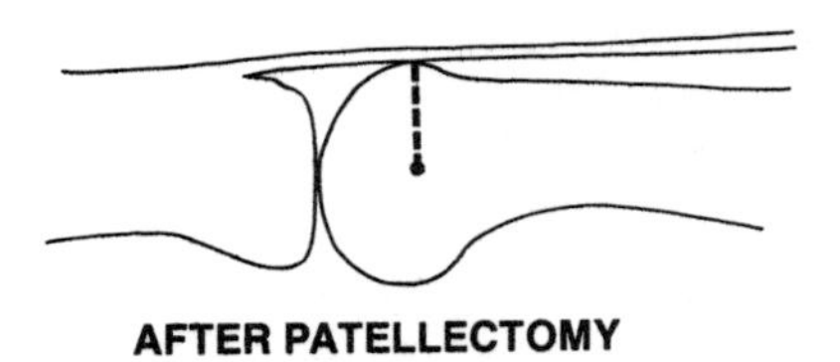

FIG. 6–20
Quadriceps muscle lever arm (represented by the broken line) in a normal knee and a knee from which the patella has been removed. The lever arm is the perpendicular distance between the force exerted by the quadriceps muscle through the patellar tendon and the instant center of the tibiofemoral joint for the last two degrees of extension. The patellar tendon lies closer to the instant center in the knee without a patella. (Adapted from Kaufer, 1971.)

STATICS AND DYNAMICS OF THE PATELLOFEMORAL JOINT

During dynamic activities, the magnitude of the muscle forces acting on a joint directly affects the magnitude of the joint reaction force. In general, the greater the muscle forces are, the greater is the joint reaction force.

In the patellofemoral joint, the quadriceps muscle force increases with knee flexion. During relaxed upright standing, minimal quadriceps muscle forces are required to counterbalance the small flexion moments about the patellofemoral joint because the center of gravity of the body above the knee is almost directly above the center of rotation of the patellofemoral joint. As knee flexion increases, however, the center of gravity shifts farther away from the center of rotation, thereby greatly increasing the flexion moments to be counterbalanced by the quadriceps muscle force. As the quadriceps muscle force rises, so does the patellofemoral joint reaction force.

Knee flexion also influences the patellofemoral joint reaction force by affecting the angle between the patellar tendon force and the quadriceps tendon force. The angle of these two force components becomes more acute with knee flexion, increasing the magnitude of the patellofemoral joint reaction force, which is their resultant (Fig. 6–21). In Figure 6–22, the patellofemoral joint reaction force is shown to increase with knee flexion even though the quadriceps muscle force remains constant.

Reilly and Martens (1972) determined the magnitude of the patellofemoral joint reaction force during several dynamic activities involving varying amounts of knee flexion. During level walking, which requires relatively little knee flexion, the reaction force was low. The peak value, in the middle of the stance phase when flexion was greatest, was one half body weight.

The joint reaction force was much greater during activities that require greater flexion. During knee bends to 90 degrees, this force reached 2.5 to 3 times body weight with the knee flexed 90 degrees

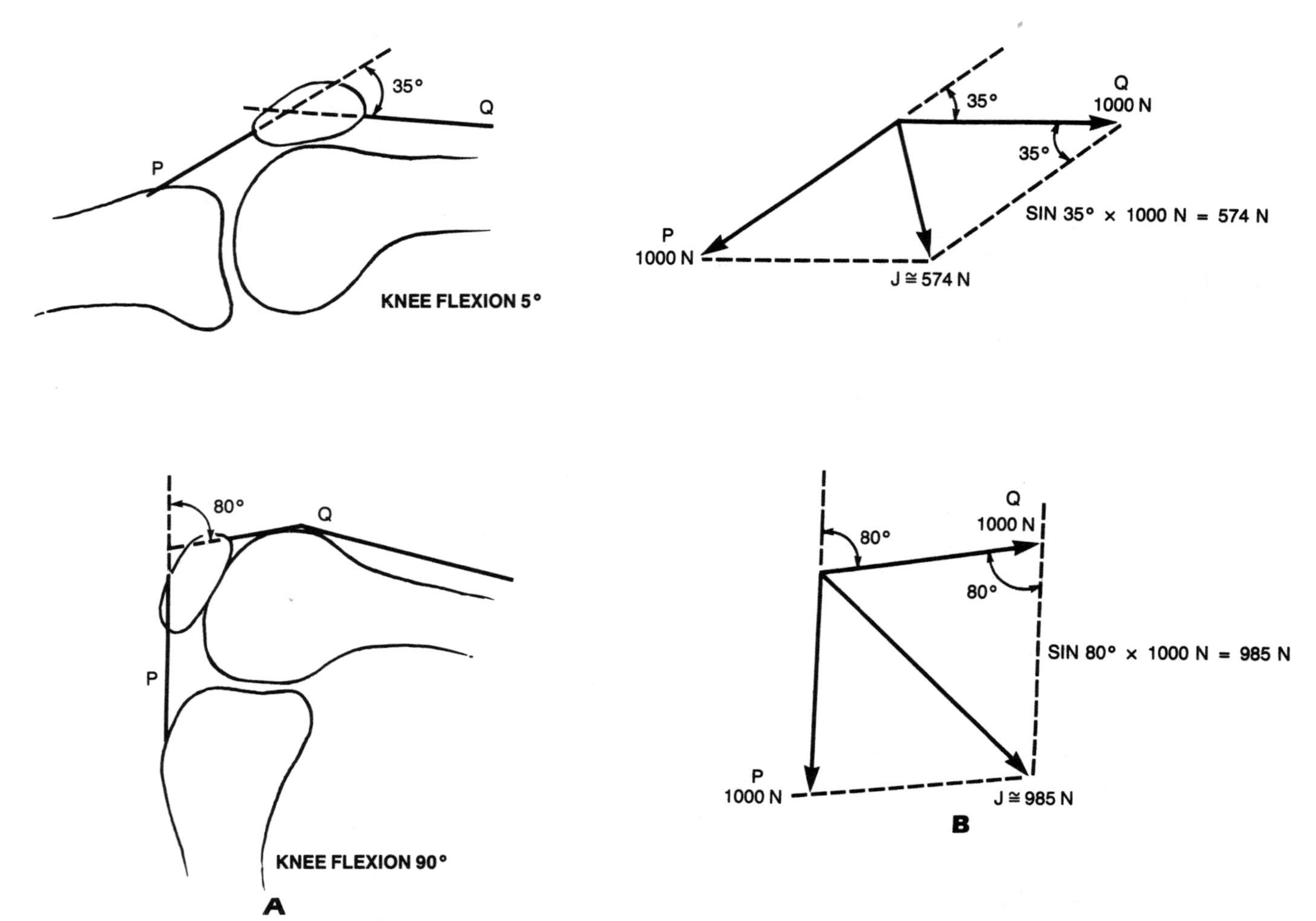

FIG. 6–21
Knee flexion influences the patellofemoral joint reaction force by changing the angle between the patellar tendon and the quadriceps tendon. **A.** The angle between the patellar tendon (P) and the quadriceps tendon (Q) is 35 degrees with the knee flexed 5 degrees (top) and 80 degrees with the knee flexed 90 degrees (bottom). Values for the tendon angles are from Matthews and associates (1977), who determined the angle roentgenographically after placing two metal wires along each of these tendons. **B.** The patellofemoral joint reaction force with the knee in 5 degrees and 90 degrees of flexion is obtained by constructing a parallelogram of forces for each situation and using trigonometric calculations. The patellofemoral joint reaction force (J) is the resultant of the two equal force components through the patellar tendon (P) and the quadriceps tendon (Q). As the angle between these force components becomes more acute with greater knee flexion, the resultant joint reaction force becomes larger. (Adapted from Wiktorin and Nordin, 1986.)

(Fig. 6–23). Throughout the knee bend the patellofemoral joint reaction force remained higher than the quadriceps muscle force. During stair climbing and descent, at the point when knee flexion reached a maximum of about 60 degrees, the peak value equaled 3.3 times body weight.

When the knee is extended, the lower part of the patella rests against the femur. As the knee is flexed to 90 degrees, the contact surface between the patella and femur shifts cranially and its size increases somewhat (Goodfellow et al., 1976). To some extent this increase in the contact surface with knee flexion compensates for the larger patellofemoral joint reaction force.

The quadriceps muscle force and the torque around the patellofemoral joint can be extremely high under certain circumstances, particularly when the knee is flexed. An extreme situation was

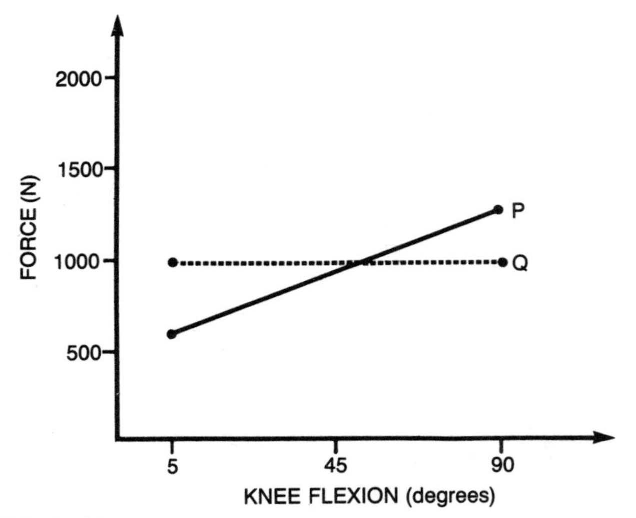

FIG. 6–22
Patellofemoral joint reaction force (P) for a theoretical situation in which the quadriceps muscle force (Q) is held constant at 1,000 N and the knee is flexed 5 degrees, 45 degrees, and 90 degrees. (Adapted from Wiktorin and Nordin, 1986.)

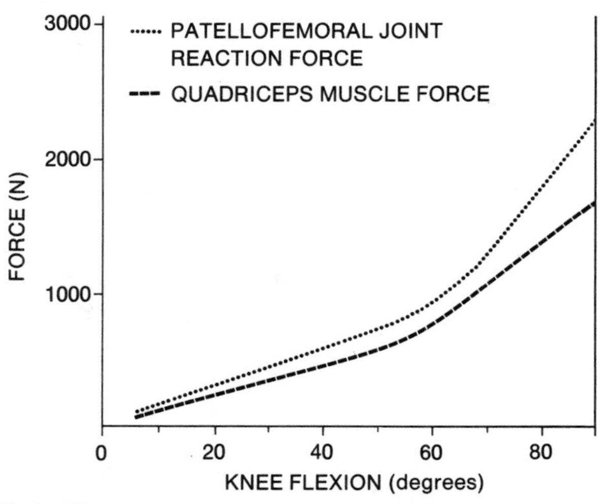

FIG. 6–23
Patellofemoral joint reaction force and quadriceps muscle force during knee bend to 90 degrees (three subjects). (Adapted from Reilly and Martens, 1972.)

observed during a study of the external torque on the knee produced by weight lifting: one subject ruptured a patellar tendon when he lifted a barbell weighing 175 kg (Zernicke et al., 1977). At the instant of tendon rupture, the knee was flexed 90 degrees; the torque on the knee joint was 550 Nm, and the quadriceps muscle force was about 10,330 N.

Because of the high magnitude of the quadriceps muscle force and joint reaction force during activities requiring a large amount of knee flexion, patients with patellofemoral joint derangements experience increased pain when performing these activities. An effective mechanism for reducing these forces is to limit the amount of knee flexion.

SUMMARY

1. The knee is a two-joint structure that is composed of the tibiofemoral joint and the patellofemoral joint.
2. In the tibiofemoral joint, surface motion occurs in three planes simultaneously but is greatest by far in the sagittal plane. In the patellofemoral joint, surface motion occurs simultaneously in two planes, the frontal and the transverse, and is greater in the frontal.
3. The surface joint motion can be described with the use of an instant center technique. When performed on a normal knee the technique reveals the following: the instant center for successive intervals of motion of the tibiofemoral joint in the sagittal plane follows a semicircular pathway, and the direction of displacement of the tibiofemoral contact points is tangential to the surface of the tibia, indicating gliding throughout the range of motion.
4. The screw-home mechanism of the tibiofemoral joint adds stability to the joint in full extension.
5. The tibiofemoral and patellofemoral joints are subjected to great forces. Muscle forces have the greatest influence on the magnitude of the joint reaction force, which can reach several times body weight in both joints. In the patellofemoral joint, knee flexion also affects the joint reaction force, greater knee flexion resulting in a higher joint reaction force.
6. Although the tibial plateaus are the main load-bearing structures in the knee, the cartilage, menisci, and ligaments also bear loads. The menisci aid in distributing the stresses imposed on the tibial plateaus.
7. The patella aids knee extension by lengthening the lever arm of the quadriceps muscle force throughout the entire range of motion and allows a wider distribution of compressive stress on the femur.

REFERENCES

Drillis, R., Contini, R., and Bluestein, M.: Body segment parameters. A survey of measurement techniques. Artif. Limbs, *8*:44, 1964.

Frankel, V. H., and Burstein, A. H.: Orthopaedic Biomechanics. Philadelphia, Lea & Febiger, 1970.

Frankel, V. H., Burstein, A. H., and Brooks, D. B.: Biomechanics of internal derangement of the knee. Pathomechanics as determined by analysis of the instant centers of motion. J. Bone Joint Surg., *53A*:945, 1971.

Goodfellow, J., Hungerford, D. S., and Zindel, M.: Patello femoral joint mechanics and pathology. I. Functional anatomy of the patello-femoral joint. J. Bone Joint Surg., *58B*:287, 1976.

Helfet, A. J.: Anatomy and mechanics of movement of the knee joint. *In* Disorders of the Knee. Edited by A. Helfet. Philadelphia, J. B. Lippincott, 1974, pp. 1–17.

Kaufer, H.: Mechanical function of the patella. J. Bone Joint Surg., *53A*:1551, 1971.

Kettelkamp, D. B., and Jacobs, A. W.: Tibiofemoral contact area—determination and implications. J. Bone Joint Surg., *54A*:349, 1972.

Kettelkamp, D. B., et al.: An electrogoniometric study of knee motion in normal gait. J. Bone Joint Surg., *52A*:775, 1970.

Laubenthal, K. N., Smidt, G. L., and Kettelkamp, D. B.: A quantitative analysis of knee motion during activities of daily living. Phys. Ther., *52*:34, 1972.

Levens, A. S., Inman, V. T., and Blosser, J. A.: Transverse rotation of the segments of the lower extremity in locomotion. J. Bone Joint Surg., *30A*:859, 1948.

Lieb, F. J., and Perry, J.: Quadriceps function. An anatomical and mechanical study using amputated limbs. J. Bone Joint Surg., *50A*:1535, 1968.

Lindahl, O., and Movin, A.: The mechanics of extension of the knee-joint. Acta Orthop. Scand., *38*:226, 1967.

Matthews, L. S., Sonstegard, D. A., and Henke, J. A.: Load bearing characteristics of the patello-femoral joint. Acta Orthop. Scand., *48*:511, 1977.

Morrison, J. B.: The mechanics of the knee joint in relation to normal walking. J. Biomech., *3*:51, 1970.

Murray, M. P., Drought, A. B., and Kory, R. C.: Walking patterns of normal men. J. Bone Joint Surg., *46A*:335, 1964.

Perry, J., Norwood, L., and House, K.: Knee posture and biceps and semimembranosis muscle action in running and cutting (an EMG study). Trans. Orthop. Res. Soc., *2*:258, 1977.

Reilly, D. T., and Martens, M.: Experimental analysis of the quadriceps muscle force and patello-femoral joint reaction force for various activities. Acta Orthop. Scand., *43*:126, 1972.

Reuleaux, F.: The Kinematics of Machinery: Outline of a Theory of Machines. London, Macmillan, 1876.

Seedhom, B. B., Dowson, D., and Wright, V.: The load-bearing function of the menisci: A preliminary study. *In* The Knee Joint. Recent Advances in Basic Research and Clinical Aspects. Edited by O. S. Ingwersen et al. Amsterdam, Excerpta Medica, 1974, pp. 37–42.

Selvik, G.: Röntgen stereophotogrammetry in Lund, Sweden. *In* Applications of Human Biostereometrics. Edited by A. M. Coblenz and R. E. Herron. Proc. SPIE 166, 1978, pp. 184–189.

Selvik, G.: Röntgen stereophotogrammetry in orthopaedics. *In* Biostereometrics '82. Edited by R. E. Herron. Proc. SPIE 361, 1983, pp. 178–185.

Smidt, G. L.: Biomechanical analysis of knee flexion and extension. J. Biomech., *6*:79, 1973.

Wiktorin, C. v. H., and Nordin, M.: Introduction to Problem Solving in Biomechanics. Philadelphia, Lea & Febiger, 1986, pp. 87–129.

SUGGESTED READING

Tibiofemoral Joint

Andriacchi, T. P., et al.: A study of lower-limb mechanics during stair-climbing. J. Bone Joint Surg., *62A*:749–757, 1980.

Brantigan, O. C., and Voshell, A. F.: The mechanics of the ligaments and menisci of the knee joint. J. Bone Joint Surg., *23A*:44–66, 1941.

Brattström, H. H., Junerfält, I., and Moritz, U.: Behandlingen av sträckdefekter i leder: Kontraktur eller fraktur? Läkartidningen, *68*:304–306, 1971.

Butler, D. L., Noyes, F. R., and Grood, E. S.: Ligamentous restraints to anterior-posterior drawer in the human knee. J. Bone Joint Surg., *62A*:259–270, 1980.

Cailliet, R.: Knee Pain and Disability. Philadelphia, F. A. Davis, 1968.

Collopy, M. C., et al.: Kinesiologic measurements of functional performance before and after geometric total knee replacement. One-year follow-up of twenty cases. Clin. Orthop., *126*:196–202, 1977.

Cox, J. S., Nye, C. E., Schaefer, W. W., and Woodstein, I. J.: The degenerative effects of partial and total resection of the medial meniscus in dogs' knees. Clin. Orthop., *109*:178–183, 1975.

Detenbeck, L. C.: Function of the cruciate ligaments in knee stability. J. Sports Med., *2*:217–221, 1974.

Drillis, R., Contini, R., and Bluestein, M.: Body segment parameters. A survey of measurement techniques. Artif. Limbs, *8*:44–66, 1964.

Ducroquet, R., Ducroquet, J., and Ducroquet, P.: Walking and Limping. A Study of Normal and Pathological Walking. Philadelphia, J. B. Lippincott, 1968.

Edholm, P., et al.: Knee instability. An orthoradiographic study. Acta Orthop. Scand., *47*:658–663, 1976.

Fairbank, T. J.: Knee joint changes after meniscectomy. J. Bone Joint Surg., *30B*:664–670, 1948.

Frankel, V. H., and Burstein, A. H.: Orthopaedic Biomechanics. Philadelphia, Lea & Febiger, 1970.

Frankel, V. H., Burstein, A. H., and Brooks, D. B.: Biomechanics of internal derangement of the knee. Pathomechanics as determined by analysis of the instant centers of motion. J. Bone Joint Surg., *53A*:945–962, 1971.

Hainaut, K.: Introduction à la biomecanique. Brussels, Presses Universitaires de Bruxelles, 1971.

Hallen, L. G., and Lindahl, O.: The "screw-home" movement in the knee-joint. Acta Orthop. Scand., *37*:97–106, 1966.

Helfet, A.: Disorders of the Knee. Philadelphia, J. B. Lippincott, 1974.

Hsieh, H.-H., and Walker, P. S.: Stabilizing mechanisms of the loaded and unloaded knee joint. J. Bone Joint Surg., *58A*:87–93, 1976.

Ingwersen, O. S., et al. (eds.): The Knee Joint. Recent Advances in Basic Research and Clinical Aspects. Amsterdam, Excerpta Medica, 1974.

Johnson, R. J., Kettelkamp, D. B., Clark, W., and Leaverton, P.: Factors affecting late results after meniscectomy. J. Bone Joint Surg., *56A*:719–729, 1974.

Kapandji, I. A.: The Physiology of the Joints. Vol. 2. Lower Limb. Edinburgh, Churchill Livingstone, 1970.

Kettelkamp, D. B., and Jacobs, A. W.: Tibiofemoral contact area—Determination and implications. J. Bone Joint Surg., *54A*:349–356, 1972.

Kettelkamp, D. B., et al.: An electrogoniometric study of knee motion in normal gait. J. Bone Joint Surg., *52A*:775–790, 1970.

Krause, W. R., Pope, M. H., Johnson, R. J., and Wilder, D. G.: Mechanical changes in the knee after meniscectomy. J. Bone Joint Surg., *58A*:599–604, 1976.

Laasonen, E. M., and Wilppula, E.: Why a meniscectomy fails. Acta Orthop. Scand., *47*:672–675, 1976.

Laubenthal, K. N., Smidt, G. L., and Kettelkamp, D. B.: A quantitative analysis of knee motion during activities of daily living. Phys. Ther., *52*:34–42, 1972.

Levens, A. S., Inman, V. T., and Blosser, J. A.: Transverse rotation of the segments of the lower extremity in locomotion. J. Bone Joint Surg., *30A*:859–872, 1948.

Lindahl, O., and Movin, A.: The mechanics of extension of the knee-joint. Acta Orthop. Scand., *38*:226–324, 1967.

Lindahl, O., Movin, A., and Ringqvist, I.: Knee extension. Acta Orthop. Scand., *40*:79–85, 1969.

Markolf, K. L., Graff-Radford, A., and Amstutz, H. C.: In vivo knee stability. J. Bone Joint Surg., *60A*:664–674, 1978.

Markolf, K. L., Mensch, J. S., and Amstutz, H. C.: Stiffness and laxity of the knee—The contributions of the supporting structures. J. Bone Joint Surg., *58A*:583–593, 1976.

Marshall, J. L., and Olsson, S.-E.: Instability of the knee. A long-term experimental study in dogs. J. Bone Joint Surg., *53A*:1561–1570, 1971.

McLeod, P. C., Kettelkamp, D. B., Srinivasan, V., and Henderson, O. L.: Measurements of repetitive activities of the knee. J. Biomech., *8*:369–373, 1975.

McLeod, W. D., Moschi, A., Andrews, J. R., and Hughston, J. C.: Tibial plateau topography. Am. J. Sports Med., *5*:13–18, 1977.

Moore, T. M., Meyers, M. H., and Harvey, P. J.: Collateral ligament laxity of the knee. J. Bone Joint Surg., *58A*:594–598, 1976.

Morrison, J. B.: Bioengineering analysis of force actions transmitted by the knee joint. Biomed. Eng., *3*:164–171, 1968.

Morrison, J. B.: The mechanics of the knee joint in relation to normal walking. J. Biomech., *3*:51–61, 1970.

Murray, M.: Gait as a total pattern of movement. Including a bibliography on gait. Am. J. Phys. Med., *46*:290–298, 1967.

Murray, M. P., Drought, A. B., and Kory, R. C.: Walking patterns of normal men. J. Bone Joint Surg., *46A*:335–360, 1964.

Nissan, M.: Review of some basic assumptions in knee biomechanics. J. Biomech., *13*:375–381, 1980.

Noyes, F. R.: Functional properties of knee ligaments and alterations induced by immobilization. Clin. Orthop., *123*:210–242, 1977.

Perry, J., Norwood, L., and House, K.: Knee posture and biceps and semimembranosis muscle action in running and cutting (an EMG study). Trans. Orthop. Res. Soc., *2*:258, 1977.

Pope, M. H., Crowninshield, R., Miller, R., and Johnson, R.: The static and dynamic behavior of the human knee *in vivo*. J. Biomech., *9*:449–452, 1976.

Reuleaux, F.: The Kinematics of Machinery: Outline of a Theory of Machines. London, Macmillan, 1876.

Roberts, E. M., Zernicke, R. F., Youm, Y., and Huang, T. C.: Kinetic parameters of kicking. *In* Biomechanics IV. Edited by R. C. Nelson and C. A. Morehouse. Baltimore, University Park Press, 1974, pp. 157–162.

Saunders, J. B. DeC. M., Inman, V. T., and Eberhardt, H. D.: The major determinants in normal and pathological gait. J. Bone Joint Surg., *35A*:543–558, 1953.

Scudder, G. N.: Torque curves produced at the knee during isometric and isokinetic exercise. Arch. Phys. Med. Rehabil., *61*:68–73, 1980.

Seedhom, B. B., Dowson, D., and Wright, V.: The load-bearing function of the menisci: A preliminary study. *In* The Knee Joint. Recent Advances in Basic Research and Clinical Aspects. Edited by O. S. Ingwersen et al. Amsterdam, Excerpta Medica, 1974, pp. 37–42.

Seireg, A., and Arvikar, R. J.: The prediction of muscular load sharing and joint forces in the lower extremities during walking. J. Biomech., *8*:89–102, 1975.

Shrive, N. G., O'Connor, J. J., and Goodfellow, J. W.: Load-bearing in the knee joint. Clin. Orthop., *131*:279–287, 1978.

Smidt, G. L.: Biomechanical analysis of knee flexion and extension. J. Biomech., 6:79–92, 1973.

Smillie, T.: Injuries of the Knee Joint. Edinburgh, Churchill Livingstone, 1970.

Stauffer, R. N., Chao, E. Y. S., and Györy, A. N.: Biomechanical gait analysis of the diseased knee joint. Clin. Orthop., *126*:246–255, 1977.

Townsend, M. A., Izak, M., and Jackson, R. W.: Total motion knee goniometry. J. Biomech., *10*:183–193, 1977.

Trent, P. S., Walker, P. S., and Wolf, B.: Ligament length patterns, strength, and rotational axes of the knee joint. Clin. Orthop., *117*:263–270, 1976.

Walker, P. S., and Erkman, M. J.: The role of the menisci in the force transmission across the knee. Clin. Orthop., *109*:184–192, 1975.

Walker, P. S., and Hajek, J. V.: The load-bearing area in the knee joint. J. Biomech., *5*:581–589, 1972.

Wang, C.-J., and Walker, P. S.: Rotatory laxity of the human knee joint. J. Bone Joint Surg., *56A*:161–170, 1974.

Wang, C.-J., Walker, P. S., and Wolf, B.: The effects of flexion and rotation on the length patterns of the ligaments of the knee. J. Biomech., *6*:587–596, 1973.

Warren, C. G., Lehmann, J. F., and Kirkpatrick, G. S.: Measurement of moments in the knee-ankle orthosis of ambulating paraplegics. *In* Biomechanics IV. Edited by R. C. Nelson and C. A. Morehouse. Baltimore, University Park Press, 1974, pp. 409–414.

Wiktorin, C.v.H., and Nordin, M.: Introduction to Problem Solving in Biomechanics. Philadelphia, Lea & Febiger, 1986, pp. 7–29, 87–129.

Williams, M., and Lissner, H.: Biomechanics of Human Motion. 2nd Ed. Edited by B. LeVeau. Philadelphia, W. B. Saunders, 1977.

Patellofemoral Joint

Bandi, W.: Die retropatellaren Kniegelenk-schäden. *In* Pathomechanik und pathologische Anatomie, Klinik und Therapie. Band 4. Aktuelle Probleme in Chirurgie und Orthopadie. Edited by M. Saegesser. Bern, Verlag Hans Huber, 1977.

Böstrom, A.: Fracture of the patella. A study of 422 patellar fractures. Acta Orthop. Scand., Suppl. *143*:1–80, 1972.

Brattström, H. H., Junerfält, I., and Moritz, U.: Behandlingen av sträckdefekter i leder: Kontraktur eller fraktur? Läkartidningen, *68*:304–306, 1971.

Cailliet, R.: Knee Pain and Disability. Philadelphia, F. A. Davis, 1968.

Drillis, R., Contini, R., and Bluestein, M.: Body segment parameters. A survey of measurement techniques. Artif. Limbs, *8*:44–66, 1964.

Frankel, V. H., and Burstein, A. H.: Orthopaedic Biomechanics. Philadelphia, Lea & Febiger, 1970.

Frankel, V. H., Burstein, A. H., and Brooks, D. B.: Biomechanics of internal derangement of the knee. J. Bone Joint Surg., *53A*:945–962, 1971.

Goodfellow, J., Hungerford, D. S., and Zindel, M.: Patellofemoral joint mechanics and pathology. I. Functional anatomy of the patello-femoral joint. J. Bone Joint Surg., *58B*:287–290, 1976.

Hainaut, K.: Introduction à la biomecanique. Brussels, Presses Universitaires de Bruxelles, 1971.

Helfet, A.: Disorders of the Knee. Philadelphia, J. B. Lippincott, 1974.

Huberti, H. H., and Hayes, W. C.: Patellofemoral contact pressures. The influence of Q-angle and tendofemoral contact. J. Bone Joint Surg., *66A*:715–724, 1984.

Kapandji, I. A.: The Physiology of the Joints. Vol. 2. Lower Limb. Edinburgh, Churchill Livingstone, 1970.

Kaufer, H.: Mechanical function of the patella. J. Bone Joint Surg., *53A*:1551–1560, 1971.

Matthews, L. S., Sonstegard, D. A., and Henke, J. A.: Load bearing characteristics of the patello-femoral joint. Acta Orthop. Scand., *48*:511–516, 1977.

Reilly, D. T., and Martens, M.: Experimental analysis of the quadriceps muscle force and patello-femoral joint reaction force for various activities. Acta Orthop. Scand., *43*:126–137, 1972.

Seireg, A., and Arvikar, R. J.: The prediction of muscular load sharing and joint forces in the lower extremities during walking. J. Biomech., *8*:89–102, 1975.

Smidt, G. L.: Biomechanical analysis of knee flexion and extension. J. Biomech., *6*:79–92, 1973.

Smillie, T.: Injuries of the Knee Joint. Edinburgh, Churchill Livingstone, 1970.

West, F. E.: End results of patellectomy. J. Bone Joint Surg., *44A*:1089–1108, 1962.

Wiktorin, C.v.H., and Nordin, M.: Introduction to Problem Solving in Biomechanics. Philadelphia, Lea & Febiger, 1968, pp. 7–29, 87–129.

Williams, M., and Lissner, H.: Biomechanics of Human Motion. 2nd Ed. Edited by B. LeVeau. Philadelphia, W. B. Saunders, 1977.

Quadriceps Muscle

Damholt, V., and Zdravkovic, D.: Quadriceps function following fractures of the femoral shaft. Acta Orthop. Scand., *43*:148–156, 1972.

Elftman, H.: Biomechanics of muscle. With particular application to studies of gait. J. Bone Joint Surg., *48A*:363–377, 1966.

Frankel, V. H., and Burstein, A. H.: Orthopaedic Biomechanics. Philadelphia, Lea & Febiger, 1970.

Grood, E. S., Suntay, W. J., Noyes, F. R., and Butler, D. L.: Biomechanics of the knee-extension exercise. Effect of cutting the anterior cruciate ligament. J. Bone Joint Surg., *66A*:725–734, 1984.

Haffajee, D., Moritz, U., and Svantesson, G.: Isometric knee extension strength as a function of joint angle, muscle length and motor unit activity. Acta Orthop. Scand., *43*:138–147, 1972.

Helfet, A.: Disorders of the Knee. Philadelphia, J. B. Lippincott, 1974.

Lieb, F. J., and Perry, J.: Quadriceps function. An anatomical and mechanical study using amputated limbs. J. Bone Joint Surg., *50A*:1535–1548, 1968.

Lieb, F. J., and Perry, J.: Quadriceps function. An electromyographic study under isometric conditions. J. Bone Joint Surg., *53A*:749–758, 1971.

Reilly, D. T., and Martens, M.: Experimental analysis of the quadriceps muscle force and patello-femoral joint reaction force for various activities. Acta Orthop. Scand., *43*:126–137, 1972.

Smillie, I. S.: Injuries of the Knee Joint. Edinburgh, Churchill Livingstone, 1970.

Williams, M., and Lissner, H.: Biomechanics of Human Motion. 2nd Ed. Edited by B. LeVeau. Philadelphia, W. B. Saunders, 1977.

Zernicke, R. F., Garhammer, J., and Jobe, F. W.: Human patellar-tendon rupture. J. Bone Joint Surg., *59A*:179–183, 1977.

7

BIOMECHANICS OF THE HIP

Margareta Nordin
Victor H. Frankel

The hip joint is one of the largest and most stable joints in the body. In contrast to the knee, the hip joint has intrinsic stability, provided by its relatively rigid ball-and-socket configuration. It also has a great deal of mobility, which allows normal locomotion in the performance of daily activities. Derangements of the hip can produce altered stress distributions in the joint cartilage and bone, leading to degenerative arthritis. Such damage is further potentiated by the large forces borne by the joint.

ANATOMIC CONSIDERATIONS

The hip joint is composed of the head of the femur and the acetabulum of the pelvis (Fig. 7–1). This articulation has a loose joint capsule and is surrounded by large, strong muscles. The construction of this stable joint allows for the wide range of motion required for normal daily activities such as walking, sitting, and squatting. Such a joint must be precisely aligned and controlled.

THE ACETABULUM

The acetabulum is the concave component of the ball-and-socket configuration of the hip joint. The acetabular surface is covered with articular cartilage that thickens peripherally (Kempson et al., 1971) and predominantly laterally (Rushfeld et al., 1979). The cavity of the acetabulum faces obliquely forward, outward, and downward. A plane through the circumference of the acetabulum at its opening would intersect with the sagittal plane at an angle of 40 degrees opening posteriorly and with the transverse plane at an angle of 60 degrees opening laterally. The acetabular cavity is deepened by the labrum, a flat rim of fibrocartilage.

THE FEMORAL HEAD

The femoral head is the convex component of the ball-and-socket configuration of the hip joint and forms two thirds of a sphere. The articular cartilage covering the femoral head is thickest on the medial-central surface and thinnest toward the periphery. The variations in the cartilage thickness result in a different strength and stiffness in various regions of the femoral head (Kempson et al., 1971). These differences in the mechanical properties from point to point on the femoral head cartilage may influence the transmission of stresses from the acetabulum through the femoral head to the femoral neck. Although it is not known just how stresses on the femoral head are distributed, the joint reaction force usually acts on the superior quadrant (Rydell, 1965).

THE FEMORAL NECK

The femoral neck has two angular relationships with the femoral shaft that are important to hip joint

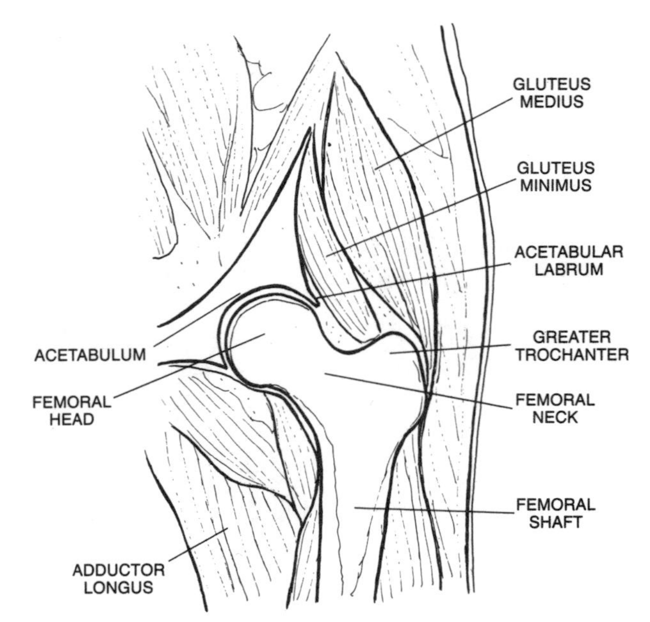

FIG. 7–1
Schematic drawing of the hip joint (front view).

function: the angle of inclination of the neck to the shaft in the frontal plane (the neck-to-shaft angle) and the angle of inclination in the transverse plane (the angle of anteversion). Freedom of motion of the hip joint is facilitated by the neck-to-shaft angle, which offsets the femoral shaft from the pelvis laterally. In most adults this angle is about 125 degrees, but it can vary from 90 to 135 degrees. An angle exceeding 125 degrees produces a condition known as coxa valga; an angle less than 125 degrees results in coxa vara (Fig. 7–2). Deviation of the femoral shaft in either way alters the force relationships about the hip joint.

The angle of anteversion is formed as a projection of the long axis of the femoral head and the transverse axis of the femoral condyles (Fig. 7–3). In adults this angle averages about 12 degrees, but it varies a great deal. Anteversion of more than 12 degrees causes a portion of the femoral head to be uncovered and creates a tendency toward internal rotation of the leg during gait to keep the femoral head in the acetabular cavity. An angle of less than 12 degrees (retroversion) produces a tendency toward external rotation of the leg during gait. Both anteversion and retroversion are fairly common in children but are usually outgrown.

The interior of the femoral neck is composed of cancellous bone with trabeculae organized into medial and lateral trabecular systems (Fig. 7–4). The fact that the joint reaction force on the femoral head parallels the trabeculae of the medial system (Frankel, 1960) indicates the importance of the system for supporting this force. The epiphyseal plates are at right angles to the trabeculae of the medial system and are thought to be perpendicular to the joint reaction force on the femoral head (Inman, 1947). It is likely that the lateral trabecular system resists the compressive force on the femoral head produced by contraction of the abductor muscles—the gluteus medius, the gluteus minimus, and the tensor fasciae latae. The thin shell of cortical bone around the superior femoral neck progressively thickens in the inferior region.

With aging the femoral neck gradually undergoes degenerative changes: the cortical bone is thinned and cancellated and the trabeculae are gradually

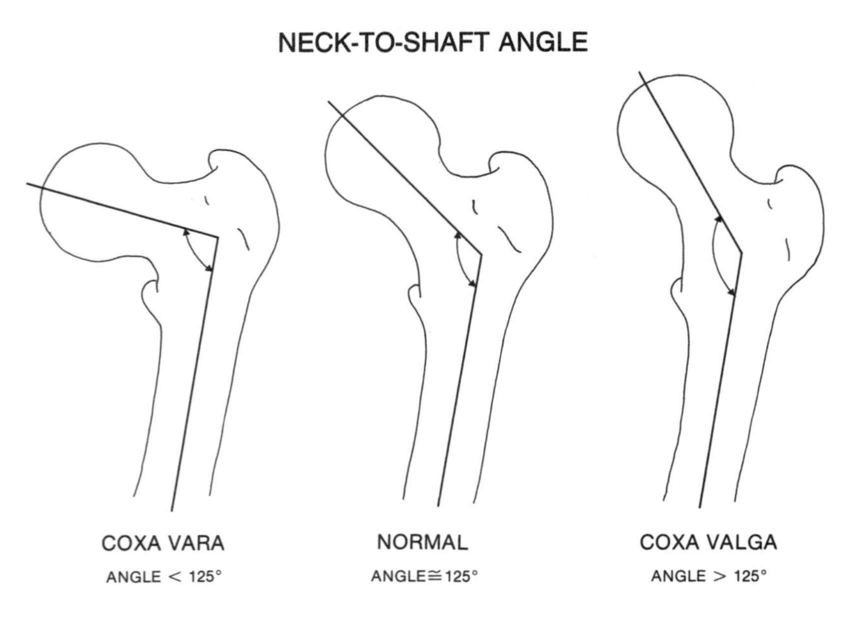

FIG. 7–2
The normal neck-to-shaft angle (angle of inclination of the femoral neck to the shaft in the frontal plane) is approximately 125 degrees. The condition wherein this angle is less than 125 degrees is called coxa vara. If the angle is greater than 125 degrees, the condition is called coxa valga.

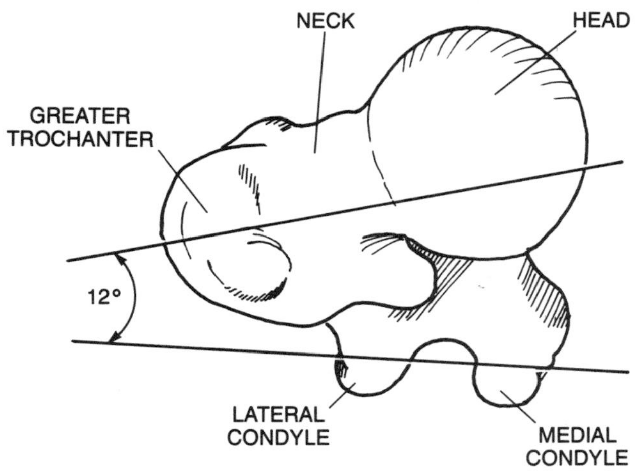

FIG. 7–3
Top view of the proximal end of the left femur showing the angle of anteversion, formed by the intersection of the long axis of the femoral head and the transverse axis of the femoral condyles. This angle averages about 12 degrees in adults.

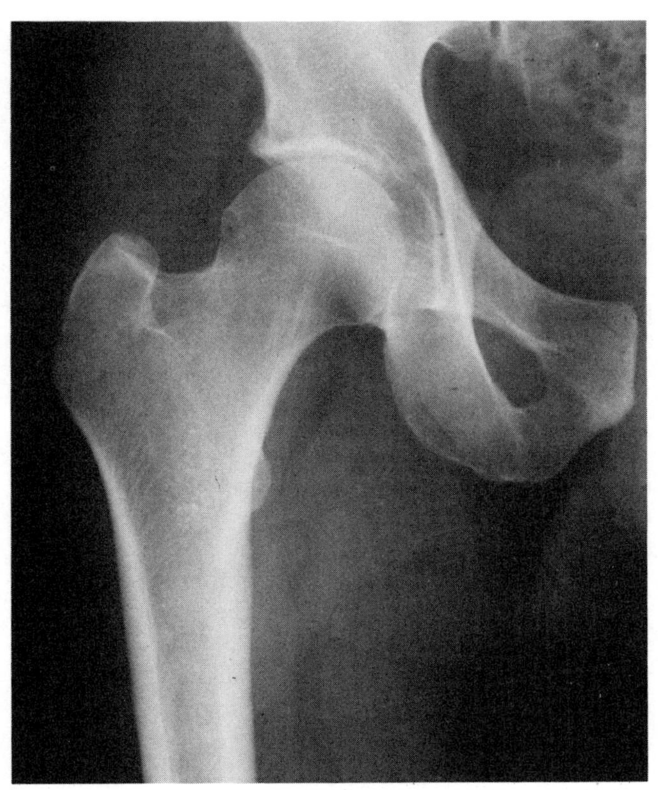

FIG. 7–4
Roentgenogram of a femoral neck showing the medial and lateral trabecular systems. The thin shell of cortical bone around the superior femoral neck progressively thickens in the inferior region.

resorbed (see Fig. 1–49). These changes may predispose the femoral neck to fracture. It is noteworthy that the femoral neck is the most common fracture site in elderly persons.

KINEMATICS

In considering the kinematics of the hip joint, it is useful to view the joint as a stable ball-and-socket configuration wherein the femoral head and acetabulum can move in all directions.

RANGE OF MOTION

Hip motion takes place in all three planes: sagittal (flexion-extension), frontal (abduction-adduction), and transverse (internal and external rotation) (Fig. 7–5). Motion is greatest in the sagittal plane, where the range of flexion is from zero to approximately 140 degrees and the range of extension is from zero to 15 degrees. The range of abduction is from zero to 30 degrees, whereas that of adduction is somewhat less, from zero to 25 degrees. External rotation ranges from zero to 90 degrees and internal rotation from zero to 70 degrees when the hip joint is flexed. Less rotation occurs when the hip joint is extended because of the restricting function of the soft tissues.

The range of motion of the hip joint during gait has been measured electrogoniometrically in all three planes. Measurements in the sagittal plane during level walking (Murray, 1967) showed that the joint was maximally flexed during late swing phase of gait, as the limb moved forward for heel strike. The joint extended as the body moved forward at the beginning of stance phase. Maximum extension was reached at heel-off. The joint reversed into flexion during swing phase and again reached maximal flexion, 35 to 40 degrees, prior to heel strike. Figure 7–6 shows the pattern of hip joint motion in the sagittal plane during a gait cycle and allows a comparison of this motion with that of the knee and ankle.

Motion in the frontal plane (abduction-adduction) and transverse plane (internal and external rotation) during gait (Johnston and Smidt, 1969) is illustrated in Figure 7–7. Abduction occurred during swing phase, reaching its maximum just after toe-off; at heel strike the hip joint reversed into adduction, which continued until late stance phase. The hip joint was externally rotated throughout the swing phase, rotat-

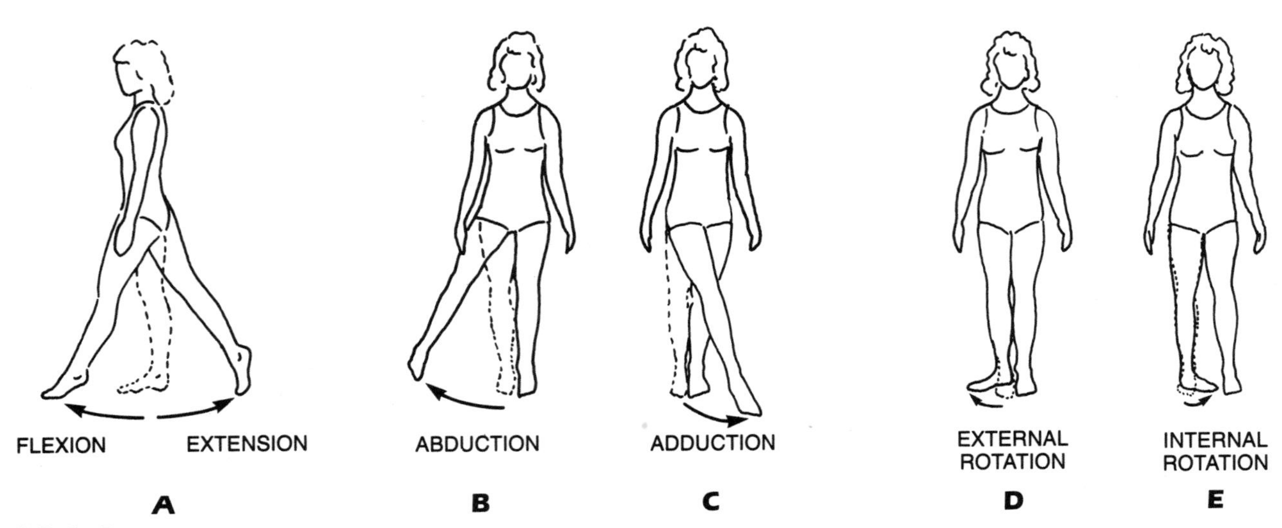

FIG. 7–5
Movements of the hip joint. **A.** Flexion-extension. **B.** Abduction. **C.** Adduction. **D.** External rotation. **E.** Internal rotation.

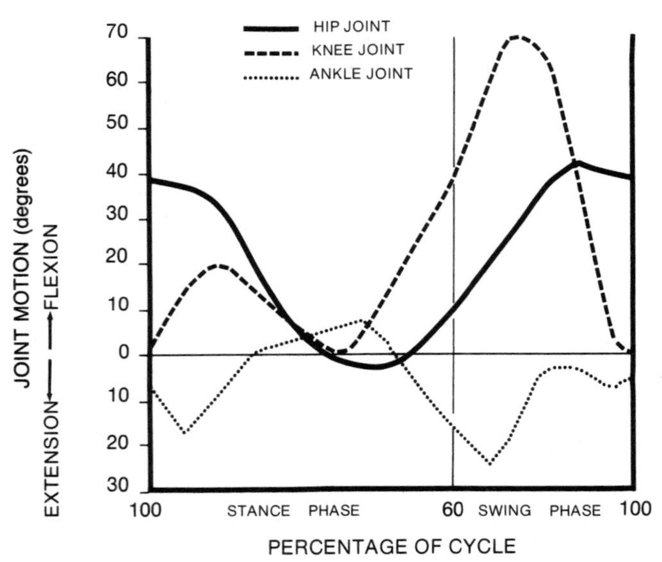

FIG. 7–6
Range of hip joint motion in the sagittal plane for 30 normal men during level walking, one gait cycle. The ranges of motion for the knee and ankle joints are shown for comparison. (Adapted from Murray, 1967.)

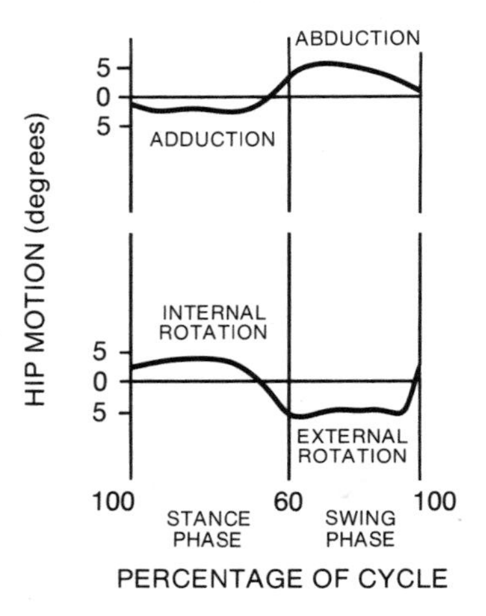

FIG. 7–7
A typical pattern for range of motion in the frontal plane (top) and transverse plane (bottom) during level walking, one gait cycle. (Adapted from Johnston and Smidt, 1969.)

ing internally just before heel strike. The joint remained internally rotated until late stance phase, when it again rotated externally. The average ranges of motion recorded for the 33 normal men in this study were 12 degrees for the frontal plane and 13 degrees for the transverse plane.

As people age they use a progressively smaller portion of the range of motion of the lower extremity joints during ambulation. Murray and coworkers (1969) studied the walking patterns of 67 normal men of similar weight and height ranging in age from 20 to 87 years and compared the gait patterns of older and younger men. The differences in the sagittal body positions of the two groups at the instant of heel strike are illustrated in Figure 7–8. The older men had shorter strides, a decreased range of hip flexion and extension, decreased plantar flexion of the ankle, and a decreased heel-to-floor angle of the tracking limb; they also showed reduced dorsiflexion of the ankle and diminished elevation of the toe of the forward limb.

The range of motion in three planes during common daily activities such as tying a shoe, sitting

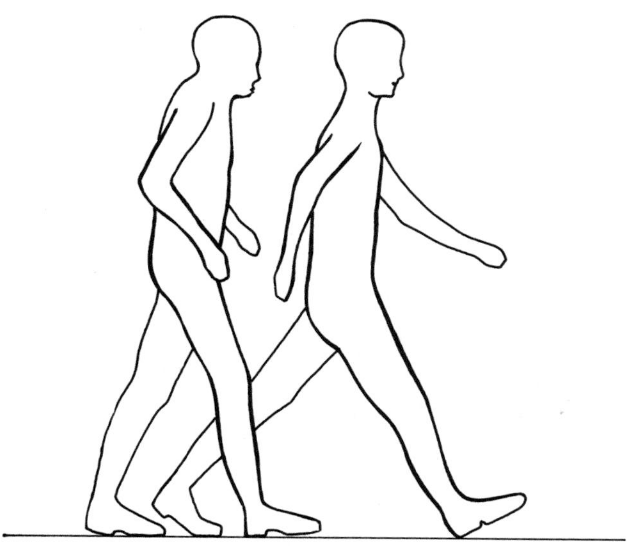

FIG. 7–8
Differences in the sagittal body positions of older men (left) and younger men (right) at the instant of heel strike. The older men showed shorter strides, a decreased range of hip flexion and extension, decreased plantar flexion of the ankle, and a decreased heel-to-floor angle of the tracking limb; they also showed less dorsiflexion of the ankle and less elevation of the toe of the forward limb. (Reprinted with permission from Murray, M. P., et al.: Walking patterns in healthy old men. J. Gerontol., *24*:169–178, 1969.)

TABLE 7–1
MEAN VALUES FOR MAXIMUM HIP MOTION IN THREE PLANES DURING COMMON ACTIVITIES

ACTIVITY	PLANE OF MOTION	RECORDED VALUE (DEGREES)
Tying shoe with foot on floor	Sagittal	124
	Frontal	19
	Transverse	15
Tying shoe with foot across opposite thigh	Sagittal	110
	Frontal	23
	Transverse	33
Sitting down on chair and rising from sitting	Sagittal	104
	Frontal	20
	Transverse	17
Stooping to obtain object from floor	Sagittal	117
	Frontal	21
	Transverse	18
Squatting	Sagittal	122
	Frontal	28
	Transverse	26
Ascending stairs	Sagittal	67
	Frontal	16
	Transverse	18
Descending stairs	Sagittal	36

(Data from Johnston and Smidt, 1970. Mean for 33 normal men.)

down on a chair, rising from it, picking up an object from the floor, and climbing stairs was measured electrogoniometrically in 33 normal men by Johnston and Smidt (1970). The mean motion during these activities is shown in Table 7–1. Maximal motion in the sagittal plane (hip flexion) was needed for tying the shoe, and bending down to squat to pick up an object from the floor. The greatest motion in the frontal and transverse planes (abduction-adduction and internal-external rotation, respectively) was recorded during squatting and during shoe tying with the foot across the opposite thigh. The values obtained for these common activities indicate that hip flexion of at least 120 degrees abduction and external rotation of at least 20 degrees are necessary for carrying out daily activities in a normal manner.

SURFACE JOINT MOTION

Surface motion in the hip joint can be considered as gliding of the femoral head on the acetabulum. The pivoting of the ball and socket in three planes around the center of rotation in the femoral head produces this gliding of the joint surfaces. If incongruity is present in the femoral head, gliding may not be parallel or tangential to the joint surface and the articular cartilage may be abnormally compressed or distracted. Instant center analysis by means of the Reuleaux method cannot be performed accurately in the hip joint because motion takes place in three planes simultaneously.

KINETICS

Kinetic studies have demonstrated that substantial forces act on the hip joint during simple activities. The factors involved in producing these forces must be understood if rational rehabilitation programs are to be developed for patients with pathologic conditions of the hip.

STATICS

During a two-leg stance the line of gravity of the superincumbent body passes posterior to the

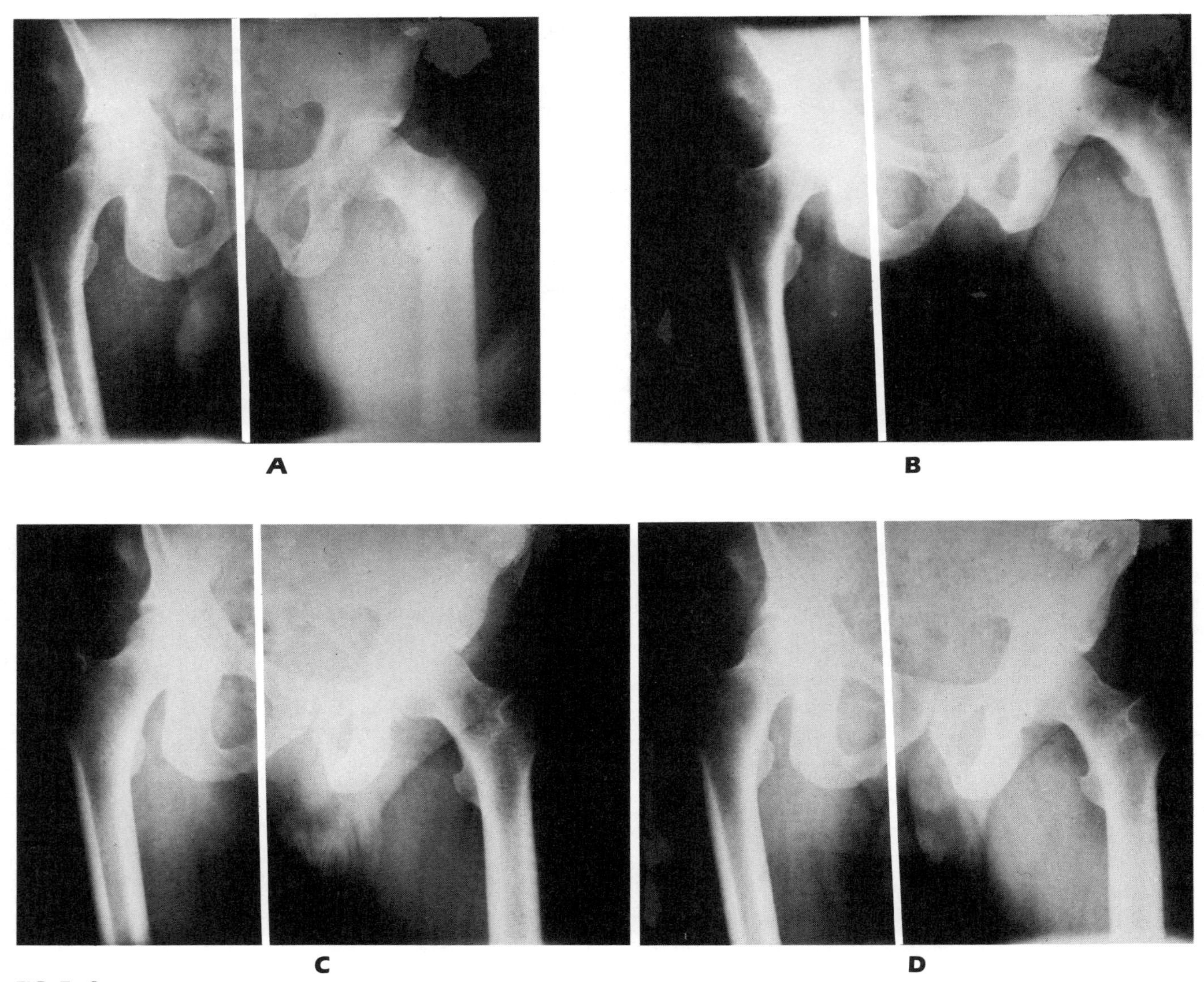

FIG. 7–9
Roentgenograms utilizing a plumb line (white line) show that the line of gravity shifts in the frontal plane with different positions of the upper body and inclinations of the pelvis. **A.** The pelvis is in a neutral position. The gravity line falls approximately through the pubic symphysis. The lever arm for the force produced by body weight (the perpendicular distance between the gravity line and the center of rotation in the femoral head) influences the moment about the hip joint and hence the joint reaction force. **B.** The shoulders are maximally tilted over the supporting hip joint. The gravity line has shifted and is now nearest the supporting hip. Because this shift minimizes the lever arm, the moment about the hip joint and the joint reaction force are also minimized. **C.** The shoulders are maximally tilted away from the supporting hip joint. Again the gravity line has shifted toward the supporting hip, thus decreasing the joint reaction force. **D.** The pelvis sags away from the supporting hip joint (Trendelenburg's test). The shift in the gravity line is similar to that in Part C. (Courtesy of John C. Baker, M.D.)

pubic symphysis, and since the hip joint is stable, an erect stance can be achieved without muscle contraction through the stabilizing effect of the joint capsule and capsular ligaments. With no muscle activity to produce moments around the hip joint, calculation of the joint reaction force becomes simple: the magnitude of this force on each femoral head during upright standing is one half the weight of the superincumbent body. Since each lower extremity is one sixth of body weight, the reaction force on each hip joint will be one half of the remaining two thirds, or one third of body weight; however, if the muscles surrounding the hip joint contract to prevent swaying and to maintain an upright position of the body (e.g., during prolonged standing), this force increases in proportion to the amount of muscle activity.

When a person changes from a two-leg to a single-leg stance, the line of gravity of the superincumbent body shifts in all three planes, producing moments around the hip joint that must be counteracted by muscle forces and thus increasing the joint reaction force. The magnitude of the moments, and hence the magnitude of the joint reaction force, depend on the posture of the spine, the position of the non-weight-bearing leg and upper extremities, and, in particular, the inclination of the pelvis (McLeish and Charnley, 1970). Figure 7–9 demonstrates how the line of gravity in the frontal plane shifts with four different positions of the upper body and inclinations of the pelvis: standing with the pelvis in a neutral position, standing with a maximum tilt of the upper body over the supporting hip joint, standing with the upper body tilting away from the supporting hip joint, and standing with the pelvis sagging away from the supporting hip joint (Trendelenburg's test). The shift in the gravity line, and hence in the length of the lever arm of the gravitational force (the perpendicular distance between the gravity line and the center of rotation in the femoral head), influences the magnitude of the moments about the hip joint and, consequently, the joint reaction force. The gravitational force lever arm and the joint reaction force are minimized when the trunk is tilted over the hip joint (see Fig. 7–9B).

In the following sections, two methods for deriving the magnitude of the joint reaction force acting on the femoral head are illustrated: the simplified free body technique for coplanar forces and a mathematical method utilizing equilibrium equations.

SIMPLIFIED FREE BODY TECHNIQUE FOR COPLANAR FORCES

The simplified free body technique for coplanar forces was described in detail in Chapter 6. In the following example this technique is used to estimate the joint reaction force in the frontal plane acting on the femoral head during a single-leg stance with the pelvis in a neutral position. The stance limb is considered as a free body, and a free body diagram is drawn. From all of the forces acting on the free body, the three main coplanar forces are identified as the force of gravity against the foot (the ground reaction force), which is transmitted through the tibia to the femoral condyles; the force produced by contraction of the abductor muscles; and the joint reaction force on the femoral head. The *ground reaction force* (W) has a known magnitude equal to five sixths of body weight, and a known sense, line of application, and point of application. The *abductor muscle force* (M) has a known sense, a known line of application and point of application estimated from the muscle origin and insertion on a roentgenogram, but an unknown magnitude. Because several muscles are involved in the action of hip abduction, simplifying assumptions are made in determining the direction of this force (McLeish and Charnley, 1970). Furthermore, forces produced by other muscles active in stabilizing the hip joint are not taken into account. The *joint reaction force* (J) has a known point of application on the surface of the femoral head, but an unknown magnitude, sense, and line of application.

The magnitudes of the abductor muscle force and the joint reaction force can be derived by designating all three forces on the free body diagram (Fig. 7–10A) and constructing a triangle of forces (Fig. 7–10B). The muscle force is found to be approximately two times body weight, whereas the joint reaction force is somewhat greater.

MATHEMATICAL METHOD UTILIZING EQUILIBRIUM EQUATIONS

The mathematical method of calculating the joint reaction force on the femoral head using equilibrium equations will be demonstrated for a single-leg stance with the pelvis level. First, the external forces acting on the body during the single-leg stance are indicated on a free body diagram (Fig. 7–11A). Since the body is in equilibrium (i.e., the sum of the moments and the sum of the forces both equal zero),

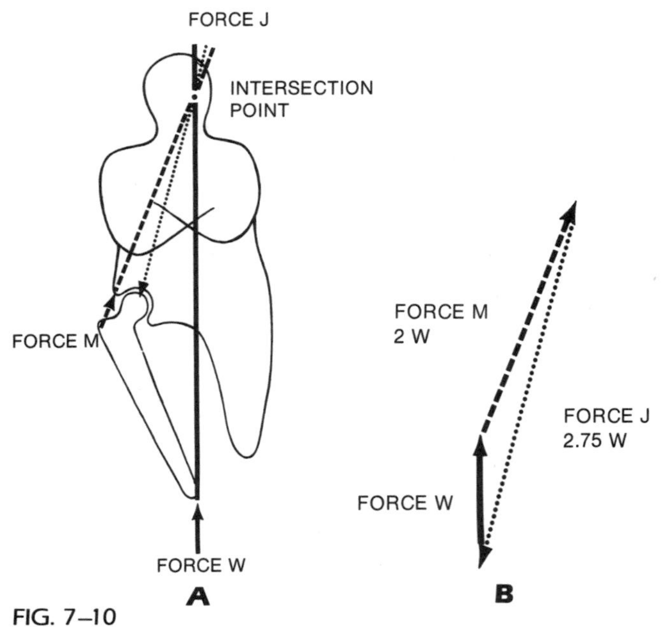

FIG. 7–10
A. On a free body diagram of the upper body and supporting lower extremity, the lines of application for W and M are extended until they meet (intersection point). The line of application for J is then determined by connecting its point of application (contact point of the acetabulum and femoral head) with the intersection point for W and M. **B.** A triangle of forces is constructed. The magnitudes of M and J can be scaled from W, which is equal to body weight. Force M is approximately two times body weight, and force J is about 2.75 times body weight.

the ground reaction force is equal to the gravitational force of the body, which can be divided into two components, the gravitational force of the stance leg (equal to one sixth body weight) and the remaining force (equal to five sixths body weight).

Next, the body is divided at the hip joint into two free bodies. The main coplanar forces and moments acting on these free bodies must be determined. The upper free body is considered first (Fig. 7–11B). In this free body two moments are required for stability. The moment arising from the superincumbent body weight (equal to 5/6 W) must be balanced by a moment arising from the force of the abductor muscles. The force produced by the superincumbent body weight (5/6 W) acts at a distance of b from the center of rotation of the hip (Q), thus producing a moment of 5/6 W times b. The force produced by the principal abductor, the gluteus medius, designated as M, acts at a distance of c from the center of rotation, producing a counterbalancing moment of M times c. For the body to remain in moment equilibrium, the sum of the moments must equal zero. In this example, the moments acting clockwise are considered to be positive and the counterclockwise moments are considered to be negative. Thus,

$$(5/6\ W \times b) - (M \times c) = 0$$

$$M = \frac{5/6\ W \times b}{c}.$$

To solve for M it is necessary to find the values of b and c. The gravitational force lever arm (b) is found roentgenographically. Since the center of gravity must lie over the base of support, a plumb line intersecting the heel can be extended upward; a perpendicular drawn from the center of rotation in the femoral head (Q) to the line represents distance b. The muscle force lever arm (c) is similarly found by identifying the gluteus medius muscle on a roentgenogram and drawing a perpendicular from the center of rotation of the femoral head to a line approximating the gluteus medius muscle tendon.

In this example a value for M of two times body weight has been chosen (Rydell, 1966; English and Kilvington, 1979). The direction of force M is found from a roentgenogram to be 30 degrees from the vertical. The horizontal and vertical components of this force are found by vector analysis (Fig. 7–11C). The horizontal component (M_x) is equal to body weight; the vertical component (M_y) is approximately 1.7 times body weight.

Attention is then directed to the lower free body (Fig. 7–12A). The gravitational forces (W and 1/6 W) are known. The joint reaction force (force J) has an unknown magnitude and direction but must pass through the center of rotation in the femoral head. The magnitude of force J is determined by finding the horizontal and vertical force components and adding them (Fig. 7–12B). The vertical and horizontal components of the forces acting on the lower free body are identified. Since the body is in force equilibrium, the sum of the forces in the horizontal direction must equal zero and so must the forces in the vertical direction. The horizontal and vertical forces are added:

$M_X - J_X = 0$	$M_Y - J_Y - 1/6\ W + W = 0$
$M_X = J_X$	$M_Y \simeq 1.7\ W$
$M_X = W$	$J_Y \simeq 1.7\ W + 5/6\ W$
$J_X = W$	$J_Y \simeq 2.5\ W$

The value of J is found by vector addition (Fig. 7–12C), and its direction is measured on the parallelogram of forces. The joint reaction force on the

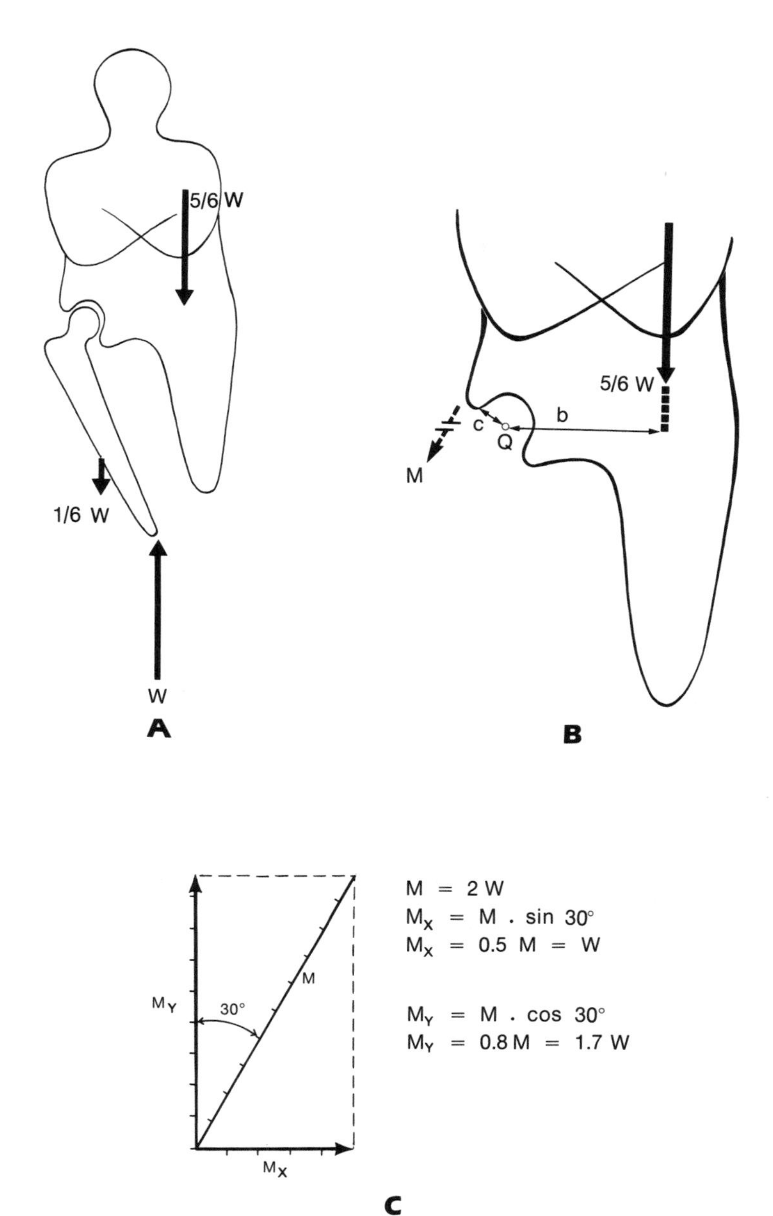

FIG. 7–11

A. External forces acting on the body in equilibrium during a single-leg stance. The ground reaction force is equal to body weight (W). The gravitational force of the stance leg is equal to one sixth of body weight; the remaining force is equal to five sixths of body weight. **B.** The internal forces acting on the hip joint are found by separating the joint into an upper and lower free body; the upper free body is considered first. Moment equilibrium is attained by the production of two equal moments. A moment arising from the force of the abductor muscle (M × c) counterbalances the moment arising from the gravitational force of the superincumbent body (5/6 W × b), which tends to tilt the pelvis away from the supporting lower extremity. Q, center of rotation in the femoral head; M, abductor muscle force; 5/6 W, gravitational force of the body above the hip joint; c, abductor force lever arm; b, gravitational force lever arm. **C.** Force M is equal to two times body weight and has a direction of 30 degrees from the vertical. The magnitudes of its horizontal (M_X) and vertical (M_Y) components are found by vector analysis. Perpendiculars are drawn from the tip of M to a horizontal and a vertical line extended from the base of the force, representing M_X and M_Y, respectively. M_X and M_Y can then be scaled off. Alternatively, trigonometry is used to find the magnitudes of the components.

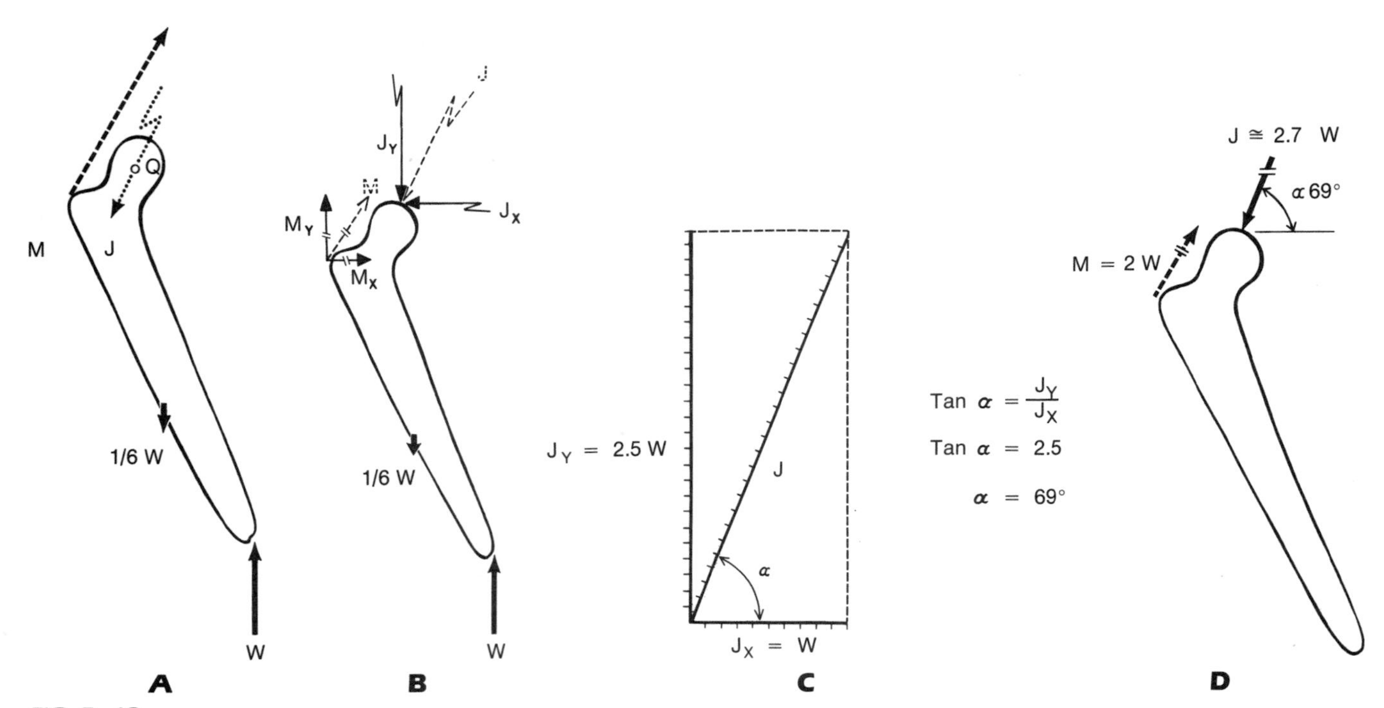

FIG. 7–12
A. The supporting lower extremity is considered as a free body, and the forces acting on the free body are identified. M, abductor muscle force; J, joint reaction force; 1/6 W, gravitational force of the limb; W, ground reaction force; Q, center of rotation in the femoral head. **B.** The forces acting on the lower free body are divided into horizontal and vertical components. The magnitudes of J_Y and J_X are found from force equilibrium equations. **C.** Addition of the horizontal and vertical components J_X and J_Y is performed graphically and the joint reaction force (J) is scaled off. Its direction is measured on the parallelogram of forces. Alternately, trigonometry is used to find the direction of J. **D.** The joint reaction force has a magnitude of approximately 2.7 times body weight and acts at an angle of 69 degrees from the horizontal.

femoral head in a single-leg stance with the pelvis level is found to be approximately 2.7 times body weight, and its direction is 69 degrees from the horizontal (Fig. 7–12D).

A key factor influencing the magnitude of the joint reaction force on the femoral head is the ratio of the abductor muscle force lever arm (c) to the gravitational force lever arm (b). Figure 7–13 illustrates the relationship of this ratio to the joint reaction force. A low ratio (i.e., a small muscle force lever arm and a large gravitational force lever arm) yields a greater joint reaction force than does a high ratio.

A short lever arm of the abductor muscle force, as in coxa valga (see Fig. 7–2), results in a small ratio and thus a somewhat elevated joint reaction force. Moving the greater trochanter laterally during total hip replacement lowers the joint reaction force, as it increases the lever arm ratio by increasing the muscle force lever arm. Inserting a prosthetic cup deeper in the acetabulum reduces the gravitational force lever arm, thereby increasing the ratio and decreasing the joint reaction force. It is difficult, however, to change the lever arm ratio in such a way as to reduce the joint reaction force significantly, because the curve formed from plotting the ratios becomes asymptotic when the ratio of c to b approaches 0.8 (see Fig. 7–13).

The ratio of the abductor muscle force lever arm (c) to the gravitational force lever arm (b) was determined during a one-leg stance before and after hip joint resurfacing in a patient with degenerative arthritis of the right hip. After joint resurfacing the line of gravity for the superincumbent body shifted laterally away from the affected hip joint as the patient assumed a more normal body position, having been relieved of hip joint pain by the surgical procedure (Fig. 7–14). This shift in the line of gravity

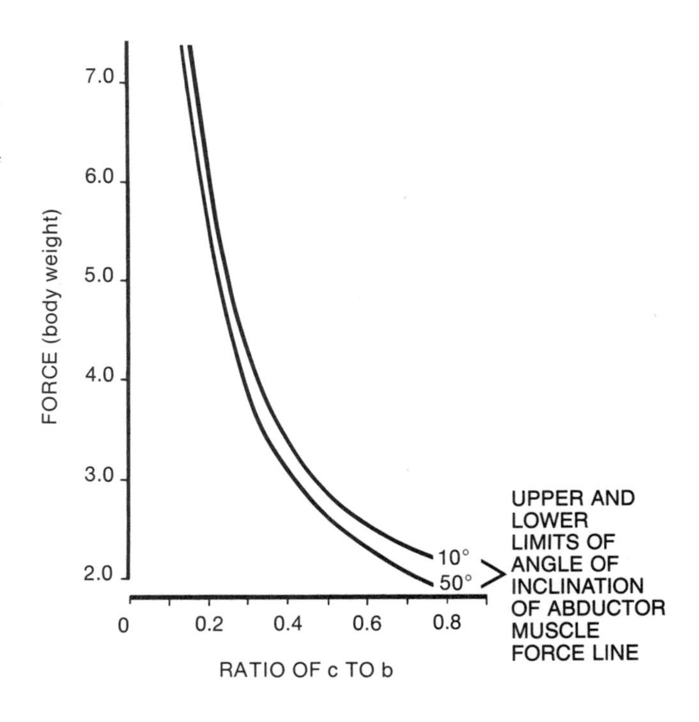

FIG. 7–13
The value of the ratio of the abductor muscle force lever arm (c) to the gravitational force lever arm (b) is plotted against the joint reaction force on the femoral head in units of body weight. Since the line of application of the abductor muscle force (its angle of inclination in the frontal plane) has finite upper and lower limits (10 degrees and 50 degrees), the force envelope is plotted. The curve can be utilized to determine the minimal force acting on the femoral head during a one-leg stance if the ratio of c to b is known. (Adapted from Frankel, 1960.)

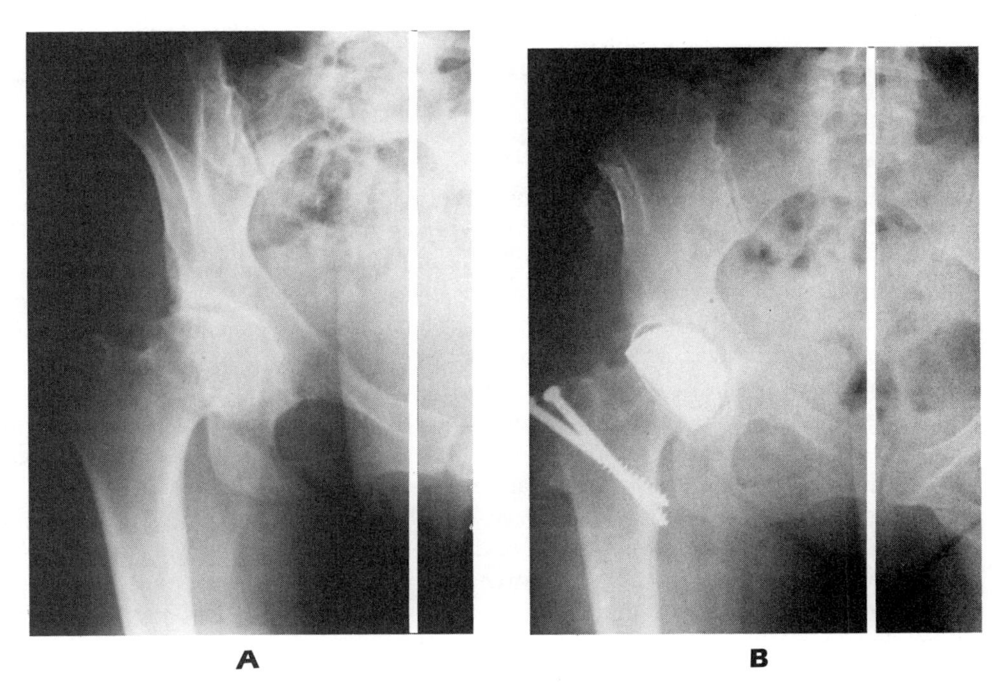

FIG. 7–14
Roentgenograms of the right hip of a patient with degenerative arthritis before and after hip joint resurfacing. The patient stood on the right foot while lifting the left foot 3.5 cm from the floor. **A.** Before surgery. The line of gravity is represented by a plumb line (white line). **B.** After surgery. The line of gravity shifted laterally away from the affected hip joint because the patient assumed a more normal body configuration. (Courtesy of William G. Boettcher, M.D.)

increased the gravitational force lever arm, thereby reducing the ratio of c to b.

With the direction of the muscle force taken to be 15 degrees to the vertical, the ratio of c to b before surgery was found to be 7.0 and the joint reaction force about 2.2 times body weight. After surgery the ratio was 6.0 and the joint reaction force about 2.5 times body weight. The patient was able to tolerate this higher joint reaction force on the femoral head because joint resurfacing relieved her pain and restored the mechanical function of the joint.

DYNAMICS

The loads on the hip joint during dynamic activities have been studied by several investigators (Rydell, 1965, 1966; Paul, 1967; Seireg and Arvikar, 1975; English and Kilvington, 1979; Draganich et al., 1980; Andriacchi et al., 1980). Using a force plate system and kinematic data for the normal hip, Paul (1967) examined the joint reaction force on the femoral head in normal men and women during gait and correlated the peak magnitudes with specific muscle activity recorded electromyographically. In the men two peak forces were produced during the stance phase when the abductor muscles contracted to stabilize the pelvis. One peak of about four times body weight occurred just after heel strike, and a large peak of about seven times body weight was reached just before toe-off (Fig. 7–15A). During foot flat, the joint reaction force decreased to about body weight because of the rapid deceleration of the center of gravity of the body. During the swing phase the joint reaction force was influenced by contraction of the extensor muscles in decelerating the thigh, and the magnitude remained relatively low, about equal to body weight.

In the women the force pattern was the same, but the magnitude was somewhat lower, reaching a maximum of only about four times body weight at late stance phase (Fig. 7–15B). The lower magnitude of the joint reaction force in the women may have been due to several factors: a wider female pelvis, a difference in the inclination of the femoral neck-to-shaft angle, a difference in footwear, and differences in the general pattern of gait.

Intravital measurements of the forces acting on an instrumented hip joint prosthesis also demonstrated that a large joint reaction force may act on the femoral head during the stance phase of gait (Rydell, 1965)

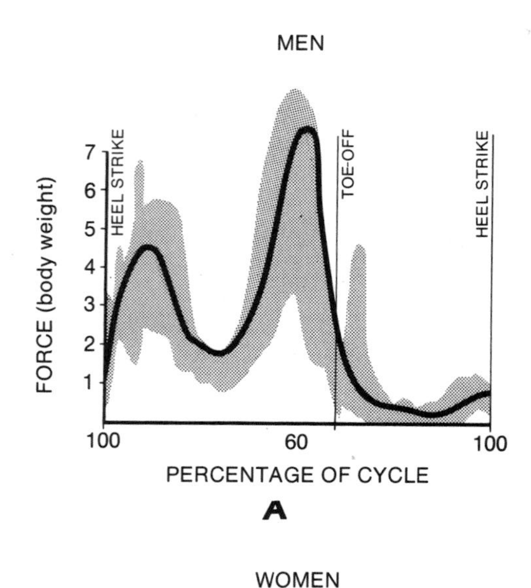

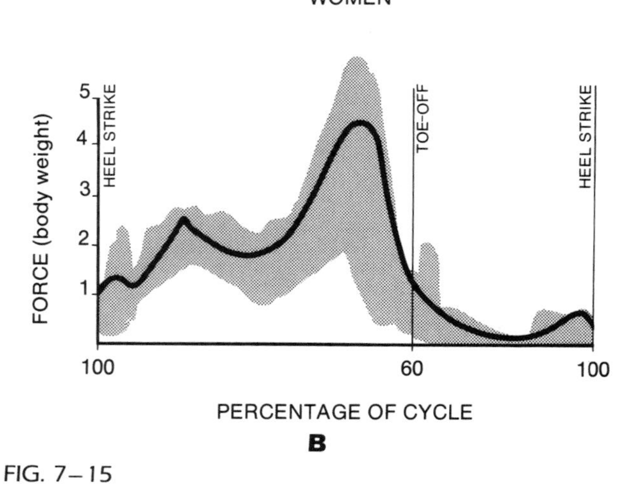

FIG. 7–15
Hip joint reaction force in units of body weight during walking, one gait cycle. Shaded area indicates variations among subjects. **A.** Force pattern for normal men. **B.** Force pattern for normal women. (Adapted from Paul, 1967.)

(Fig. 7–16A). At a faster cadence the forces acting on the prosthesis greatly increased because of an increase in muscle activity (Fig. 7–16B). At both cadences the magnitude of the forces during swing phase was about half that during stance phase.

Insertion of an instrumented nail plate in the proximal femur after osteotomy or during fixation of a femoral neck fracture allowed a subsequent determination of the forces acting on the implant during the activities of daily living (Fig. 7–17) (Frankel et al., 1971; Lygre, 1970; Milde, 1974). Although the device measured forces on the implant and not on the hip joint, it was possible to determine the proportion of the load transmitted through the device and to calculate the total load acting on the hip joint by

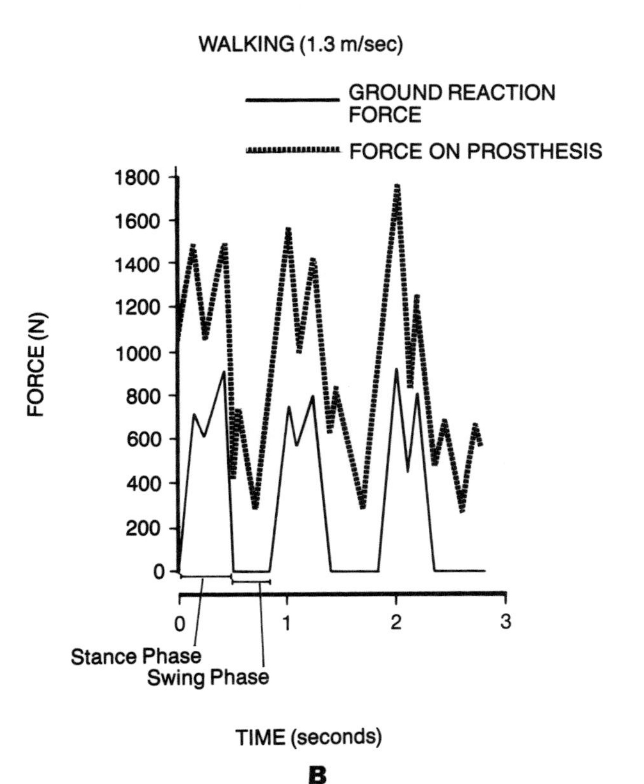

FIG. 7–16
Forces on an instrumented hip prosthesis during walking. The broken line represents the force on the prosthesis, and the solid line represents the ground reaction force. **A.** Walking speed 0.9 m per second. **B.** Walking speed 1.3 m per second. An increase in muscle activity at the faster cadence resulted in higher forces on the prosthesis. (Adapted from Rydell, 1967.)

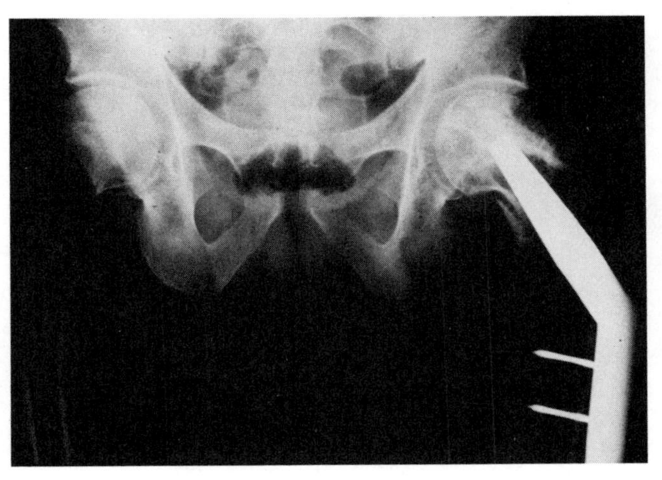

FIG. 7–17
An instrumented nail plate in the proximal end of the femur was used to determine the forces acting on the implant during the activities of daily living following fracture of the femoral neck. In this case the nail plate was found to transmit one fourth of the total load on the hip joint.

means of static analysis. In the case illustrated in Figure 7–17, the nail plate transmitted one fourth of the total load.

Strong forces acting on the nail plate were encountered during such diverse activities as moving onto a bedpan, transferring to a wheel chair, and walking. The magnitude of the forces was greatly modified by skillful assistance from the nurse or therapist to control the patient's movement. Forces of up to four times body weight acted on the hip joint when the patient used the elbows and heels to elevate the hips while being placed on a bedpan (Fig. 7–18A), but these forces were greatly reduced through the use of a trapeze and assistance from an attendant (Fig. 7–18B). The use of a hip spica cast reduced the forces acting on the hip by about two thirds for all activities, but since the spica cast could not prevent muscle contraction during the activities, some force was still produced. A 5-kg extension traction on the hip had little effect in modifying the forces acting on the hip joint. Exercises of the foot and ankle increased these forces.

Use of the instrumented nail plate demonstrated that, for a bedridden patient with a fractured femoral neck, the forces on the femoral head during the activities of daily living approached those during walking with external supports. The magnitude of the moments acting on the nail-plate junction in the transverse plane (i.e., during internal and external rotation) was only approximately one half the magnitude of the moments acting in the frontal plane

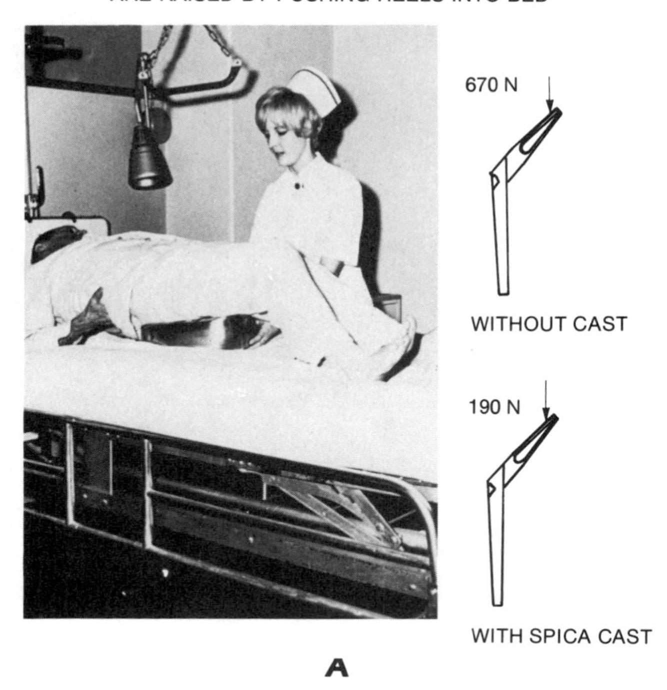

FORCE ON TIP OF NAIL
DURING TRAPEZE MANEUVER TO RAISE HIPS

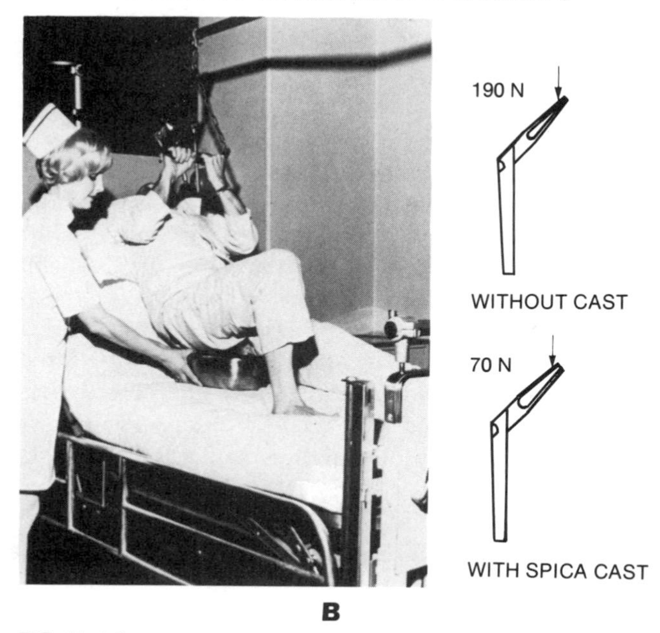

FIG. 7–18
A. When the patient used elbows and heels to elevate the hips while being placed on a bedpan, the force on the tip of the instrumented nail was 670 N. With a spica cast the force on the tip of the nail was 190 N. **B.** The use of a trapeze and assistance from an attendant reduced the force to 190 N without a cast, and to 70 N with a spica cast. (Reprinted with permission from Frankel, V. H.: Biomechanics of the hip. *In* Surgery of the Hip Joint. Edited by R. G. Tronzo. Philadelphia, Lea & Febiger, 1973, pp. 105–125.)

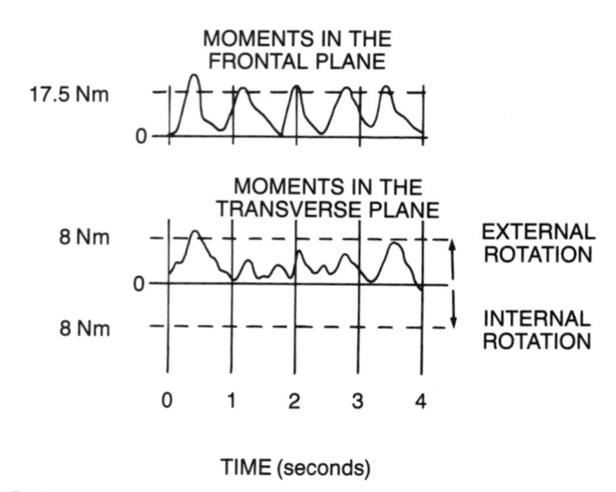

FIG. 7–19
Moments acting on the nail-plate junction in the frontal and transverse planes while the patient walked independently.

(i.e., during abduction) for many activities (Fig. 7–19).

EFFECT OF EXTERNAL SUPPORT ON THE HIP JOINT REACTION FORCE

Static analysis of the joint reaction force on the femoral head during walking with a cane demonstrates that the cane should be used on the side opposite the painful or operated hip (Pauwels, 1936; Blount, 1956; Denham, 1959). Such use reduces the force on the femoral head of the painful joint without necessitating an antalgic body position. The support offered by the cane greatly reduces the amount of contraction of the abductor muscles needed to support the body weight. Because a cane used on the side opposite the painful hip works through a large lever arm, a moderate force on the cane greatly decreases the abductor muscle force, and consequently the joint reaction force, on the painful hip. A cane used on the side of the painful hip works through a shorter lever arm, and thus a greater push on the cane is needed to decrease the joint reaction force.

The use of a brace on the leg may alter the forces on the hip joint but may not always reduce the joint reaction force on the femoral head. An ischial long-leg brace used in the treatment of Perthes' disease raises the joint reaction force during late swing phase because the large mass moment of inertia of the brace results in a higher extensor muscle force during this part of the gait cycle.

Kinematic data from stroboscopic studies were used to determine the joint reaction force acting on the femoral head in late swing phase of the gait cycle for an 8-year-old boy weighing 24 kg and wearing a long-leg brace. The main muscle force (M) was produced by contraction of the gluteus maximus muscle.

The torque about the hip joint was calculated according to the formula

$$T = I\alpha,$$

where T is the torque expressed in newton meters (Nm)
I is the mass moment of inertia expressed in newton meters times seconds squared (Nm sec^2)
α is the angular acceleration in late swing phase, expressed in radians per second squared (r/sec^2).

In the case of the braced side,

$$I = I_L + I_B,$$

where I_L is the mass moment of inertia of the leg and I_B is the mass moment of inertia of the brace.

On the normal side,	On the braced side,
$I = 0.45$ Nm sec^2	$I = 0.45$ Nm $sec^2 + 0.35$ Nm sec^2
$\alpha = 24$ r/sec^2.	$\alpha = 24$ r/sec^2.
Thus,	Thus,
$T = 0.45$ Nm $sec^2 \times 24$ r/sec^2	$T = (.45$ Nm $sec^2 + .35$ Nm $sec^2) \times 24$ r/sec^2
$T = 10.8$ Nm.	$T = 19.2$ Nm.

The extensor muscle force (M) was then found from the moment relationship

$$T = Fd,$$

where F is the extensor muscle force
d is the perpendicular distance from the center of rotation of the femur to the middle of the gluteus maximus muscle.

Distance d was measured from a roentgenogram and found to be 3.2 cm. From the equation $M = \frac{T}{d}$, the muscle force on the normal side was calculated to be 338 N, and on the braced side, 600 N.

The joint reaction force on the femoral head (J) is equal to the muscle force (M) minus the gravitational force produced by the weight of the limb (W_L). In this example W_L was estimated to be 40 N.

On the normal side,	On the braced side,
$J = M - W_L$	$J = M - W_L$
$J = 338\ N - 40\ N$	$J = 600\ N - 40\ N$
$J = 298\ N.$	$J = 560\ N.$

Thus, the joint reaction force on the femoral head in the braced limb was over 80% higher than the force in the nonbraced limb, reaching more than two times body weight.

SUMMARY

1. The hip joint is a ball-and-socket joint composed of the acetabulum and femoral head.
2. The thickness and mechanical properties of the cartilage on the femoral head and acetabulum vary from point to point.
3. Hip flexion of at least 120 degrees, abduction of at least 20 degrees, and external rotation of at least 20 degrees are necessary for carrying out daily activities in a normal manner.
4. A joint reaction force of approximately three times body weight acts on the hip joint during a single-leg stance with the pelvis in a neutral position; its magnitude varies as the position of the upper body changes.
5. The magnitude of the hip joint reaction force is influenced by the ratio of the abductor muscle force and gravitational force lever arms. A low ratio yields a greater joint reaction force than does a high ratio.
6. The hip joint reaction force during gait reaches levels of six times body weight or more in stance phase and is approximately equal to body weight during swing phase.
7. An increase in gait velocity increases the magnitude of the hip joint reaction force in both swing and stance phase.
8. The forces acting on an internal fixation device during the activities of daily living vary greatly depending on the nursing care and the therapeutic activities undergone by the patient.
9. The use of a brace on the leg can alter the magnitude of the hip joint reaction force.

REFERENCES

Andriacchi, T. P., et al. A study of lower-limb mechanics during stair-climbing. J. Bone Joint Surg., *62A*:749, 1980.

Blount, W. P.: Don't throw away the cane. J. Bone Joint Surg., *38A*:695, 1956.

Denham, R. A.: Hip mechanics. J. Bone Joint Surg., *41B*:550, 1959.

Draganich, L. F., Andriacchi, T. P., Strongwater, A. M., and Galante, J. O.: Electronic measurement of instantaneous floot-floor contact patterns during gait. J. Biomech., *13*:875, 1980.

English, T. A., and Kilvington, M.: In vivo records of hip loads using a femoral implant with telemetric output (a preliminary report). J. Biomed. Eng., *1*:111, 1979.

Frankel, V. H.: The Femoral Neck: Function, Fracture Mechanisms, Internal Fixation. Springfield, Charles C Thomas, 1960.

Frankel, V. H.: Biomechanics of the hip. *In* Surgery of the Hip Joint. Edited by R. G. Tronzo. Philadelphia, Lea & Febiger, 1973, pp. 105–125.

Frankel, V. H., Burstein, A. H., Lygre, L., and Brown, R. H.: The telltale nail. J. Bone Joint Surg., *53A*:1232, 1971.

Inman, V. T.: Functional aspects of the abductor muscles of the hip. J. Bone Joint Surg., *29*:607, 1947.

Johnston, R. C., and Smidt, G. L.: Measurement of hip-joint motion during walking. Evaluation of an electrogoniometric method. J. Bone Joint Surg., *51A*:1083, 1969.

Johnston, R. C., and Smidt, G. L.: Hip motion measurements for selected activities of daily living. Clin. Orthop., *72*:205, 1970.

Kempson, G. E., Spivey, C. J., Swanson, S. A. V., and Freeman, M. A. R.: Patterns of cartilage stiffness on normal and degenerate human femoral heads. J. Biomech., *4*:597, 1971.

Lygre, L.: The loads produced on the hip joint by nursing procedures: A telemeterization study. M.S. thesis, Case Western Reserve University, 1970.

McLeish, R. D., and Charnley, J.: Abduction forces in the one-legged stance. J. Biomech., *3*:191, 1970.

Milde, F. K.: Loads on femoral head during nursing care activities as measured by a telemeterized nail-plate. M.S. thesis, Case Western Reserve University, 1974.

Murray, M. P.: Gait as a total pattern of movement. Am. J. Phys. Med., *46*:290, 1967.

Murray, M. P., Kory, R. C., and Clarkson, B. H.: Walking patterns in healthy old men. J. Gerontol., *24*:169, 1969.

Paul, J. P.: Forces at the human hip joint. Ph.D. thesis, University of Chicago, 1967.

Pauwels, F.: Der Schenkelhalsbruch, ein mechanisches Problem: Grundlagen des heilungsvorganges Prognose und kausale Therapie. Stuttgart, Ferdinand Enke, 1936.

Rydell, N.: Forces in the hip-joint. Part II, intravital measurements. *In* Biomechanics and Related Bio-Engineering Topics. Edited by R. M. Kenedi. Oxford, Pergamon Press, 1965, pp. 351–357.

Rydell, N. W.: Forces acting on the femoral head prosthesis. A study on strain gauge supplied prostheses in living persons. Acta Orthop. Scand., Suppl. *88*:1–132, 1966.

Seireg, A., and Arvikar, R. J.: The prediction of muscular load sharing and joint forces in the lower extremities during walking. J. Biomech., *8*:89, 1975.

SUGGESTED READING

Andriacchi, T. P., et al.: A study of lower-limb mechanics during stair-climbing. J. Bone Joint Surg., *62A*:749–757, 1980.

Blount, W. P.: Don't throw away the cane. J. Bone Joint Surg., *38A*:695–708, 1956.

Charnley, J., and Pusso, R.: The recording and analysis of gait in relation to the surgery of the hip joint. Clin. Orthop., *58*:153–164, 1968.

Crowninshield, R. D., Johnston, R. C., Andrews, J. G., and Brand, R. A.: The effects of walking velocity and age on hip kinematics and kinetics. Clin. Orthop., *132*:140–144, 1978.

Denham, R. A.: Hip mechanics. J. Bone Joint Surg., *41B*:550–557, 1959.

Draganich, L. F., Andriacchi, T. P., Strongwater, A. M., and Galante, J. O.: Electronic measurement of instantaneous foot-floor contact patterns during gait. J. Biomech., *13*:875–880, 1980.

English, T. A., and Kilvington, M.: In vivo records of hip loads using a femoral implant with telemetric output (a preliminary report). J. Biomed. Eng., *1*:111–115, 1979.

Frankel, V. H.: The Femoral Neck: Function, Fracture Mechanisms, Internal Fixation. Springfield, Charles C Thomas, 1960.

Frankel, V. H.: Biomechanics of the hip. *In* Surgery of the Hip Joint. Edited by R. G. Tronzo. Philadelphia, Lea & Febiger, 1973, pp. 105–125.

Frankel, V. H., Burstein, A. H., Lygre, L., and Brown, R. H.: The telltale nail. J. Bone Joint Surg., *53A*:1232, 1971.

Gore, D. R., Murray, M. P., Sepic, S. B., and Gardner, G. M.: Walking patterns of men with unilateral surgical hip fusion. J. Bone Joint Surg., *57A*:759–765, 1975.

Inman, V. T.: Functional aspects of the abductor muscles of the hip. J. Bone Joint Surg., *29*:607–619, 1947.

Johnston, R. C.: Detailed analysis of hip joint during gait. *In* The Hip. Proceedings of the Second Open Scientific Meeting of the Hip Society, 1974. Edited by W. H. Harris. St. Louis, C. V. Mosby, 1974, pp. 94–110.

Johnson, R. C., and Smidt, G. L.: Measurement of hip-joint motion during walking. Evaluation of an electrogoniometric method. J. Bone Joint Surg., *51A*:1083–1094, 1969.

Johnson, R. C., and Smidt, G. L.: Hip motion measurements for selected activities of daily living. Clin. Orthop., *72*:205–215, 1970.

Kempson, G. E., Spivey, C. J., Swanson, S. A. V., and Freeman, M. A. R.: Patterns of cartilage stiffness on normal and degenerate human femoral heads. J. Biomech., *4*:597–609, 1971.

Lygre, L.: The loads produced on the hip joint by nursing procedures: A telemeterization study. M.S. thesis, Case Western Reserve University, 1970.

McLeish, R. D., and Charnley, J.: Abduction forces in the one-legged stance. J. Biomech., *3*:191–209, 1970.

Merchant, A. C.: Hip abductor muscle force. An experimental study of the influence of hip position with particular reference to rotation. J. Bone Joint Surg., *47A*:462–476, 1965.

Milde, F. K.: Loads on femoral head during nursing care

activities as measured by a telemeterized nail-plate. M.S. thesis, Case Western Reserve University, 1974.

Morris, J. M.: Biomechanical aspects of the hip joint. Orthop. Clin. North Am., *2*:33–54, 1971.

Murray, M. P.: Gait as a total pattern of movement. Am. J. Phys. Med., *46*:290–333, 1967.

Murray, M. P., and Peterson, R. M.: Weight distribution and weight-shifting activity during normal standing posture. Phys. Ther., *53*:741–748, 1973.

Murray, M. P., and Sepic, S. B.: Maximum isometric torque of hip abductor and adductor muscles. J. Am. Phys. Ther. Assoc., *48*:1327–1335, 1968.

Murray, M. P., Gore, D. R., and Clarkson, B. H.: Walking patterns of patients with unilateral hip pain due to osteo-arthritis and avascular necrosis. J. Bone Joint Surg., *53A*:259–274, 1971.

Murray, M. P., Kory, R. C., and Clarkson, B. H.: Walking patterns in healthy old men. J. Gerontol., *24*:169–178, 1969.

Murray, M. P., Seireg, A. A., and Scholz, R. C.: Center of gravity, center of pressure, and supportive forces during human activities. J. Appl. Physiol., *23*:831–838, 1967.

Murray, M. P., Seireg, A. A., and Scholz, R. C.: A survey of the time, magnitude and orientation of forces applied to walking sticks by disabled men. Am. J. Phys. Med., *48*:1–13, 1969.

Paul, J. P.: Forces at the hip joint. Ph.D. thesis, University of Chicago, 1967.

Paul, J. P., and McGrouther, D. A.: Forces transmitted at the hip and knee joint of normal and disabled persons during a range of activities. Acta Orthop. Belg., Suppl. *41*:78–88, 1975.

Pauwels, F.: Der Schenkelhalsbruch, ein mechanisches Problem: Grundlagen des heilungsvorganges Prognose und kausale Therapie. Stuttgart, Ferdinand Enke, 1936.

Rydell, N.: Forces in the hip-joint. Part II, intravital measurements. *In* Biomechanics and Related Bio-Engineering Topics. Edited by R. M. Kenedi. Oxford, Pergamon Press, 1965, pp. 351–357.

Rydell, N. W.: Forces acting on the femoral head prosthesis. A study on strain gauge supplied prostheses in living persons. Acta Orthop. Scand., Suppl. *88*:1–132, 1966.

Seireg, A., and Arvikar, R. J.: The prediction of muscular load sharing and joint forces in the lower extremities during walking. J. Biomech., *8*:89–102, 1975.

Wiktorin, C. v. H., and Nordin, M.: Introduction to Problem Solving in Biomechanics. Philadelphia, Lea & Febiger, 1986, pp. 58–86.

8

BIOMECHANICS OF THE ANKLE

Victor H. Frankel
Margareta Nordin

The role of the ankle joint, like that of other major joints in the lower extremity, is to participate in kinematic functions and load bearing. This joint complex consists of the tibiotalar, fibulotalar, and distal tibiofibular joints (Fig. 8–1). The anatomic configuration of the ankle joint is more like that of the hip, which is inherently stable, than that of the knee, which requires ligamentous and muscular restraints for stability. The ankle mortise is maintained by the shape of the three articulations, the medial (deltoid) and lateral collateral ligament systems, the joint capsule, and the interosseous ligaments.

The ankle joint, again like the hip joint, responds poorly to small changes in its anatomic configuration. Loss of kinematic and structural restraints due to severe sprains can seriously affect ankle stability and can produce malalignment of the ankle joint surfaces. Even a slight malalignment may result in profound pathologic changes.

KINEMATICS

The ankle joint is basically a uniplanar hinge joint. Motion of the talus takes place primarily in the sagittal plane about a transverse axis that deviates posteriorly from the frontal plane on the lateral side (Barnett and Napier, 1952; Inman, 1976) (Fig. 8–2). This motion allows dorsiflexion (flexion) and plantar flexion (extension) of the foot. The talus in the mortise may also rotate a few degrees around a sagittal axis. Deviations in the inclination of the axes of the ankle joint due to epiphyseal injury, ligamentous injury, or malunion of a fractured tibia can result in severe pathologic alterations in the joint. Movement of the tibiofibular joint is limited to a few degrees because the tibia and fibula are tightly bound together by the interosseous membrane (see Fig. 8–1).

RANGE OF MOTION

The total range of motion of the ankle joint in the sagittal plane is approximately 45 degrees, but it can vary widely among individuals and with age. Ten to 20 degrees of this motion is defined as dorsiflexion and the remaining 25 to 35 degrees as plantar flexion.

The normal pattern of ankle joint motion during gait has been studied extensively (Murray et al., 1964; Wright et al., 1964; Lamoreaux, 1971; Stauffer et al., 1977) (Fig. 8–3A). At heel strike the ankle is in slight plantar flexion. Plantar flexion increases until foot flat, but the motion rapidly reverses to dorsiflexion during midstance as the body passes over the supporting foot. The motion then returns to plantar flexion after heel-off at the end of the stance phase. At toe-off at the beginning of the swing phase the ankle is in plantar flexion; the motion reverses to dorsiflexion in the middle of the swing phase and changes again to slight plantar flexion at heel strike. The amount of plantar flexion from heel strike to foot

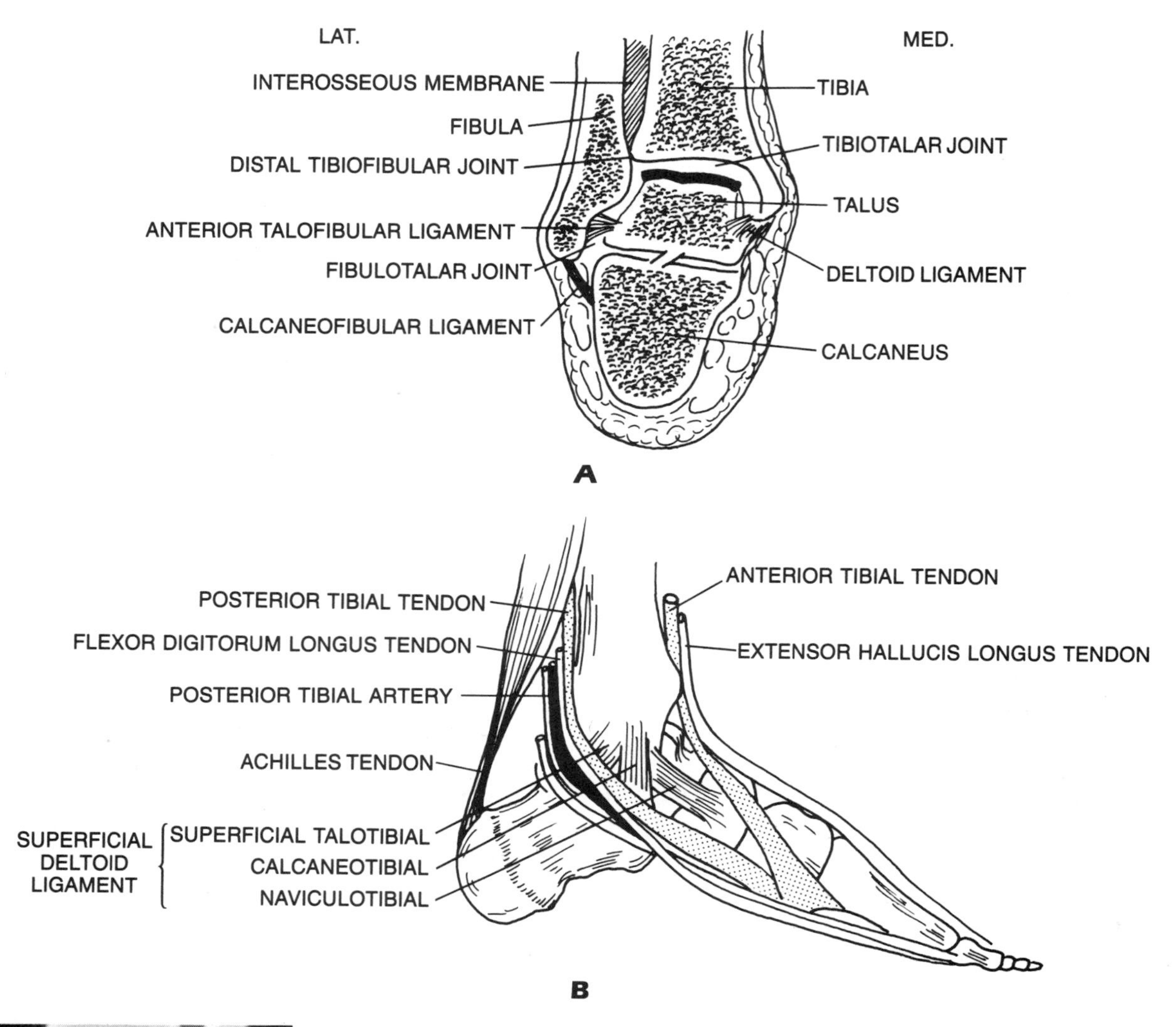

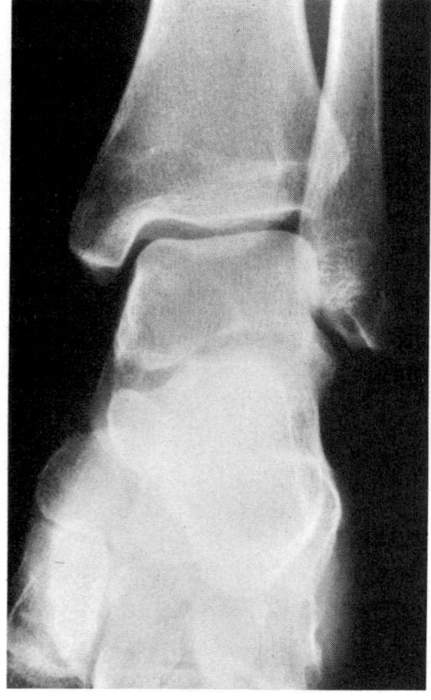

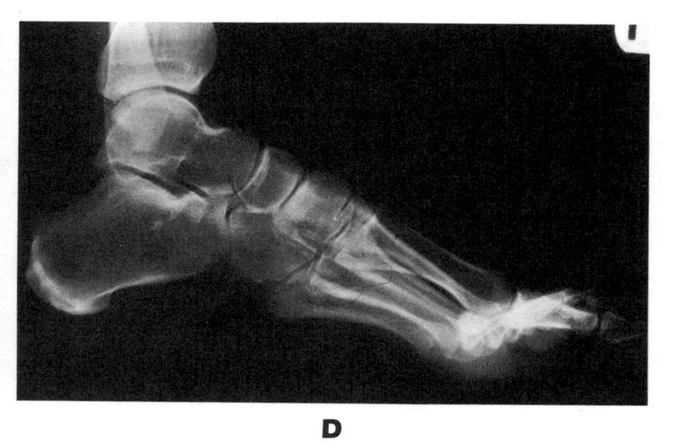

FIG. 8–1
Ankle joint complex composed of the tibiotalar, fibulotalar, and distal tibiofibular joints. Posterior cross-sectional view **(A)** and medial view **(B)** of the left ankle. Anteroposterior **(C)** and lateral **(D)** roentgenograms of the left ankle. (Courtesy of Paul Brisson, M.D.)

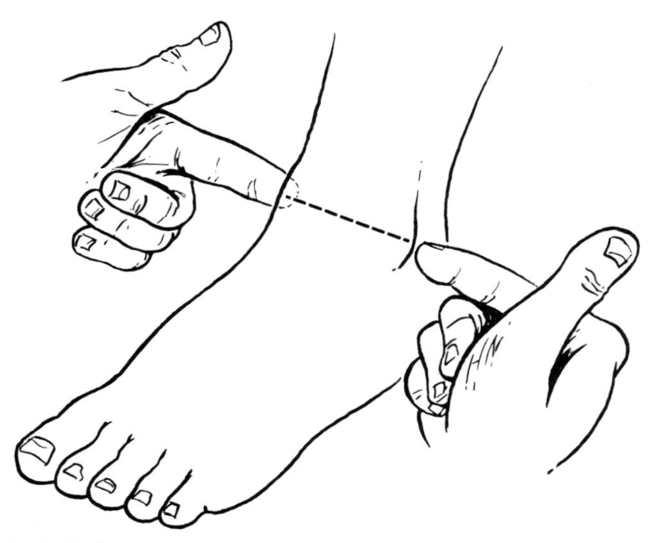

FIG. 8–2
The axis of dorsiflexion (flexion) and plantar flexion (extension) of the ankle (tibiotalar) joint deviates posteriorly from the frontal plane on the lateral side. This transverse axis is not rigid but changes slightly with ankle motion. Its position may be estimated clinically by palpation of the malleoli. (Adapted from Inman, 1973.)

flat depends on the height of the shoe heels. The higher the heel is, the more plantar flexion (Stauffer et al., 1977) (Fig. 8–4). The total amount of ankle joint motion during the gait cycle decreases as heel height increases (Murray et al., 1970).

Sammarco and coworkers (1973) studied total ankle joint motion roentgenographically and recorded the average range of motion in the sagittal plane during gait for 245 normal subjects ranging in age from 20 to 60 years. The total range varied from 24 to 75 degrees (average 43±12.7 degrees) and tended to decrease with age. The average amounts of dorsiflexion and plantar flexion were nearly the same (21 and 23 degrees, respectively).

In a study of normal gait at two velocities in five men, Stauffer and associates (1977) found that plantar flexion at heel strike was diminished at the higher velocity and reached its peak earlier in the midstance phase; dorsiflexion remained essentially unchanged in the two cadences (see Fig. 8–3B). This finding contrasts with that for both the hip and knee joints, where joint motion increases directly as a person's pace accelerates (Pauwels, 1936; Rydell, 1966; Perry et al., 1977).

Ankle joint disease affects total ankle joint motion to varying degrees in different individuals (Sammarco et al., 1973; Stauffer et al., 1977). Measurements of joint motion in the sagittal plane during gait in 10 patients with abnormal ankle joints revealed an overall reduction in range of motion compared with that in a control group (Sammarco et al., 1973). The greatest decrease took place in dorsiflexion. The same pattern was found in a later study of nine patients with abnormal ankles (Stauffer et al., 1977).

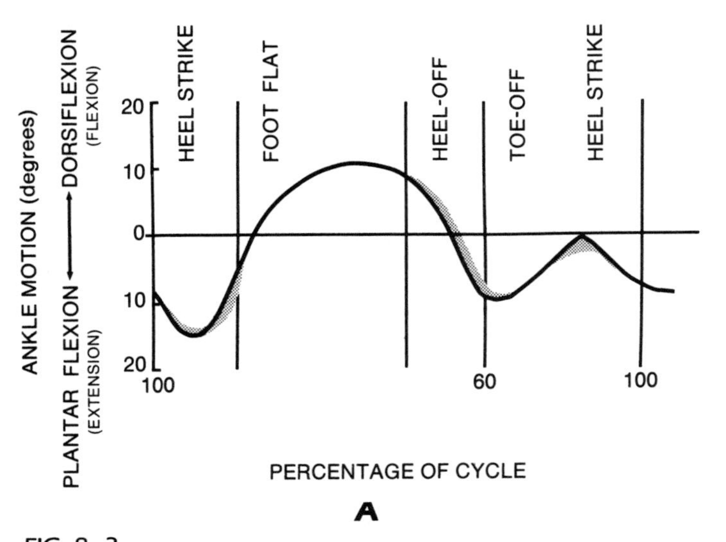

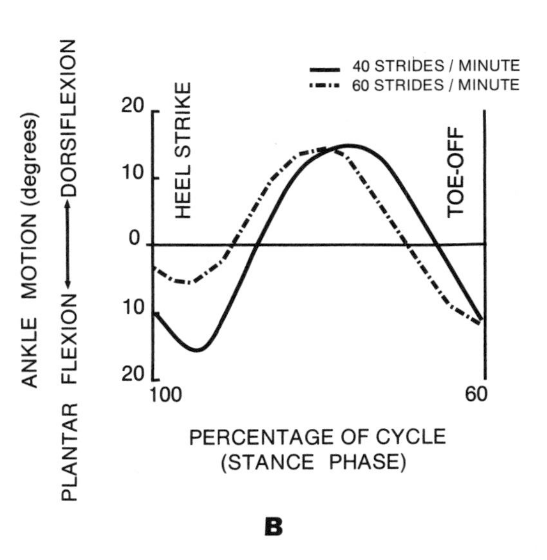

FIG. 8–3
A. Range of ankle joint motion in the sagittal plane during level walking, one gait cycle. Shaded area indicates variation among 60 subjects (age range 20 to 65 years). (Adapted from Murray et al., 1964.) **B.** Range of ankle joint motion in a normal ankle joint during the stance phase of gait at two velocities. The amount of plantar flexion at heel strike was smaller in the faster cadence, and the peak plantar flexion occurred earlier in the midstance phase. (Adapted from Stauffer et al., 1977.)

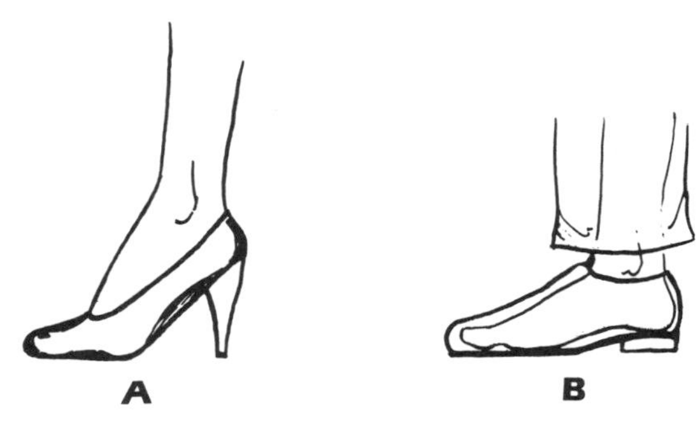

FIG. 8–4
The amount of plantar flexion at foot flat is much greater when a high-heeled shoe is worn **(A)** than when a flatter shoe is chosen **(B)**.

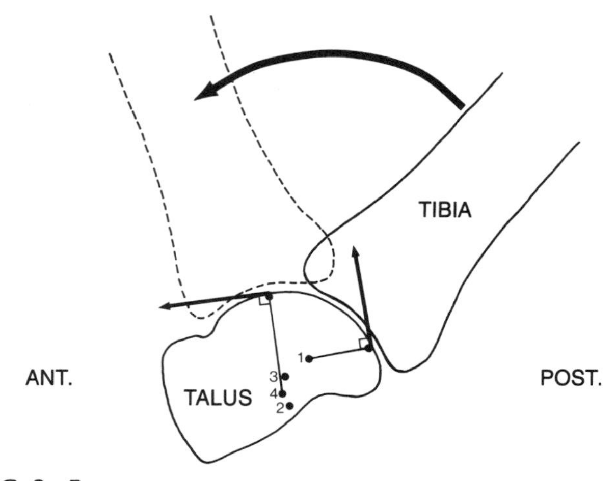

FIG. 8–5
Instant center pathway for surface joint motion at the tibiotalar joint in a normal ankle from full plantar flexion to full dorsiflexion. All instant centers fall within the talus. The direction of displacement of the contact points shows distraction of the joint surfaces at the beginning of the motion (points 1 and 2) and gliding thereafter (points 3 and 4). (Adapted from Sammarco et al., 1973.)

SURFACE JOINT MOTION

In the ankle joint, surface motion occurs primarily in the tibiotalar and fibulotalar articulations. A few degrees of motion also takes place in the distal tibiofibular joint during plantar flexion to accommodate the narrowing of the posterior talus. Sammarco and coworkers (1973) performed instant center analyses on 24 normal weight-bearing ankles and on 11 diseased ankles. Using multiple roentgenographic exposures and the instant center technique described for the knee in Chapter 6, they determined the instant center pathway for surface motion of the tibiotalar joint for a range of ankle motion from full plantar flexion to full dorsiflexion and found that all instant centers fell within the talus. The direction of displacement of the contact points was then determined.

In the normal ankles the joint surfaces distracted at the beginning of the motion, and gliding then took place (Fig. 8–5); the motion ended with jamming of the surfaces. During the reverse motion the compressed surfaces distracted at the beginning of motion, demonstrated gliding throughout the range of motion, and then jammed again. It is possible that distraction and jamming of the tibiotalar joint surfaces at the extremes of motion play an important role in lubrication of the joint.

In the abnormal ankles the direction of displacement of the contact points showed no consistent pattern. The tibiotalar joint surfaces distracted in an unpredictable manner, and they jammed when the joint was in a neutral position rather than at the end of dorsiflexion.

A study of 152 tali showed that the medial and lateral profiles of the talus have different curvatures (Barnett and Napier, 1952). This finding indicates that the inclination of the axis of flexion-extension of the tibiotalar joint is not rigid but changes slightly with motion. This change in the axis was confirmed by Inman (1976) and by the instant center analyses of Sammarco and coworker (1973). Clinically, the position of the axis can be estimated by palpating the malleoli (Inman, 1973) (see Fig. 8–2).

ANKLE JOINT STABILITY

The mortise of the ankle joint is maintained primarily by the shape of the talus, its tight fit between the fibula and tibia, and the interosseous membrane between these two bones. This stabilizing effect is greatest in dorsiflexion. The mortise is further maintained by the anterior and posterior talofibular ligaments and by the ankle joint capsule. The anterior talofibular ligament and the calcaneofibular ligament are important for stabilizing the lateral side of the ankle joint, while the deltoid ligament secures the medial side (see Fig. 8–1A). Because the talus narrows posteriorly, mortise stability in plantar flexion is provided mainly by the tautness of the ligaments. The musculotendinous apparatus surrounding the ankle joint on the medial and lateral sides plays a small role in stabilizing the joint; its primary function is to move the foot (see Fig. 8–1B).

KINETICS

The reaction force on the ankle joint during gait is equivalent to, or somewhat greater than, the reaction forces determined for the knee and hip joints in Chapters 6 and 7, respectively. Since the ankle joint has a larger weight-bearing surface area, lower loads per unit area (stresses) may be transmitted across the ankle. The following static and dynamic analyses give an estimate of the magnitude of the reaction forces acting on the ankle joint during standing on tiptoe on one leg and during level walking.

STATICS

When an individual stands on both feet, each ankle joint supports approximately one half the body weight. The line of gravity of the body passes a few centimeters anterior to the transverse axis of the ankle joint; hence, the body weight produces a dorsiflexing torque on the joint, which varies between 3 and 24 newton meters (Nm) as a result of body oscillations (Smith, 1957). Therefore, standing with the body weight distributed evenly on both feet requires some activity in the plantar flexor muscles. When these and other muscles in the lower leg are involved in balancing the body, the joint reaction force on the ankle increases in proportion to the amount of muscle force used for these balancing activities.

In a static analysis of the forces acting on the ankle joint, the magnitude of the force produced by contraction of the gastrocnemius and the soleus muscles through the Achilles tendon, and consequently the magnitude of the joint reaction force, can be calculated through the use of the simplified free body technique for coplanar forces described for the knee in Chapter 6. In the following example, the muscle force transmitted through the Achilles tendon and the reaction force on the ankle joint are calculated for a subject standing on tiptoe on one leg. For the body to maintain equilibrium, the line of gravity for the body must pass through the ball of the foot to the sole of the shoe, and then to the ground.

In this example the foot, including the talus, is considered as a free body. From all the forces acting on this free body the three main coplanar forces are identified as the ground reaction force, the tensile muscle force through the Achilles tendon, and the reaction force on the dome of the talus. The *ground reaction force* (W) has a known magnitude (equal to body weight), sense, line of application, and point of application. The *Achilles tendon force* (A), which holds the foot in plantar flexion, has a known sense, line of application (along the Achilles tendon), and point of application (point of insertion of the Achilles tendon into the calcaneus) but an unknown magnitude. The *joint reaction force* (J) has a known point of application on the dome of the talus (estimated from a roentgenogram) but an unknown magnitude, sense, and line of application.

The magnitude of A and J can be derived by designating the forces on a free body diagram and constructing a triangle of forces (Fig. 8–6). Not surprisingly, these forces are found to be quite large. The joint reaction force is about 2.1 times body weight, and the Achilles tendon force reaches about 1.2 times body weight. The great force required for rising up on tiptoe explains why the patient with weak gastrocnemius and soleus muscles has diffi-

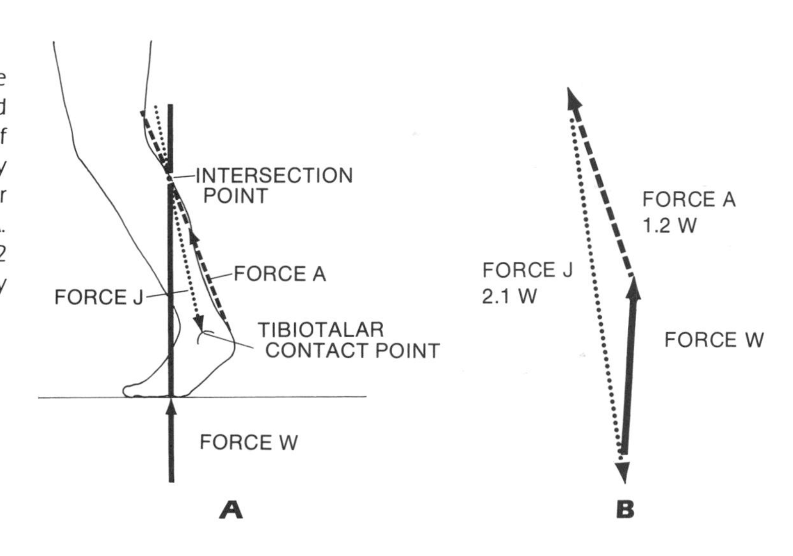

FIG. 8–6
A. On a free body diagram of the foot, including the talus, the lines of application for W and A are extended until they intersect (intersection point). The line of application for J (dotted line) is then determined by connecting its point of application, the tibiotalar contact point, with the intersection point for W and A. **B.** A triangle of forces is constructed. Force A is 1.2 times body weight, and force J is 2.1 times body weight.

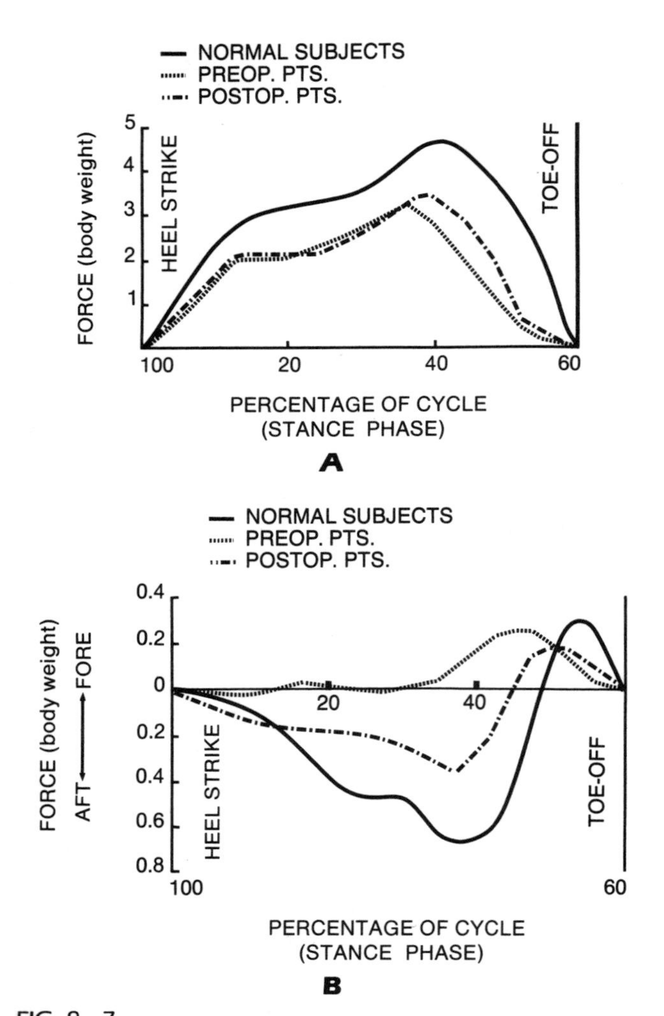

FIG. 8–7

A. The compressive component of the ankle joint reaction force expressed in multiples of body weight during the stance phase of level walking for five normal subjects and for nine patients with joint disease before and after prosthetic ankle replacement. (Adapted from Stauffer et al., 1977.) **B.** The shear component (in the "fore" and "aft" directions) produced in the ankle during the stance phase of level walking for the same subjects. (Adapted from Stauffer et al., 1977.)

culty performing this exercise 10 times in rapid succession. It also explains why a patient with degenerative arthritis of the ankle joint will have pain on rising up on tiptoe.

In a study of the weight-bearing function of the fibula, Lambert (1971) demonstrated that a portion of the compressive component of the ankle joint reaction force is transmitted from the knee to the ankle through the fibula. Using a static model of the ankle joint, including the fibula, he determined that approximately one sixth of the load of the leg was borne by the fibula; it was generated by the articulation with the talus of the fibula and possibly also by the inferior tibiofibular ligaments. This load on the fibula was taken up by the proximal tibiofibular joint, and little of it was transmitted through the interosseous membrane.

DYNAMICS

Dynamic studies of the ankle joint are important for understanding the magnitude of the loads on the normal ankle during exercise, the loads on the damaged ankle during normal activities, and the loads expected to act on an ankle joint prosthesis. The loads on the joint during level walking have been studied by Stauffer and coworkers (1977) and by Proctor and Paul (1982). Using a force plate, high-speed photography, roentgenograms, and free body calculations, Stauffer's group determined both the compressive and shear components of the reaction force acting on the ankle joint during the stance phase of gait. These force components were calculated for normal subjects and for patients with joint disease before and after prosthetic ankle replacement.

Stauffer's group found that the main compressive force across the ankle during gait in the normal subjects was produced by contraction of the gastrocnemius and soleus muscles and was transmitted through the Achilles tendon. They noted that the force produced by contraction of the anterior tibial muscle group acted only during early stance phase and was mild (<20% of body weight). The Achilles tendon force was great during late stance phase when the muscles acting through the Achilles tendon began to produce a plantar flexion moment at push-off. The compressive force on the joint was highest at this point in the gait cycle: about five times body weight (Fig. 8–7A). The shear force for these subjects reached its maximum value, about 0.8 times body weight, just after the middle of the stance phase during heel-off (Fig. 8–7B).

Similar results were obtained in a later study by Proctor and Paul (1982), who used a force plate and tracing system to develop a three-dimensional model of the ankle. The peak compressive forces acting on normal ankles during gait averaged about four times body weight. In contrast to Stauffer and coworkers' results, Proctor and Paul found substantial activity in the anterior tibial muscle group, with mean peak forces equal to body weight.

Stauffer's group's results for patients with ankle joint disease showed a decreased compressive force on the joint, about three times body weight (see Fig. 8–7A). This force reached its peak slightly earlier in these patients than in the normal subjects. The shear forces were also lower (see Fig. 8–7B). A follow-up study 1 year after ankle joint replacement in these patients showed no change in the compressive force patterns. The shear forces, however, showed a pattern and magnitude nearly equal to those in the normal subjects.

Stauffer and associates' study also showed the effect of two walking cadences on the ankle joint reaction force in normal subjects. The patterns were somewhat different for the two cadences, but the magnitudes of the peak forces were the same (Fig. 8–8). In the faster cadence the pattern showed two peak forces of three to five times body weight, one in early stance phase and the other in late stance phase. In the slower cadence only one peak force of approximately five times body weight was reached during late stance phase.

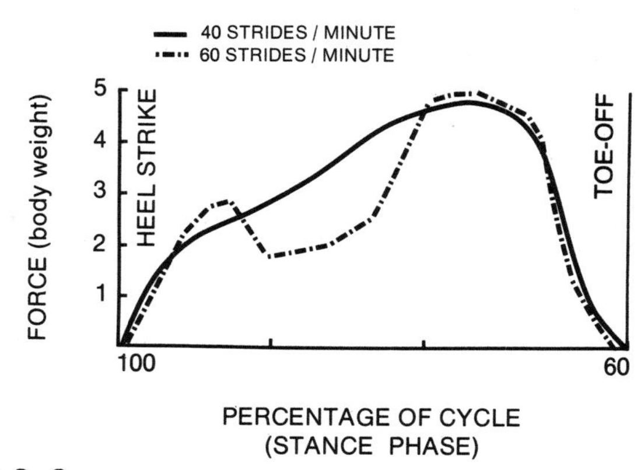

FIG. 8–8
Ankle joint reaction force expressed in multiples of body weight in a normal ankle during the stance phase of gait at two velocities. Although the patterns varied somewhat for the two cadences, the magnitudes of the peak forces were the same. In the faster cadence there were two peak forces of three to five times body weight, one in early stance phase and one in late stance phase. In the slower cadence only one peak force of approximately five times body weight was reached during late stance phase. (Adapted from Stauffer et al., 1977.)

Owing to the large load-bearing surface of the ankle (11 to 13 cm^2) (Greenwald, 1977), lower stresses can result across this joint than in the knee or hip. A small deviation in the anatomic configuration of the ankle joint can produce gross changes in the weight-bearing pattern and consequent peak loads. Ramsey and Hamilton (1976) noted changes in the tibiotalar contact area produced by lateral talar shift (Fig. 8–9), which is a frequent sequela of major sprains and fractures of the ankle. If not corrected, talar shift can lead to significant biomechanical alterations in the joint. In the case of a shift of only 1 to 2 mm, the contact surface stresses rise precipitously, and early degenerative changes in the ankle joint could result.

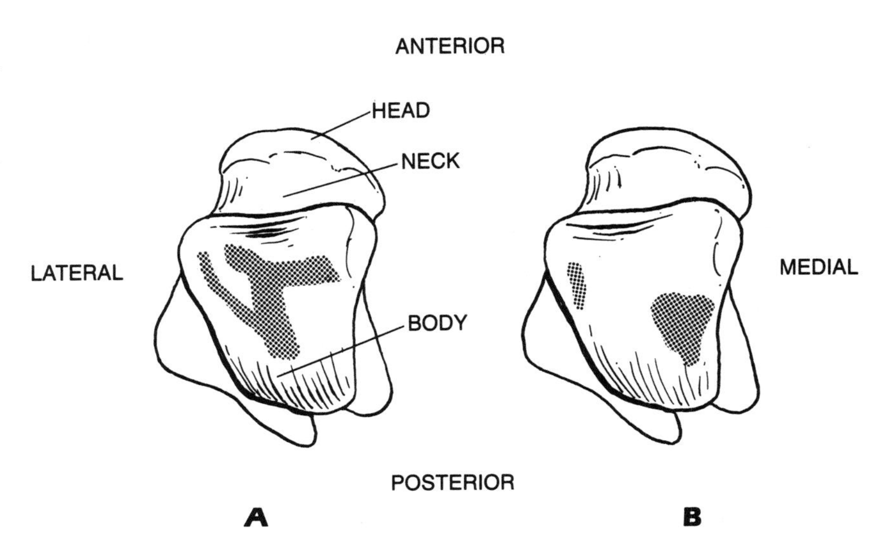

FIG. 8–9
Top view of the left talus showing the tibiotalar contact area (shaded). **A.** No talar shift (displacement). The major contact area is on the lateral side. **B.** Two mm of lateral talar shift. The total contact area has markedly decreased, and the major contact area is now on the medial side. (Adapted from Ramsey and Hamilton, 1976.)

SUMMARY

1. The ankle joint is composed of the tibiotalar, fibulotalar, and distal tibiofibular joints.
2. Motion of the talus takes place primarily in the sagittal plane about a transverse axis that deviates posteriorly from the frontal plane on the lateral side.
3. The instant center of motion falls within the talus throughout the full range of ankle flexion-extension.
4. The forces acting on the ankle joint during gait can rise to levels exceeding five times body weight.
5. The fibulotalar joint transmits approximately one sixth of the force exerted through the leg.
6. Small anatomic deviations in the tibiotalar and fibulotalar articulations can result in significant changes in the magnitude and direction of the stresses on the talus.

REFERENCES

Barnett, C. H., and Napier, J. R.: The axis of rotation at the ankle joint in man. Its influence upon the form of the talus and the mobility of the fibula. J. Anat., *86*:1, 1952.

Greenwald, S.: Unpublished data. Cited in Stauffer, R. N., Chao, E. Y. S., and Brewster, R. C.: Force and motion analysis of the normal, diseased, and prosthetic ankle joint. Clin. Orthop., *127*:189,1977.

Inman, V. T.: DuVries' Surgery of the Foot. St. Louis, C. V. Mosby, 1973.

Inman, V. T.: The Joints of the Ankle. Baltimore, Williams & Wilkins, 1976.

Lambert, L. L.: The weight-bearing function of the fibula. A strain gauge study. J. Bone Joint Surg., *53A*:507, 1971.

Lamoreaux, L. W.: Kinematic measurements in the study of human walking. Bull. Prosthet. Res., *10*:3, 1971.

Murray, M. P., Drought, A. B., and Kory, R. C.: Walking patterns in normal men. J. Bone Joint Surg., *46A*:335, 1964.

Murray, M. P., Kory, R. C., and Sepic, S. B.: Walking patterns of normal women. Arch. Phys. Med. Rehabil., *51*:637, 1970.

Pauwels, F.: Der Schenkelhalsbruch, ein mechanisches Problem: Grundlagen des heilungsvorganges Prognose und kausale Therapie. Stuttgart, Ferdinand Enke, 1936.

Perry, J., Norwood, L., and House, K.: Knee posture and biceps and semimembranosis muscle action in running and cutting (an EMG study). Trans. Orthop. Res. Soc., *2*:258, 1977.

Proctor, P., and Paul, J. P.: Ankle joint biomechanics. J. Biomech., *15*:627, 1982.

Ramsey, P. L., and Hamilton, W.: Changes in tibiotalar area of contact caused by lateral talar shift. J. Bone Joint Surg., *58A*:356, 1976.

Rydell, N. W.: Forces acting on the femoral head prosthesis. A study on strain gauge supplied prostheses in living persons. Acta Orthop. Scand., Suppl. *88*, pp. 1–132, 1966.

Sammarco, G. J., Burstein, A. H., and Frankel, V. H.: Biomechanics of the ankle: A kinematic study. Orthop. Clin. North Am., *4*:75, 1973.

Smith, J. W.: The forces operating at the human ankle joint during standing. J. Anat., *91*:545, 1957.

Stauffer, R. N., Chao, E. Y. S., and Brewster, R. C.: Force and motion analysis of the normal, diseased, and prosthetic ankle joint. Clin. Orthop., *127*:189, 1977.

Wright, D. G., Desai, S. M., and Henderson, W. H.: Action of the subtalar and ankle-joint complex during the stance phase of walking. J. Bone Joint Surg., *46A*:361, 1964.

SUGGESTED READING

Andriacchi, T. P., et al.: A study of lower-limb mechanics during stair-climbing. J. Bone Joint Surg., *62A*:749–757, 1980.

Barnett, C. H., and Napier, J. R.: The axis of rotation at the ankle joint in man. Its influence upon the form of the talus and the mobility of the fibula. J. Anat., *86*:1–9, 1952.

Bresler, B., and Frankel, J. P.: The forces and moments in the leg during level walking. Trans. Am. Soc. Mech. Eng., *72*:27–36, 1950.

Close, J. R.: Some applications of the functional anatomy of the ankle joint. J. Bone Joint Surg., *38A*:761–781, 1956.

Close, J. R.: Motor Function in the Lower Extremity. Analyses by Electronic Instrumentation. American Lecture Series, No. 551. Springfield, Charles C Thomas, 1964.

Inman, V. T.: The influence of the foot-ankle complex on the proximal skeletal structures. Artif. Limbs, *13*:59–65, 1969.

Inman, V. T.: DuVries' Surgery of the Foot. St. Louis, C. V. Mosby, 1973.

Inman, V. T.: The Joints of the Ankle. Baltimore, Williams & Wilkins, 1976.

Isman, R. E., and Inman, V. T.: Anthropometric studies of the human foot and ankle. Bull. Prosthet. Res., *10–11*:97–129, 1969.

Kapandji, I. A.: The Physiology of the Joints, Vol. 2. Lower Limb. Edinburgh, Churchill Livingstone, 1975.

Kempson, G. E., Freeman, M. A. R., and Tuke, M. A.: Engineering considerations in the design of an ankle joint. Biomed. Eng., *10*:166–171, 180, 1975.

Lambert, L. L.: The weight-bearing function of the fibula. A strain gauge study. J. Bone Joint Surg., *53*:507–513, 1971.

Lamoreaux, L. W.: Kinematic measurements in the study of human walking. Bull. Prosthet. Res., *10–11*:3–84, 1971.

Miura, M., Miyashita, M., Matsui, H., and Sodeyama, H.: Photographic method of analyzing the pressure distribution of the foot against the ground. J. Biomech., *7*:482–487, 1974.

Murray, M. P., Drought, A. B., and Kory, R. C.: Walking patterns in normal men. J. Bone Joint Surg., *46A*:335–360, 1964.

Murray, M. P., Kory, R. C., and Sepic, S. B.: Walking patterns of normal women. Arch. Phys. Med. Rehabil., *51*:637–650, 1970.

Murray, M. P., Guten, G. N., Baldwin, J. M., and Gardner, G. M.: A comparison of plantar flexion torque with and without the triceps surae. Acta Orthop. Scand., *47*:122–124, 1976.

Murray, M. P., et al.: Function of the triceps surae during gait. Compensatory mechanisms for unilateral loss. J. Bone Joint Surg., *60A*:473–476, 1978.

Paul, J. P.: Bio-engineering studies of the forces transmitted by joints. II. Engineering analysis. *In* Biomechanics and Related Bioengineering Topics. Edited by R. N. Kenedi. Edinburgh, Pergamon Press, 1965, pp. 368–380.

Pauwels, F.: Der Schenkelhalsbruch, ein mechanisches Problem: Grundlagen des heilungsvorganges Prognose und kausale Therapie. Stuttgart, Ferdinand Enke, 1936.

Proctor, P., and Paul, J. P.: Ankle joint biomechanics. J. Biomech., *15*:627–634, 1982.

Ramsey, P. L., and Hamilton, W.: Changes in tibiotalar area of contact caused by lateral talar shift. J. Bone Joint Surg., *58A*:356–357, 1976.

Rasmussen, O., Kromann-Andersen, C., and Boe, S.: Deltoid ligament. Functional analysis of the medial collateral ligamentous apparatus of the ankle joint. Acta Orthop. Scand., *54*:36–44, 1983.

Rydell, N. W.: Forces acting on the femoral head prosthesis. A study on strain gauge supplied prostheses in living persons. Acta Orthop. Scand., Suppl. *88*:1–132, 1966.

Sammarco, G. J., Burstein, A. H., and Frankel, V. H.: Biomechanics of the ankle: A kinematic study. Orthop. Clin. North Am., *4*:75–96, 1973.

Seireg, A., and Arvikar, R. J.: The prediction of muscular load sharing and joint forces in the lower extremities during walking. J. Biomech., *8*:89–102, 1975.

Smith, J. W.: The forces operating at the human ankle joint during standing. J. Anat., *91*:545–564, 1957.

Stauffer, R. N., Chao, E. Y. S., and Brewster, R. C.: Force and motion analysis of the normal, diseased, and prosthetic ankle joint. Clin. Orthop., *127*:189–196, 1977.

Wiktorin, C. v. H., and Nordin, M.: Introduction to Problem Solving in Biomechanics. Philadelphia, Lea & Febiger, 1986, pp. 32–35.

Wright, D. G., Desai, S. M., and Henderson, W. H.: Action of the subtalar and ankle-joint complex during the stance phase of walking. J. Bone Joint Surg., *46A*:361–464, 1964.

9

BIOMECHANICS OF THE FOOT

G. James Sammarco

The biomechanics of the foot are complex and may be considered distinctly different from those of the ankle. However, several authors have considered portions of the foot and ankle as a single unit (Isman and Inman, 1968). While the foot must be considered as an intricate mechanical part of the entire lower extremity, the necessity for foot-to-floor stability dictates that the foot itself must independently adapt to a variety of conditions.

The unique qualities of the foot allow it to be rigid when necessary, converting the 26 bones (28 with sesamoids) and 57 joints into a solid unit, as in ballet dancing on point, or quite flexible when necessary, as in climbing barefoot (Fig. 9–1). Between these two extremes of rigidity and flexibility is the motion of the foot during walking. The need for a range of motions and stability in the structure of the foot arises from the fact that the surfaces on which we stand and move vary considerably from soft, smooth, and slippery to firm, rough, and sticky. In addition, foot coverings vary from nothing at all to the sophisticated modern ski boot, which is thoroughly rigid and snug fitting below the ankle.

This chapter discusses the motion that occurs in the foot during various phases of gait as well as at the extremes of motion. The location of forces as they pass from the tibiofibular complex into the dome of the talus, and then into the foot, are also covered. The ground (foot-to-floor) reaction force, the resultant of all the forces in the body, is discussed as well. Further, the muscles of the foot required for control are described.

A discussion of sophisticated electromyographic activity during walking is not within the scope of this text; however, the activity of certain extrinsic and intrinsic muscles is by necessity presented to allow a better understanding of foot control during various activities. In addition to the dynamics of the lower limb, the chapter includes a description of motion in certain joints of the foot as determined by analysis of instant centers of motion and the direction of displacement of the contact points. The necessity of understanding muscle and joint function in order to effect the proper rehabilitation of the foot following injury, surgery, or immobilization is also illustrated.

Examples of certain disease entities are presented to illustrate the difference between the function of the normal foot and that of a diseased or deformed foot. In Western society the foot is more often than not protected by a semirigid covering, the shoe, and certain disease conditions develop simply because of this circumstance. When the pathomechanics of such externally restricting materials are understood, a rationale for treatment of foot disorders such as bunions can be appreciated.

GROWTH OF THE FOOT

The foot is formed at the time that the limb buds develop during the eighth week of gestation. Following birth, the growth of the foot in both girls and boys proceeds at a steady rate, although it is recognized that the child as a whole has two periods of fast growth, namely the first 2 years of life and puberty. Blais and associates (1956) showed that the foot appears to be closer to the adult size at all times

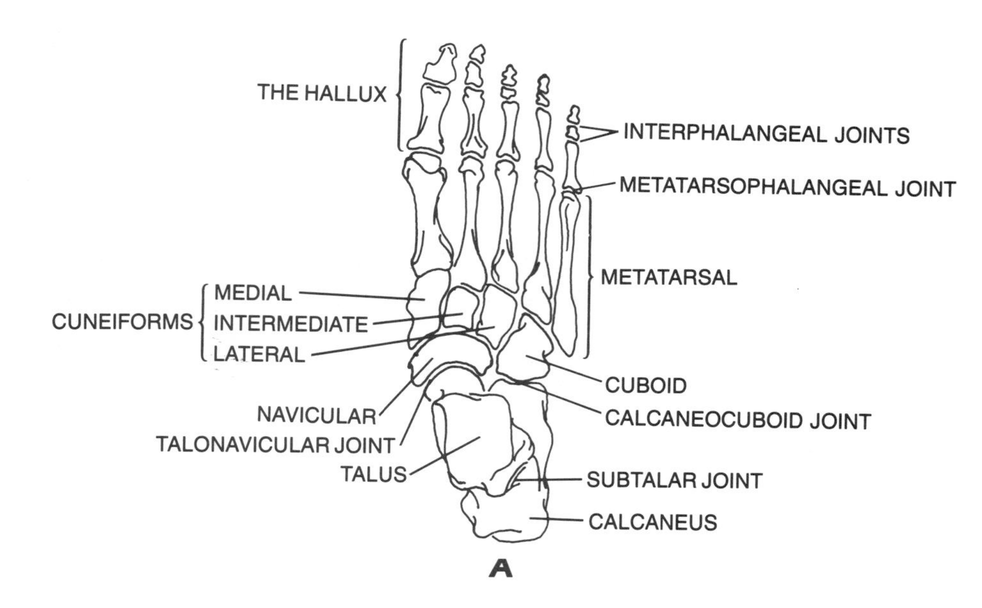

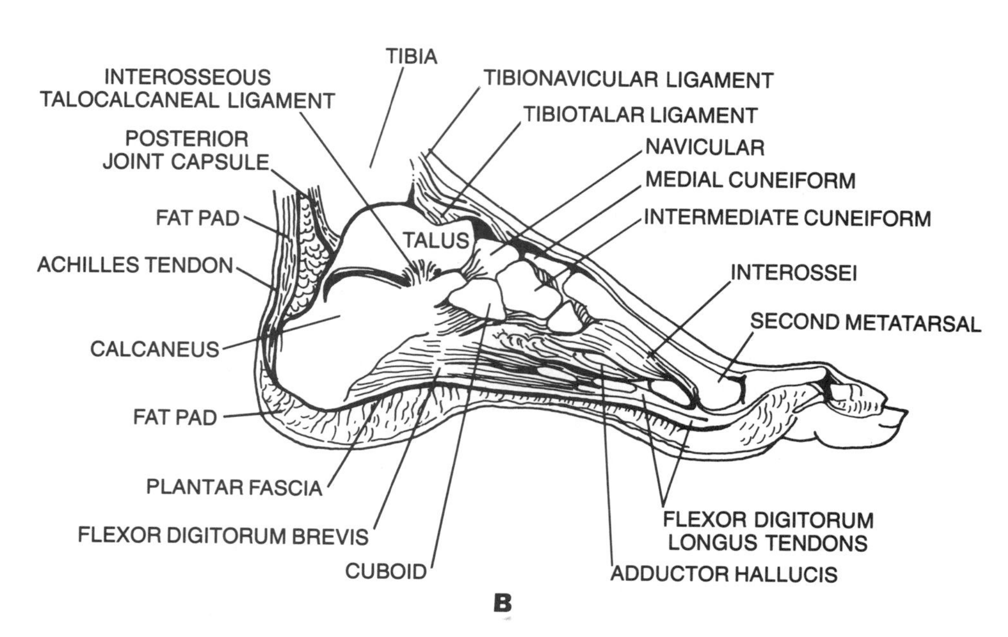

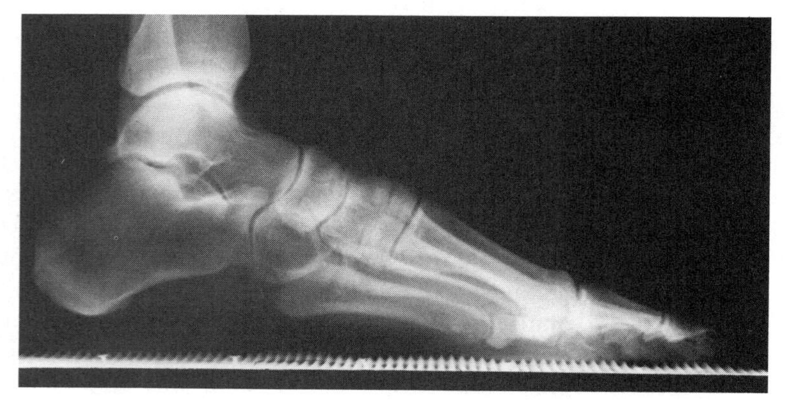

FIG. 9 – 1
The 26 bones (28 with sesamoids) and 57 joints of the foot allow the foot either to act as a solid unit or to be quite flexible, depending on the function it must perform. **A.** Top view. (Adapted from Basmajian, 1982.) **B.** Midsagittal section. (Adapted from Jaffe and Laitman, 1982.) **C.** Lateral view.

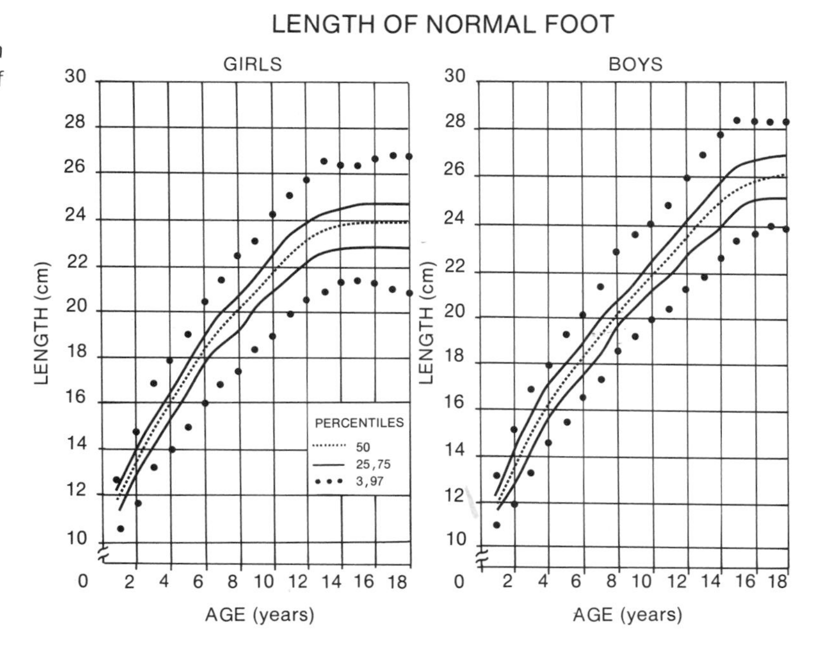

FIG. 9–2
Lengths of normal girls' and boys' feet derived from serial measures on 512 children from 1 to 18 years of age. (Adapted from Blais et al., 1956.)

during normal development of the child than are other parts of the limb. On the average, at age 1 year in girls and 18 months in boys, the length of the foot is one half the length of the respective adult foot (Fig. 9–2). This situation contrasts with that in the femur and the tibia, which do not attain half their mature length until 3 years later in both girls and boys. The relatively large size of the foot, then, is important for providing a broad base on which the child's body is supported, and this base may at times compensate for the child's lack of muscle strength and coordination.

KINEMATICS

Gross motion of the foot is complex and occurs simultaneously in several planes. During inversion of the foot the sole faces medially, and during eversion it faces laterally. Circumduction of the foot involves flexion and extension of the ankle and foot and inversion and eversion of the foot itself. Toe motion includes flexion, extension, abduction, and adduction.

For practical purposes foot motion can be considered to be of two distinct types: non-weight-bearing and weight-bearing. Passive, non-weight-bearing motion, achieved by holding the leg and moving the heel during examination of the foot, permits inversion or eversion of the hindfoot through the subtalar joint. Abduction and adduction of the forefoot can be tested if the heel is held immobile. Supination and pronation (twist) of the forefoot may also be tested with the heel fixed, as may flexion and extension of the tarsometatarsal joints and the toes.

Active, weight-bearing motion of the foot differs from passive motion because the forces produced by body weight and by muscle contraction act to stabilize the joints. When one stands on the ball of the foot, the hindfoot inverts slightly as the midfoot is in plantar flexion, and the forefoot exhibits some pronator twist, creating an arch. Standing flat-footed on an externally rotating leg also raises the arch by moving the heel into slight inversion and causing the forefoot to twist into pronation. Rotating the leg internally has the opposite effect: it lowers the arch.

FOOT MOTION DURING GAIT

At the speed of normal walking (about 5.63 km per hour), a person averages 60 cycles per minute and spends 62% of each cycle in stance phase (Fig. 9–3). Part of this time is a period of double support when both feet are on the ground, each foot in a different part of the stance phase. As the speed of gait increases, less time is spent in the period of double support and more time is spent in swing phase.

During normal walking, the entire lower extremity (including the pelvis, femur, and tibia) tends to rotate internally through the swing phase of gait into the initial 15% of stance phase. In the middle of stance phase and at push-off, the entire lower extremity begins to reverse and rotate externally down to and

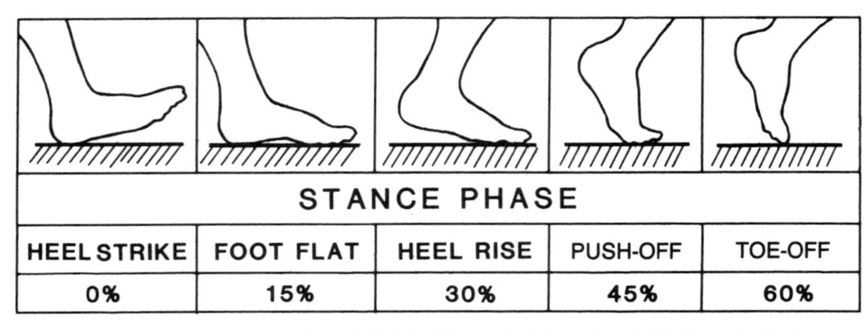

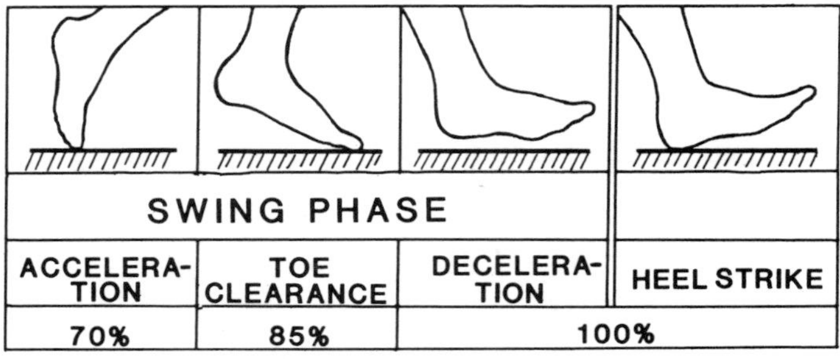

FIG. 9–3
About 60% of the normal gait cycle is spent in stance phase and about 40% in swing phase. After heel strike the center of the load progresses forward rapidly to the great toe, where toe-off occurs at about 60% of the gait cycle.

including the talus, so that at toe-off, the end of stance phase, maximum external rotation is achieved in the lower extremity, including the foot. As this external rotation occurs, a degree of increased stability is reached along the medial aspect of the hip, knee, ankle, and foot. Since muscles are also contracting during this portion of the gait cycle, both ligaments and muscles stabilize the foot until it is lifted off the ground.

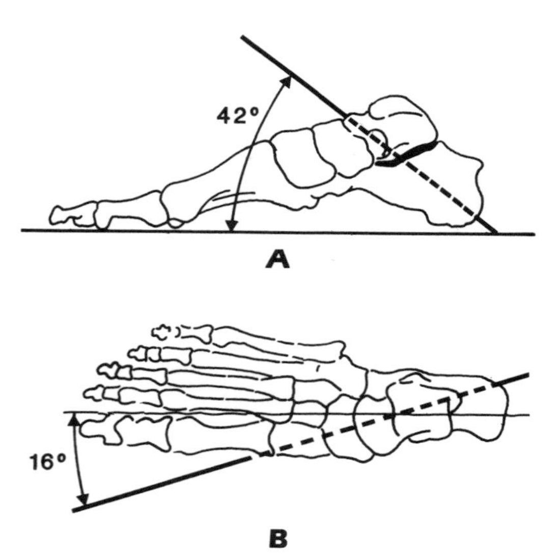

FIG. 9–4
Simplified axis of rotation of the subtalar joint. **A.** Sagittal plane (lateral view). The axis rises up at a 42-degree angle from the plantar surface. **B.** Transverse plane (top view). The axis is oriented 16 degrees medial to the midline of the foot. (Adapted from Manter, 1941.)

MOTION OF THE TARSAL BONES

The motion of the separate joints within the foot varies considerably with the phase of gait as well as with the type of surface beneath the foot. At times the foot functions as a single unit and at other times as a supple grasping appendage (Crelin, 1983). Indeed, the potential of the normal foot as a prehensile limb has been well demonstrated by persons with complete amelia of the upper extremities who dress, feed themselves, and even write with their feet and toes (Swinyard, 1969).

Subtalar Joint Motion

Motion in the subtalar joint, located between the talus and calcaneus, has been described by several investigators. The talus itself is a bone with no muscle attachments and can be likened to the carpal scaphoid. It sits atop the calcaneus, stabilized only by ligaments and cradled by all tendons passing from the leg to the foot (Sarrafian, 1983). Sarrafian (1983), Wright and coworkers (1964), and Inman (1976) conceived of the ankle and subtalar joint complex as a universal joint functioning as a single unit. Manter (1941), in investigating the subtalar axis of rotation, showed that on the average the axis extends upward and forward at an angle of 42 degrees from the floor at the heel. This axis deviates 16 degrees medially from the midline of the foot (Fig. 9–4). The subtalar facets closely resemble segments of a "spiral of Archimedes," a right-handed screw in the right foot

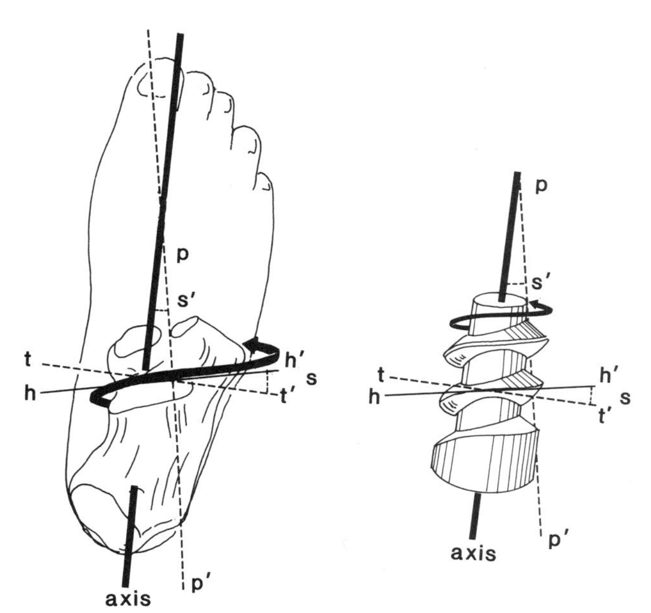

FIG. 9–5
Comparison of the posterior calcaneal facet of the right subtalar joint with a right-handed screw. Arrow represents the path of a body following the screw. The horizontal plane in which motion is occurring is hh'; tt' is a plane perpendicular to the axis of the screw; s is the helix angle of the screw, equal to the angle s', which is obtained by dropping a perpendicular (pp') from the axis. (Adapted from Manter, 1941.)

and a left-handed screw in the left foot, passing from the posterior to the anterior facets (Fig. 9–5).

On average, the subtalar joint can be inverted 20 degrees and everted about 5 degrees. Wright and coworkers (1964) found, however, that throughout the stance phase of gait the average range of motion of the subtalar joint was only 6 degrees, a so-called functional range of motion. The foot rests in slight supination (varus) during swing phase. At heel strike it rotates into slight pronation, or valgus (Mann, 1979). The muscles of the calf and foot actively stabilize the subtalar joint, controlling the heel within this small range of motion.

Transverse Tarsal Joint Motion

The transverse tarsal joint, Chopart's joint, lies just anterior to the talus and calcaneus and is closely associated with the subtalar joint. It represents motion between the talus and navicular and between the calcaneus and cuboid. Manter (1941) noted that two types of motion are achieved at this complex joint through two axes: the axis of mediolateral rotation and the axis of flexion and extension. The former is a longitudinal axis of rotation rising up from the floor anterodorsally at an angle of 15 degrees and directed away from the midline of the foot, pointing anteromedially at an angle of 9 degrees (Fig. 9–6). Internal and external rotation of the midfoot occurs around this axis during gait while the foot accommodates various plantar surfaces. Flexion and extension of the midfoot occur around the latter axis. This axis is a more oblique axis, which rises up from the floor anterodorsally about 52 degrees and is directed away from the midline of the foot, pointing anteromedially 57 degrees (Fig. 9–7). The axis varies considerably

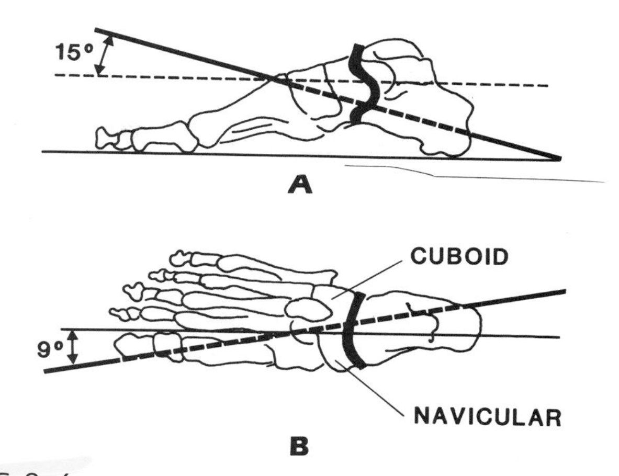

FIG. 9–6
Longitudinal axis of rotation of the transverse tarsal joint. **A.** Lateral view. **B.** Top view. (Adapted from Manter, 1941.)

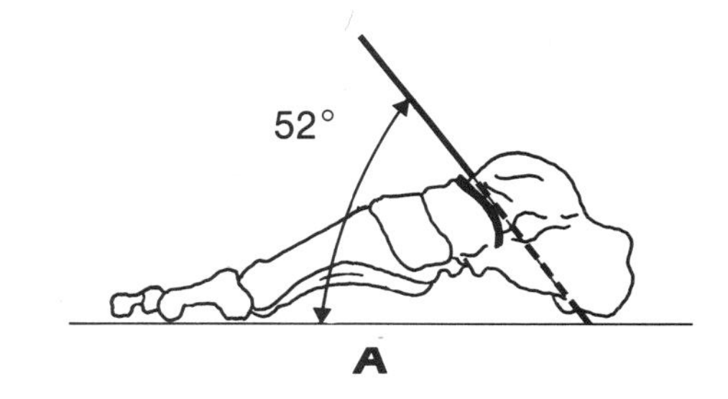

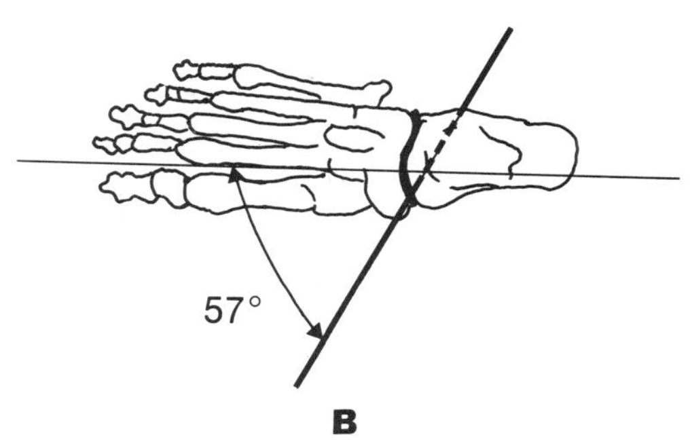

FIG. 9–7
Oblique axis of flexion and extension of the transverse tarsal joint. **A.** Lateral view. **B.** Top view. (Adapted from Manter, 1941.)

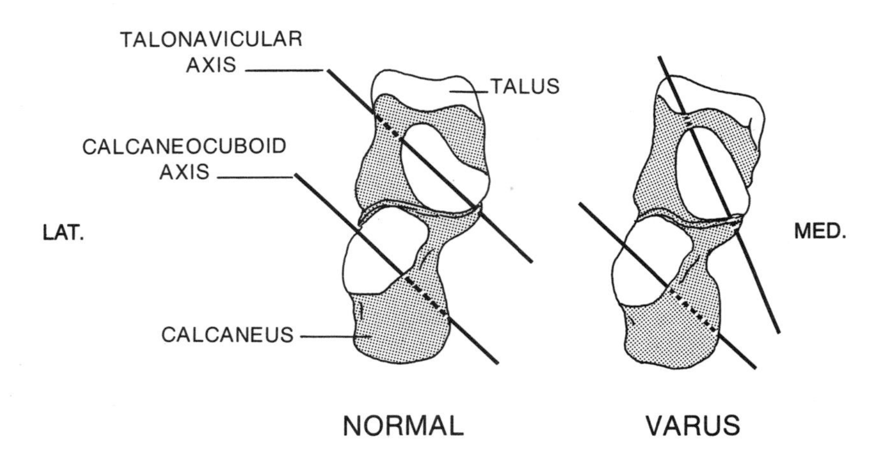

FIG. 9–8
Anteroposterior view of the transverse tarsal joint of the right foot. The anterior articulations of the talar head and calcaneus are shown. The axes of the talonavicular and calcaneocuboid joints are shown in the neutral position (parallel) and with the heel in varus (convergent). (Adapted from Mann and Inman, 1964.)

with both the age of the person and the shape of the foot; therefore, this description of "normal" subjects is, at best, a composite picture.

Mann and Inman (1964) analyzed flexion and extension of the midfoot in terms of parallel axes through the talus and calcaneus (Fig. 9–8) and described the transverse tarsal joint as functioning in a unique manner. These two axes are positioned in the frontal plane; the superior axis passes through the talar neck, and the inferior axis passes through the calcaneal body. These axes are concerned with motion at the talonavicular and calcaneocuboid joints, respectively. When the foot is in eversion, that is, with a tendency to be rolled flat or pronated, these two axes fall into parallel alignment. With two axes parallel in the same plane, the midfoot is able to flex and extend with ease in relation to the hindfoot (heel); however, when the heel is inverted, that is, with the arch elevated or the foot supinated, the axes diverge. With two axes crossing one another in the same plane, flexion and extension of the midfoot are significantly restricted with respect to the hindfoot. This limitation of flexion and extension may be one reason why patients are able to tolerate a pronated foot or flatfoot more easily than they tolerate a varus or supinated foot as in clubfoot.

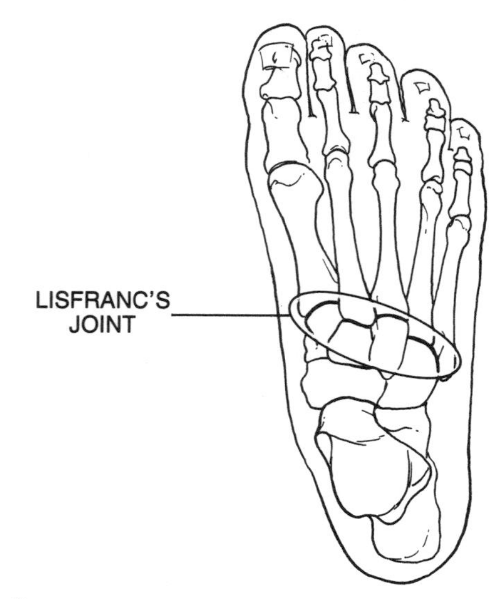

FIG. 9–9
The tarsometatarsal joints, known as Lisfranc's joint (circled).

Intertarsal and Tarsometatarsal Joint Motion

Motion on the surfaces of the intertarsal and tarsometatarsal joints is restricted by several factors—the shapes of the bones, the many restrictive ligaments (the foot contains 108), and the contracting muscles. Gliding motion occurs between the cuneiform and cuboid and also within the tarsometatarsal joints during gait. Since the total excursion of any two bones of the intertarsal joints is small, for practical purposes motion may be considered as translation, or parallel motion (gliding), of one surface across another. Total motion in the midfoot ranges from just a few degrees of dorsiflexion to about 15 degrees of plantar flexion; this motion is shared by all the tarsal bones.

The shape of the arch is affected by motion of all the tarsal joints. The arch may be raised by passive means through external rotation of the tibia during standing and also through extension of the toes and tightening of the plantar fascia. Of particular importance to the shape of the arch are the tarsometatarsal joints, which together are known as Lisfranc's joint (Fig. 9–9). Hicks (1953) showed that with upward and downward movement of the first metatarsal ray (the metatarsal and phalanges of the hallux), the second through the fifth rays moved successively less. Conversely, with the same movement of the fourth and fifth rays, the medial rays of the foot tended to move less.

The second tarsometatarsal joint is recessed into the midfoot, forming a keylike configuration with the intermediate cuneiform. This configuration restricts motion of the second ray, making it more stable than the first ray. The lateral rays also have a greater range of motion than does the second ray. This relative rigidity of the second ray is important during the later stages of stance, when the load is transferred into the forefoot for toe-off, because it permits an increased load to be transmitted through the second metatarsal. A dramatic example of the consequence of this increased loading can be seen in roentgenograms of a ballet dancer's foot (Fig. 9–10). Excessive repeated loading through the ray produced hypertrophy of the second metatarsal (Sammarco, 1982).

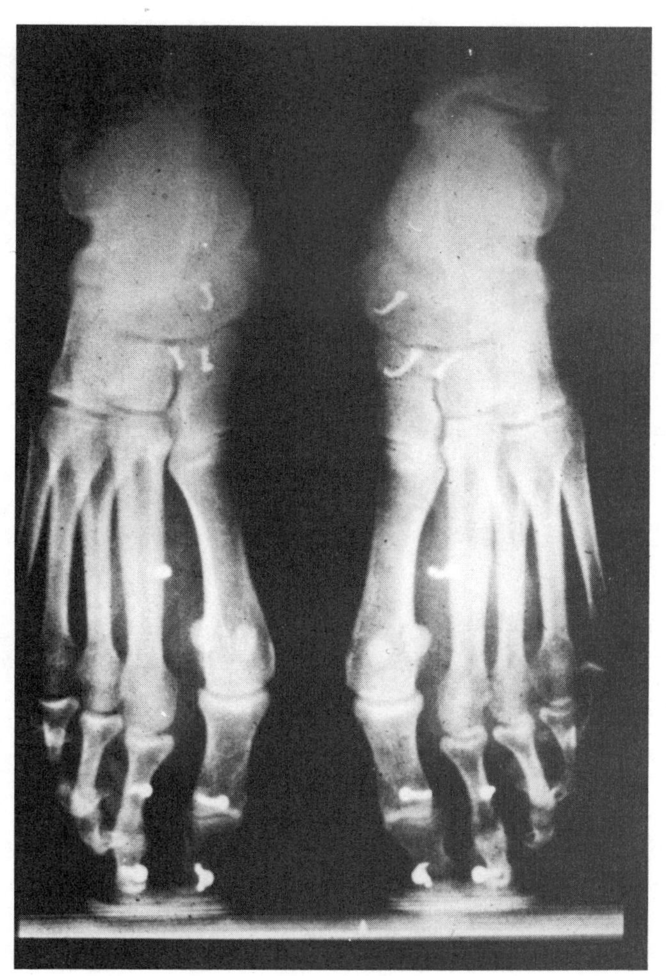

FIG. 9–10
Roentgenograms of the feet of a ballerina standing on point. The thickened second metatarsal and the recessed second tarsometatarsal joint form a keylike lock that stabilizes the second ray.

The Metatarsal Break

In the later part of stance phase, as the weight is transferred to the forefoot, there is an axis through which all toes extend at the metatarsophalangeal joints. This oblique axis, which overlies the metatarsophalangeal joints, is called the metatarsal break. It varies considerably among individuals in its orientation to the long axis of the foot, from 50 to 70 degrees (Fig. 9–11), and is a generalization of the instant centers of rotation of all five metatarsophalangeal joints. The metatarsal break has been used to analyze shoe wear and fit.

Motion of the Hallux

Motion of the forefoot can be illustrated by the metatarsophalangeal joint motion of the hallux. The hallux must accommodate a wide range of motion for the foot to perform a great variety of tasks. The hallux

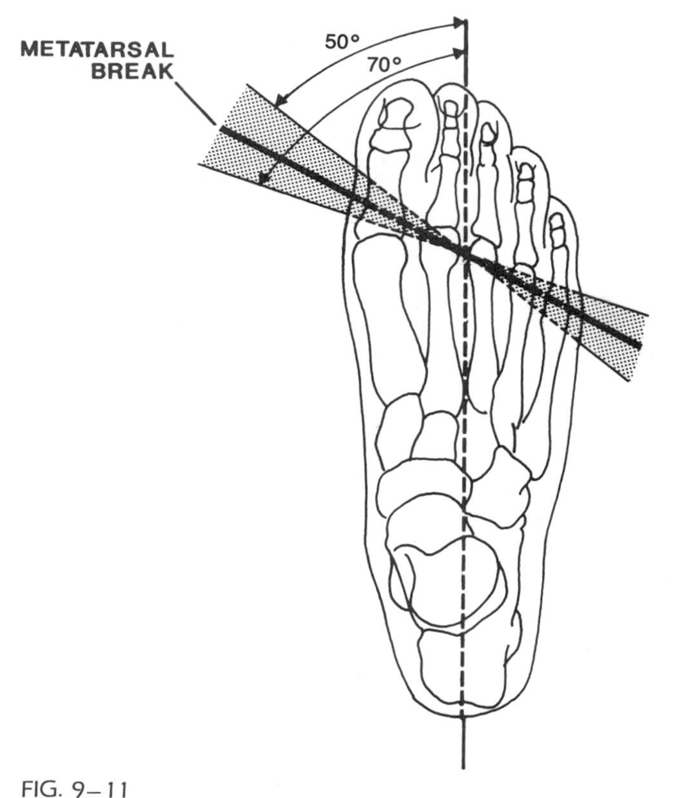

FIG. 9–11
The metatarsal break (top view), a generalization of the instant centers of rotation of all five metatarsophalangeal joints, may vary among individuals from 50 to 70 degrees in its orientation to the long axis of the foot. (Adapted from Mann, 1975.)

FIG. 9–12
Extension of the hallux metatarsophalangeal joint approaches 90 degrees when a crouched position, such as that of a sprinter poised on starting blocks, is assumed.

metatarsophalangeal joint has a range of motion from 30 degrees of flexion to 90 degrees of extension. A hooking action with the hallux flexed to support the body, as when one stands on the edge of a step supported by the toes alone, is an example of flexion. For this action to be accomplished, the toe is flexed and all muscles must contract. It is interesting to note that modern boots used for climbing rock and timber have rough soles that "hook" onto a crag or limb, a stiff shank for support, and a thick upper portion for protection of the foot. This design helps substitute boot function for foot function. One must also be able to extend the joint in order to accommodate a crouched position, best illustrated by a sprinter poised at the start of a race on starting blocks (Fig. 9–12). In this situation extension of the hallux metatarsophalangeal joint approaches 90 degrees. Extension of almost 90 degrees also occurs at the toe-off phase of walking.

Analysis of motion of the great toe in the sagittal plane reveals that instant centers often fall within the head of the first metatarsal (Fig. 9–13). The direction of displacement of the contact points in the sagittal plane reflected by these instant centers tends to show gliding, as in all diarthrodial joints, through most of the range of motion, that is, the joint motion occur-

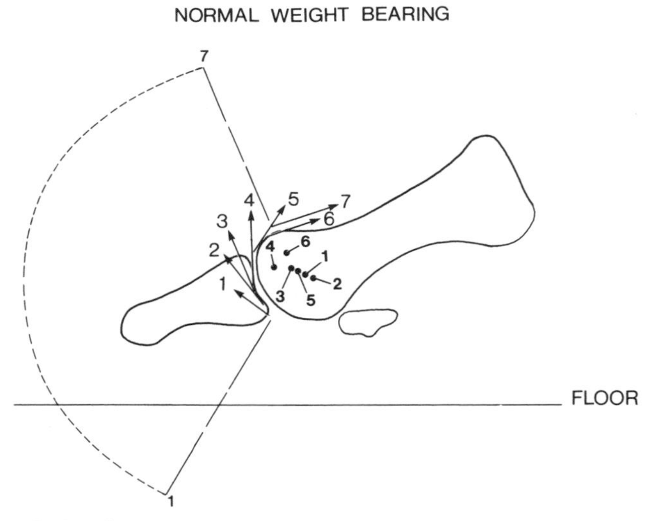

FIG. 9–13
Instant center analysis of the metatarsophalangeal joint of the hallux in the sagittal plane during normal weight bearing. Each arrow denoting direction of displacement of the contact points corresponds to the similarly numbered instant center. Gliding takes place throughout most of the motion except at the limit of extension, which occurs at toe-off in the gait cycle and with squatting. Here compression takes place. The range of motion of the hallux is indicated by the arc.

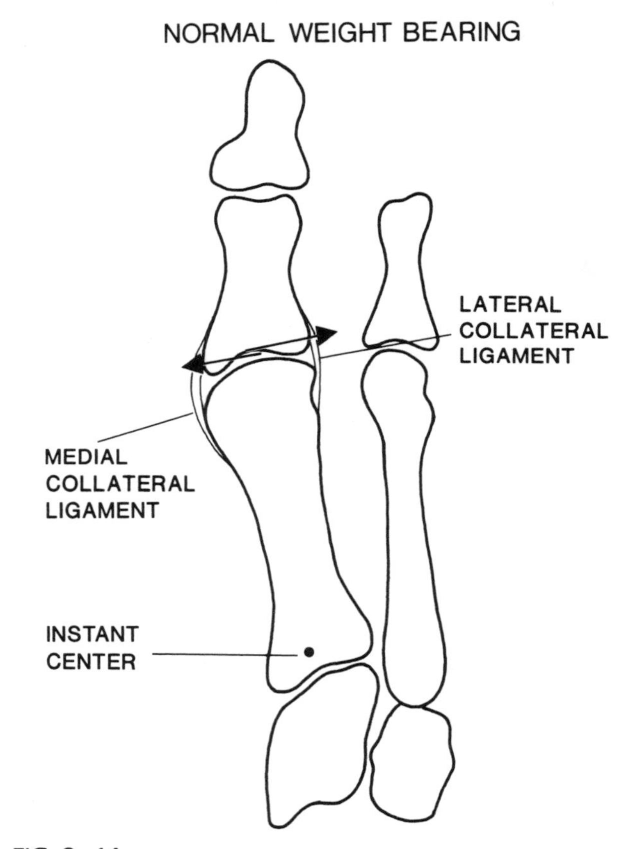

FIG. 9–14
Instant center analysis of the metatarsophalangeal joint of the hallux in the transverse plane during normal weight bearing. Gliding (denoted by the arrows) occurs at the joint surface even though the range of motion is small.

ring during normal activity from the few degrees of flexion through almost 90 degrees of extension. During walking the direction of displacement is parallel to the joint surfaces and motion occurs between the metatarsals and the proximal phalanges. At the extreme of extension, however, as in the case of a baseball catcher crouched behind home plate, the direction of displacement shows that the joint is being compressed or jammed. In this crouched position, a similar action occurs in the ankle (see Fig. 8–5).

The collateral ligaments of the hallux metatarsophalangeal joint are somewhat loose and provide a certain degree of play, both medially and laterally. Instant centers for analysis of medial and lateral motion lie far from the joint surface (Fig. 9–14), and for practical purposes these motions can be considered to be translation between the two bones.

In the development of hallux valgus, the hallux rotates internally and drifts into valgus angulation. The lateral collateral ligaments shorten, and the medial collateral ligaments stretch out. The metatarsal head develops a medial osteophyte (a bunion) (Fig. 9–15). Since the hallux lies in an unnatural position, the direction of displacement of the contact points shows an altered pattern. Now jamming and distraction take place where gliding normally occurred (Fig. 9–16).

Motion of the Small Toes

The lateral four toes have three phalanges each (see Fig. 9–1A). The second toe may be shorter or longer than or the same length as the hallux. Motion at the metatarsophalangeal joint is approximately 90 degrees of extension and 50 degrees of flexion, or

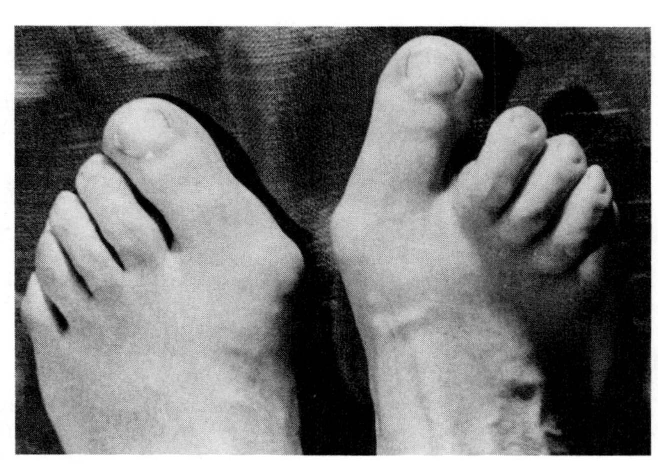

FIG. 9–15
Feet of a patient with hallux valgus and bunions. The hallux valgus is represented by the position of the great toes, which are pointed outward on both feet. Just proximal to the toes, bony protuberances on each foot—the bunions—are visible.

slightly greater than the hallux motion. The mechanism by which the small toes move is similar to that of the hand. Muscles that control the metatarsophalangeal and interphalangeal joints originate within the foot (intrinsic muscles) and in the calf (extrinsic muscles). During the later part of stance phase, in the period up to toe-off, the muscles stop functioning to allow the toes to extend greatly during push-off. This extension also passively stiffens the tarsals and metatarsals by tightening the plantar fascia.

Motion of the toes in flexion is brought about at the metatarsophalangeal joint by the weak lumbricals and interossei (Sarrafian and Topouzian, 1969) (Figs. 9–17 and 9–18), but the flexor digitorum brevis and flexor digitorum longus, which insert on the middle and distal phalanx, respectively, contribute their

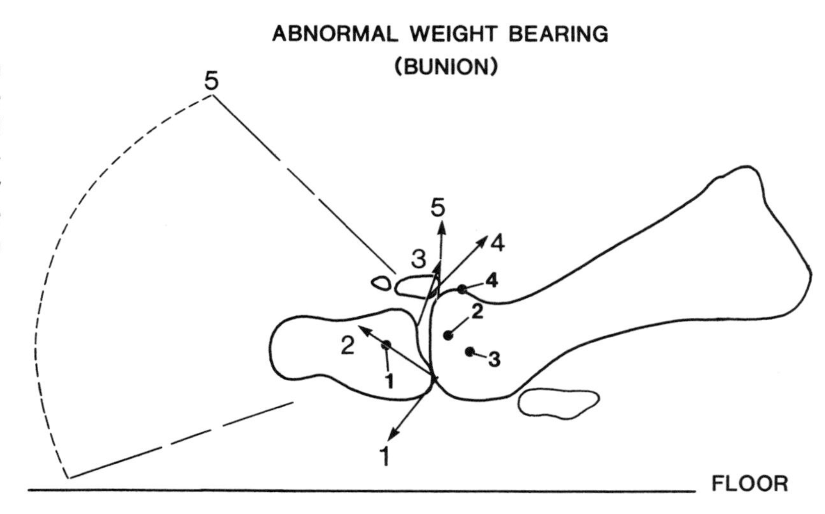

FIG. 9–16
Altered instant centers caused by the presence of a bunion, seen through analysis of motion of the metatarsophalangeal joint of the hallux in the sagittal plane. Each arrow denoting direction of displacement of the contact points corresponds to a similarly numbered instant center. The arc indicates the range of motion of the hallux, which is more limited than that in a normal foot (compare Fig. 9–13).

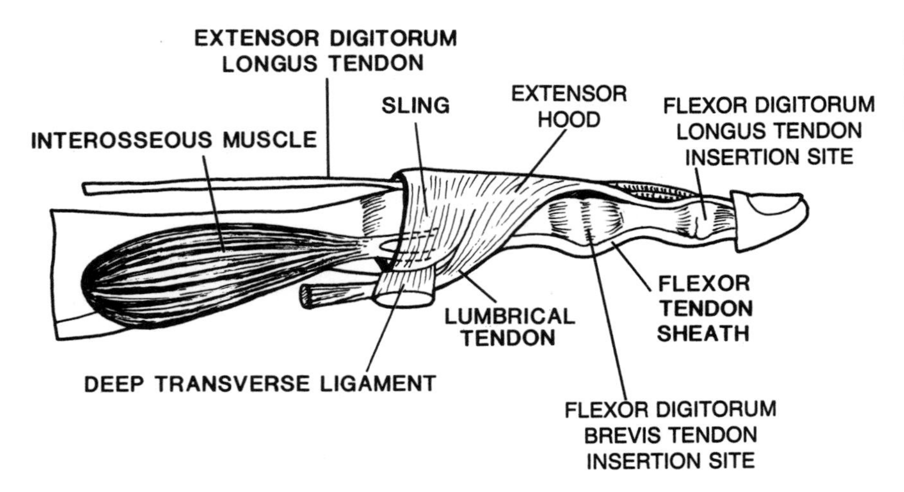

FIG. 9–17
Simplified lateral diagram of a single toe illustrating muscles and ligaments required for motion.

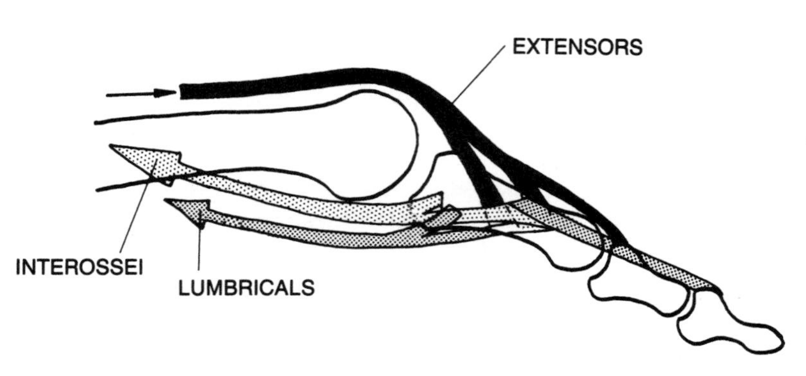

FIG. 9–18
Diagram demonstrating the pull of the lumbricals and interossei causing flexion of the metatarsophalangeal joint and extension of the interphalangeal joint. (Adapted from Viladot, 1982.)

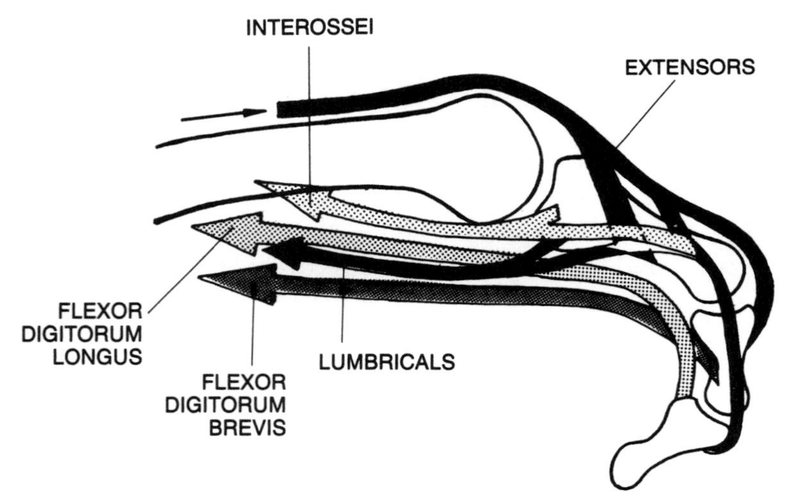

FIG. 9–19
Diagram demonstrating the pull of the lumbricals and interossei along with the pull of the long and short flexors of the toe. This pull causes clawing of the toe. (Adapted from Viladot, 1982.)

power and provide a stronger flexion force (Fig. 9–19). The metatarsophalangeal joints extend by the action of the extensor digitorum longus through the extensor sling, which supports the proximal portion of the proximal phalanges (Fig. 9–20). The extensor digitorum longus, a strong muscle, acts as the principal motor for the action. Extension of the middle and distal phalanges is mediated through an extensor hood (see Fig. 9–17), much as in the hand, and is controlled by the lumbricals and interossei.

FUNCTION OF THE PLANTAR FASCIA

The function of the plantar fascia is complex. From its attachment to the calcaneus it extends forward to span all the tarsal and metatarsopha-

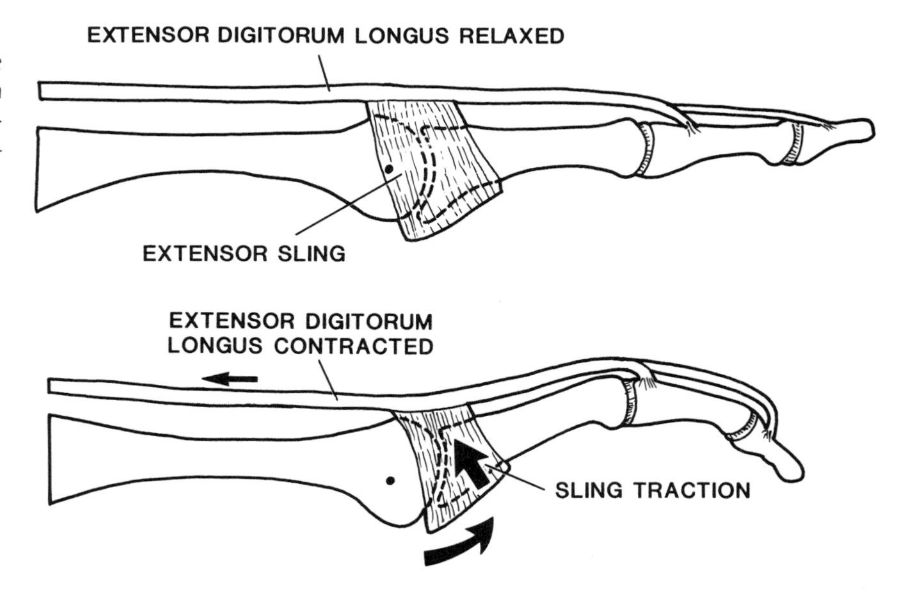

FIG. 9–20
Lateral diagram showing the action of the extensor sling. When the extensor digitorum longus contracts (bottom), the proximal phalanx is lifted into extension through this mechanism.

langeal joints and to attach to the plantar aspect of the proximal phalanges. The result is a trusslike structure whose links are the tarsal bones and ligaments of the foot, which are held at their base by a tether, the plantar fascia (Fig. 9–21). Although Wright and Rennels (1964) showed that this heavy ligamentous structure elongates slightly with increasing loads, this elongation is more of a shock-absorbing mechanism than a mechanism for obtaining motion in the foot. Since the plantar fascia spans the entire longitudinal arch and has relatively little intrinsic ability to lengthen, it acts as a cable between the heel and toes.

The subtalar, intertarsal, and tarsometatarsal joints all have small ranges of motion because of the irregular shape of each bone and the multifaceted, curved articular surfaces, which are bound by tight ligaments. The stability created ensures a good platform for standing; however, the combined motion of these joints is great because of the multitude of moving parts. There must be a mechanism for stiffening the foot during such actions as running and climbing. This mechanism is provided actively by the contraction of the calf muscles such as the triceps surae, peroneus longus, peroneus brevis, tibialis anterior, and tibialis posterior, all of which insert into tarsal bones. The intrinsic foot muscles also have their origin on the tarsal bones. When all contract, the foot becomes more rigid and acts as a stiff spring.

The mechanism by which all the tarsal joints are passively locked together is the action of the plantar fascia. A Spanish-windlass mechanism is formed at the metatarsophalangeal attachment of the fascia (Fig. 9–22). As the metatarsophalangeal joints are extended passively, when one stands on the ball of the foot, the plantar fascia is pulled distally across the metatarsophalangeal joints, shortening the distance from the calcaneus to the metatarsal heads. This process makes the base of the truss shorter. As the tether is shortened and the distance between the heel and the ball of the foot is reduced, the tarsal joints are locked into a forced flexed position and the height of the longitudinal arch of the foot is increased (Fig. 9–23).

The plantar fascia also prevents the collapse of the longitudinal arch of the foot during standing because it prevents passive flexion of the toes, which would allow the fascia to relax and the arch to flatten slightly. Thus, when a welder, sprinter, or baseball catcher squats, the 12 tarsal and metatarsal bones of the foot are forced by this mechanism to create a solid structure to support the rest of the body.

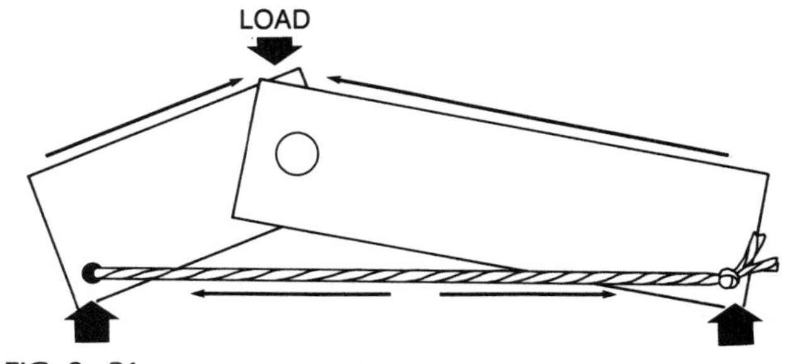

FIG. 9–21
Diagram of a truss. The wooden members are analogous to the bony structures of the foot. The plantar fascia is represented by a tether between the ends of the bones. The shorter the tether, the higher the truss is raised.

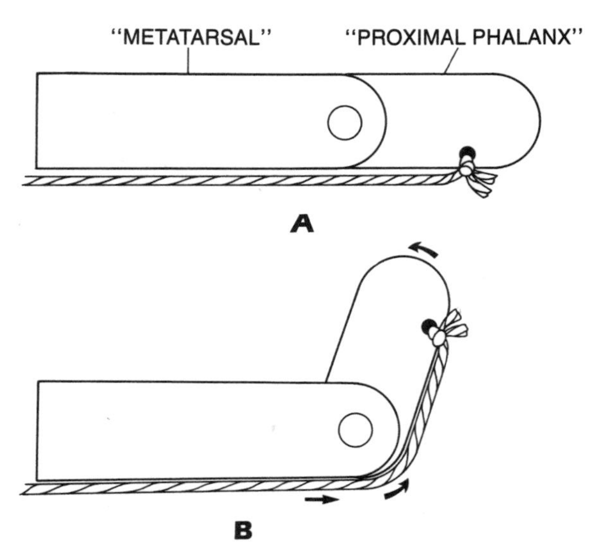

FIG. 9–22
A. Diagram of a Spanish windlass. The metatarsal is represented by the fixed wooden member, and the proximal phalanx is represented by the moving member. The rope attached to the moving member represents the attachment of the plantar fascia to the proximal phalanx. **B.** As the moving member turns, the rope advances.

The passive function of the plantar fascia complements the active function of the muscles in standing, walking, running, squatting, and other activities. Surgical removal of the plantar fascia is seldom indicated and usually creates a need for an arch support for postoperative care.

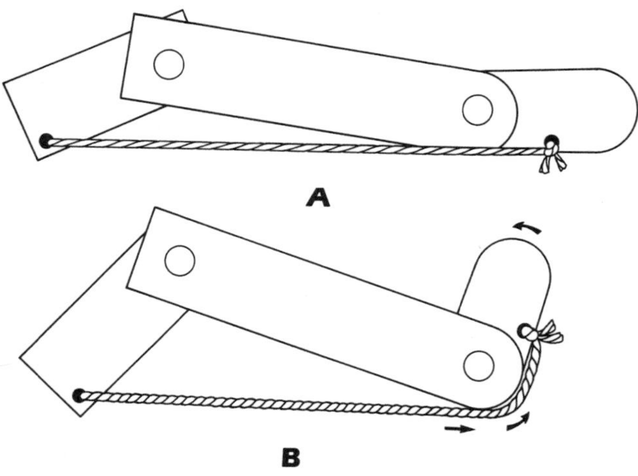

FIG. 9–23
Diagram of a combined truss and Spanish windlass **(A)**, which illustrates the function of the plantar fascia in raising the arch of the foot and at the same time locking the joints and making a single unit from multiple individual bones and joints **(B)**.

MUSCLE CONTROL IN THE FOOT

Thirteen extrinsic and 19 intrinsic muscles control the foot. The extrinsic muscles provide the active control. In the calf, the gastrocnemius and soleus muscles, which combine to form the Achilles tendon inserting on the calcaneus, are the strongest flexors of the ankle. Their function is mediated through the subtalar joint, which is rather stiff in the sagittal plane and transmits the force directly to the talus and foot. The peroneus longus muscle controls, to a great extent, the downward pressure of the first metatarsal head, and it also controls the fine movements of the hallux. Injury or paralysis of this muscle results in a great decrease in the loads transmitted to the head of the first metatarsal and often leads to the development of a dorsal bunion. Alpine skiers use the peroneus longus muscle a great deal to control the inner edges of the skis while turning. Along with the tibialis posterior muscle, the peroneus longus acts as a sling to actively support the longitudinal arch of the foot. The peroneus brevis muscle acts as an accessory ankle flexor and stabilizes the foot laterally.

Of the muscles of the posteromedial compartment of the ankle, the tibialis posterior, relatively active in normal standing, is most important in controlling the

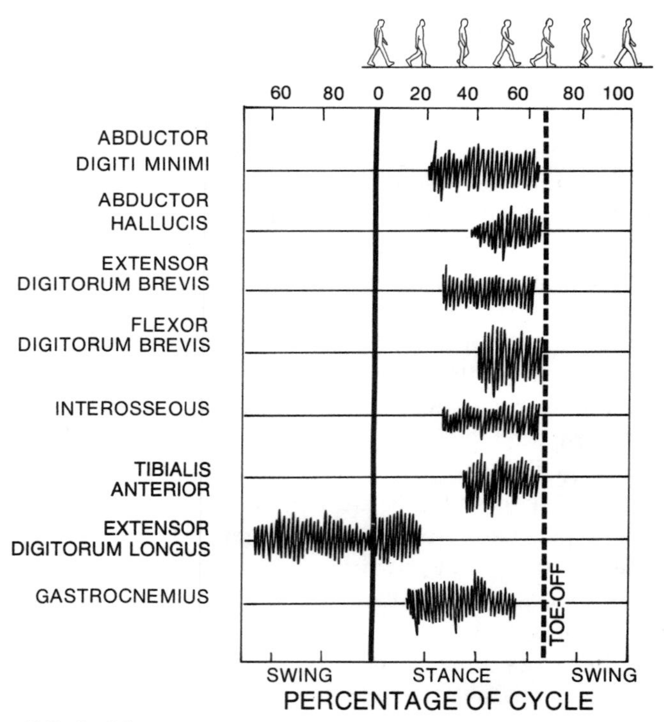

FIG. 9–24
Electromyographic tracing of calf and foot muscles during walking. (Adapted from Mann and Inman, 1964.)

medial stability of the foot and functioning as an accessory flexor of the ankle. This muscle is particularly important because of its special function in actively supporting the longitudinal arch. It flexes the midfoot through its pull on a broad insertion of several tarsal bones and also supports the talar head and neck through a slinglike action. Isolated loss of this muscle due to a severed tendon leads to collapse of the arch.

The flexor hallucis longus and flexor digitorum longus are strong muscles that control flexion of the toes during walking. The tibialis anterior and the extensor hallucis longus and brevis decelerate the foot as it strikes the ground, preventing the foot from slapping at heel strike; flaccid paralysis of these muscles from disease causes a slapping gait (Fig. 9–24).

Control of the great toe is mediated through the tendons of both the intrinsic and extrinsic muscles of the foot. A cross section of the proximal phalanx reveals that the relative positions of the flexors, extensors, abductors, and adductors are such that these muscles move the hallux in any direction within the confines of its ligaments (Fig. 9–25).

Within the tendons of the flexor hallucis brevis are the two sesamoid bones of the hallux located directly beneath the head of the first metatarsal. The sesamoids have several functions. They act to transfer loads from the ground through the soft tissues of the forefoot to the metatarsal heads. Their position within the tendons of the flexor hallucis brevis permits them to increase the length of the moment arm, giving the muscle a greater mechanical advantage as a flexor, similar to the mechanical advantage that the patella provides at the knee (see Fig. 6–20). The sesamoids are at a fixed distance from the proximal phalanx, into which the flexor hallucis brevis inserts. Therefore, as the hallux is extended at the end of the stance phase, the sesamoids are pulled forward, so an optimal relationship is maintained among the ground reaction forces, the soft tissue fat pads of the hallux, and the metatarsal heads.

KINETICS

The distribution of loads under the foot during stance has been the subject of intensive investigation for the last half century. Morton (1935), in his investigation of weight bearing during normal stance, showed that all metatarsal heads are in contact with the floor. This finding contrasts with those of some investigators who felt that a "transverse metatarsal arch" existed in the foot. There may indeed be a curvature of the metatarsal heads while the foot is held in the non-weight-bearing position. However, the mobility of the metatarsal heads permits them all to fall into contact with the floor as soon as load is applied during standing.

In the portions of the foot that have contact with the floor during normal stance, approximately 50% of the load is borne by the heel and 50% is transmitted across the metatarsal heads. The load on the metatarsal head of the hallux is twice that on each of the lateral four metatarsal heads (Fig. 9–26). The first metatarsal head thus transmits twice the load of each individual lateral metatarsal head, and each of these bears an equal amount of the remaining portion of the load in the forefoot. A slight change in the foot structure alters the load distribution. It may also be changed with only slight modification of weight bearing, such as rocking slightly from side to side or forward and backward while standing; thus, when one stands for long periods, as a soldier at attention does, a slight shift of the weight relieves pressure on the plantar soft tissues and lessens the burning and pain of fatigue in the soft tissues of the foot.

During the stance phase of gait, the center of the load progresses forward rapidly to the great toe. During the later part of stance, increased loads tend to be transmitted across the second metatarsal head (Figs. 9–3 and 9–27), as noted by Collis and Jayson (1972) and by Klenerman (1976). The reason for this

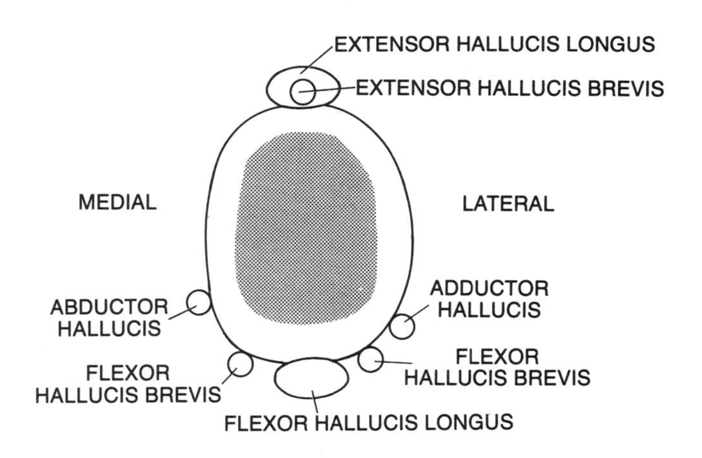

FIG. 9–25
Diagrammatic cross section of the proximal phalanx of the hallux, showing normal positions of the various tendons in relation to the bone.

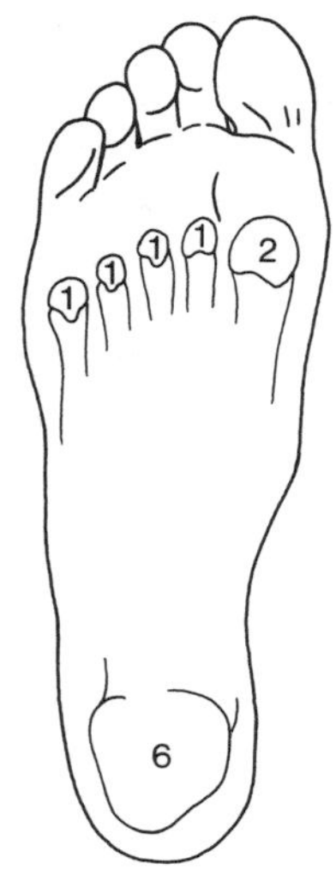

FIG. 9–26
Loads beneath the foot during normal stance. The numbers represent the proportion of the total load borne by each structure. Approximately half of the load is borne by the heel and the other half by the heads of the metatarsals. The load on the head of the hallux is twice that on each of the lateral four metatarsal heads.

increase is twofold. The second metatarsal tends to be longer than the others; thus, as the heel rises and the load is shifted forward, the second metatarsal head, being more distal than the others, tends to concentrate the ground reaction force. In addition, the second tarsometatarsal joint is quite stiff compared with the other midfoot joints and thus tends to yield less under loading.

During the stance portion of gait at heel strike, the ground reaction force is slightly medial to the center part of the heel pad. When the foot rolls into slight valgus shortly thereafter and the force is applied in the foot flat portion of stance, the ground reaction force progresses slightly laterally to beneath the cuboid and then toward the base of the first metatarsal (Figs. 9–28 and 9–29). Toward the end of the stance phase, the reaction force courses medially to beneath the second metatarsal head and then to the hallux at toe-off. Variations in magnitude, position, and direction of ground reaction forces at each portion of stance phase are shown in Figure 9–30.

Loads beneath the foot vary during sitting and standing because of the difference in the position of the foot in relation to the torso. Jones (1941) examined the loads through the foot in a seated subject onto whose knees external loads were applied. With all the muscles of the foot relaxed, 80% of the load was distributed to the heel and only 20% was borne across the metatarsal heads. During normal standing the load produced by the body is balanced over the center of the foot, anterior to the ankle. Slight weight shifts caused by the action of the calf muscles (the tibialis anterior and the triceps surae) adjust deviations to equilibrium.

Little muscle activity occurs in the foot during relaxed standing, as evidenced by electromyographic activity (Basmajian, 1974). In the cadaver foot, where no active muscle contraction or variation of loading position accompanies slight changes in posture, loads borne by the forefoot are distributed equally across all metatarsal heads.

In analyzing the load transmitted across the individual tarsal joints during stance, Manter (1946)

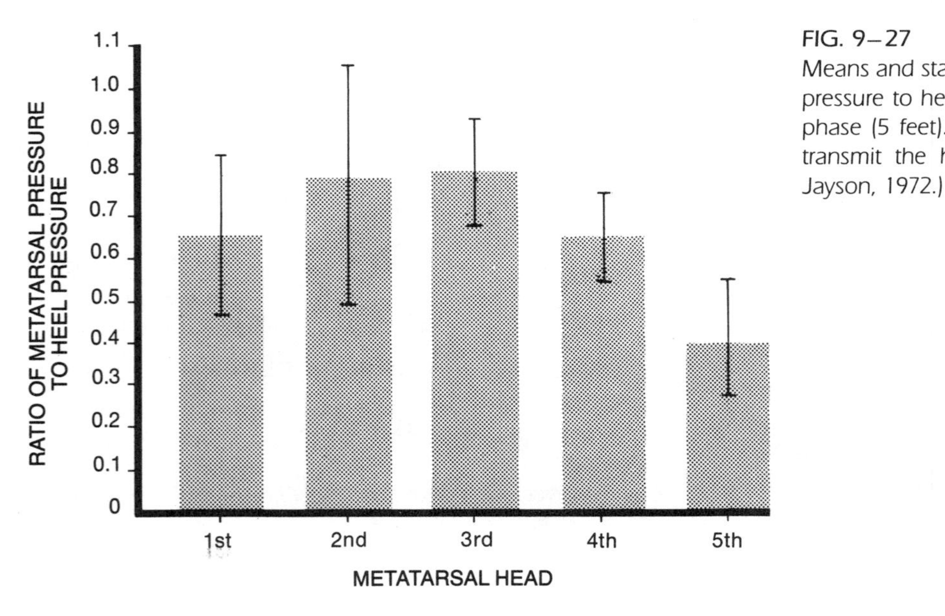

FIG. 9–27
Means and standard deviations of the ratio of metatarsal pressure to heel pressure during the later part of stance phase (5 feet). The second metatarsal heads tended to transmit the highest loads. (Adapted from Collis and Jayson, 1972.)

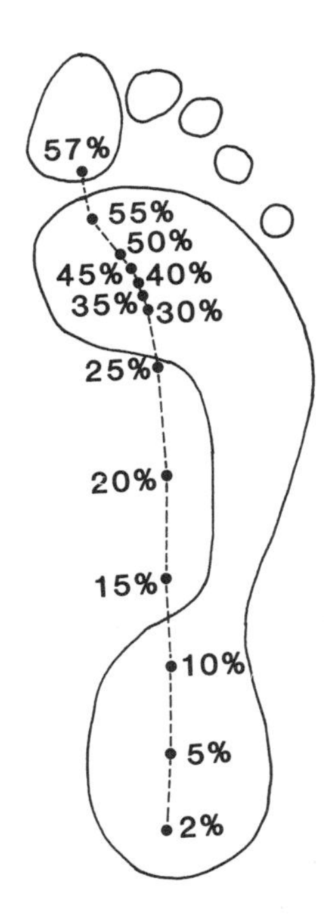

FIG. 9–28
The ground reaction force (broken line) progresses rapidly through the foot during gait. It passes slightly medial to the midline of the heel and along the medial border of the foot. Percentages of the gait cycle are indicated. At 30% of the cycle the force is beneath the second metatarsal. From 30% to 50% of the cycle the progression slows, and the force shifts medially beneath the first metatarsal head. It then progresses rapidly along the plantar aspect of the big toe. (Adapted from Hutton et al., 1976.)

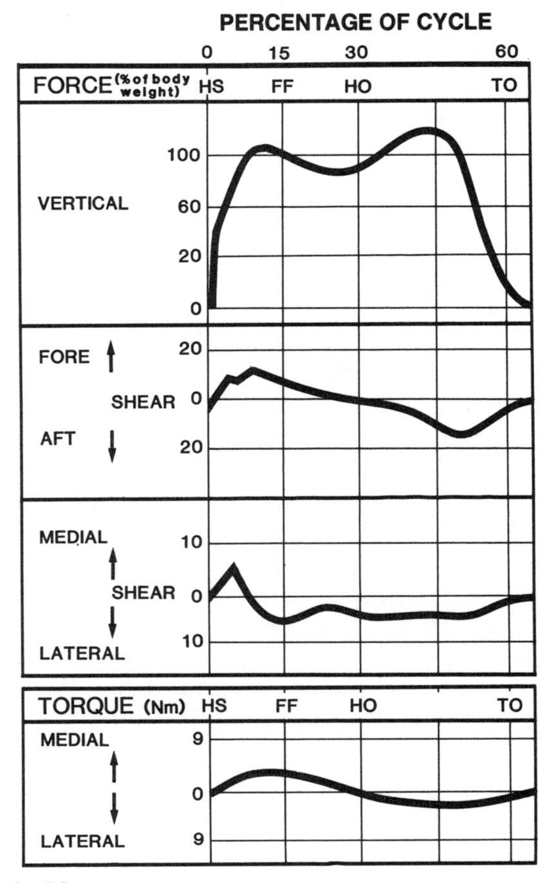

FIG. 9–29
Forces acting on the foot during the gait cycle: HS, heel strike; FF, foot flat; HO, heel-off; TO, toe-off. (Adapted from Mann, 1982.)

concluded that, although one cannot trace an individual longitudinal pathway through which all the forces pass, the load should be greatest at the height of the longitudinal arch. Therefore, since the talonavicular and naviculocuneiform joints are the highest on the longitudinal arch, they bear the greatest loads. The load is distributed toward the heel and is transmitted to the forefoot by two general routes. The greater load is borne across the highest portion of the longitudinal arch and it then passes from the tarsal joints into the first and second metatarsals and metatarsal heads. The lesser load is transmitted somewhat laterally through the cuboid into the base and then to the head of the fifth metatarsal. These routes are described for the peak forces within the foot, however, and it should be recognized that loads are transmitted through all tarsal and metatarsal bones.

The orientation of the trabeculae of the bones within the foot indicates major stress patterns (Mann, 1975) (see Fig. 9–1C). The trabeculae orient themselves into stress lamellae in the region where loads pass from the dome of the talus into the foot (Harty,

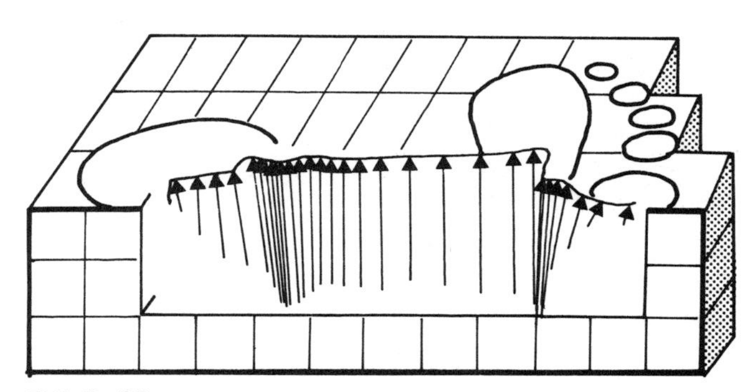

FIG. 9–30
Depiction of a force plate showing the distribution, magnitude, and direction of the ground reaction forces during the stance phase of the gait cycle. The point of application of the force passes through the heel and the medial border of the foot. It reaches the ball of the foot at the level of the second metatarsal head, shifts medially toward the first metatarsal head, and progresses rapidly to the big toe. The forces are directed downward and anteriorly during heel strike, become progressively vertical at midstance, and are directed downward and posteriorly at push-off. (Adapted from Elftman, 1939.)

1973). In the calcaneus some trabeculae are directed toward the posterior portion of the bone and others toward the anterior portion, leaving a triangle in the center with few trabeculae. This area should not be mistaken for a cyst lying in the calcaneus. The bony trabeculae are oriented longitudinally in the tarsal bones and the metatarsals and transversely across the top of the arch in the midfoot, indicating the direction of intraosseous force distribution.

The soft tissues of the foot are modified to provide strength and resistance to hostile external forces on the sole. The skin of the dorsum of the foot is thin and supple, allowing the underlying structures to slide easily beneath it during motion. By contrast, the skin of the sole of the foot is fixed to deeper tissue. It is thicker at the heel and at the lateral margin of the foot and ball, and it is thinner medially. Cleavage lines on the sole of the foot tend to be somewhat convex laterally, curving around the heel. These lines give some indication of the subcutaneous structure that provides cushioning during normal gait as well as during running (Fig. 9–31). The skin of the sole is affixed tightly to deeper tissue at the heel and beneath the ball of the foot. The fixation of the metatarsal fat pad is important in the later part of stance phase during push-off, when the toes extend to 50 degrees at the metatarsophalangeal joints, locking the fat pad beneath the metatarsal heads (Bojsen-Møller and Lamoreux, 1979).

To absorb the energy of the ground reaction force at heel strike, the skin and soft tissues are compartmentalized into a unique shock-absorbing mechanism. Under the thick skin are slanted spiral chambers that contain adipose tissue beneath the calcaneus. These chambers are so placed as to take their deep origin from the calcaneus and, passing posteriorly and curving laterally, to attach to the skin. A second spiral system, located more superficial to the calcaneus, converges on itself as it spirals laterally. The two systems represent a right-handed spiral that opens laterally beneath the calcaneus and a second spiral that closes as it passes laterally, superficial to the first (Fig. 9–32). The cushioning provided

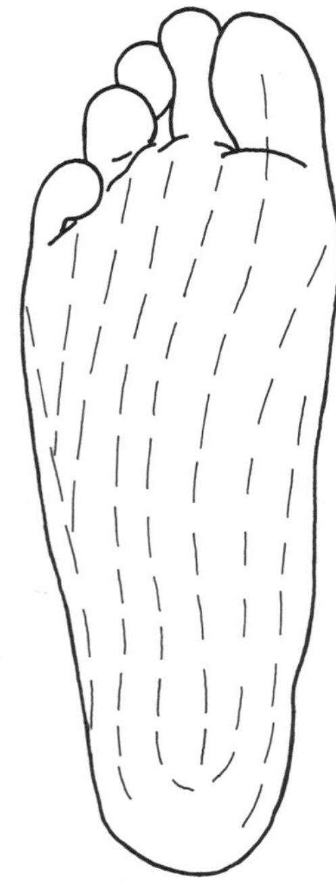

FIG. 9–31
On the plantar aspect of the foot, the skin flexion lines are longitudinally oriented with a slight curvature, convex to the fibular side. At the level of the heel the lines are arciform and parallel to the plantar border of the heel. Incisions placed in the cleavage lines leave fine linear scars, whereas incisions at right angles to these lines may leave wide scars. (Adapted from Cox, 1941.)

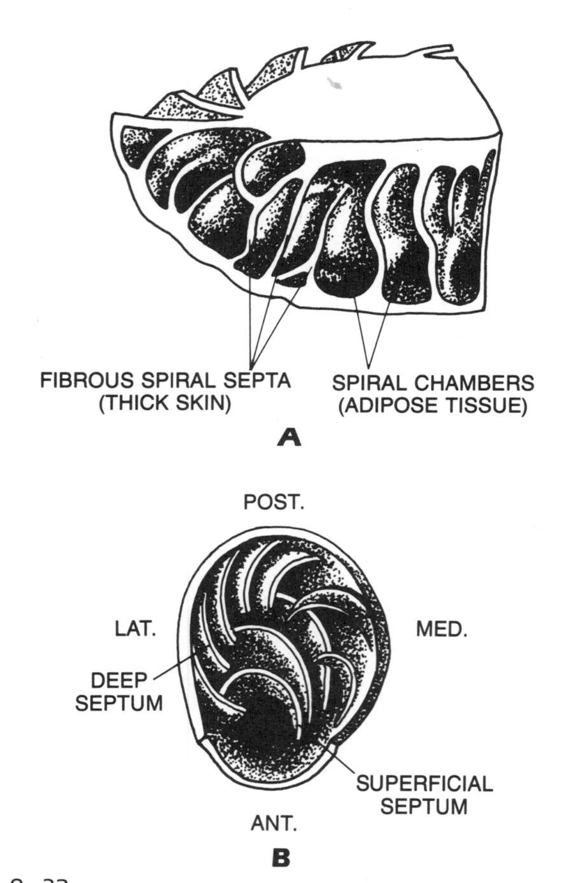

FIG. 9–32
Structure of the heel pad. **A.** Lateral view. **B.** Top view of the right foot showing the septa of the deep spiral system, which originate on the lateral calcaneus, and the septa of the superficial spiral system, which originate beneath the heel. (Adapted from Blechschmidt, 1933.)

by this system accounts for the ability of the heel to resist repeated loads for long periods, as during hiking.

COMMON DEFORMITIES CAUSED BY SHOE WEAR

Since in Western cultures shoes are commonly worn, an external force is normally present along the medial side of the hallux. Valgus angulation of less than 15 degrees at the hallux metatarsophalangeal joint is normal; however, the external force applied when a tight-fitting "stylish" shoe is worn over a period of years tends to increase the angle, producing hallux valgus and a bunion (see Fig. 9–15). This type of bunion differs from that caused by a disease process such as rheumatoid arthritis. The tight tip of the "stylish" shoe limits hallux motion in two planes. The medial ligaments of the hallux metatarsophalangeal joint stretch and the lateral ligaments contract. The extensor tendons drift laterally as the hallux is pushed laterally. The flexor tendons, now changed in position, begin to pull the first phalanx laterally and to rotate it internally (Fig. 9–33). When the first phalanx shifts, the abductor hallucis, which normally abducts the hallux, begins to function as a flexor, causing rotation of the proximal phalanx; the adductor hallucis pulls the toe laterally, unchecked by medial muscles or ligaments. Balance in the toe is then lost. The first metatarsal begins to drift medially, assuming a characteristic metatarsus primus varus position. (Of course, certain types of bunions are hereditary. Reference is made here only to those types caused by poorly fitting shoes.)

Shoes are a combination of design, materials, and fabrication. Despite attempts to obtain a correct fit, a shoe may be too small and the foot may be forced into a position in which the toes are crammed together with the metatarsophalangeal joints extended and the interphalangeal joints flexed. In this situation, normal muscle balance is lost. This posturing of the toes allows the stronger calf muscles to function, causing clawing of the toes while at the same time preventing the smaller, and weaker, intrinsic muscles (lumbricals and interossei) and flexor digitorum quinti from flexing the metatarsophalangeal joints and extending the interphalangeal joints. If such conditions persist for several years, for example, from continual wear of poorly fitted work shoes, a deformity known as hammertoe can develop with fixed flexion contractures of the interphalangeal joints. Dorsal dislocation at the metatarsophalangeal joint can also occur (Fig. 9–34). The deformity also occurs with muscle imbalance (caused, for example, by neurologic disease or by arthritis), with a resultant loss of joint stability. Since a small change in the loading of the foot can cause a significant alteration in load distribution, a slight change in muscle contraction can also cause a significant variation.

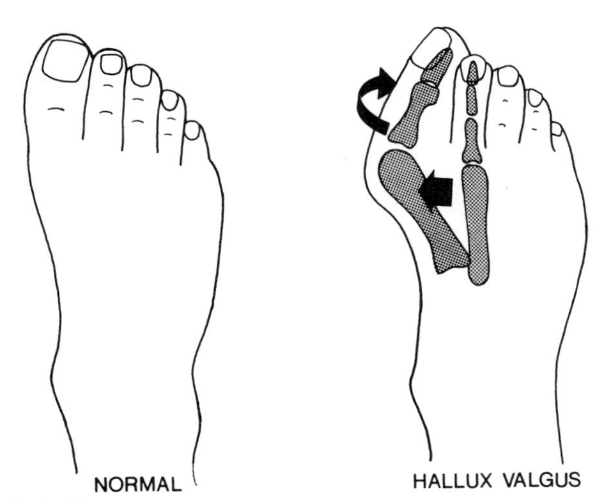

FIG. 9–33
Diagram demonstrating the formation of hallux valgus and a bunion. The pressure from the shoe forces the hallux laterally. The result is medial drift of the first metatarsal, which increases the mediolateral angular relation of the first metatarsal and phalanx. When this happens, the tendons that normally flex and adduct the hallux begin to rotate it, in turn forcing the head of the first metatarsal more medially. Osteophytes then form on the medial aspect of the head of the first metatarsal.

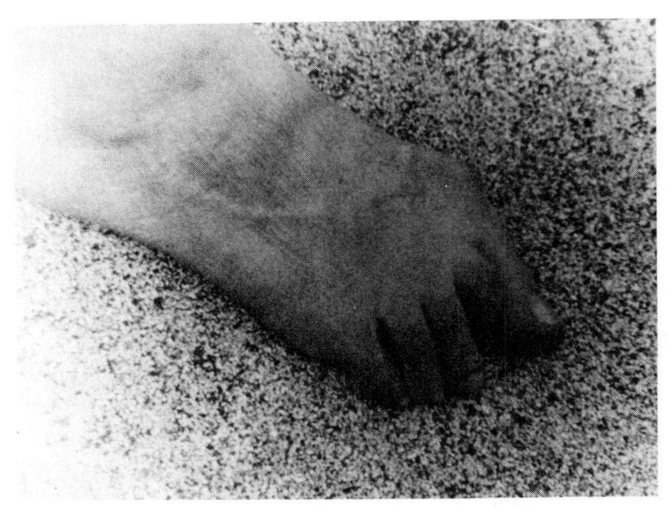

FIG. 9–34
Hammertoe. The second toe is in extension at the metatarsophalangeal joint and in fixed flexion at the proximal interphalangeal joint.

SUMMARY

1. Motion of the tarsal bones during gait is screwlike at the subtalar joint and takes place through a double axis at the transverse tarsal joint.
2. The keylike Lisfranc's joint (composed of the tarsometatarsal joints) in the midfoot stabilizes the second metatarsal, making it the most rigid forefoot bone and allowing it to carry most of the load during walking.
3. Instant center analysis of the hallux shows that gliding occurs throughout most of gait at the metatarsophalangeal joint with jamming at extreme extension; this surface motion is not unlike that in the ankle joint.
4. Motion in the small toes is controlled by both intrinsic and extrinsic foot muscles. The control mechanism is similar to that in the hand.
5. The plantar fascia functions through a complex truss or windlass mechanism to help passively stabilize tarsal and metatarsal bones.
6. Active control of the foot is provided by the extrinsic muscles. Anterior leg muscles decelerate the foot at heel strike, whereas posterior calf muscles propel the foot toward toe-off.
7. During standing, loads are distributed equally on the heel and ball of the foot; the hallux bears twice the load of any other metatarsal. During walking, however, most of the load is transmitted through the second metatarsal.
8. During walking, ground reaction forces act slightly lateral to the center of the heel at heel strike; they then shift laterally toward the cuboid, finally moving beneath the second metatarsal and hallux at toe-off.
9. Loads within the foot are generally transmitted from the talus backward to the calcaneus and forward to the navicular and cuneiforms, then through the second metatarsal; only a minimal force is transmitted laterally.
10. Soft tissues of the sole are designed to absorb shock at heel strike and protect the bony structures through the stance phase of gait and during push-off.

REFERENCES

Basmajian, J. V.: Muscles Alive. Their Function Revealed by Electromyography. 3rd Ed. Baltimore, Williams & Wilkins, 1974.

Blais, M. M., Green, W. T., and Anderson, M.: Lengths of the growing foot. J. Bone Joint Surg., *38A*:988, 1956.

Blechschmidt, E.: Die Architektur des Fersenpolsters. *In* Gegenbaurs morphologisches Jahrbuch, Vol. 73, Heft 1. Leipzig, Akademische Verlagsgesellschaft M.B.H., 1933, pp. 44, 66.

Bojsen-Møller, F., and Lamoreux, L.: Significance of free dorsiflexion of the toes in walking. Acta Orthop. Scand., *50*:471, 1979.

Brand, P. W.: Personal communication, 1973.

Collis, W. J. M. F., and Jayson, M. I. V.: Measurement of pedal pressures. An illustration of a method. Ann. Rheum. Dis., *31*:215, 1972.

Cox, H. T.: The cleavage lines of the skin. Br. J. Surg., *29*:234, 1941.

Crelin, E. S.: The development of the human foot as a resumé of its evolution. Foot Ankle, *3*:305, 1983.

Elftman, H.: Forces and energy changes in the leg during walking. Am. J. Physiol., *125*:339, 1939.

Harty, M.: Anatomic considerations in injuries of the calcaneus. Orthop. Clin. North Am., *4*:179, 1973.

Hicks, J. H.: The mechanics of the foot. I. The joints. J. Anat., *87*:345, 1953.

Hutton, W. C., Stott, J. R. R., and Stokes, I. A. F.: The mechanics of the foot. *In* The Foot and Its Disorders. Edited by L. Klenerman. Oxford, Blackwell Scientific Publications, 1976, p. 41.

Inman, V. T.: The Joints of the Ankle. Baltimore, Williams & Wilkins, 1976, p. 37.

Isman, R. E., and Inman, V. T.: Anthropometric studies of the human foot and ankle. Biomechanics Laboratory, University of California, San Francisco and Berkeley. Technical Report 58. San Francisco, The Laboratory, 1968.

Jones, R. L.: The human foot. An experimental study of its mechanics and the role of its muscles and ligaments in the support of the arch. Am. J. Anat., *68*:1, 1941.

Klenerman, L. (ed.): The Foot and Its Disorders. Oxford, Blackwell Scientific Publications, 1976.

Klenerman, L.: Biomechanics. *In* Disorders of the Foot, Vol. 1. Edited by M. H. Jahss. Philadelphia, W. B. Saunders, 1982, pp. 47, 49.

Mann, R. A.: Biomechanics of the foot. *In* Atlas of Orthotics: Biomechanical Principles and Application. American Academy of Orthopaedic Surgeons. St. Louis, C. V. Mosby, 1975, pp. 257–266.

Mann, R. A.: Personal communication, 1984.

Mann, R. A., and Inman, V. T.: Phasic activity of intrinsic muscles of the foot. J. Bone Joint Surg., *46A*:469, 1964.

Manter, J. T.: Distribution of compression forces in the joints of the human foot. Anat. Rec., *96*:313, 1946.

Manter, J. T.: Movements of the subtalar and transverse tarsal joints. Anat. Rec., *80*:397, 1941.

Morton, D. J.: The Human Foot. Its Evolution, Physiology and Functional Disorders. New York, Columbia University Press, 1935.

Sammarco, G. J.: The foot and ankle in classical ballet and modern dance. *In* Disorders of the Foot, Vol. 2. Edited by M. H. Jahss. Philadelphia, W. B. Saunders, 1982, pp. 1629–1659.

Sarrafian, S. K.: Anatomy of the Foot and Ankle. Philadelphia, J. B. Lippincott, 1983.

Sarrafian, S. K., and Topouzian, L. K.: Anatomy and physiology of the extensor apparatus of the toes. J. Bone Joint Surg., *51A*:669, 1969.

Swinyard, C. A.: Limb Development and Deformity: Problems of Evaluation and Rehabilitation. Springfield, Charles C Thomas, 1969.

Viladot, A.: The metatarsals. *In* Disorders of the Foot, Vol. 1. Edited by M. H. Jahss. Philadelphia, W. B. Saunders, 1982, pp. 659–710.

Wright, D. G., and Rennels, D. C.: A study of the elastic properties of plantar fascia. J. Bone Joint Surg., *46A*:482, 1964.

Wright, D. G., DeSai, S. M., and Henderson, W. H.: Action of the subtalar and ankle-joint complex during the stance phase of walking. J. Bone Joint Surg., *46A*:361, 1964.

SUGGESTED READING

Basmajian, J. V.: Muscles Alive. Their Function Revealed by Electromyography. 3rd Ed. Baltimore, Williams & Wilkins, 1974.

Blais, M. M., Green, W. T., and Anderson, M.: Lengths of the growing foot. J. Bone Joint Surg., *38A*:998–1000, 1956.

Collis, W. J. M. F., and Jayson, M. I. V.: Measurement of pedal pressures. An illustration of a method. Ann. Rheum. Dis., *31*:215–217, 1972.

Eyring, E. J., and Murray, W. R.: The effect of joint position on the pressure of intra-articular effusion. J. Bone Joint Surg., *46A*:1235–1241, 1964.

Giannestras, N. J., and Sammarco, G. J.: Fractures and dislocations in the foot. *In* Fractures, Vol. 2. Edited by C. A. Rockwood and D. P. Green. Philadelphia, J. B. Lippincott, 1975, pp. 1400–1490.

Hicks, J. H.: The mechanics of the foot. I. The joints. J. Anat., *87*:345–357, 1953.

Hicks, J. H.: The mechanics of the foot. II. The plantar aponeurosis and the arch. J. Anat., *88*:25–31, 1954.

Hollinshead, W. H.: Knee, leg, ankle, and foot. *In* Anatomy for Surgeons, Vol. 3. Back and Limbs. New York, Harper and Row, 1969, pp. 752–874.

Inman, V. T.: The Joints of the Ankle. Baltimore, Williams & Wilkins, 1976.

Isman, R. E., and Inman, V. T.: Anthropometric studies of the human foot and ankle. Biomechanics Laboratory, University of California, San Francisco and Berkeley. Technical Report 58. San Francisco, The Laboratory, 1968.

Jones, R. L.: The human foot. An experimental study of its mechanics and the role of its muscles and ligaments in the support of the arch. Am. J. Anat., *68*:1–39, 1941.

Mann, R. A.: Biomechanics of the foot. *In* Atlas of Orthotics: Biomechanical Principles and Application. American Academy of Orthopaedic Surgeons. St. Louis, C. V. Mosby, 1975, pp. 257–266.

Mann, R. A., and Inman, V. T.: Phasic activity of intrinsic muscles of the foot. J. Bone Joint Surg., *46A*:469–480, 1964.

Manter, J. T.: Distribution of compression forces in the joints of the human foot. Anat. Rec., *96*:313–321, 1946.

Manter, J. T.: Movements of the subtalar and transverse tarsal joints. Anat. Rec., *80*:397–410, 1941.

Morton, D. J.: The Human Foot. Its Evolution, Physiology and Functional Disorders. New York, Columbia University Press, 1935.

Sammarco, G. J.: The foot and ankle in classical ballet and modern dance. *In* Disorders of the Foot, Vol. 2. Edited by M. H. Jahss. Philadelphia, W. B. Saunders, 1982, pp. 1629–1659.

Sammarco, G. J.: Surgical treatment of disorders of the skin of the foot and toenails. *In* Skin Surgery. Edited by M. Harap. St. Louis, Warren H. Green, 1985, pp. 685–708.

Sarrafian, S. K.: Anatomy of the Foot and Ankle. Philadelphia, J. B. Lippincott, 1983.

Shephard, E.: Tarsal movements. J. Bone Joint Surg., *33B*:258–263, 1951.

Swinyard, C. A.: Limb Development and Deformity: Problems of Evaluation and Rehabilitation. Springfield, Charles C Thomas, 1969.

Viladot, A.: The metatarsals. *In* Disorders of the Foot, Vol. 1. Edited by M. H. Jahss. Philadelphia, W. B. Saunders, 1982, pp. 659–710.

Wright, D. G., and Rennels, D. C.: A study of the elastic properties of plantar fascia. J. Bone Joint Surg., *46A*:482–492, 1964.

Wright, D. G., DeSai, S. M., and Henderson, W. H.: Action of the subtalar and ankle-joint complex during the stance phase of walking. J. Bone Joint Surg., *46A*:361–382, 1964.

10

BIOMECHANICS OF THE LUMBAR SPINE

Margareta Lindh

The human spine is a complex structure whose principal functions are to protect the spinal cord and transfer loads from the head and trunk to the pelvis. Each of the 24 vertebrae articulates with the adjacent ones to permit motion in three planes. The spine gains stability from the intervertebral discs and from the surrounding ligaments and muscles; the discs and ligaments provide intrinsic stability and the muscles give extrinsic support.

This chapter describes the basic characteristics of the various structures of the spine and the interaction of these structures during normal spine function. Kinematics and kinetics of the spine are also examined. The discussion of kinematics covers both the thoracic and lumbar spine, but that of kinetics involves only the lumbar spine because it is subjected to significantly greater loads than the rest of the spine and has received more attention clinically and experimentally. The information in the chapter has been selected to provide an understanding of some fundamental aspects of lumbar spine biomechanics that can be put to practical use.

THE MOTION SEGMENT: THE FUNCTIONAL UNIT OF THE SPINE

The functional unit of the spine, the motion segment, consists of two vertebrae and their intervening soft tissues (Fig. 10–1). The anterior portion of the segment is composed of two superimposed vertebral bodies, the intervertebral disc, and the longitudinal ligaments (Fig. 10–2). The corresponding vertebral arches, the intervertebral joints formed by the facets, the transverse and spinous processes, and various ligaments make up the posterior portion. The arches and vertebral bodies form the vertebral canal, which protects the spinal cord (Fig. 10–3).

THE ANTERIOR PORTION OF THE MOTION SEGMENT

The vertebral bodies are designed to bear mainly compressive loads and they are progressively larger caudally as the superimposed weight of the upper body increases. The vertebral bodies in the lumbar region are thicker and wider than those in the thoracic and cervical regions; their greater size allows them to sustain the larger loads to which the lumbar spine is subjected.

The intervertebral disc, which bears and distributes loads and restrains excessive motion, is of great mechanical and functional importance. It is well suited for its dual role because of its location between the vertebrae and the unique composition of its inner and outer structures (Fig. 10–4). The inner portion of the disc, the nucleus pulposus, is a gelatinous mass. Rich in hydrophilic (water-binding) glycosaminoglycans in the young adult, it diminishes in glycosaminoglycan content with age and becomes progressively less hydrated (Urban and McMullin, 1985).

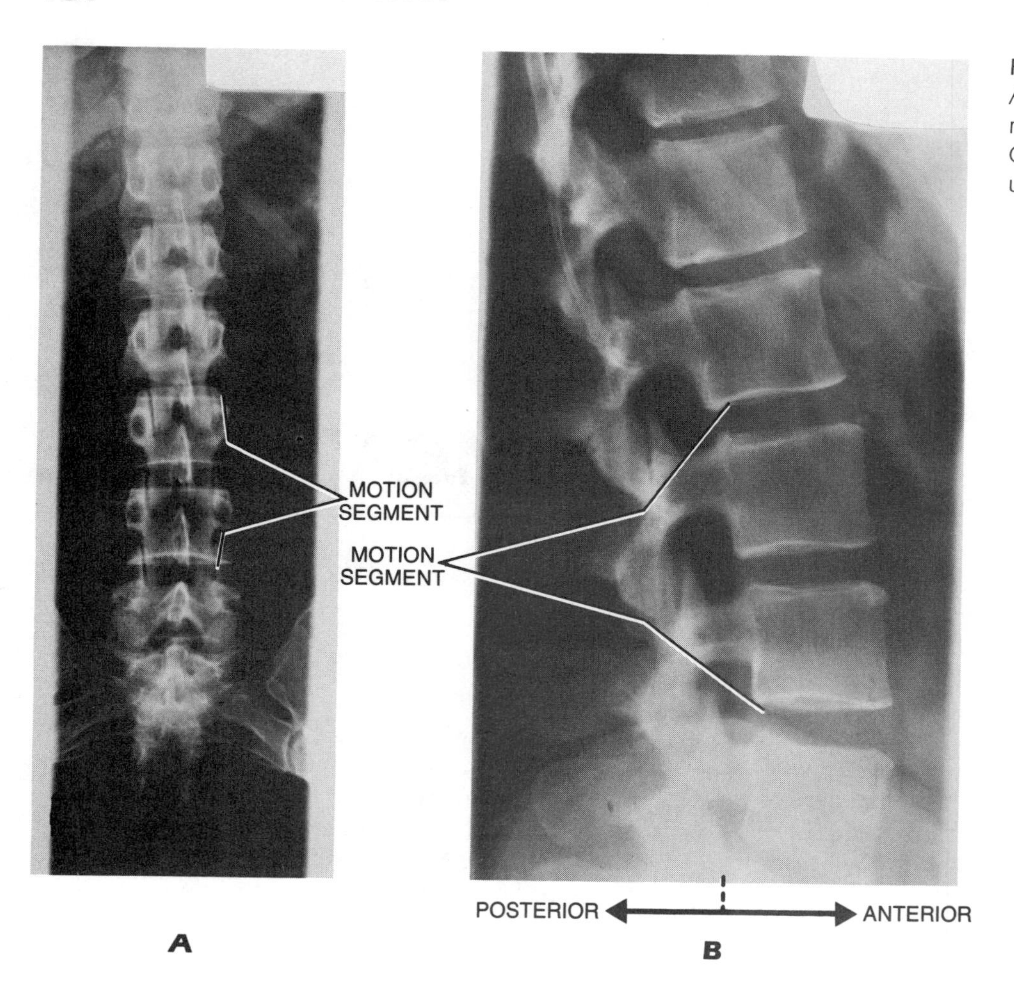

FIG. 10–1
Anteroposterior **(A)** and lateral **(B)** roentgenograms of the lumbar spine. One motion segment, the functional unit of the spine, is indicated.

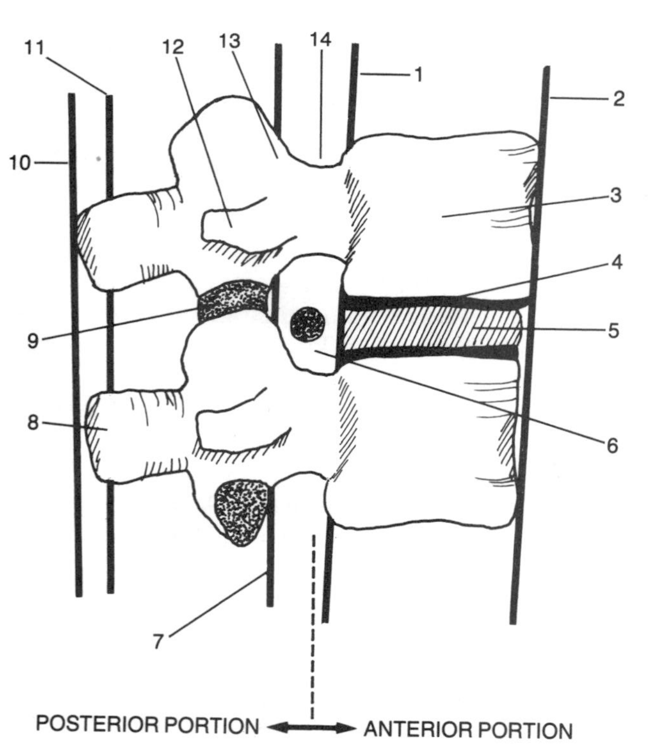

FIG. 10–2
Schematic representation of a motion segment in the lumbar spine (sagittal view). Anterior portion: 1, posterior longitudinal ligament; 2, anterior longitudinal ligament; 3, vertebral body; 4, cartilaginous end-plate; 5, intervertebral disc; 6, intervertebral foramen with nerve root. Posterior portion: 7, ligamentum flavum; 8, spinous process; 9, intervertebral joint formed by the superior and inferior facets (the capsular ligament is not shown); 10, supraspinous ligament; 11, interspinous ligament; 12, transverse process (the intertransverse ligament is not shown); 13, arch; 14, vertebral canal (the spinal cord is not depicted).

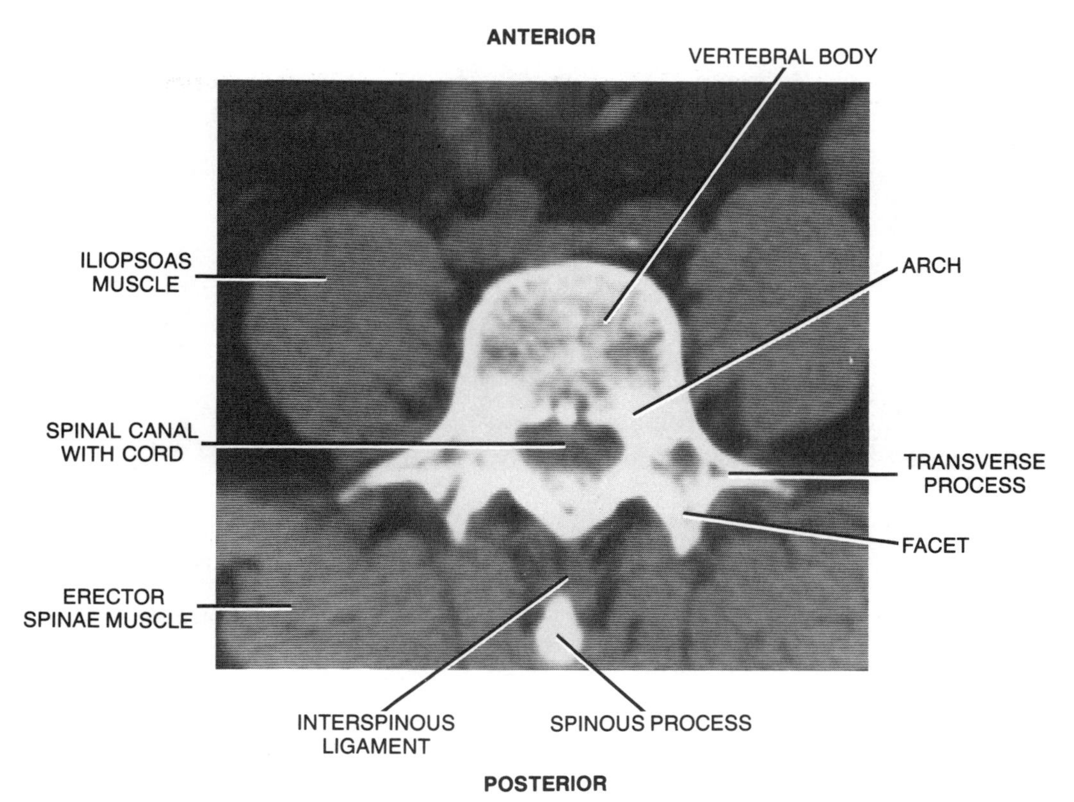

FIG. 10–3
Transverse section of a motion segment at the L4 level viewed by computed tomography. The vertebral body, arch, spinal canal with spinal cord, and transverse processes are clearly seen. The view is taken at a level that depicts only the tip of the spinous process, with the interspinous ligament visible between the spinous process and the facets of the intervertebral joints. Directly anterior to the transverse processes and adjacent to the vertebral body are the iliopsoas muscles. Posterior to the vertebral body the erector spinae muscles can be seen.

The nucleus pulposus lies directly in the center of all discs except those in the lumbar segments, where it has a slightly posterior position. This inner mass is surrounded by a tough outer covering, the annulus fibrosus, composed of fibrocartilage. The crisscross arrangement of the coarse collagen fiber bundles within the fibrocartilage allows the annulus fibrosus to withstand high bending and torsional loads (see Fig. 11–14). The end-plate, composed of hyaline cartilage, separates the disc from the vertebral body (see Fig. 10–2). The disc composition is similar to that of articular cartilage, described in detail in Chapter 2.

During daily activities the disc is loaded in a complex manner and is usually subjected to a combination of compression, bending, and torsion. Flexion, extension, and lateral flexion of the spine produce mainly tensile and compressive stresses in the disc, whereas rotation produces mainly shear stress.

When a motion segment is transected vertically, the nucleus pulposus of the disc protrudes, indicating that it is under pressure. Measurement of the intradiscal pressure in normal and slightly degenerated cadaver lumbar nuclei pulposi has shown an intrinsic pressure in the unloaded disc of about 10 N per square centimeter (Nachemson, 1960). This intrinsic pressure, or prestress, in the disc results from forces exerted by the longitudinal ligaments and the ligamenta flava. During loading of the spine, the nucleus pulposus acts hydrostatically (Nachemson, 1960), allowing a uniform distribution of pressure throughout the disc; hence, the entire disc serves a hydrostatic function in the motion segment, acting as a cushion between the vertebral bodies to store energy and distribute loads.

In a disc loaded in compression, the pressure is about 1.5 times the externally applied load per unit area. Because the nuclear material is only slightly compressible, a compressive load makes the disc

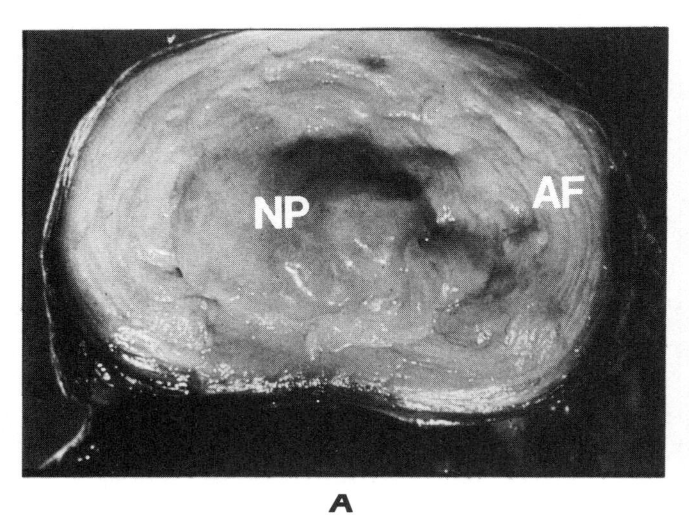

A

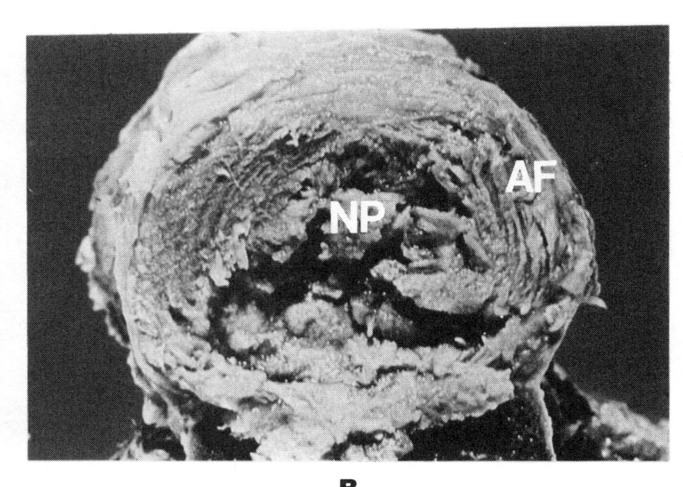

B

FIG. 10–4
Intervertebral disc composed of an inner gelatinous mass, the nucleus pulposus (NP), and a tough outer covering, the annulus fibrosus (AF). **A.** Normal disc. The gelatinous nucleus pulposus is 80 to 88% water (Gower and Pedrini, 1969) and is easy to distinguish from the firmer annulus fibrosus. **B.** Severely degenerated disc. The nucleus pulposus has become dehydrated and has lost its gel-like character. The boundary between the nucleus and annulus is difficult to distinguish, as the degree of hydration is now about the same in both structures.

bulge laterally and circumferential tensile stress is sustained by the annular fibers. In the lumbar spine the tensile stress in the posterior part of the annulus fibrosus has been estimated to be four to five times the applied axial compressive load (Nachemson, 1960, 1963; Galante, 1967) (Fig. 10–5). The tensile stress in the annulus fibrosus in the thoracic spine is less than that in the lumbar spine because of differences in disc geometry. The higher ratio of disc diameter to height in the thoracic discs reduces the circumferential stress in these discs (Kulak et al., 1975).

Degeneration of a disc reduces its proteoglycan content and thus its hydrophilic capacity (see Fig. 10–4). As the disc becomes drier, its elasticity and its ability to store energy and distribute loads gradually decrease; these changes render it less capable of resisting loads. Studies of the mechanical behavior of lumbar motion segments from cadavers have shown that age does not affect the loading response of the discs to any significant extent, nor does gender or disc level (Nachemson et al., 1979).

THE POSTERIOR PORTION OF THE MOTION SEGMENT

The posterior portion of the motion segment guides its movement. The type of motion possible at any level of the spine is determined by the orientation of the facets of the intervertebral joints to the transverse and frontal planes. This orientation changes throughout the spine.

Except for the facets of the two uppermost cervical vertebrae (C1 and C2), which are parallel to the transverse plane, the facets of the cervical intervertebral joints are oriented at a 45-degree angle to the transverse plane and are parallel to the frontal plane (Fig. 10–6A). This alignment of the joints of C3 to C7 allows flexion, extension, lateral flexion, and rotation. The facets of the thoracic joints are oriented at a 60-degree angle to the transverse plane and a 20-degree angle to the frontal plane (Fig. 10–6B); this orientation allows lateral flexion, rotation, and some flexion and extension. In the lumbar region the facets are oriented at right angles to the transverse plane and at a 45-degree angle to the frontal plane (Fig. 10–6C) (White and Panjabi, 1978). This alignment allows flexion, extension, and lateral flexion, but almost no rotation. The lumbosacral joints differ from the other lumbar intervertebral joints in that the oblique orientation of the facets allows appreciable rotation (Lumsden and Morris, 1968). The above-cited values for facet orientation are only approximations, as considerable variation is found within and among individuals.

In the past it was thought that the facets mainly guided movement of the motion segment and had only a modest load-bearing function. Recent studies

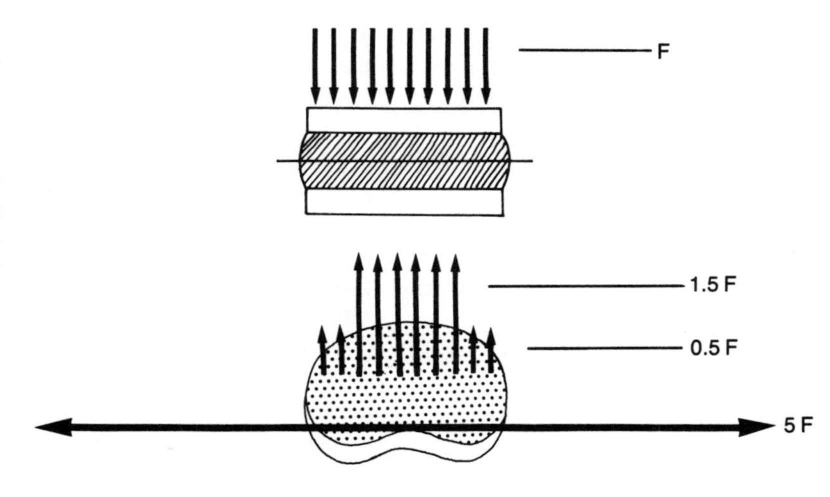

FIG. 10–5
Distribution of stress in a cross section of a lumbar disc under compressive loading. The compressive stress is highest in the nucleus pulposus, 1.5 times the externally applied load (F) per unit area. By contrast, the compressive stress on the annulus fibrosus is only about 0.5 times the externally applied load. This part of the disc bears predominantly tensile stress, which is four to five times greater than the externally applied load per unit area. (Adapted from Nachemson, 1975.)

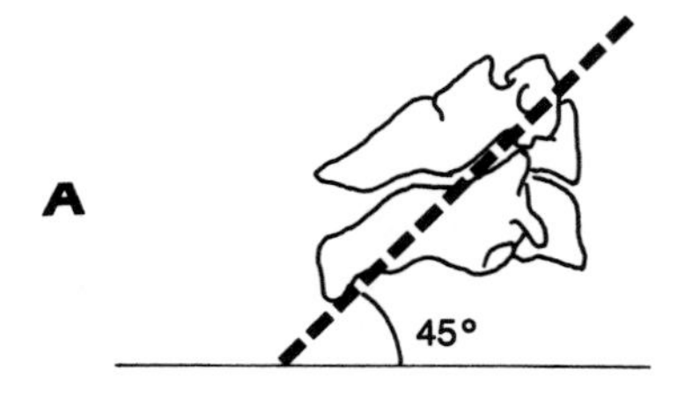

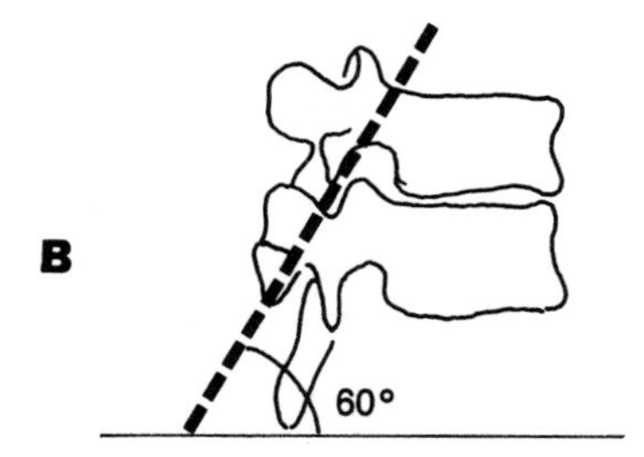

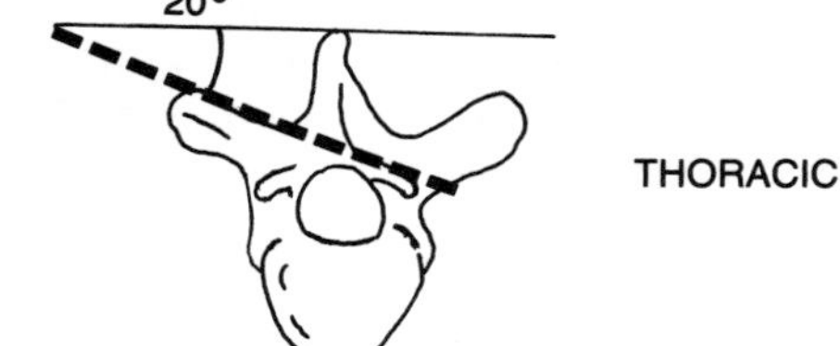

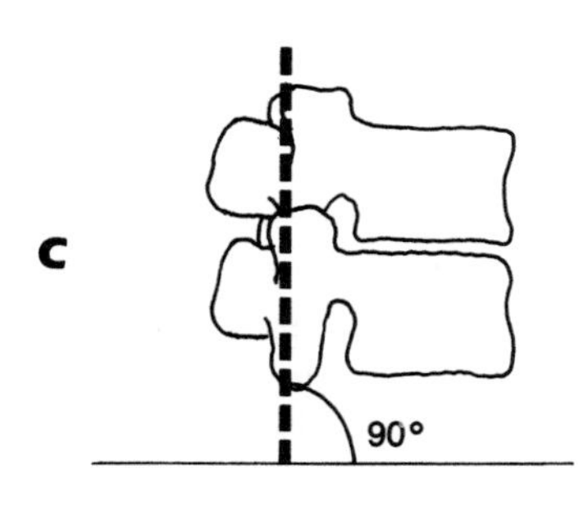

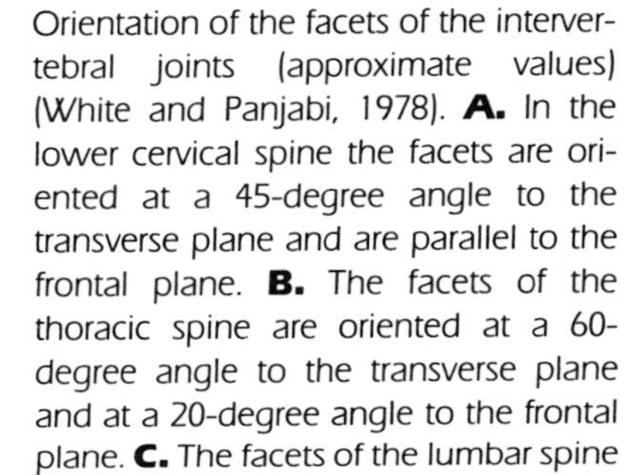

FIG. 10–6
Orientation of the facets of the intervertebral joints (approximate values) (White and Panjabi, 1978). **A.** In the lower cervical spine the facets are oriented at a 45-degree angle to the transverse plane and are parallel to the frontal plane. **B.** The facets of the thoracic spine are oriented at a 60-degree angle to the transverse plane and at a 20-degree angle to the frontal plane. **C.** The facets of the lumbar spine are oriented at a 90-degree angle to the transverse plane and at a 45-degree angle to the frontal plane.

have shown that their load-bearing function is significant. Load-sharing between the facets and the disc varies with the position of the spine. The loads on the facets are greatest (about 30% of the total load) when the spine is hyperextended (King et al., 1975) and are also quite high during forward bending coupled with rotation (El-Bohy and King, 1986).

The vertebral arches and intervertebral joints play an important role in resisting shear forces. This function is demonstrated by the fact that patients with deranged arches or defective joints (e.g., from spondylolysis) are at increased risk for forward displacement of the vertebral body (Miller et al., 1983; Adams and Hutton, 1983). The transverse and spinous processes serve as sites of attachment for spinal muscles, whose activity initiates spine motion and provides extrinsic stability.

THE LIGAMENTS OF THE SPINE

The ligamentous structures surrounding the spine contribute to its intrinsic stability (see Fig. 10–2). Most have a high collagen content, which limits their extensibility during spine motion. The ligamentum flavum, which connects two adjacent vertebral arches longitudinally, is an exception, having a large percentage of elastin. The elasticity of this ligament allows it to contract during extension of the spine and to elongate during flexion. Even when the spine is in a neutral position the ligamentum flavum is under constant tension as a result of its elastic properties. Because it is located at a distance from the center of motion in the disc, it prestresses the disc; that is, along with the longitudinal ligaments, it creates an intradiscal pressure and thus helps provide intrinsic support to the spine (Roland, 1966; Nachemson and Evans, 1968).

The amount of strain on the various ligaments differs with the type of motion of the spine. During flexion the interspinous ligaments are subjected to the greatest strain, followed by the capsular ligaments and the ligamenta flava. During extension the anterior longitudinal ligament bears the greatest strain. During lateral flexion the contralateral transverse ligament sustains the highest strains, followed by the ligamenta flava and the capsular ligaments. The capsular ligaments bear the most strain during rotation (Panjabi et al., 1982).

KINEMATICS

Motion of the spine is produced by the coordinated action of nerves and muscles. Agonistic muscles (prime movers) initiate and carry out motion, and antagonistic muscles often control and modify it. The range of motion differs at various levels of the spine and depends on the orientation of the facets of the intervertebral joints (see Fig. 10–6). Motion between two vertebrae is small and does not occur independently; all spine movements involve the combined action of several motion segments. The skeletal structures that influence motion of the spine are the rib cage, which limits thoracic motion, and the pelvis, which augments trunk movements by tilting.

SEGMENTAL MOTION OF THE SPINE

The vertebrae have six degrees of freedom: rotation about and translation along a transverse, a sagittal, and a longitudinal axis. The motion produced during flexion, extension, lateral flexion, and axial rotation of the spine is a complex combined motion resulting from simultaneous rotation and translation.

Range of Motion

Various investigations using autopsy material or radiographic measurements in vivo have shown divergent values for the range of motion of individual motion segments, but there is agreement on the relative amount of motion at different levels of the spine. Representative values from White and Panjabi (1978) are presented in Figure 10–7 to allow a comparison of motion at various levels of the thoracic and lumbar spine. (Representative values for motion in the cervical spine are included for the sake of completeness.)

Investigations of the thoracic and lumbar spine show that the range of flexion and extension is approximately 4 degrees in the upper thoracic motion segments, about 6 degrees in the midthoracic region, and about 12 degrees in the two lower thoracic segments. This range progressively increases in the lumbar motion segments, reaching a maximum of 20 degrees at the lumbosacral level.

Lateral flexion shows the greatest range in the lower thoracic segments, reaching 8 to 9 degrees. In the upper thoracic segments the range is uniformly 6 degrees. Six degrees of lateral flexion is also found in all lumbar segments except the lumbosacral segment, which demonstrates only 3 degrees of motion.

Rotation is greatest in the upper segments of the thoracic spine, where the range is 9 degrees. The range of rotation progressively decreases caudally, reaching 2 degrees in the lower segments of the lumbar spine. It then increases to 5 degrees in the lumbosacral segment.

Surface Joint Motion

Motion between the surfaces of two adjacent vertebrae during flexion-extension or lateral flexion may be analyzed by means of the instant center method of Reuleaux. The procedure is essentially the same as that described for the cervical spine in Chapter 11 (see Figs. 11–11 and 11–12). The instantaneous center of flexion-extension and lateral flexion in a motion segment of the lumbar spine lies

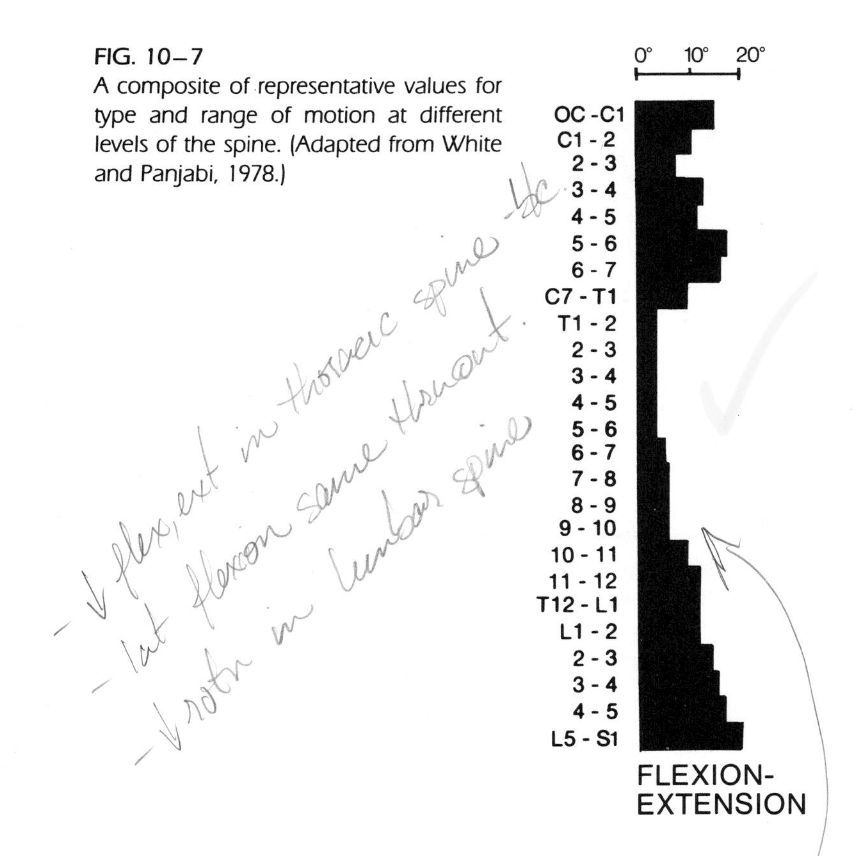

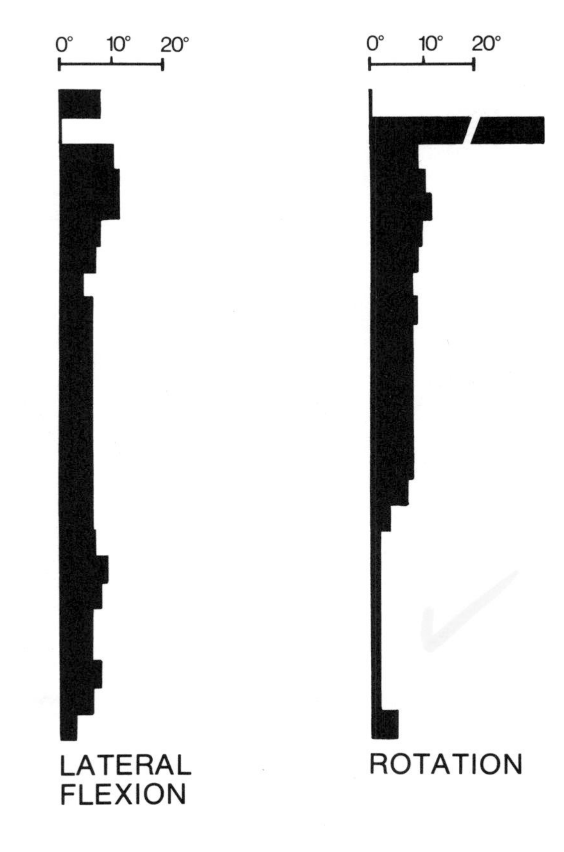

FIG. 10–7
A composite of representative values for type and range of motion at different levels of the spine. (Adapted from White and Panjabi, 1978.)

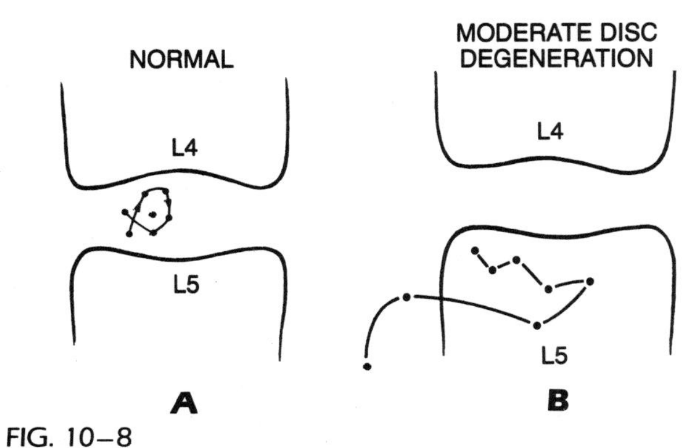

FIG. 10–8
Instant center pathway for a normal cadaver spine **(A)** and a cadaver spine with moderate disc degeneration **(B).** Instant centers were determined for 3-degree intervals of motion from maximum extension to maximum flexion. In the normal spine all instant centers fell within a small area in the disc. In the degenerated spine the centers were displaced, and hence the surface motion was abnormal. (Adapted from Gertzbein et al., 1985.)

within the disc under normal conditions (Fig. 10–8A) (Rolander, 1966; Cossette et al., 1971), but some abnormal circumstances such as pronounced disc degeneration may cause it to lie outside the disc (Fig. 10–8B) (Reichmann et al., 1972; Gertzbein et al., 1985).

FUNCTIONAL MOTION OF THE SPINE

Because of its complexity, the motion of a single motion segment cannot be measured clinically. Not even approximate values for the normal functional range of motion of the spine can be given because the variations among individuals are so great; in fact, the range of motion in each of the three planes shows a Gaussian distribution. The range of motion is strongly age-dependent, decreasing by about 50% from youth to old age. Differences have also been noted between the sexes: men have greater mobility in flexion and extension, whereas women are more mobile in lateral flexion (Moll and Wright, 1971; Biering-Sørensen, 1984).

Flexion and Extension

The first 50 to 60 degrees of spine flexion occurs in the lumbar spine, mainly in the lower motion segments (Farfan, 1975). Tilting the pelvis forward allows further flexion (Fig. 10–9). The thoracic spine contributes little to flexion of the total spine because of the oblique orientation of the facets (see Fig. 10–6), the nearly vertical orientation of the spinous processes, and the limitation of motion imposed by the rib cage.

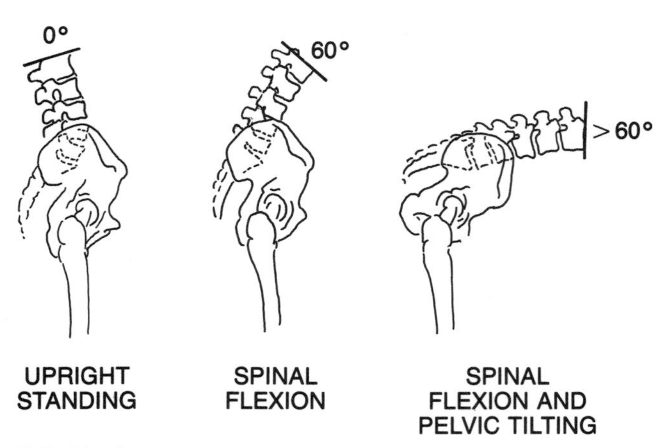

FIG. 10–9
The first 50 to 60 degrees of trunk flexion occurs in the lumbar spine. Additional flexion is accomplished primarily through forward tilting of the pelvis. (Adapted from Farfan, 1975.)

Flexion is initiated by the abdominal muscles and the vertebral portion of the psoas muscle. The weight of the upper body produces further flexion, which is controlled by the gradually increasing activity of the erector spinae muscles as the forward-bending moment acting on the spine increases. The posterior hip muscles are active in controlling the forward tilting of the pelvis as the spine is flexed (Carlsöö, 1961). In full flexion the erector spinae muscles become inactive and are fully stretched; in this position the forward-bending moment may be counteracted passively by these muscles and also by the posterior ligaments, which are initially slack but become taut at this point because the spine has fully elongated (Farfan, 1975).

From full flexion to upright positioning of the trunk, a reverse sequence is observed. The pelvis tilts backward and the spine then extends. Some studies have shown that the concentric work performed by the muscles involved in raising the trunk is greater than the eccentric work performed by the muscles involved in lowering the trunk (Friedebold, 1958; Joseph, 1960). The mechanism responsible for this difference has not yet been elucidated. When the trunk is extended from the upright position, the extensor muscles are active during the initial phase. This initial burst of activity decreases during further extension, and the abdominal muscles become active to control and modify the motion. In extreme or forced extension, extensor activity is again required.

Lateral Flexion and Rotation

During lateral flexion of the trunk, motion may predominate in either the thoracic or the lumbar spine. In the thoracic spine the facet orientation allows lateral flexion, but the rib cage restricts it (to varying degrees in different people); in the lumbar spine the wedge-shaped spaces between the intervertebral joint surfaces show variations during this motion (Reichmann, 1970/71). The spinotransversal and transversospinal systems of the erector spinae muscles and the abdominal muscles are active during lateral flexion; ipsilateral contractions of these muscles initiate the motion and contralateral contractions modify it.

Significant axial rotation occurs at the thoracic and lumbosacral levels but is quite limited at other levels of the lumbar spine, being restricted by the vertical orientation of the facets (see Fig. 10–6C). In the thoracic region rotation is consistently associated with lateral flexion. During this coupled motion, which is most marked in the upper thoracic region, the vertebral bodies generally rotate toward the concavity of the lateral curve of the spine (White, 1969). Coupling of rotation and lateral flexion also takes place in the lumbar spine, with the vertebral bodies rotating toward the convexity of the curve (Miles and Sullivan, 1961). During axial rotation, back and abdominal muscles are active on both sides of the spine, as both ipsilateral and contralateral muscles cooperate to produce this movement (Fig. 10–10).

Functional trunk movements not only involve a combined motion of different parts of the spine but also require the cooperation of the pelvis, since pelvic motion is essential for increasing the range of functional motion of the trunk (Fig. 10–11; see also Fig. 10–9). During walking, spinal and pelvic movements

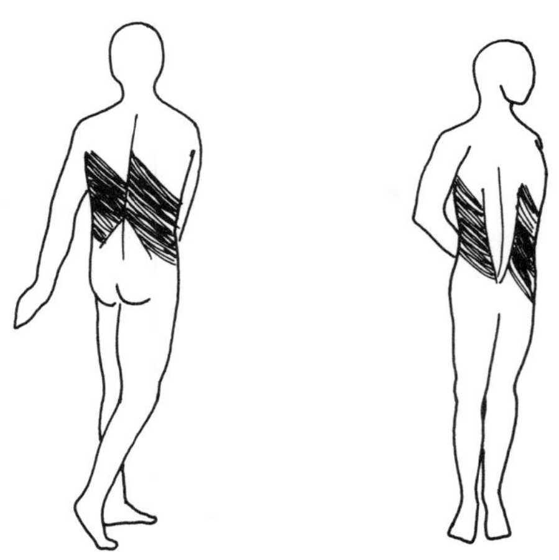

FIG. 10–10
When the trunk is rotated, the back muscles are active on both sides of the spine, with ipsilateral contractions of the spinotransversal muscles and contralateral contractions of the transversospinal muscles. The abdominal muscles are also active, but to a lesser extent; the ipsilateral obliquus internus muscle and the contralateral obliquus externus muscle contract.

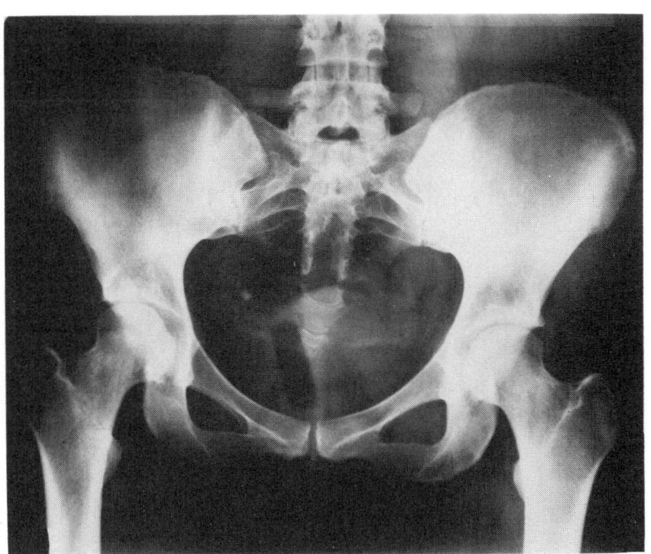

FIG. 10–11
The pelvic ring with its linkage to the spine and the lower extremities. The anteroposterior view of these structures on film gives a hint of the irregular shape of the sacroiliac joint surfaces, but an oblique projection is required for an accurate view of the joints.

in all three planes—sagittal, frontal, and transverse—are necessary for a smooth gait cycle and a minimal expenditure of energy. The pattern of pelvic motion in these planes is complex, and the relation between the movements of the pelvis and those of the lumbar spine is not completely understood. The responses of the pelvis and the spine to the action of the lower limbs during locomotion vary among individuals (Thurston and Harris, 1983).

Restriction of motion at any level of the spine may increase motion at another level. Thus, a brace worn to restrict thoracic and lumbar motion may result in compensatory motion at the lumbosacral level (Norton and Brown, 1957; Lumsden and Morris, 1968). Braces and corsets also can affect muscle activity. If a tight brace or corset is worn, the activity of the abdominal muscles is decreased because the brace or corset assumes the function of these muscles when the force it exerts on the anterior aspect of the trunk is great enough to support the abdomen (Waters and Morris, 1970).

The relationship between pelvic movements and spinal motion is generally analyzed in terms of motion of the lumbosacral joints, the hip joints, or both (see Fig. 10–11). The function of the sacroiliac joints is currently under discussion. These joints are thought to produce very little clinically significant motion because they are surrounded by a dense ligamentous structure and their joint surfaces are irregular. Biomechanical analyses of the sacroiliac joints suggest that these joints function mainly as shock absorbers and are important in protecting the intervertebral joints (Wilder et al., 1980).

KINETICS

Loads on the spine are produced primarily by body weight, muscle activity, prestress exerted by the ligaments, and externally applied loads. Rough calculations of the loads at various levels of the spine can be made with the use of the simplified free body technique for coplanar forces. Direct information regarding loads on the spine at the level of individual intervertebral discs can be obtained by measuring the pressure within the discs both in vitro and in vivo. Because this method is too complex for general application, however, a semidirect measuring method is often used that involves measuring myoelectric activity of the trunk muscles and correlating this activity with calculated values for muscle contraction forces. The values obtained correlate well with those obtained through intradiscal pressure measurement, so the method can be used to predict the loads on the spine (Örtengren et al., 1981; Schultz et al., 1982b). Another method is the use of a mathematical model for force estimation that allows the loads on the lumbar spine and the contraction forces in the trunk muscles to be calculated for various physical activities; the calculated values agree well with those obtained by means of the direct and semidirect methods (Schultz and Andersson, 1981; Schultz et al., 1982a, 1982b, 1982c).

Because the lumbar spine is the main load-bearing area and the most common site of pain, studies of spine loads have focused on this region and lumbar spine loading will be emphasized here. In the following section, static loads on the lumbar spine are examined for common postures such as standing and sitting and also for lifting, a common activity involving external loads. The dynamic loads are generally higher than the static loads, since almost all motion in the body produces a rise in lumbar spine loads from a slight increase during slow walking to a great increase during vigorous physical activity. In the final section, the dynamic loads on the lumbar spine during walking and during common strengthening exercises for the back and abdominal muscles are discussed.

STATICS

The spine can be considered as a modified elastic rod because of the flexibility of the spinal column, the shock-absorbing behavior of the discs and vertebrae, the stabilizing function of the longitudinal ligaments, and the elasticity of the ligamenta flava. The two curvatures of the spine in the sagittal plane—kyphosis and lordosis—also contribute to the springlike capacity of the spine and allow the vertebral column to withstand higher loads than if it were straight. A study of the capacity of cadaver thoracolumbar spines devoid of muscles to resist vertical loads showed that the critical load (the point at which buckling occurred) was about 20 N (Lucas and Bresler, 1961). The critical load is much higher in vivo and varies greatly among individuals. The extrinsic support provided by the trunk muscles helps stabilize and modify the loads on the spine in both dynamic and static situations.

Loading of the Spine during Standing

When a person stands, the postural muscles are constantly active. This activity is minimized when the body segments are well aligned. During standing the line of gravity of the trunk usually passes ventral to the center of the fourth lumbar vertebral body (Asmussen and Klausen, 1962). Thus, it falls ventral to the transverse axis of motion of the spine, and the motion segments are subjected to a forward-bending moment, which must be counterbalanced by ligament forces and erector spinae muscle forces (Fig. 10–12). Any displacement of the line of gravity alters the magnitude and direction of the moment on the spine. For the body to return to equilibrium, the moment must be counteracted by increased muscle activity, which causes intermittent postural sway. In addition to the erector spinae muscles, the abdominal muscles are often intermittently active in maintaining the upright position of the trunk. The vertebral portion of the psoas muscles is also involved in producing postural sway (Basmajian, 1958; Nachemson, 1966). The level of activity in these muscles varies considerably among individuals and depends to some extent on the shape of the spine, for example, the magnitude of habitual kyphosis and lordosis.

The pelvis also plays a role in the muscle activity and resulting loads on the spine during standing (Fig. 10–13). The base of the sacrum is inclined forward and downward. The angle of inclination, or sacral angle, is about 30 degrees to the transverse plane during relaxed standing (Fig. 10–13B). Tilting of the pelvis about the transverse axis between the hip joints changes the angle. When the pelvis is tilted backward, the sacral angle decreases and the lumbar lordosis flattens (Fig. 10–13A). This flattening affects the thoracic spine, which extends slightly to adjust the center of gravity of the trunk so that energy expenditure, in terms of muscle work, is minimized. When the pelvis is tilted forward the sacral angle increases, accentuating the lumbar lordosis and the thoracic kyphosis (Fig. 10–13C). Forward and backward tilting of the pelvis influences the activity of the

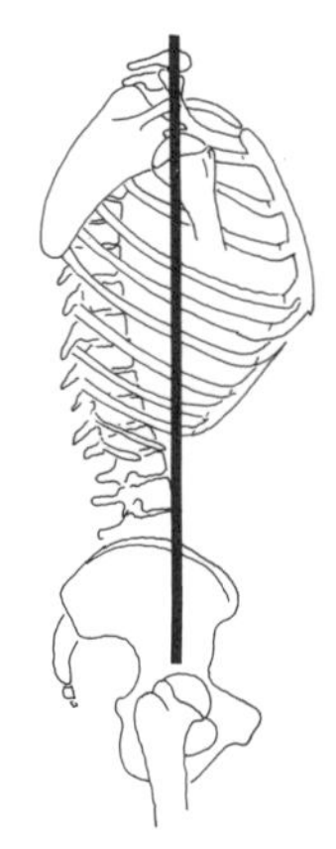

FIG. 10–12
The line of gravity for the trunk (solid line) is usually ventral to the transverse axis of motion of the spine, and thus the spine is subjected to a constant forward-bending moment.

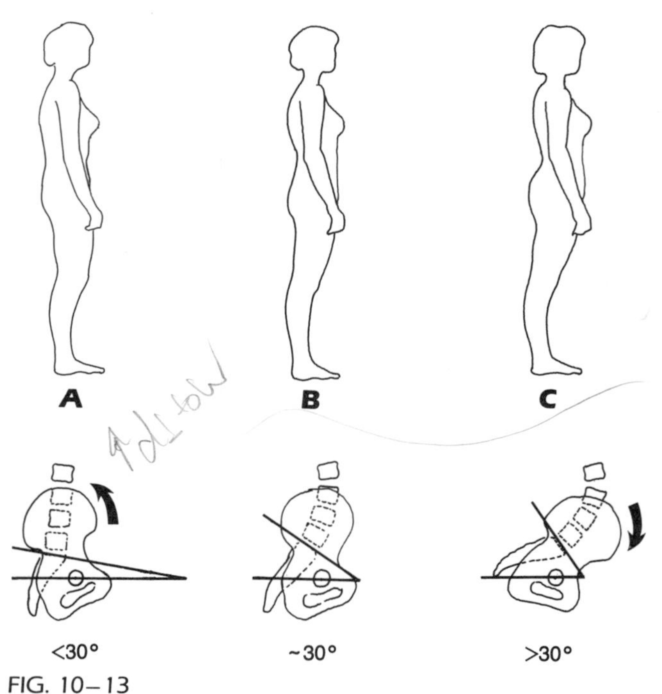

FIG. 10–13
Effect of pelvic tilting on the inclination of the base of the sacrum to the transverse plane (sacral angle) during upright standing. **A.** Tilting the pelvis backward reduces the sacral angle and flattens the lumbar spine. **B.** During relaxed standing the sacral angle is about 30 degrees. **C.** Tilting the pelvis forward increases the sacral angle and accentuates the lumbar lordosis.

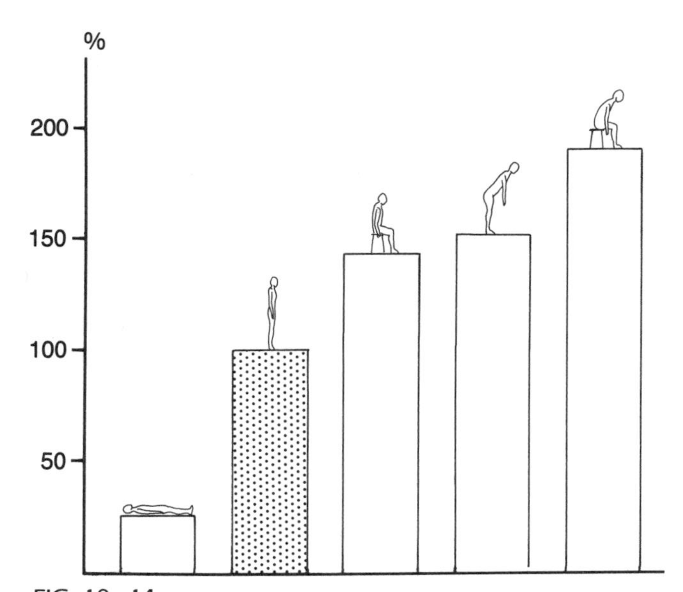

FIG. 10–14
The relative loads on the third lumbar disc for living subjects in various body positions are compared with the load during upright standing, depicted as 100%. (Adapted from Nachemson, 1975.)

postural muscles by affecting the static loads on the spine (Floyd and Silver, 1955).

Comparative Loads on the Lumbar Spine during Standing, Sitting, and Reclining

Body position affects the magnitude of the loads on the spine. These loads are minimal during well-supported reclining, remain low during relaxed upright standing, and rise during sitting. The relative loads on the spine during various body postures are presented in Figure 10–14.

During relaxed upright standing, the load on the third lumbar disc, calculated from the disc pressure, has been shown to be 70 kg in a 70-kg man; the load is almost twice the weight of the body above the measured level, which is approximately 60% of the total body weight, or about 40 kg (Nachemson and Morris, 1964; Nachemson and Elfström, 1970; Nachemson, 1975). Trunk flexion increases the load by increasing the forward-bending moment on the spine. The forward inclination of the spine makes the disc bulge on the concave side of the spinal curve and retract on the convex side (Fig. 10–15). Hence, when the spine is flexed the disc protrudes anteriorly and is retracted posteriorly. Both compressive and tensile stresses on the disc increase. The addition of rotary motion and accompanying torsional loads further increases the stresses on the disc (Andersson et al., 1977).

During relaxed unsupported sitting the loads on the lumbar spine are greater than during relaxed upright standing (Nachemson and Elfström, 1970; Andersson et al., 1974). In this sitting position the pelvis is tilted backward and the lumbar lordosis straightens out. The line of gravity for the upper body, already ventral to the lumbar spine, shifts further ventrally, creating a longer lever arm for the force exerted by the weight of the trunk (Fig. 10–16A, B). This longer lever arm produces an increased moment, or torque, on the lumbar spine that increases further if the trunk is bent forward. Psoas muscle activity also contributes to the loads on the lumbar region during sitting (Nachemson, 1968).

During erect sitting, forward tilting of the pelvis and an increase in the lumbar lordosis reduce the loads on the lumbar spine, but these loads still exceed those produced during relaxed upright standing (Fig. 10–16C). During this sitting position, particularly if the knees are somewhat extended, tight hamstring

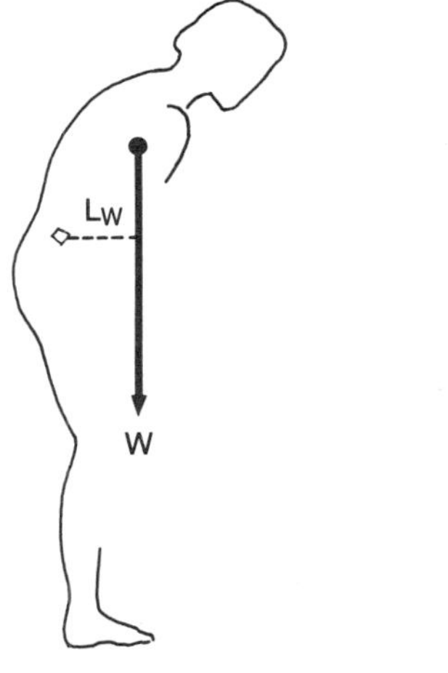

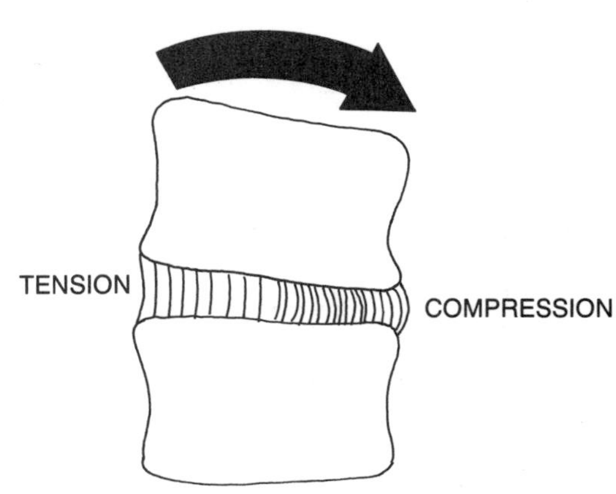

FIG. 10–15
The forward-bent position produces a bending moment on the lumbar spine. The moment is a product of the force produced by the weight of the upper body (W) and the lever arm of the force (L_W). The forward inclination of the upper body subjects the disc to increased tensile and compressive stresses. The disc bulges on the compressive side and retracts on the tensile side.

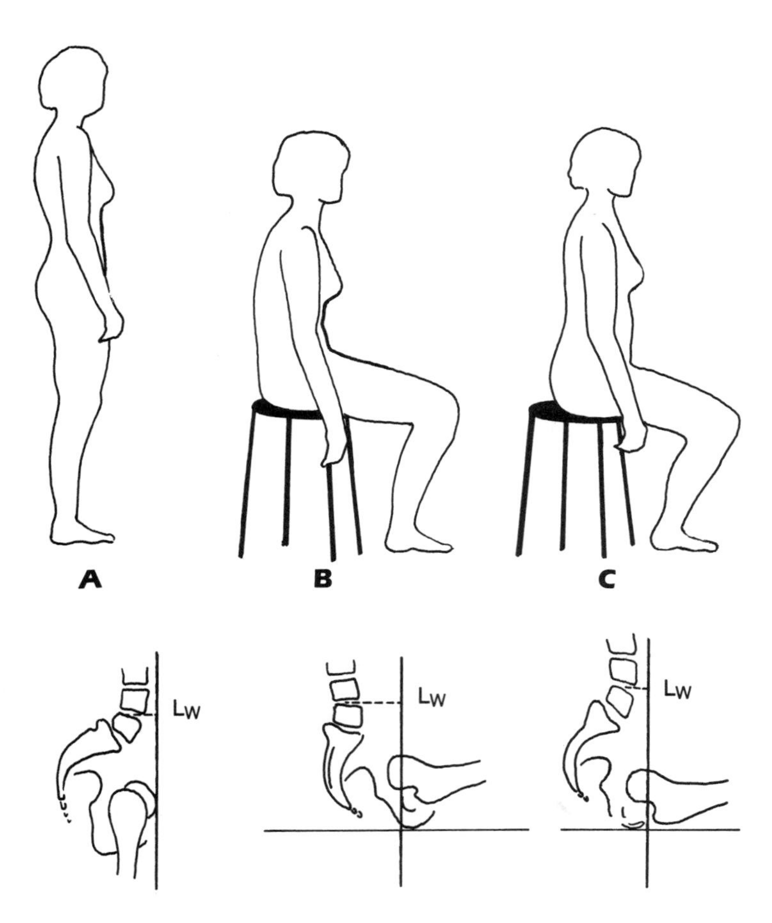

FIG. 10–16
Compared with relaxed upright standing **(A)**, the line of gravity for the upper body, already ventral to the lumbar spine, shifts further ventrally during relaxed unsupported sitting as the pelvis is tilted backward and the lumbar lordosis flattens **(B)**. This shift creates a longer lever arm (L_W) for the force exerted by the weight of the upper body. During erect sitting the backward pelvic tilt is reduced and the lever arm shortens **(C)**, but it is still slightly longer than during relaxed upright standing.

muscles may restrict forward tilting of the pelvis and thus may cause an increase in the loads on the lumbar spine (Stokes and Abery, 1980).

The loads on the lumbar spine are lower during supported sitting than during unsupported sitting because part of the weight of the upper body is supported by the backrest. A backward inclination of the backrest and the use of a lumbar support further reduce the loads. The use of a support in the thoracic region, however, which pushes the thoracic spine and the trunk forward, makes the lumbar spine move toward kyphosis to remain in contact with the backrest, increasing the loads on the lumbar spine (Andersson et al., 1974) (Fig. 10–17).

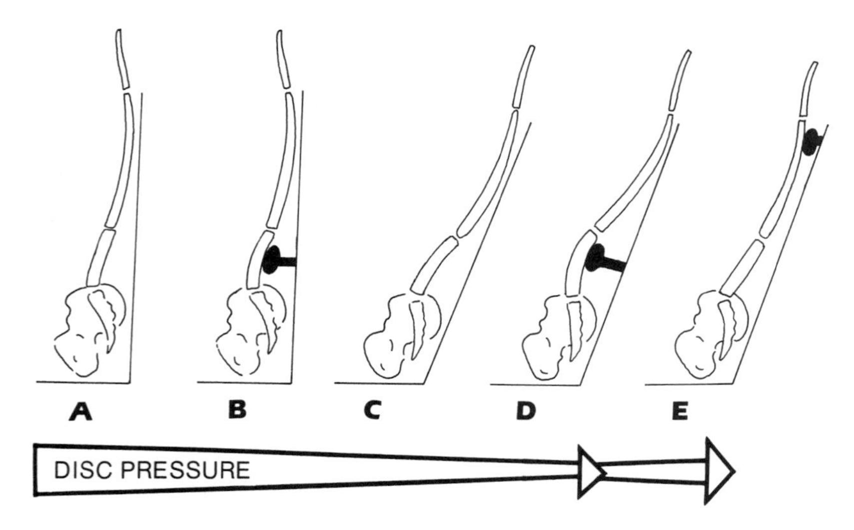

FIG. 10–17
Influence of backrest inclination and back support on loads on the lumbar spine, in terms of pressure in the third lumbar disc, during supported sitting. **A.** Backrest inclination is 90 degrees, and disc pressure is at a maximum. **B.** Addition of a lumbar support decreases the disc pressure. **C.** Backward inclination of the backrest to 110 degrees, but with no lumbar support, produces less disc pressure. **D.** Addition of a lumbar support with this degree of backrest inclination further decreases the pressure. **E.** Shifting the support to the thoracic region pushes the upper body forward, moving the lumbar spine toward kyphosis and increasing disc pressure. (Adapted from Andersson et al., 1974.)

Loads on the spine are minimal when an individual assumes a supine position because the loads produced by the body's weight are eliminated (see Fig. 10–14). With the body supine and the knees extended, the pull of the vertebral portion of the psoas muscle produces some loads on the lumbar spine. With the hips and knees bent and supported, however, the lumbar lordosis straightens out as the psoas muscle relaxes and the loads decrease (Fig. 10–18). A further decrease in these loads is achieved by the application of traction. When traction is applied to a patient in the supine position with the hips and knees bent and supported to flatten the spine, the traction forces are distributed more evenly throughout the spine than if the legs are kept straight to maintain the lumbar lordosis (Fig. 10–19).

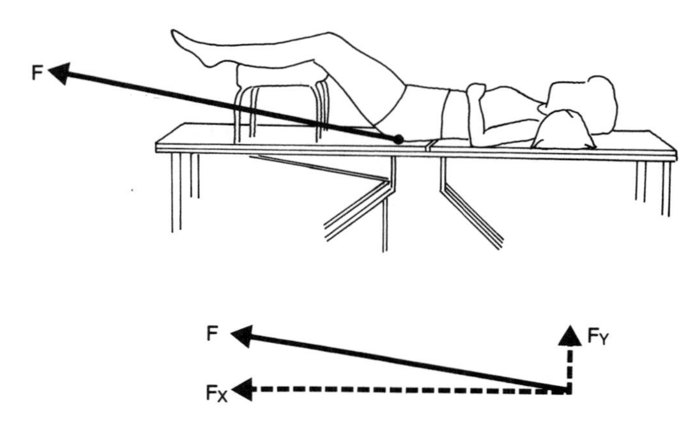

FIG. 10–19
The semiflexed position, which flattens the lumbar spine, favors a more even distribution of the force produced by traction (F). The pull must be directed diagonally to keep the spine flat. This diagonal arrangement means that not all of the pull exerts a horizontal force. Resolving the traction force (F) into horizontal (F_X) and vertical (F_Y) components demonstrates that part of the pull produces lifting. In this example friction forces are not taken into account.

Static Loads on the Lumbar Spine during Lifting

The highest loads on the spine are generally external loads, such as that produced by a heavy object that is being lifted. Just how much load can be sustained by the spine before damage occurs is still under investigation. Pioneering studies by Eie (1966) of lumbar vertebral specimens from adult humans showed that the compressive load to failure ranged from about 5,000 to 8,000 N. On the whole, values reported subsequently by other authors correspond to those of Eie, although values above 10,000 N and below 5,000 N have been documented (Hutton and Adams, 1982). Both age and degree of disc degeneration influence this range.

Eie (1966) observed that during compressive testing the fracture point was reached in the vertebral body or end-plate before the intervertebral disc sustained damage. This finding shows that the bone is less capable of resisting compression than is an intact disc. During the testing a yield point was reached before the vertebra or end-plate fractured. When the load was removed at this point, the vertebral body recovered but was more susceptible to damage when reloaded.

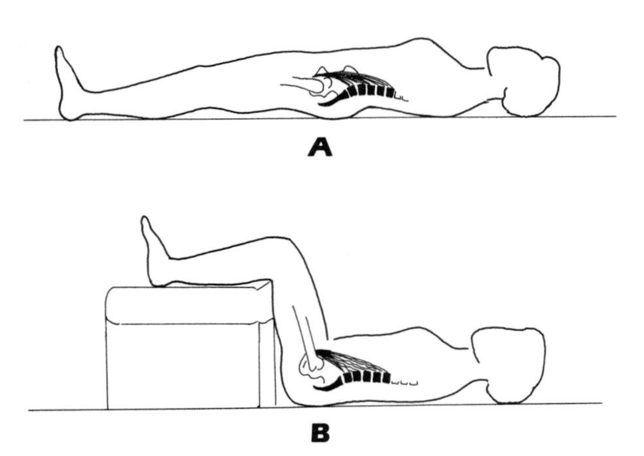

FIG. 10–18
A. When a person assumes a supine position with legs straight, the pull of the vertebral portion of the psoas muscle produces some loads on the lumbar spine. **B.** When the hips and knees are bent and supported, the psoas muscle relaxes and the loads on the lumbar spine decrease.

More recent studies have provided evidence that the spine may incur microdamage as a result of high loads in vivo. Hansson (1977) observed microfractures in specimens from "normal" human lumbar vertebrae and interpreted this microdamage to be fatigue fractures resulting from stresses and strains on the spine in vivo. Correspondingly, Hansson suggested that radiating ruptures observed in the posterior part of the annulus fibrosus were the result of excessive tension in vivo.

Lifting an object and carrying an object over a horizontal distance are common situations wherein loads applied to the vertebral column may be so high as to damage the spine. Several factors influence the loads on the spine during these activities: (1) the position of the object relative to the center of motion in the spine; (2) the size, shape, weight, and density of the object; and (3) the degree of flexion or rotation of the spine.

Holding the object close to the body instead of away from it reduces the bending moment on the lumbar spine because the distance from the center of gravity of the object to the center of motion in the spine (the lever arm) is minimized. The shorter the lever arm is for the force produced by the weight of a given object, the lower the magnitude of the bending moment, and thus the lower the loads on the lumbar spine (Andersson et al., 1976b; Németh, 1984). The geometry of the object and its weight and density influence the loads on the spine. If objects of the same weight, shape, and density but of different sizes are held, the lever arm for the force produced by the weight of the object is longer for the larger object, and thus the bending moment on the lumbar spine is greater (Fig. 10–20).

When a person holding an object bends the body forward, the force produced by the weight of the object plus that produced by the weight of the upper body create a bending moment on the disc, increasing the loads on the spine. This bending moment is greater than that produced when the person stands erect while holding the object (Fig. 10–21).

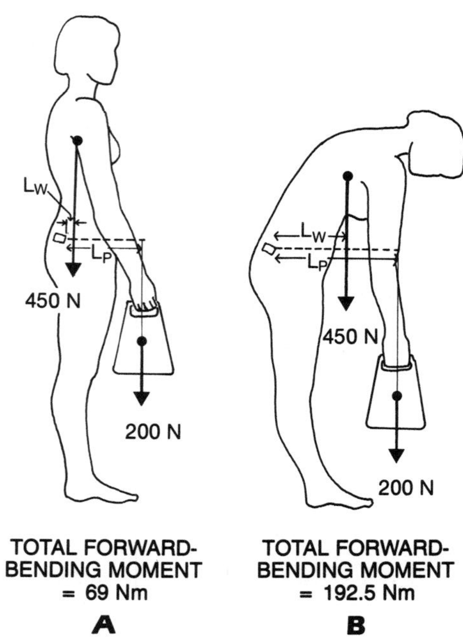

FIG. 10–21
The position of the upper body during lifting influences the loads on the lumbar spine. In these two situations an identical object weighing 20 kg is lifted. In case **A** (upright standing), the lever arm of the force produced by the weight of the object (L_p) is 30 cm, creating a forward-bending moment of 60 Nm (200 N × 0.3 m). The forward-bending moment created by the upper body is 9 Nm; the length of the lever arm (L_W) is estimated to be 2 cm, and the force produced by the weight of the upper body is 450 N. Thus, the total forward-bending moment in case **A** is equal to 69 Nm (60 Nm + 9 Nm). In case **B** (upper body flexed forward), the lever arm of the force produced by the weight of the object (L_p) is increased to 40 cm, creating a forward-bending moment of 80 Nm (200 N × 0.4 m). Furthermore, the force of 450 N produced by the weight of the upper body increases in importance, as it acts with a lever arm (L_W) of 25 cm, creating a forward-bending moment of 112.5 Nm (450 N × 0.25 m). Thus, the total forward-bending moment in case **B** is 192.5 Nm (112.5 Nm + 80 Nm).

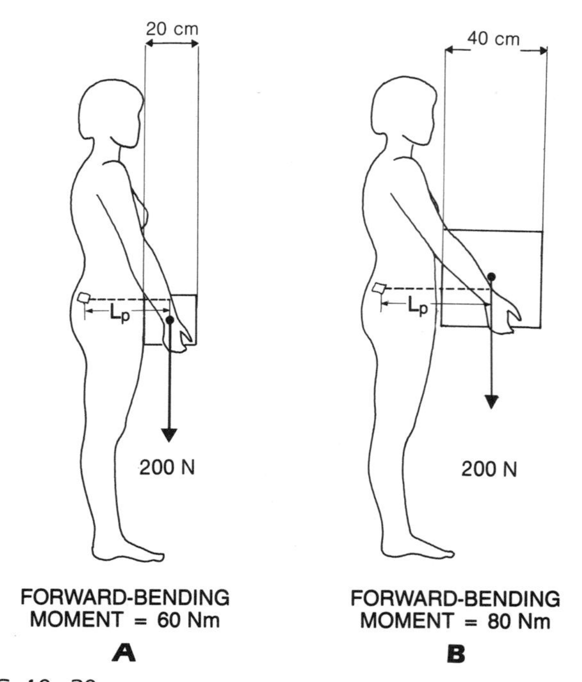

FIG. 10–20
The size of the object held influences the loads on the lumbar spine. In these two situations the distance from the center of motion in the disc to the front of the abdomen is 20 cm. In both cases the object has a uniform density and weighs 20 kg. In case **A** the width of the cubic object is 20 cm; in case **B** the width is 40 cm. Thus, in case **A** the forward-bending moment acting on the lowest lumbar disc is 60 Nm, as the force of 200 N produced by the weight of the object acts with a lever arm (L_p) of 30 cm (200 N × 0.3 m). In case B the forward-bending moment is 80 Nm, as the lever arm (L_p) is 40 cm (200 N × 0.4 m).

It is generally recommended that lifting be done with the knees bent and the back relatively straight to reduce the loads on the spine. This recommendation is valid only if this technique is used correctly. Lifting with bent knees and a straight back allows the object, if it is not too big, to be held closer to the trunk and to the center of motion in the spine than does lifting with the trunk flexed forward and the knees straight (Fig. 10–22A, B). However, the loads are not reduced by the use of this technique if the object is held out in front of the knees, that is, farther away from the center of motion; despite the bent knees, lifting the load in this way increases the bending moment rather than decreasing it (Fig. 10–22C) (Ekholm et al., 1982).

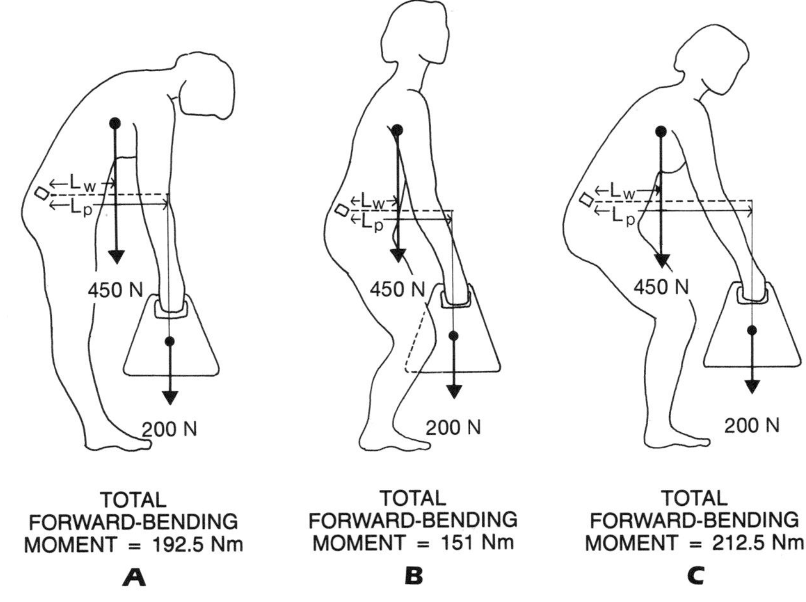

FIG. 10–22
The technique employed during lifting influences the loads on the lumbar spine. In these three situations an identical object weighing 20 kg is lifted. Case **A** (upper body flexed forward) is identical to case **B** in Figure 10–21; the total forward-bending moment is 192.5 Nm. In case **B**, lifting with knees bent and back straight places the object closer to the trunk, decreasing the forward-bending moments. The lever arms of the forces produced by the weight of the object (L_p) and upper body (L_W) are shortened to 35 and 18 cm, respectively, at this point in the lifting process. The result is a total forward-bending moment of 151 Nm ([200 N × 0.35 m] + [450 N × 0.18 m]). Case **C** shows that bent knees per se do not decrease the forward-bending moments. If the object lifted is held out in front of the knees, the lever arm of the force produced by the weight of the object (L_p) increases to 50 cm, and the lever arm of the force produced by the weight of the upper body (L_W) increases to 25 cm. Thus, the total forward-bending moment created is 212.5 Nm ([200 N × 0.5 m] + [450 N × 0.25 m]).

In the following example the simplified free body technique for coplanar forces will be used to make a rough calculation of the static loads on the spine as an object is lifted. The loads on a lumbar disc will be calculated for one point in time when a person who weighs 70 kg lifts a 20-kg object. The spine is flexed about 35 degrees. In this example (Fig. 10–23A), the three principal forces acting on the lumbar spine at the lumbosacral level are: (1) the force produced by the weight of the upper body (W), calculated to be 450 N (approximately 65% of the force exerted by the total body weight); (2) the force produced by the weight of the object (P), 200 N; and (3) the force produced by contraction of the erector spinae muscles (M), which has a known direction and point of application but an unknown magnitude.

Because these three forces act at a distance from the center of motion in the spine, they create moments in the lumbar spine. Two forward-bending moments (WL_w and PL_p) are the products of W and P and the perpendicular distance of these forces from the instant center (their lever arms). A counterbalancing moment (ML_m) is the product of M and its lever arm. The lever arm for W is 0.25 m, for P, 0.4 m, and for M, 0.05 m. The magnitude of M can be found through the use of the equilibrium equation for moments. For the body to be in moment equilibrium, the sum of the moments acting on the lumbar spine must be zero. (In this example, clockwise moments are considered to be positive, and counterclockwise moments are considered to be negative.) Thus,

$$\Sigma \text{ moments} = 0$$
$$(W \times L_W) + (P \times L_P) - (M \times L_M) = 0$$
$$(450 \text{ N} \times 0.25 \text{ m}) + (200 \text{ N} \times 0.4 \text{ m}) - (M \times 0.05 \text{ m}) = 0$$
$$M \times 0.05 \text{ m} = 112.5 \text{ Nm} + 80 \text{ Nm}.$$

Solving this equation for M yields 3,850 N.

The total compressive force exerted on the disc (C) can now be calculated trigonometrically (Fig. 10–23B). In the example, C is the sum of the compressive forces acting over the disc, which is inclined 35 degrees to the transverse plane. These forces are:

1. The compressive force produced by the weight of the upper body (W), which acts on the disc inclined 35 degrees (W × cos 35°).
2. The force produced by the weight of the object (P), which acts on the disc inclined 35 degrees (P × cos 35°).
3. The force produced by the erector spinae muscles (M), which acts approximately at a right angle to the disc inclination.

The total compressive force acting on the disc (C) has a known sense, point of application, and line of application, but an unknown magnitude. The magnitude of C can be found through the use of the equilibrium equation for forces. For the body to be in force equilibrium, the sum of the forces must equal zero. Thus,

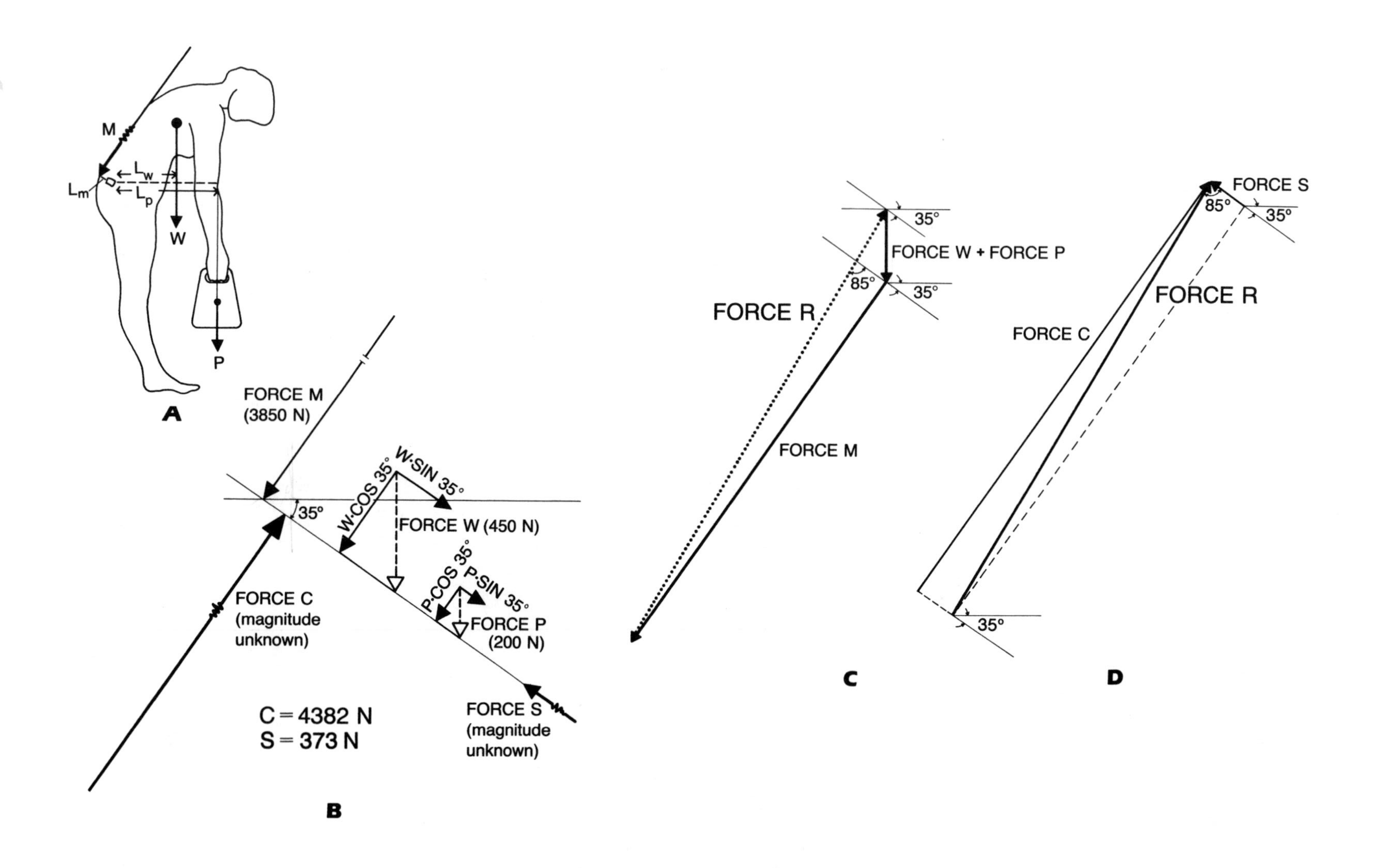

M
L_m
L_w
L_p
W
P
A
FORCE M
(3850 N)
35°
W·COS 35°
W·SIN 35°
FORCE W (450 N)
P·COS 35°
P·SIN 35°
FORCE P
(200 N)
FORCE C
(magnitude
unknown)
FORCE S
(magnitude
unknown)
C = 4382 N
S = 373 N
B
FORCE W + FORCE P
35°
85°
FORCE R
FORCE M
C
FORCE S
85°
35°
FORCE R
FORCE C
D

FIG. 10–23

A. Three main forces act on the lumbar spine at the lumbosacral level:

1. The force produced by the weight of the upper body (W), 450 N
2. The force produced by the weight of the object (P), 200 N
3. The force produced by the erector spinae muscles (M), magnitude unknown

Two forward-bending moments, WL_W and PL_P, are produced by W and P and their distances from the instant center. The lever arm (L_P) for P is 40 cm, and the lever arm (L_W) for W is 25 cm. The counterbalancing moment, ML_M, is produced by M and its distance from the instant center. The lever arm (L_M) is 5 cm. The magnitude of M is found by solving the equilibrium equation that states that the sum of all the moments must equal zero:

$$(450\text{ N} \times 0.25\text{ m}) + (200\text{ N} \times 0.4\text{ m}) - (M \times 0.05\text{ m}) = 0$$
$$M \times 0.05\text{ m} = 112.5\text{ Nm} + 80\text{ Nm}$$
$$M = 3850\text{ N.}$$

B. W and P are resolved into a compressive component, W × cos 35° and P × cos 35° respectively, and a shear component, W × sin 35° and P × sin 35° respectively. The total compressive reaction force acting on the disc (C) can now be solved by means of trigonometric functions. The magnitude of C is found by solving the equilibrium equation that states that the sum of all the forces must equal zero:

$$(450\text{ N} \times \cos 35°) + (200\text{ N} \times \cos 35°) + 3850\text{ N} - C = 0$$
$$C = 4382\text{ N.}$$

The shear component for the reaction force on the disc (S) is found in the same way:

$$(450\text{ N} \times \sin 35°) + (200\text{ N} \times \sin 35°) - S = 0$$
$$S = 373\text{ N.}$$

Since C and S form a right angle, the Pythagorean theorem can be used to find the total reaction force on the disc (R):

$$R = \sqrt{J^2 + S^2}$$
$$R = 4398\text{ N.}$$

The direction of R ($\sqrt{\ }$) is determined by means of a trigonometric function:

$$\sin \sqrt{\ } = \frac{C}{R}$$
$$\sqrt{\ } = 85°.$$

Thus, the line of application of R forms an 85-degree angle with the disc inclination.

C. The problem can be graphically solved by constructing a vector diagram based on the known values. A vertical line representing W + P is drawn first; M is added at a right angle to the disc inclination, and R closes the triangle. The direction of R in relation to the disc is measured.

D. When the magnitude and direction of the reaction force R are known, R can be resolved into its compressive force component C, and its shearing component S, by drawing a parallelogram with R as the diagonal. S, which mainly represents the shear-resisting effect of the disc and the posterior elements of the motion segment, is proportionately rather small in this example but will rise with an increasing inclination of the disc.

$$\Sigma \text{ forces} = 0$$
$$(W \times \cos 35°) + (P \times \cos 35°) + M - C = 0$$
$$(450 \text{ N} \times \cos 35°) + (200 \text{ N} \times \cos 35°) + 3850 \text{ N} - C = 0$$
$$C = 368.5 \text{ N} + 163.8 \text{ N} + 3850 \text{ N}.$$

Solving the equation for C yields 4,382 N. Solutions for the shear force (S) and total reaction force (R) acting on the disc are given in Figures 10–23B, C, D.

Calculations made in this way for one point in time during lifting are valuable for demonstrating how the lever arms of the forces produced by the weight of the upper body and by the weight of the object affect the loads imposed on the spine. The use of the same calculations to compute the loads produced when an 80-kg object is lifted (representing a force of 800 N) yields an approximate load of 10,000 N on the disc, which is likely to exceed the fracture point of the vertebra.

Since athletes who lift weights can easily reach such calculated loads without sustaining fractures, other factors, such as intra-abdominal pressure, may be involved in reducing the loads on the spine in vivo. The effect of intra-abdominal pressure in reducing the loads on the lumbosacral spine has been demonstrated through calculations based on data from intra-abdominal pressure measurements (Bartelink, 1957; Morris et al., 1961; Eie and Wehn, 1962). Using calculations of the abdominal cross-sectional area and the perpendicular distance from the muscular wall to the spinous process of L5, Eie and Wehn (1962) converted the abdominal pressure to a moment to be substituted in an equation for the various moments acting on the disc. They concluded that in athletes the spinal loads produced by erector spinae contraction were reduced up to 40% by the intra-abdominal pressure. It is difficult to find support for these calculations in later studies, however.

During moderate lifting, reduction of the loads on the spine due to intra-abdominal pressure has not been demonstrated to reach values of such magnitude. On the contrary, the effect of intra-abdominal pressure has been believed to be negligible when static positions are involved or when the spine is moderately flexed (Asmussen and Poulsen, 1968; Andersson et al., 1976b). There is agreement in most studies, however, that the intra-abdominal pressure rises with increasing forward bending of the trunk and with increased weight of the lifted load. Also, higher values are found during dynamic lifting, when acceleration forces are involved, than at one static point during the lift. Particularly in the initial phase of lifting, just at the point when the weight of the load is about to be overcome, the intra-abdominal pressure increases greatly and probably plays a major part in stabilizing the spine and moderating the intervertebral compressive forces. It has been suggested that the transverse and oblique abdominal muscles provide the muscle activity needed to build up the pressure; the role of these muscles may be overemphasized, however, as the pressure-altering mechanism is unclear (Ekholm et al., 1982).

During lifting, the erector spinae muscles are active to a greater or lesser extent, depending on the degree of trunk flexion. The activity of these muscles rises as the bending moment increases within the range of motion from the upright standing position to about 60 degrees of flexion. In any lifting situation, however, the bending moment must be counterbalanced by the activity of the back muscles.

The muscles and the ligaments are involved to varying degrees when different lifting techniques are used. For example, a static pull of a given magnitude from a short distance from the floor requires less back muscle activity when the back-lifting method is used (knees straight and trunk flexed, in this case more than 60 degrees) (Fig. 10–24A) than when the leg-lifting method is used (knees bent and back straight)

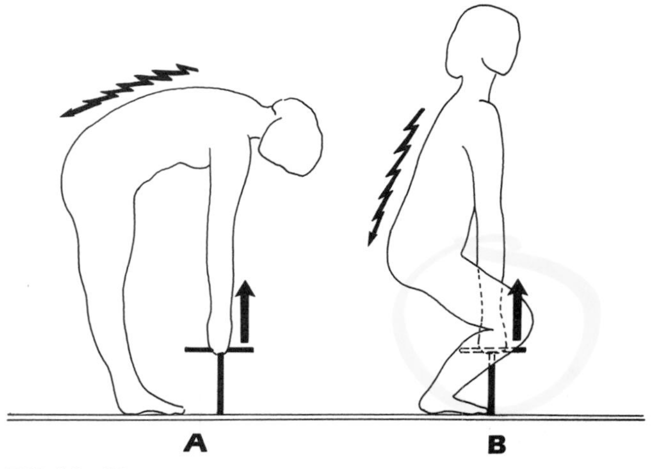

FIG. 10–24
In two static pulls of the same magnitude from a short distance from the floor, the activity of the erector spinae muscles (designated by the arrows) varies with the lifting technique employed. In case **A** (knees straight and upper body flexed more than 60 degrees), the back muscle activity is less than in case **B** (knees bent and back straight). However, this does not mean that less load on the spine is produced in case **A**, because in this situation the forward-bending moment is also counterbalanced by ligament forces. In case **B**, the ligaments are slack and the counterbalancing force is mainly provided by muscle contraction; thus, the ligaments are protected from excessive strain. On the other hand, the deep bending of the knees subjects the knee joints to heavy loads and increases the demand on the leg muscles, which are in an unfavorable position for exerting the force needed to raise the body.

(Fig. 10–24B) (Andersson et al., 1976a). This does not mean, however, that the back-lifting method creates less load on the lumbar spine at this point in the lifting process. Although electromyographic studies have shown that less muscle activity is needed to counterbalance the upper body when the trunk is flexed more than 60 degrees, the ligaments must sustain increased tension at this point to maintain this body position (Farfan, 1975; Adams et al., 1980). The stretched erector spinae muscles and the posterior hip muscles also help maintain this position.

At the beginning of the lifting process, the pelvis tilts backward and lumbar extension is delayed; thus, back muscle activity is low (Davis et al., 1965). When the spine starts to extend, however, the muscles increase their activity to counteract the forward-bending moment. Since the forces produced by the ligaments generally have shorter lever arms than do muscle forces, the load on the ligaments may become extremely high if the spine is not stabilized by muscle contractions. Thus, to protect the ligaments and the spine, back muscles should be active at the beginning of the lift, but no motion within the spine should occur until the initial phase of lifting is complete, that is, until the inertia of the load is overcome. Motion of the spine at the very start of lifting increases the stresses on the motion segment.

The lumbar spine provides less resistance to bending forces than to compressive forces (Lin et al., 1978). This finding suggests that a vertical position of the spine during lifting, which prevents wedging of the discs, is preferred. Measurements in vivo have also demonstrated that disc pressure in the lumbar spine increases when the trunk is loaded in lateral flexion or in flexion combined with axial rotation (Andersson et al., 1977).

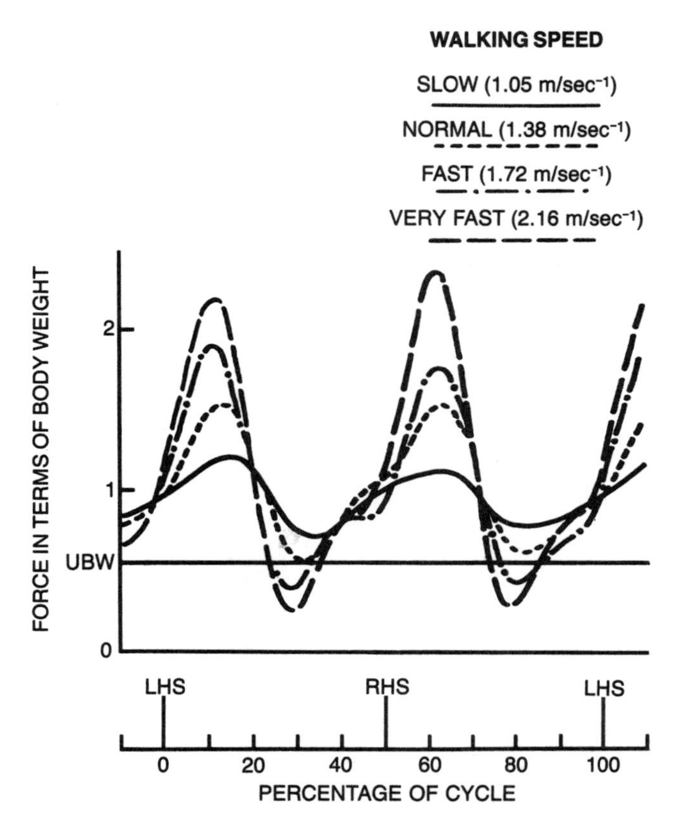

FIG. 10–25
Axial load on the L3–L4 motion segment in terms of body weight for one subject during walking at four speeds. The horizontal line (UBW) denotes the weight of the upper body, which represents the gravitational component of this load. Loads were predicted using experimental data from photogrammetric measurements along with a biomechanical model of the trunk. LHS, left heel strike; RHS, right heel strike. (Adapted from Cappozzo, 1984.)

DYNAMICS

Almost all motion in the body increases the loads on the lumbar spine. This increase is modest during such activities as slow walking or easy twisting but becomes more marked during various exercises (Nachemson and Elfström, 1970). In a study of normal walking at four speeds, the compressive loads at the L3–L4 motion segment ranged from 0.2 to 2.5 times body weight (Fig. 10–25) (Cappozzo, 1984). The loads were maximal around toe-off and increased approximately linearly with walking speed. Muscle action was mainly concentrated in the trunk extensors. Individual walking traits, particularly the amount of forward flexion of the trunk, influenced the loads. The greater this flexion was, the larger the muscle forces and hence the compressive load.

During strengthening exercises for the erector spinae and abdominal muscles, the loads on the spine can be high. While such exercises must be effective for strengthening the muscles concerned, they should be performed in such a way that the loads on the spine are adjusted to suit the condition of the individual's back.

The erector spinae muscles are intensely activated by arching the back in the prone position (Fig. 10–26A) (Pauly, 1966). Loading the spine in extreme positions such as this one produces high stresses on spine structures, so this hyperextended position should be avoided. An initial position that keeps the vertebrae in a more parallel alignment is preferable when strengthening exercises for the erector spinae muscles are performed (Fig. 10–26B).

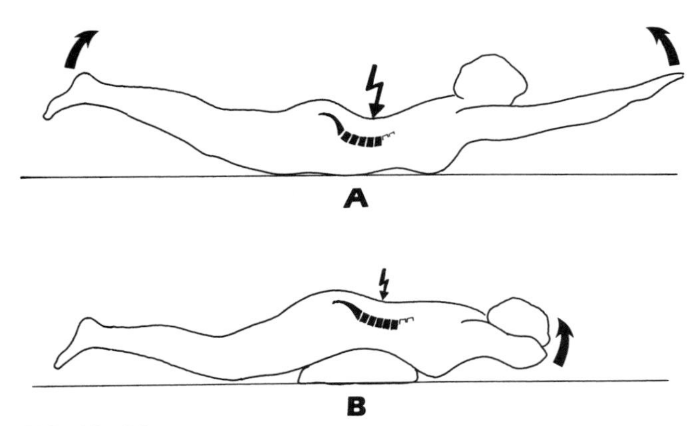

FIG. 10–26
A. Arching the back in the prone position greatly activates the erector spinae muscles but also produces high stresses on the lumbar discs, which are loaded in an extreme position. **B.** Decreasing the arch of the back by placing a pillow under the abdomen allows the discs to better resist stresses, since the vertebrae are aligned with each other. Isometric exercise in this position is preferable.

FIG. 10–27
Performing a curl to the point where only the shoulder blades clear the table minimizes lumbar motion, and hence the load on the lumbar spine is less than when a full sit-up is performed. A greater moment is produced if the arms are raised above the head or the hands are clasped behind the neck, as the center of gravity of the upper body then shifts farther away from the center of motion in the spine.

Bilateral straight-leg raising is commonly used as an abdominal muscle–strengthening exercise, but this exercise does little to activate the abdominal muscles. Instead, the vertebral portion of the psoas muscle produces the most activity and tends to pull the lumbar spine into lordosis. Performing a sit-up from the supine position, with hips and knees bent to limit the psoas activity, effectively activates the abdominal muscles but greatly increases the lumbar disc pressure (Nachemson and Elfström, 1970). The load on the lumbar spine is lessened if the range of motion of the exercise is limited; such a limitation can be accomplished through a trunk curl, wherein the head and shoulders are raised only to the point where the shoulder blades clear the table and lumbar spine motion is minimized (Fig. 10–27). This modification of the exercise has been shown to be quite effective in terms of motor unit recruitment in the muscles (Partridge and Walters, 1959; Flint, 1965; Ekholm et al., 1979); all portions of the external oblique and rectus abdominis muscles are activated.

A reverse curl, wherein the knees are brought toward the chest and the buttocks are raised from the table, activates the internal and external oblique muscles and the rectus abdominis muscle (Partridge and Walters, 1959). If the reverse curl is performed isometrically, the disc pressure is lower than that produced during a sit-up, but the exercise is just as effective for strengthening the abdominal muscles (Fig. 10–28).

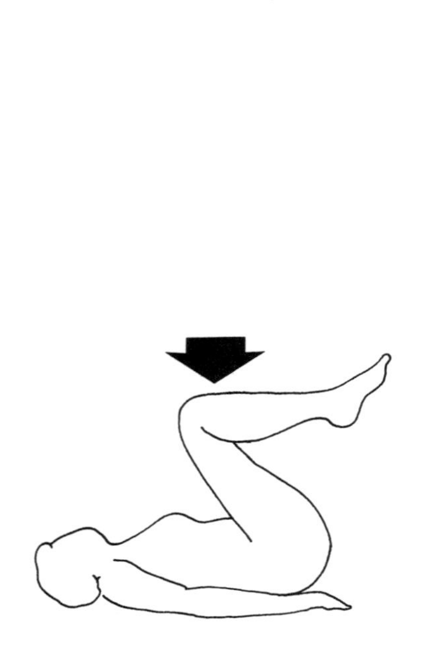

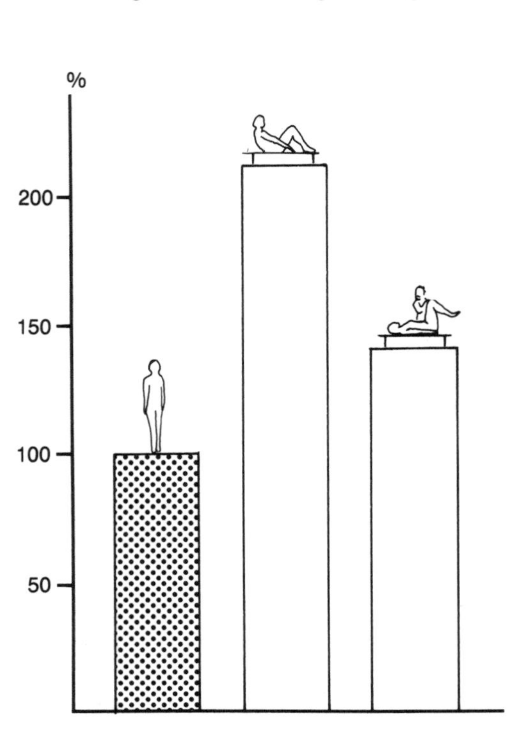

FIG. 10–28
A reverse curl, isometrically performed, provides efficient training of the abdominal muscles and produces moderate stresses on the lumbar discs. The relative loads on the third lumbar disc during a full sit-up and an isometric curl are compared with the load during upright standing, depicted as 100%. (Adapted from Nachemson, 1975.)

SUMMARY

1. A vertebra-disc-vertebra unit constitutes a motion segment, the functional unit of the spine.
2. In addition to acting as a kinematic restraint, the intervertebral disc serves a hydrostatic function in the motion segment, storing energy and distributing loads. This function is reduced with disc degeneration.
3. When compressive loads are imposed on the disc, compressive stress predominates in the inner portion of the disc, the nucleus pulposus, and tensile stress predominates in the outer covering, the annulus fibrosus.
4. The primary function of the facets is to guide the motion of the motion segment. The orientation of the facets determines the type of motion possible at any level of the spine. The facets may also sustain compressive loads, particularly during hyperextension.
5. Motion between two vertebrae is small and does not occur independently in vivo. Thus, functional motion of the spine is always a combined action of several motion segments. The range of motion varies among individuals and decreases with age. The instantaneous center of motion for the motion segments of the lumbar spine usually lies within the disc.
6. Trunk motion involves pelvic motion. The first 50 to 60 degrees of flexion occurs in the lumbar spine, and further flexion is produced primarily by tilting the pelvis forward.
7. The trunk muscles involved in motion also play an important role in providing extrinsic stability to the spine; the ligaments and discs provide intrinsic stability. Maintaining an upright standing position requires muscle activity.
8. The loads on the lumbar disc are relatively high in vivo. During relaxed upright standing the compressive load is about twice the weight of the superimposed body.
9. Body position affects the loads on the lumbar spine. Any deviation from upright relaxed standing, such as a forward bent or twisted position, produces higher stresses on the lumbar spine. Unsupported sitting also produces higher loads than standing upright does.
10. When the spine is flexed the disc bulges on the concave side of the spinal curve and retracts from the convex side. Bending and torsional loads create higher stresses on the disc than do axial compressive loads.
11. Externally applied loads produced, for example, by lifting or carrying objects may subject the lumbar spine to very high loads, although intra-abdominal pressure may help to support the spine and thus reduce these loads. For the loads on the spine to be minimized during lifting, the distance between the trunk and the object lifted should be as short as possible.
12. When a vertebra is subjected to compressive loads that exceed its failure point, the end-plate or the vertebral body fractures before the disc is damaged.

REFERENCES

Adams, M. A., and Hutton, W. C.: The mechanical functions of the lumbar apophyseal joints. Spine, *8*:327, 1983.

Adams, M. A., Hutton, W. C., and Stott, J. R. R.: The resistance to flexion of the lumbar intervertebral joint. Spine, *5*:245, 1980.

Andersson, G. B. J., Herberts, P., and Örtengren, R.: Myoelectric back muscle activity in standardized lifting postures. *In* International Series on Biomechanics. Vol. 1A. Edited by P. V. Komi. Baltimore, University Park Press, 1976, pp. 520–529.

Andersson, G. B. J., Örtengren, R., and Nachemson, A.: Quantitative studies of back loads in lifting. Spine, *1*:178, 1976.

Andersson, G. B. J., Örtengren, R., and Nachemson, A.: Intradiskal pressure, intra-abdominal pressure and myoelectric back muscle activity related to posture and loading. Clin. Orthop., *129*:156, 1977.

Andersson, G. B. J., Örtengren, R., Nachemson, A., and Elfström, G.: Lumbar disc pressure and myoelectric back muscle activity during sitting. I. Studies on an experimental chair. Scand. J. Rehabil. Med., *6*:104, 1974.

Asmussen, E., and Klausen, K.: Form and function of the erect human spine. Clin. Orthop., *25*:55, 1962.

Asmussen, E., and Poulsen, E.: On the role of the intra-abdominal pressure in relieving the back muscles while holding weights in a forward inclined position. Comm. Dan. Nat. Assoc. Infant. Paral., *28*:3, 1968.

Bartelink, D. L.: The role of abdominal pressure in relieving the pressure on the lumbar intervertebral discs. J. Bone Joint Surg., *39B*:718, 1957.

Basmajian, J. V.: Electromyography of iliopsoas. Anat. Rec., *132*:127, 1958.

Biering-Sørensen, F.: Physical measurements as risk indicators for low-back trouble over a one-year period. Spine, *9*:106, 1984.

Cappozzo, A.: Compressive loads in the lumbar vertebral column during normal level walking. J. Orthop. Res., *1*:292, 1984.

Carlsöö, S.: The static muscle load in different work positions: An electromyographic study. Ergonomics, *4*:193, 1961.

Cossette, J. W., Farfan, H. F., Robertson, G. H., and Wells, R. V.: The instantaneous center of rotation of the third lumbar intervertebral joint. J. Biomech., *4*:149, 1971.

Davis, P. R., Troup, J. D. G., and Burnard, J. H.: Movements of the thoracic and lumbar spine when lifting: A chrono-cyclophotographic study. J. Anat. Lond., *99*:13, 1965.

Eie, N.: Load capacity of the low back. J. Oslo City Hosp., *16*:73, 1966.

Eie, N., and Wehn, P.: Measurements of the intra-abdominal pressure in relation to weight bearing of the lumbosacral spine. J. Oslo City Hosp., *12*:205, 1962.

Ekholm, J., Arborelius, U. P., and Németh, G.: The load on the lumbo-sacral joint and trunk muscle activity during lifting. Ergonomics, *25*:145, 1982.

Ekholm, J., et al.: Activation of abdominal muscles during some physiotherapeutic exercises. Scand. J. Rehab. Med., *11*:75, 1979.

El-Bohy, A. A., and King, A. I.: Intervertebral disc and facet contact pressure in axial torsion. *In* 1986 Advances in Bioengineering. Edited by S. A. Lantz and A. I. King. New York, American Society of Mechanical Engineers, 1986, pp. 26–27.

Farfan, H. F.: Muscular mechanism of the lumbar spine and the position of power and efficiency. Orthop. Clin. North Am., *6*:135, 1975.

Flint, M. M.: Abdominal muscle involvement during the performance of various forms of sit-up exercise. An electromyographic study. Am. J. Phys. Med., *44*:224, 1965.

Floyd, W. F., and Silver, P. H. S.: The function of the erectores spinae muscles in certain movements and postures in man. J. Physiol., *129*:184, 1955.

Friedebold, G.: Die Aktivität normaler Rückenstreckmuskulatur im Elektromyogramm unter verschiedenen Haltungsbedingungen; eine Studie zur Skelettmuskelmechanik. Z. Orthop., *90*:1, 1958.

Galante, J. O.: Tensile properties of the human lumbar annulus fibrosus. Acta Orthop. Scand., Suppl. *100*:1–91, 1967.

Gertzbein, S. D., et al.: Centrode patterns and segmental instability in degenerative disc disease. Spine, *10*:257, 1985.

Gower, W. E., and Pedrini, V.: Age related variations in protein-polysaccharides from human nucleus pulposus, annulus fibrosus, and costal cartilage. J. Bone Joint Surg., *51A*:1154, 1969.

Hansson, T.: The bone mineral content and biomechanical properties of lumbar vertebrae. An *in vitro* study based on dual photon absorptiometry. Thesis, University of Gothenburg, Sweden, 1977.

Hutton, W. C., and Adams, M. A.: Can the lumbar spine be crushed in heavy lifting? Spine, *7*:586, 1982.

Joseph, J.: Man's Posture: Electromyographic Studies. Springfield, Charles C Thomas, 1960.

King, A. I., Prasad, P., and Ewing, C. L.: Mechanism of spinal injury due to caudocephalad acceleration. Orthop. Clin. North Am., *6*:19, 1975.

Kulak, R. F., Schultz, A. B., Belytschko, T., and Galante, J.: Biomechanical characteristics of vertebral motion segments and intervertebral discs. Orthop. Clin. North Am., *6*:121, 1975.

Lin, H. S., Liu, Y. K., and Adams, K. H.: Mechanical response of the lumbar intervertebral joint under physiological (complex) loading. J. Bone Joint Surg., *60A*:41, 1978.

Lucas, D. B., and Bresler, B.: Stability of the ligamentous spine. Biomechanics Laboratory, University of California, San Francisco and Berkeley. Technical Report 40. San Francisco, The Laboratory, 1961.

Lumsden, R. M., and Morris, J. M.: An *in vivo* study of axial rotation and immobilization at the lumbosacral joint. J. Bone Joint Surg., *50A*:1591, 1968.

Miles, M., and Sullivan, W. E.: Lateral bending at the lumbar and lumbosacral joints. Anat. Rec., *139*:387, 1961.

Miller, J. A. A., Haderspeck, K. A., and Schultz, A. B.: Posterior element loads in lumbar motion segments. Spine, *8*:331, 1983.

Moll, J. M. H., and Wright, V.: Normal range of spinal mobility. An objective clinical study. Ann. Rheum. Dis., *30*:381, 1971.

Morris, J. M., Lucas, D. B., and Bresler, B.: Role of the trunk in stability of the spine. J. Bone Joint Surg., *43A*:327, 1961.

Nachemson, A.: Lumbar intradiscal pressure. Acta Orthop. Scand., Suppl. *43*:1–140, 1960.

Nachemson, A.: The influence of spinal movements on the lumbar intradiscal pressure and on the tensile stresses in the annulus fibrosus. Acta Orthop. Scand., *33*:183, 1963.

Nachemson, A.: Electromyographic studies on the vertebral portion of the psoas muscle. With special reference to its stabilizing function of the lumbar spine. Acta Orthop. Scand., *37*:177, 1966.

Nachemson, A.: The possible importance of the psoas muscle for stabilization of the lumbar spine. Acta Orthop. Scand., *39*:47, 1968.

Nachemson, A.: Towards a better understanding of back pain; a review of the mechanics of the lumbar disc. Rheumatol. Rehabil., *14*:129, 1975.

Nachemson, A., and Elfström, G.: Intravital Dynamic Pressure Measurements in Lumbar Discs: A Study of Common Movements, Maneuvers and Exercises. Stockholm, Almqvist & Wiksell, 1970.

Nachemson, A. L., and Evans, J. H.: Some mechanical properties of the third human lumbar interlaminar ligament (ligamentum flavum). J. Biomech., *1*:211, 1968.

Nachemson, A., and Morris, J. M.: *In vivo* measurements of intradiscal pressure. Discometry, a method for the determination of pressure in the lower lumbar discs. J. Bone Joint Surg., *46A*:1077, 1964.

Nachemson, A. L., Schultz, A. B., and Berkson, M. H.: Mechanical properties of human lumbar spine motion segments. Influence of age, sex, disc level, and degeneration. Spine, *4*:1, 1979.

Németh, G.: On hip and lumbar biomechanics. A study of joint load and muscular activity. Scand. J. Rehab. Med., Suppl. *10*, 1984.

Norton, P. L., and Brown, T.: The immobilizing efficiency of back braces. Their effect on the posture and motion of the lumbosacral spine. J. Bone Joint Surg., *39A*:111, 1957.

Örtengren, R., Andersson, G. B. J., and Nachemson, A. L.: Studies of relationships between lumbar disc pressure, myoelectric back muscle activity, and intra-abdominal (intragastric) pressure. Spine, *6*:98, 1981.

Panjabi, M. M., Goel, V. K., and Takata, K.: Physiologic strains in the lumbar spinal ligaments. An in vitro biomechanical study. Spine, *7*:192, 1982.

Partridge, M. J., and Walters, C. E.: Participation of the abdominal muscles in various movements of the trunk in man. An electromyographic study. Phys. Ther. Rev., *39*:791, 1959.

Pauly, J. E.: An electromyographic analysis of certain movements and exercises. I. Some deep muscles of the back. Anat. Rec., *155*:223, 1966.

Reichmann, S.: Motion of the lumbar articular processes in flexion-extension and lateral flexion of the spine. Acta Morphol. Neerl. Scand., *8*:261, 1970/71.

Reichmann, S., Berglund, E., and Lundgren, K.: Das Bewegungszentrum in der Lendenwirbelsäule bei Flexion und Extension. Z. Anat. Entwicklungsgesch., *138*:283, 1972.

Rolander, S. D.: Motion of the lumbar spine with special reference to the stabilizing effect of posterior fusion. An experimental study on autopsy specimens. Acta Orthop. Scand., Suppl. *90*:1–144, 1966.

Schultz, A. B., and Andersson, G. B. J.: Analysis of loads on the lumbar spine. Spine, *6*:76, 1981.

Schultz, A., et al.: Analysis and quantitative myoelectric measurements of loads on the lumbar spine when holding weights in standing postures. Spine, *7*:390, 1982a.

Schultz, A., et al.: Loads on the lumbar spine. Validation of a biomechanical analysis by measurements of intradiscal pressures and myoelectric signals. J. Bone Joint Surg., *64A*:713, 1982b.

Schultz, A. B., et al.: Analysis and measurement of lumbar trunk loads in tasks involving bends and twists. J. Biomech., *15*:669, 1982c.

Stokes, I. A. F., and Abery, J. M.: Influence of the hamstring muscles on lumbar spine curvature in sitting. Spine, *5*:525, 1980.

Thurston, A. J., and Harris, J. D.: Normal kinematics of the lumbar spine and pelvis. Spine, *8*:199, 1983.

Urban, J. P. G., and McMullin, J. F.: Swelling pressure of the intervertebral disc: Influence of proteoglycan and collagen contents. Biorheology, *22*:145, 1985.

Waters, R. L., and Morris, J. M.: Effect of spinal supports on the electrical activity of muscles of the trunk. J. Bone Joint Surg., *52A*:51, 1970.

White, A. A.: Analysis of the mechanics of thoracic spine in man. An experimental study of autopsy specimens. Acta Orthop. Scand., Suppl. *127*:1–105, 1969.

White, A. A., and Panjabi, M. M.: Clinical Biomechanics of the Spine. Philadelphia, J. B. Lippincott, 1978.

Wilder, D. G., Pope, M. H., and Frymoyer, J. W.: The functional topography of the sacroiliac joint. Spine, *5*:575, 1980.

SUGGESTED READING

Adams, M. A., and Hutton, W. C.: The mechanical functions of the lumbar apophyseal joints. Spine, *8*:327–330, 1983.

Adams, M. A., Hutton, W. C., and Stott, J. R. R.: The resistance to flexion of the lumbar intervertebral joint. Spine, *5*:245–253, 1980.

Andersson, G. B. J.: On myoelectric back muscle activity and lumbar disc pressure in sitting postures. Scand. J. Rehabil. Med., Suppl. *3*, 1974.

Andersson, G. B. J., Herberts, P., and Örtengren, R.: Myoelectric back muscle activity in standardized lifting postures. *In* International Series on Biomechanics. Vol. 1A. Edited by P. V. Komi. Baltimore, University Park Press, 1976, pp. 520–529.

Andersson, G. B. J., Örtengren, R., and Nachemson, A.: Quantitative studies of back loads in lifting. Spine, *1*:178–185, 1976.

Andersson, G. B. J., Örtengren, R., and Nachemson, A.: Intradiskal pressure, intra-abdominal pressure and myoelectric back muscle activity related to posture and loading. Clin. Orthop., *129*:156–164, 1977.

Andersson, G. B. J., Örtengren, R., Nachemson, A., and Elfström, G.: Lumbar disc pressure and myoelectric back muscle activity during sitting. I. Studies on an experimental chair. Scand. J. Rehabil. Med., *6*:104–114, 1974.

Asmussen, E., and Klausen, K.: Form and function of the erect human spine. Clin. Orthop., *25*:55–63, 1962.

Asmussen, E., and Poulsen, E.: On the role of the intraabdominal pressure in relieving the back muscles while holding weights in a forward inclined position. Comm. Dan. Nat. Assoc. Infant. Paral., *28*:3–11, 1968.

Asmussen, E., Poulsen, E., and Rasmussen, B.: Quantitative evaluation of the activity of the back muscles in lifting. Comm. Dan. Nat. Assoc. Infant. Paral., *21*:3–14, 1965.

Bartelink, D. L.: The role of abdominal pressure in relieving the pressure on the lumbar intervertebral discs. J. Bone Joint Surg., *39B*:718–725, 1957.

Basmajian, J. V.: Electromyography of iliopsoas. Anat. Rec., *132*:127–132, 1958.

Berkson, M. H., Nachemson, A. L., and Schultz, A. B.: Mechanical properties of human lumbar spine motion segments. Part II: Responses in compression and shear; influence of gross morphology. J. Biomech. Eng., *101*:53–57, 1979.

Biering-Sørensen, F.: Physical measurements as risk indicators for low-back trouble over a one-year period. Spine, *9*:106–119, 1984.

Brown, T., Hansen, R. J., and Yorra, A. J.: Some mechanical tests on the lumbosacral spine with particular reference to the intervertebral discs. A preliminary report. J. Bone Joint Surg., *39A*:1135–1164, 1957.

Cappozzo, A.: Compressive loads in the lumbar vertebral column during normal level walking. J. Orthop. Res., *1*:292–301, 1984.

Carlsöö, S.: The static muscle load in different work positions: An electromyographic study. Ergonomics, *4*:193–211, 1961.

Carlsöö, S.: Att lyfta i jobbet. Uddevalla, Sweden, Pa-radet, Bohusläningens AB, 1975.

Chaffin, D. B., and Baker, W. H.: A biomechanical model for analysis of symmetric sagittal plane lifting. Am. Inst. Indus. Eng. Trans., *2*:16–27, 1970.

Cossette, J. W., Farfan, H. F., Robertson, G. H., and Wells, R. V.: The instantaneous center of rotation of the third lumbar intervertebral joint. J. Biomech., *4*:149–153, 1971.

Davis, P. R., and Troup, J. D. G.: Pressures in the trunk cavities when pulling, pushing and lifting. Ergonomics, *7*:465–474, 1964.

Davis, P. R., and Troup, J. D. G.: Effects on the trunk of erecting pit props at different working heights. Ergonomics, *9*:475–484, 1966.

Davis, P. R., Troup, J. D. G., and Burnard, J. H.: Movements of the thoracic and lumbar spine when lifting: A chrono-cyclophotographic study. J. Anat. Lond., *99*:13–26, 1965.

Donisch, E. W., and Basmajian, J. V.: Electromyography of deep back muscles in man. Am. J. Anat., *133*:25–36, 1972.

Eie, N.: Load capacity of the low back. J. Oslo City Hosp., *16*:73–98, 1966.

Eie, N., and Wehn, P.: Measurements of the intra-abdominal pressure in relation to weight bearing of the lumbosacral spine. J. Oslo City Hosp., *12*:205–217, 1962.

Ekholm, J., Arborelius, U. P., and Németh, G.: The load on the lumbo-sacral joint and trunk muscle activity during lifting. Ergonomics, *25*:145–161, 1982.

Ekholm, J., Arborelius, U., Fahlcrantz, A., Larsson, A.-M., and Mattsson, G.: Activation of abdominal muscles during some physiotherapeutic exercises. Scand. J. Rehab. Med., *11*:75–84, 1979.

El-Bohy, A. A., and King, A. I.: Intervertebral disc and facet contact pressure in axial torsion. *In* 1986 Advances in Bioengineering. Edited by S. A. Lantz and A. I. King. New York, American Society of Mechanical Engineers, 1986, pp. 26–27.

Evans, F. G., and Lissner, H. R.: Biomechanical studies on the lumbar spine and pelvis. J. Bone Joint Surg., *41A*:278–290, 1959.

Farfan, H. F.: Mechanical Disorders of the Low Back. Philadelphia, Lea & Febiger, 1973.

Farfan, H. F.: Muscular mechanism of the lumbar spine and the position of power and efficiency. Orthop. Clin. North Am., *6*:135–144, 1975.

Flint, M. M.: Abdominal muscle involvement during the performance of various forms of sit-up exercise. An electromyographic study. Am. J. Phys. Med., *44*:224–234, 1965.

Floyd, W. F., and Silver, P. H. S.: Electromyographic study of patterns of activity of the anterior abdominal wall muscles in man. J. Anat., *84*:132–145, 1950.

Floyd, W. F., and Silver, P. H. S.: The function of the erectores spinae muscles in certain movements and postures in man. J. Physiol., *129*:184–203, 1955.

Friedebold, G.: Die Aktivität normaler Rückenstreckmuskulatur im Elektromyogramm unter verschiedenen Haltungsbedingungen; eine Studie zur Skelettmuskelmechanik. Z. Orthop., *90*:1–18, 1958.

Galante, J. O.: Tensile properties of the human lumbar annulus fibrosus. Acta Orthop. Scand., Suppl. *100*:1–91, 1967.

Gertzbein, S. D., et al.: Centrode patterns and segmental instability in degenerative disc disease. Spine, *10*:257–261, 1985.

Gower, W. E., and Pedrini, V.: Age related variations in protein-polysaccharides from human nucleus pulposus, annulus fibrosus, and costal cartilage. J. Bone Joint Surg., *51A*:1154–1162, 1969.

Gregersen, G. G., and Lucas, D. B.: An *in vivo* study of the axial rotation of the human thoracolumbar spine. J. Bone Joint Surg., *49A*:247–262, 1967.

Hansson, T.: The bone mineral content and biomechanical properties of lumbar vertebrae. An *in vitro* study based on dual photon absorptiometry. Thesis, University of Gothenburg, Sweden, 1977.

Hirsch, C., and Lewin, T.: Lumbosacral synovial joints in flexion-extension. Acta Orthop. Scand., *39*:303–311, 1968.

Hutton, W. C., and Adams, M. A.: Can the lumbar spine be crushed in heavy lifting? Spine, *7*:586–590, 1982.

Jayson, M.: The Lumbar Spine and Back Pain. New York, Grune and Stratton, 1976.

Jonsson, B.: The functions of individual muscles in the lumbar part of the spinae muscle. Electromyography, *10*:5–21, 1970.

Jørgensen, K.: Back muscle strength and body weight as limiting factors for work in the standing slightly stooped position. Comm. Dan. Nat. Assoc. Infant. Paral., *30*:3–9, 1970.

Joseph, J.: Man's Posture: Electromyographic Studies. Springfield, Charles C Thomas, 1960.

Joseph, J., and McColl, I.: Electromyography of muscles of posture: Posterior vertebral muscles in males. J. Physiol., *157*:33–37, 1961.

Kazarian, L. E.: Creep characteristics of the human spinal column. Orthop. Clin. North Am., *6*:3–18, 1975.

Kazarian, L., and Graves, G. A.: Compressive strength characteristics of the human vertebral centrum. Spine, *2*:1–14, 1977.

Keegan, J. J.: Alterations of the lumbar curve related to posture and seating. J. Bone Joint Surg., *35A*:589–603, 1953.

King, A. I., Prasad, P., and Ewing, C. L.: Mechanism of spinal injury due to caudocephalad acceleration. Orthop. Clin. North Am., *6*:19–31, 1975.

Kraus, H.: Effect of lordosis on the stress in the lumbar spine. Clin. Orthop., *117*:56–58, 1976.

Kulak, R. F., Schultz, A. B., Belytschko, T., and Galante, J.: Biomechanical characteristics of vertebral motion segments and intervertebral discs. Orthop. Clin. North Am., *6*:121–133, 1975.

LeVeau, B. F.: Axes of joint rotation of the lumbar vertebrae during abdominal strengthening exercises. *In* Biomechanics IV. Edited by R. C. Nelson and C. A. Morehouse. Baltimore, University Park Press, 1974, pp. 361–364.

Lin, H. S., Liu, Y. K., and Adams, K. H.: Mechanical response of the lumbar intervertebral joint under physiological (complex) loading. J. Bone Joint Surg., *60A*:41–55, 1978.

Lucas, D. B.: Mechanics of the spine. Bull. Hosp. Joint Dis., *31*:115–131, 1970.

Lucas, D. B., and Bresler, B.: Stability of the ligamentous spine. Biomechanics Laboratory, University of California, San Francisco and Berkeley. Technical Report 40. San Francisco, The Laboratory, 1961.

Lumsden, R. M., and Morris, J. M.: An *in vivo* study of axial rotation and immobilization at the lumbosacral joint. J. Bone Joint Surg., *50A*:1591–1602, 1968.

Markolf, K. L.: Deformation of the thoracolumbar intervertebral joints in response to external loads. A biomechanical study using autopsy material. J. Bone Joint Surg., *54A*:511–533, 1972.

Miles, M., and Sullivan, W. E.: Lateral bending at the lumbar and lumbosacral joints. Anat. Rec., *139*:387–398, 1961.

Miller, J. A. A., Haderspeck, K. A., and Schultz, A. B.: Posterior element loads in lumbar motion segments. Spine, *8*:331–337, 1983.

Moll, J. M. H., and Wright, V.: Normal range of spinal mobility. An objective clinical study. Ann. Rheum. Dis., *30*:381–386, 1971.

Morris, J. M., Benner, G., and Lucas, D. B.: An electromyographic study of the intrinsic muscles of the back in man. J. Anat. Lond., *96*:509–520, 1962.

Morris, J. M., Lucas, D. B., and Bresler, B.: Role of the trunk in stability of the spine. J. Bone Joint Surg., *43A*:327–351, 1961.

Nachemson, A.: Lumbar intradiscal pressure. Acta Orthop. Scand., Suppl. *43*:1–104, 1960.

Nachemson, A.: The influence of spinal movements on the lumbar intradiscal pressure and on the tensile stresses in the annulus fibrosus. Acta Orthop. Scand., *33*:183–207, 1963.

Nachemson, A.: Electromyographic studies on the vertebral portion of the psoas muscle. With special reference to its stabilizing function of the lumbar spine. Acta Orthop. Scand., *37*:177–190, 1966.

Nachemson, A.: The possible importance of the psoas muscle for stabilization of the lumbar spine. Acta Orthop. Scand., *39*:47–57, 1968.

Nachemson, A.: Towards a better understanding of back pain; a review of the mechanics of the lumbar disc. Rheumatol. Rehabil., *14*:129–143, 1975.

Nachemson, A. L.: The lumbar spine: An orthopaedic challenge. Spine, *1*:59–71, 1976.

Nachemson, A., and Elfström, G.: Intravital Dynamic Pressure Measurements in Lumbar Discs: A Study of Common Movements, Maneuvers and Exercises. Stockholm, Almqvist & Wiksell, 1970.

Nachemson, A. L., and Evans, J. H.: Some mechanical properties of the third human lumbar interlaminar ligament (ligamentum flavum). J. Biomech., *1*:211–220, 1968.

Nachemson, A., and Morris, J. M.: *In vivo* measurements of intradiscal pressure. Discometry, a method for the determination of pressure in the lower lumbar discs. J. Bone Joint Surg., *46A*:1077–1092, 1964.

Nachemson, A. L., Schultz, A. B., and Berkson, M. H.: Mechanical properties of human lumbar spine motion segments. Influence of age, sex, disc level, and degeneration. Spine, *4*:1–8, 1979.

Naylor, A.: Intervertebral disc prolapse and degeneration. The biochemical and biophysical approach. Spine, *1*:108–114, 1976.

Németh, G.: On hip and lumbar biomechanics. A study of joint load and muscular activity. Med., Scand. J. Rehab. Med., Suppl. *10*, 1984.

Norton, P. L., and Brown, T.: The immobilizing efficiency of back braces. Their effect on the posture and motion of the lumbosacral spine. J. Bone Joint Surg., *39A*:111–139, 1957.

Okada, M., Kogi, K., and Ishii, M.: Enduring capacity of the erectores spinae muscles in static work. J. Anthrop. Soc. Nippon, *78*:99–110, 1970.

Örtengren, R., and Andersson, G. B. J.: Electromyographic studies of trunk muscles, with special reference to the functional anatomy of the lumbar spine. Spine, *2*:44–52, 1977.

Örtengren, R., Andersson, G. B. J., and Nachemson, A. L.: Studies of relationships between lumbar disc pressure, myoelectric back muscle activity, and intra-abdominal (intragastric) pressure. Spine, *6*:98–103, 1981.

Panjabi, M. M., Goel, V. K., and Takata, K.: Physiologic strains in the lumbar spinal ligaments. An in vitro biomechanical study. Spine, *7*:192–203, 1982.

Panjabi, M. M., Krag, M. H., White, A. A., and Southwick, W. O.: Effects of preload on load displacement curves of the lumbar spine. Orthop. Clin. North Am., *8*:181–192, 1977.

Park, K. S., and Chaffin, D. B.: A biomechanical evaluation of two methods of manual load lifting. Am. Inst. Indus. Eng. Trans., *6*:105–113, 1974.

Partridge, M. J., and Walters, C. E.: Participation of the abdominal muscles in various movements of the trunk in man. An electromyographic study. Phys. Ther. Rev., *39*:791–800, 1959.

Pauly, J. E.: An electromyographic analysis of certain movements and exercises. I. Some deep muscles of the back. Anat. Rec., *155*:223–234, 1966.

Portnoy, H., and Morin, F.: Electromyographic study of postural muscles in various positions and movements. Am. J. Physiol., *186*:122–126, 1956.

Posner, I., White, A. A. III, Edwards, W. T., and Hayes, W. C.: A biomechanical analysis of the clinical stability of the lumbar and lumbosacral spine. Spine, *7*:374–389, 1982.

Reichmann, S.: Motion of the lumbar articular processes in flexion-extension and lateral flexion of the spine. Acta Morphol. Neerl. Scand., *8*:261–272, 1970/71.

Reichmann, S., Berglund, E., and Lundgren, K.: Das Bewegungszentrum in der Lendenwirbelsäule bei Flexion und Extension. Z. Anat. Entwicklungsgesch., *138*:283–287, 1972.

Rolander, S. D.: Motion of the lumbar spine with special reference to the stabilizing effect of posterior fusion. An experimental study on autopsy specimens. Acta Orthop. Scand., Suppl. *90*:1–144, 1966.

Schultz, A. B., and Andersson, G. B. J.: Analysis of loads on the lumbar spine. Spine, *6*:76–82, 1981.

Schultz, A. B., Warwick, D. N., Berkson, M. H., and Nachemson, A. L.: Mechanical properties of human lumbar spine motion segments. Part 1: Responses in flexion, extension, lateral bending, and torsion. J. Biomech. Eng., *101*:46–52, 1979.

Schultz, A., et al.: Analysis and quantitative myoelectric measurements of loads on the lumbar spine when holding weights in standing postures. Spine, *7*:390–397, 1982.

Schultz, A., et al.: Loads on the lumbar spine. Validation of a biomechanical analysis by measurements of intradiscal pressures and myoelectric signals. J. Bone Joint Surg., *64A*:713–720, 1982.

Schultz, A. B., et al.: Analysis and measurement of lumbar trunk loads in tasks involving bends and twists. J. Biomech., *15*:669–675, 1982.

Stokes, I. A. F., and Abery, J. M.: Influence of the hamstring muscles on lumbar spine curvature in sitting. Spine, *5*:525–528, 1980.

Tkaczuk, H.: Tensile properties of human lumbar longitudinal ligaments. Acta Orthop. Scand., Suppl. *115*:1–68, 1968.

Thurston, A. J., and Harris, J. D.: Normal kinematics of the lumbar spine and pelvis. Spine, *8*:199–205, 1983.

Urban, J. P. G., and McMullin, J. F.: Swelling pressure of the intervertebral disc: Influence of proteoglycan and collagen contents. Biorheology, *22*:145–157, 1985.

Virgin, W. J.: Experimental investigations into the physical properties of the intervertebral disc. J. Bone Joint Surg., *33B*:607–611, 1951.

Waters, R. L., and Morris, J. M.: Effect of spinal supports on the electrical activity of muscles of the trunk. J. Bone Joint Surg., *52A*:51–60, 1970.

Weis, E. B.: Stresses of the lumbosacral junction. Orthop. Clin. North Am., *6*:83–91, 1975.

White, A. A.: Analysis of the mechanics of the thoracic spine in man. An experimental study of autopsy specimens. Acta Orthop. Scand., Suppl. *127*:1–105, 1969.

White, A. A., and Hirsch, C.: The significance of the vertebral posterior elements in the mechanics of the thoracic spine. Clin. Orthop., *81*:2–14, 1971.

White, A. A., and Panjabi, M. M.: Clinical Biomechanics of the Spine. Philadelphia, J. B. Lippincott, 1978.

White, A. A., Southwick, W. O., Panjabi, M. M., and Johnson, R. M.: Practical biomechanics of the spine for the orthopaedic surgeon. American Academy of Orthopaedic Surgeons Instructional Course Lectures, *23*:62–78, 1974.

Wiktorin, C. v. H., and Nordin, M.: Introduction to Problem Solving in Biomechanics. Philadelphia, Lea & Febiger, 1986, pp. 130–170.

Wilder, D. G., Pope, M. H., and Frymoyer, J. W.: The functional topography of the sacroiliac joint. Spine, *5*:575–579, 1980.

11

BIOMECHANICS OF THE CERVICAL SPINE

Ilan Shapiro
Victor H. Frankel

The intricacy of cervical architecture has long defied attempts to define the precise function of each component of the cervical spine, but several roles have been defined for this structure as a whole. It must give strong support to the skull, protect the neural components and vascular structures, and provide muscle attachments; yet it must have the flexibility afforded by an extensive range of motion to integrate the head with the body and environment. It must also act as a shock absorber to protect the brain in its rock-hard vault, and it must provide portals of entry, exit, and passage for neurovascular structures. Although these roles may appear to conflict, the cervical spine is structured so as to allow their mutual functioning.

That man survived his early savage environment and in the present day survives ejection from jet aircraft is a tribute to the spine's structural and functional integrity and to nature's foresight. Indeed, this structure functions so well that much of the earlier work on the biomechanics of the cervical spine, sponsored by the Victorian enthusiasm of the time, centered on finding ways to fracture it efficiently (Duff, *A Handbook on Hanging,* London, 1938). Enthusiasm today comes from other sources. These include defense agencies, space agencies, manufacturers of automobiles and football helmets, and traumatologists, all of whom are interested in preserving the function and stability of the cervical spine and in defining its limits of endurance and mechanisms of failure. The fact that extensive resources are allocated for this pursuit is understandable, as injuries to the cervical spine have always been associated with a dismal prognosis and death. The ancient Egyptians cautioned against treating cervical fractures (Breasted, 1930). More recently, at the battle of Trafalger in 1805, Lord Nelson assured those around him that his fate was sealed when he felt a sharpshooter's bullet pierce his spine (Pope, 1959).

This chapter outlines in some detail the anatomy of the cervical spine to provide an understanding of this most intricate region of the spinal column. Motion in the cervical spine, which is less restrained in range and direction than motion in the lumbar and thoracic regions, is also described. The discussion of kinetics begins with a comparison of the static loads on the cervical spine with the head in different positions. Static loads during traction of the cervical spine, a treatment used since the earliest recordings of medical history, are then examined. Finally, dynamic loading resulting in injury of the cervical spine is discussed.

ANATOMY OF THE CERVICAL SPINE

The supporting structures of the cervical spine are the vertebrae, ligaments, intervertebral discs, and surrounding muscles. Knowledge of these structures and their interrelationships is essential for understanding spine biomechanics and the injury mechanisms specific to the cervical spine.

THE OSSEOUS STRUCTURES

The cervical spine consists of seven vertebrae (Fig. 11–1): five typical vertebrae C3 to C7, similar in structure and function to the vertebrae of the thoracic and lumbar spine, and two atypical vertebrae, C1 (the atlas) and C2 (the axis), each having a unique structure and role. The articulation between C1 and the occipital bone of the skull, the atlanto-occipital joint, is also a functional part of the cervical spine.

The Typical Cervical Vertebrae—C3 to C7

Each typical cervical vertebra is composed of a body, two pedicles, two laminae, and a spinous process (Fig. 11–2). Between the pedicle and lamina on each side is an articular process that supports a superior and an inferior facet. In spine literature, reference is often made to the motion segment, the functional unit of the spine, consisting of two adjacent vertebrae, the interposed vertebral disc, and adjoining ligaments. For the purpose of biomechanical analysis, the motion segment is commonly discussed in terms of anterior and posterior components. The anterior components are the vertebral bodies, the disc, the pedicles, and the attached ligaments; all the remaining structures are posterior components (Fig. 11–3) (White and Panjabi, 1978).

The cervical vertebral body is elliptical with a transversely concave upper surface. The laterally raised lips constitute the uncinate processes. The inferior surface has a protruding anterior lip that overlaps a bevel on the underlying vertebral body's superior surface. The transverse processes are positioned anteriorly on each side of the vertebral body. Each is pierced by a transverse foramen, a structure unique to the cervical spine. The vertebral artery passes up the cervical spine bilaterally through these foramina, bypassing those of C7. The transverse processes are grooved superiorly to accommodate the exiting spinal nerve roots (see Fig. 11–2B). This grooving weakens the transverse process and predisposes it to fracture.

The joints formed by the facets of adjacent vertebrae (termed facet, intervertebral, or apophyseal joints) are true synovial joints with cartilaginous surfaces. The cervical facets, oriented at a 45-degree angle to the transverse plane (Fig. 11–4), guide motion but do not limit it. In this respect they are unlike the lumbar facets, whose 90-degree orientation to the transverse plane significantly restricts lumbar spine motion (see Fig. 10–6). The ball-shaped uncinate processes also play a role in guiding and limiting

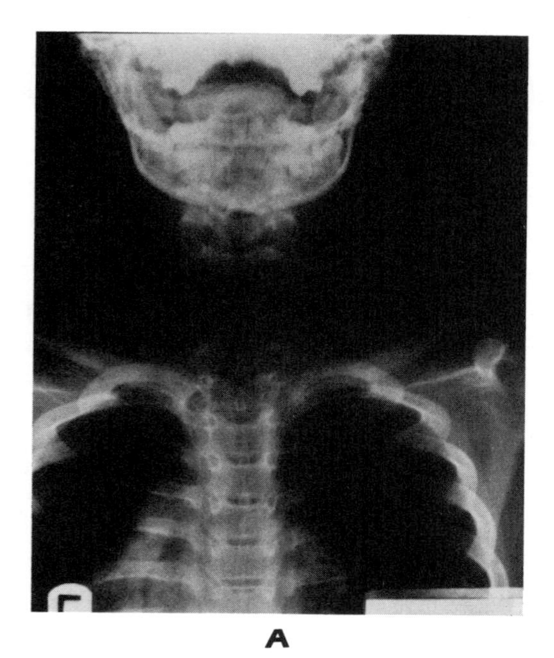

A

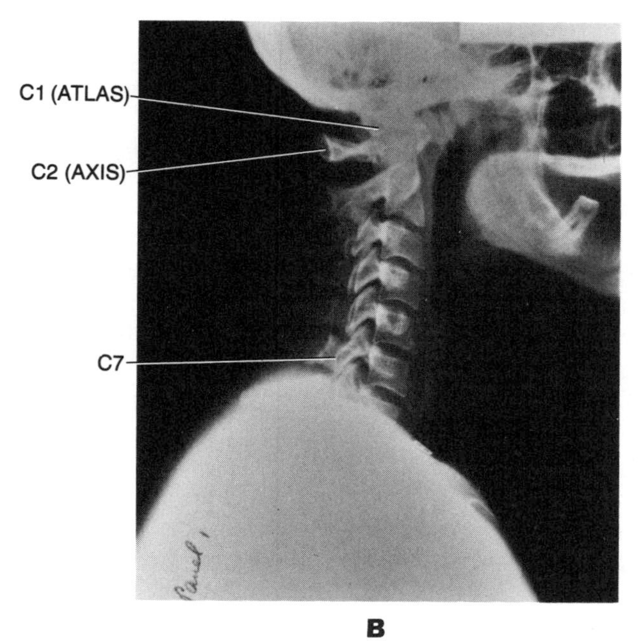

B

FIG. 11–1
Anteroposterior **(A)** and lateral **(B)** roentgenograms of the cervical spine. (Courtesy of Alex Norman, M.D.)

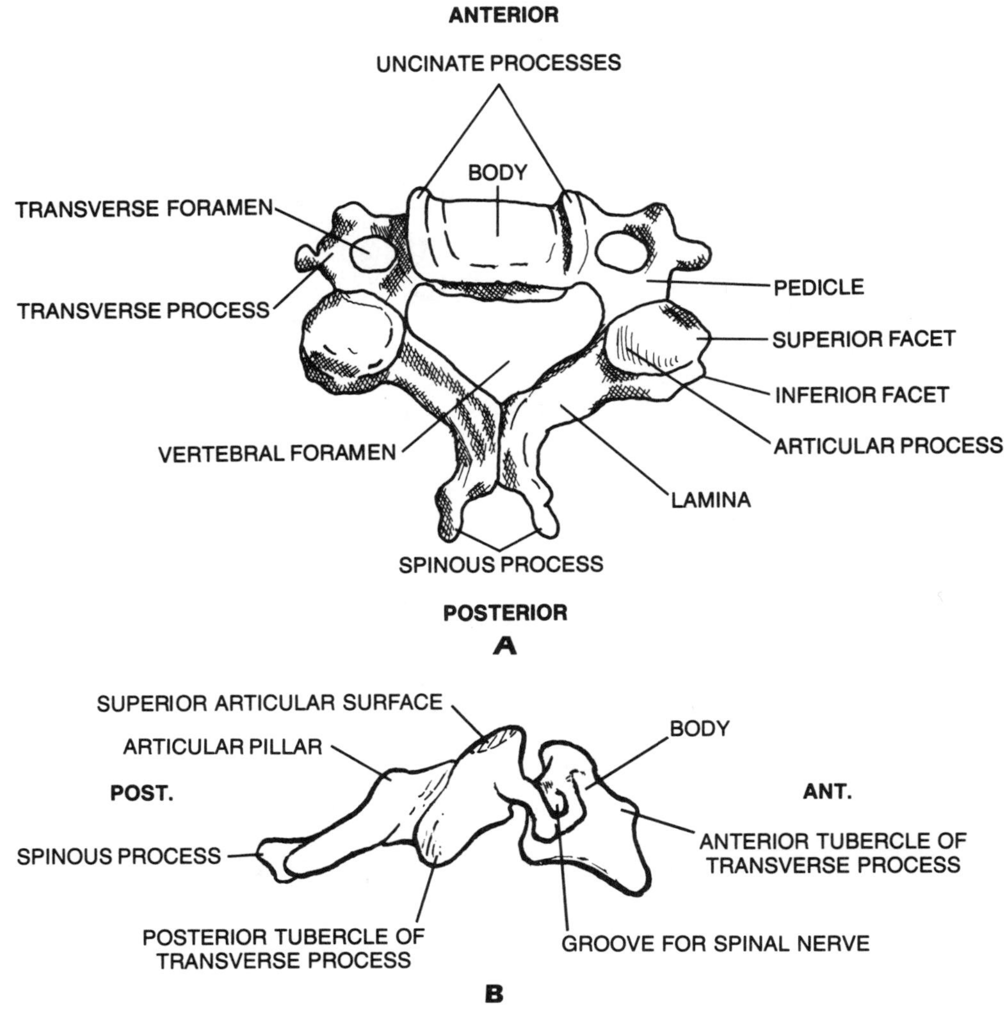

FIG. 11–2
Schematic drawings of a typical cervical vertebra, representative of C3–C6 (C7, the vertebra prominens, differs slightly in that it has a prominent nonbifid spinous process). The grooving of each transverse process superiorly accommodates the exiting spinal nerve (not pictured). **A.** Superior view. **B.** Lateral view.

cervical motion, but this role and its significance await further definition. For many years the joints of Luschka, situated between each uncinate process and the inferolateral surface of the superimposed vertebral body, were thought to be true synovial joints; however, numerous embryologic and developmental studies have shown that these joints are not present early in life and appear to be the product of degenerative fibrotic changes associated with aging (Hirsch et al., 1967).

The vertebral foramen has a rounded triangular shape, and its anteroposterior diameter is two thirds larger than that of the spinal cord, allowing motion of the neurovascular structures without compression. At the cervical level the canal also contains an epidural fat sheath (scanty at this level), the spinal cord sheaths, spinal nerves, and vascular plexuses (Fig. 11–5).

The Atypical Cervical Vertebrae—C1 and C2

The atypical vertebrae differ from the others in height, shape, and function (Fig. 11–6). C2 derives the name "axis" from the dens, an articular process that protrudes superiorly from the vertebral body and around which C1 rotates. Near the anterior tip of the dens lies an oval facet that articulates with C1 as a small synovial joint. At the top of the dens lie attachments for the apical ligament and two alar ligaments, which extend to the occipital bone and

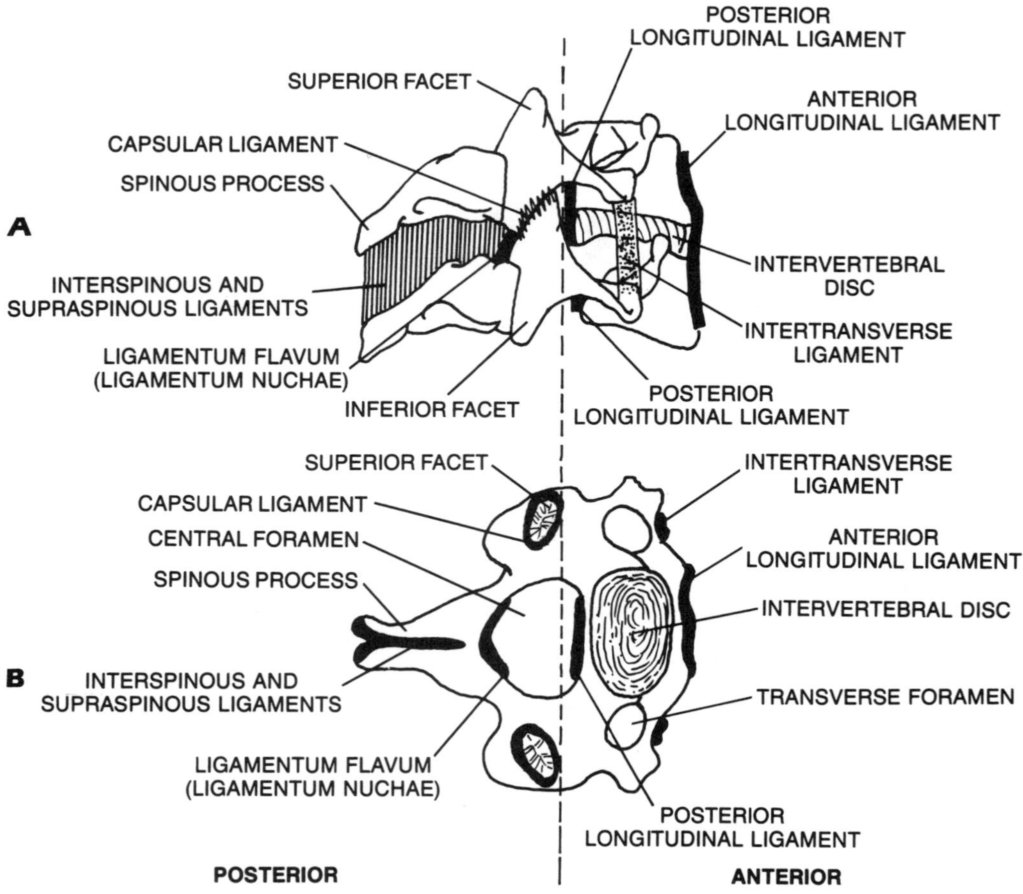

FIG. 11–3
Schematic representations of a cervical motion segment composed of two typical cervical vertebrae (C6 and C7), the intervertebral disc, and surrounding ligaments. The broken line divides the motion segment into anterior and posterior components. **A.** Lateral view. **B.** Superior view. (Adapted from White et al., 1975.)

limit rotation of the head. The inferior facets are similar to those of C3 to C7, but the superior facets have large oval convex surfaces. This convexity causes longitudinal (vertical) distraction of C1 and the head during rotation of C2 in either direction from the neutral position.

The transverse processes of C2 are small, and the two transverse foramina are oriented so that each vertebral artery is directed laterally as it extends to the protruding transverse process of C1. As each artery emerges from the corresponding C1 transverse foramen, it turns and runs posteromedially, entering the skull at the foramen magnum.

C1, the atlas, has no body but is composed of a ring within which an oval fossa articulates anteriorly with the dens of C2. The inferior facets correspond reciprocally to the superior facets of C2; both sets are parallel to the transverse plane, allowing significant rotation around a longitudinal axis. The superior facets of C1, which form the base of the atlanto-occipital joint, bear the weight of the skull, lending the atlas its name. These large, oval, semicircular facets limit the motion of the skull, particularly in rotation.

THE LIGAMENTS

The strong anterior longitudinal ligament traverses the central spine and attaches superiorly to the basilar skull (see Fig. 11–3). It is firmly attached to the ventral periosteum of the vertebral bodies and is more loosely attached to the intervertebral discs. The posterior longitudinal ligament lies on the dorsal aspect of the vertebral bodies inside the spinal canal; it continues above the axis as the tectorial membrane (see Figs. 11–3 and 11–6C). Firmly attached to the discs, it is separated from the vertebral bodies by the networks of veins and arteries that enter and leave each body.

The ligamentum flavum, composed of bands of elastic fibers attaching to each lamina, lies within the posterior spinal canal (see Fig. 11–3). It begins at the sacrum and ends between the laminae of C2 and C3. In the cervical spine the ligamentum flavum is called the ligamentum nuchae. Because of its high elastic fiber content it is considered the most elastic ligament in the body. Other posterior ligaments include the capsular ligaments, arranged at right angles to the surface of each facet joint; the interspinous ligaments; and the intertransverse ligaments, which are sparse in the cervical region. As it ascends the cervical spine, the interspinous ligament blends into the supraspinous ligament, which becomes very dense, forming a tough band called the nuchal ligament just beneath the skin.

The cruciform ligament is essential to stability at the C1–C2 articulation. Located within the spinal canal and ventral to the posterior longitudinal liga-

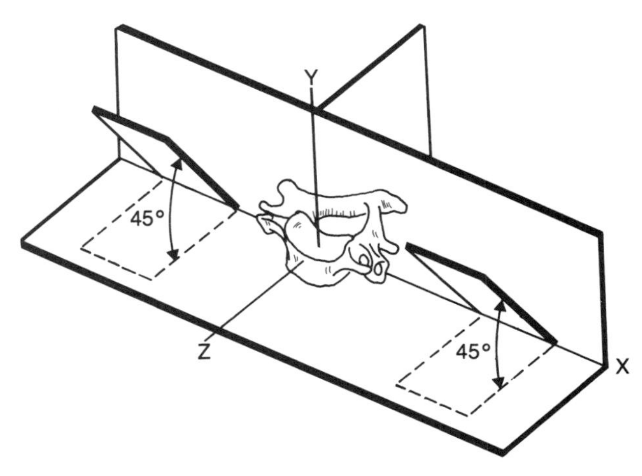

FIG. 11–4
Orientation of the facets of a typical cervical vertebra in three planes. The facets are oriented at a 45-degree angle to the transverse plane, are parallel to the frontal plane, and are at right angles to the sagittal plane. Y indicates the craniocaudal axis, z the anteroposterior axis, and x the mediolateral axis. (Adapted from White and Panjabi, 1978.)

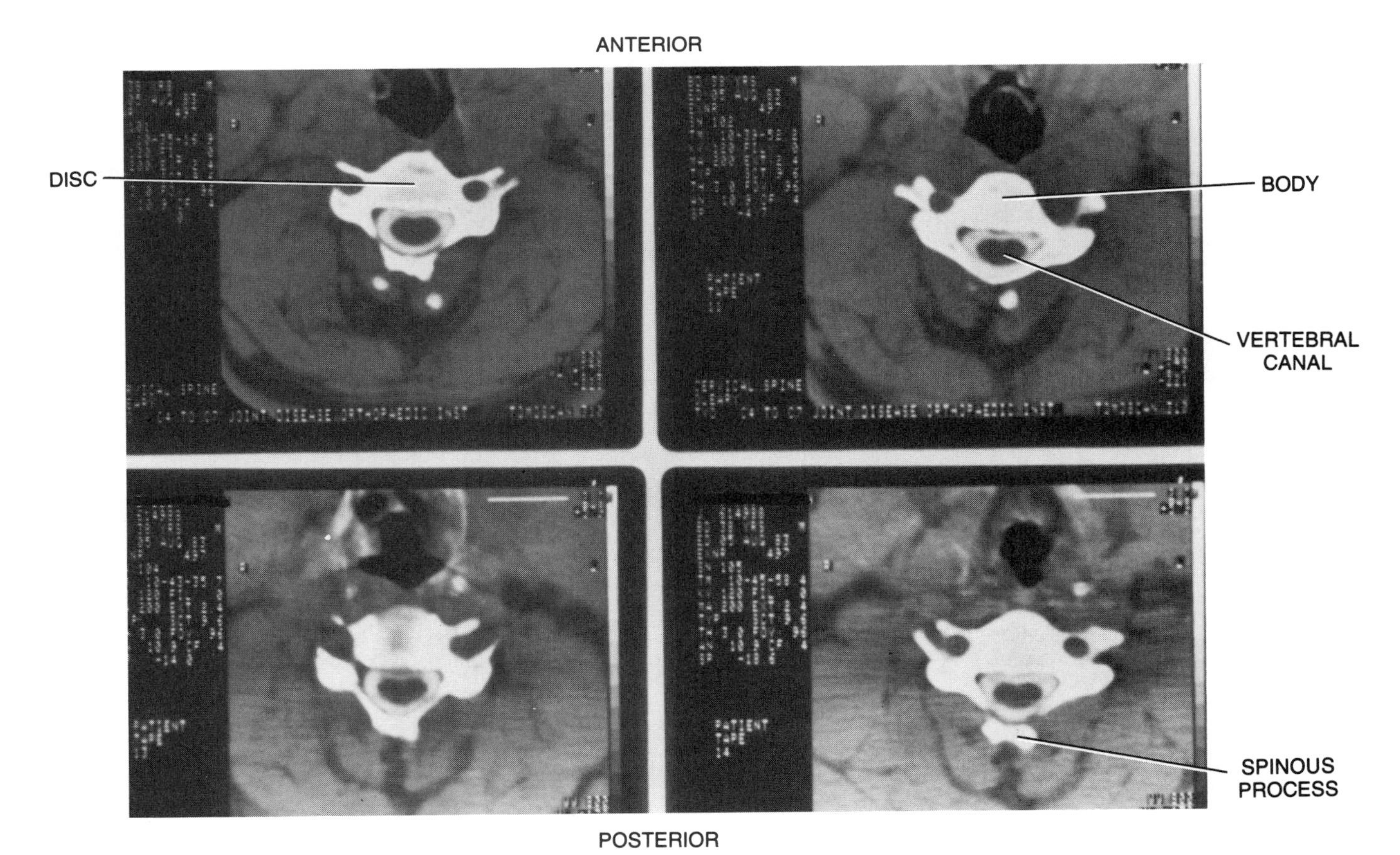

FIG. 11–5
Computed tomograms of the C5 level of the spine, showing the structures in and surrounding the vertebral canal. (Courtesy of Alex Norman, M.D.)

C1—THE ATLAS
ANT.
ANTERIOR TUBERCLE
ANTERIOR ARCH
FOSSA FOR DENS OF AXIS
SUPERIOR FACET
OUTLINE OF DENS OF AXIS
TRANSVERSE PROCESS
TRANSVERSE FORAMEN
VERTEBRAL FORAMEN
OUTLINE OF TRANSVERSE LIGAMENT
POSTERIOR ARCH
POST.
A

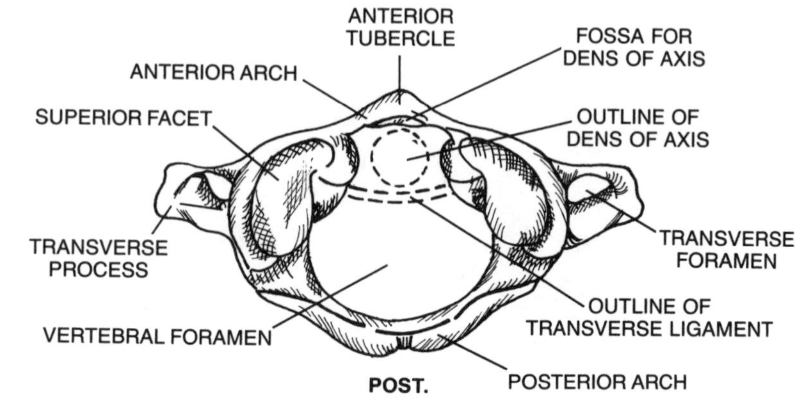

ANTERIOR
POSTERIOR
SUPERFICIAL LAYER OF TECTORIAL MEMBRANE
APICAL LIGAMENT OF DENS
TECTORIAL MEMBRANE
ANTERIOR ARCH OF ATLAS
OCCIPITAL BONE
POSTERIOR ATLANTO-OCCIPITAL MEMBRANE
DENS OF AXIS
TRANSVERSE LIGAMENT
VERTEBRAL ARTERY
POSTERIOR ARCH OF ATLAS
ANTERIOR ATLANTOAXIAL LIGAMENT
POSTERIOR ATLANTOAXIAL LIGAMENT
AXIS
INTERVERTEBRAL FIBROCARTILAGE
ARCH OF ATLAS
ANTERIOR LONGITUDINAL LIGAMENT
POSTERIOR LONGITUDINAL LIGAMENT
BODY OF C3
C

FIG. 11–6
The atypical cervical vertebrae. **A.** Superior view of C1, the atlas. The positions of the dens of C2 and the transverse ligament are outlined. **B.** Superior view of C2, the axis. **C.** Median sagittal section through the occipital bone and the first three cervical vertebrae showing the articulations and surrounding ligaments.

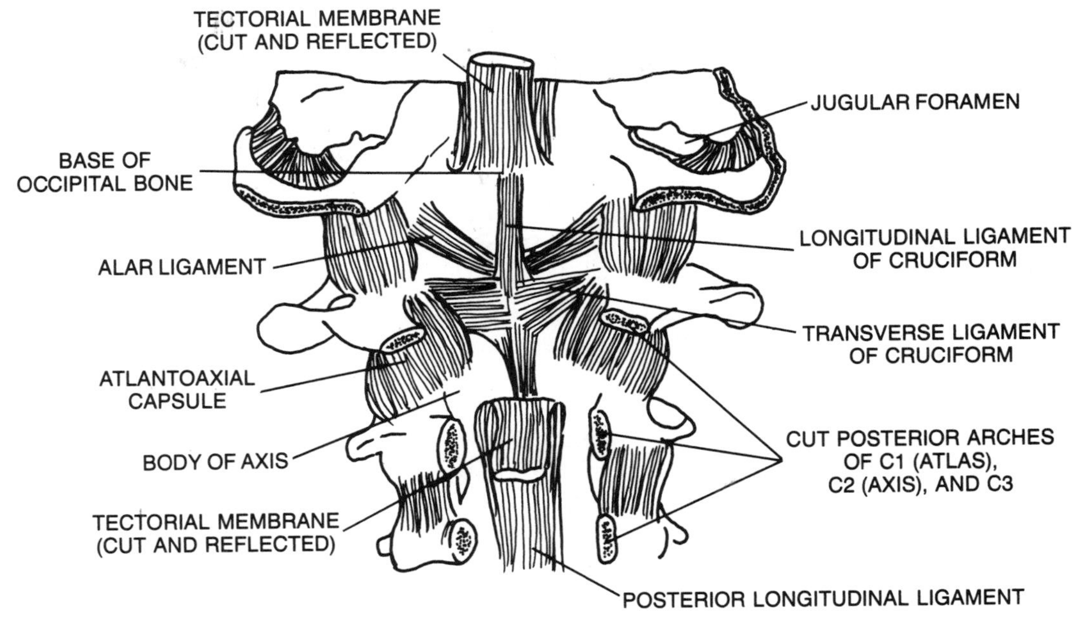

FIG. 11–7
Posterior view of the cervical spine from the occipital bone to C3 showing the cruciform ligament after the posterior arches of C1 and C2 have been removed and the tectorial membrane has been cut and reflected. The cruciform ligament consists of a longitudinal and a transverse ligament, both of which lie posterior to the dens and its alar and apical (hidden) ligaments. (Adapted from Gray, 1985.)

ment, it consists of two structures: a strong transverse ligament that attaches bilaterally to C1 and a longitudinal ligament that attaches inferiorly to the body of C2 and superiorly to the rim of the foramen magnum (Fig. 11–7). As may be surmised, the orientation of the cruciform ligament prevents posterior translation of the dens within the ring of C1.

THE INTERVERTEBRAL DISCS

The intervertebral discs form specialized symphysis joints between the cartilaginous end-plates of adjacent vertebral bodies. Each disc consists of a central gelatinous mass, the nucleus pulposus, surrounded by a tough outer covering, the annulus fibrosus. The discs of the cervical spine are slightly smaller laterally than their corresponding vertebral bodies, in contrast to those in other regions, which generally conform to the size of adjacent bodies. Like the lumbar discs, the cervical discs are thicker ventrally than dorsally; their wedge shape contributes to the lordotic curvature of the cervical spine. (The structure and function of the discs are discussed extensively in Chapter 10.)

KINEMATICS

The cervical spine is the most mobile region of the spine, affording the head a large range of motion, which in general is greater in children than in adults. Increased ligamentous laxity in adults suffering from rheumatoid arthritis makes the range of motion greater in these patients than in the general population.

RANGE OF MOTION

Several investigators have analyzed normal cervical motion in vivo and in vitro by means of roentgenography and cineroentgenography (Fielding, 1957; Lysell, 1969; White and Panjabi, 1978). From these findings some general statements can be made about the type and range of motion at each level of the cervical spine. Motion at the atlanto-occipital articulation consists of 10 to 15 degrees of flexion and extension and 8 degrees of lateral flexion. The joint configuration precludes axial rotation, and any rotational pull exerted on the skull by the neck muscles is transformed into motion at the C1–C2 articulation (the atlantoaxial joint). At C1–C2, the most mobile segment of the spine, 47 degrees of axial rotation can occur, representing 50% of the axial rotation in the cervical spine as a whole. About 10 degrees of flexion and extension takes place but little or no lateral flexion (White and Panjabi, 1978).

Below C2, flexion, extension, lateral flexion, and axial rotation occur between each vertebra. Representative values for the ranges of these motions can be given for C3–C7 as a whole, but it should be noted that wide variations exist among individuals. Approximately 90 degrees of axial rotation takes place in C3–C7, about 45 degrees to each side of neutral. Even greater lateral flexion is possible: about 49 degrees to each side of neutral, giving a total of approximately 98 degrees. The range of flexion and extension is approximately 64 degrees, about 24 degrees of extension and 40 degrees of flexion. The motion in each plane is fairly evenly distributed throughout the motion segments. Degenerative changes have little effect on the range of motion of C3–C7 in any plane (Lysell, 1969).

The combined motion of all segments of the cervical spine produces a remarkably large range of motion—about 145 degrees of flexion and extension, about 180 degrees of axial rotation, and approximately 90 degrees of lateral flexion. The great flexibility of the cervical spine allows the head to be positioned in a wide variety of ways, permitting one, with equal ease, to gaze at an airplane overhead, glance over one's shoulder, or look for an object under a table. (The range of motion in three planes for each level of the spine is presented in Fig. 10–7.)

COUPLED MOTION

C1 can move independently of the remaining cervical spine, but motion below C1 involves the entire cervical spine because the vertebrae are attached to each other functionally in motion segments (see Fig. 11–3). The facets guide the motion, referred

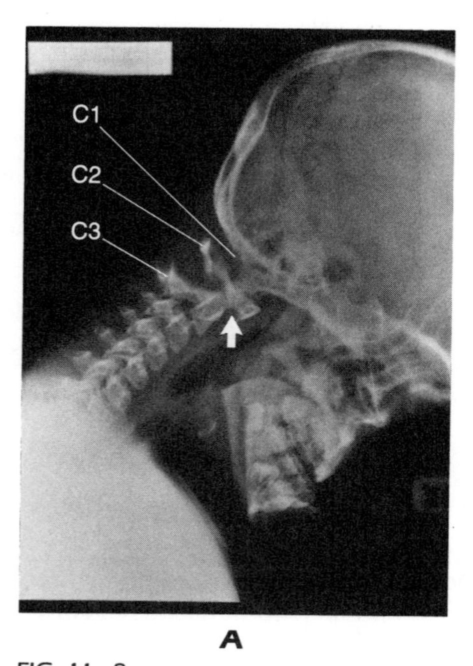

A

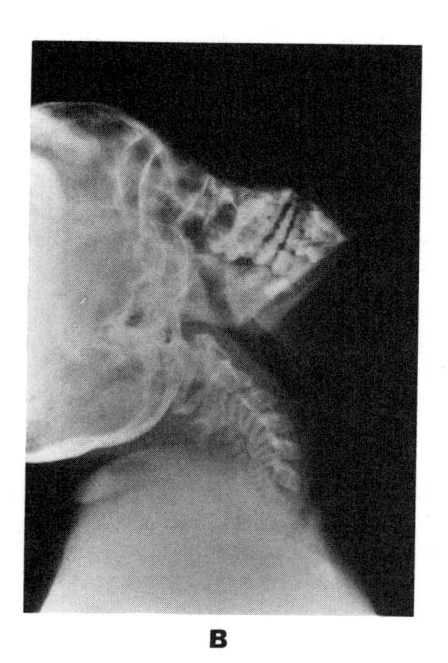

B

FIG. 11–8
Coupling of flexion-extension with transverse translation of the cervical spine is visible roentgenographically. During flexion the vertebral body shifts forward; the facets glide up and over one another with subluxation at full flexion (arrow) **(A).** During extension the reverse occurs, and the spinous processes limit motion as they touch at full extension **(B).** The size of the intervertebral foramina increases with flexion and decreases with extension (Fielding, 1957). Approximately 2.5 to 4.0 mm of transverse translation occurs at the C1–C2 articulation during flexion-extension; an appreciation of the substantial range of this motion under normal conditions is important for a proper evaluation of the integrity of the cruciform ligament following trauma. (Courtesy of Alex Norman, M.D.)

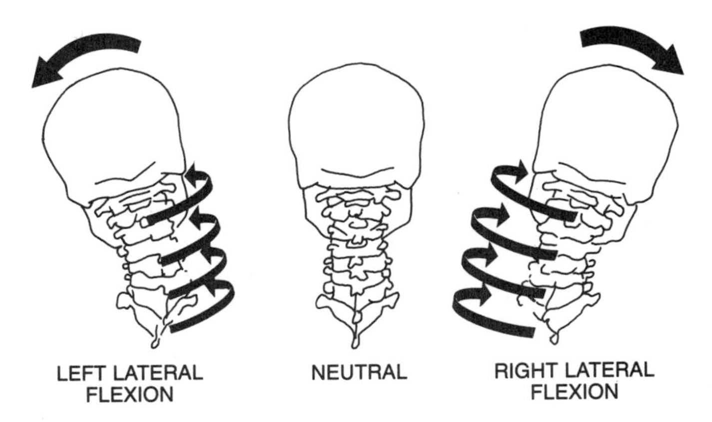

FIG. 11–9
Coupled motion during lateral flexion is depicted schematically. When the head and neck are flexed to the left the spinous processes shift to the right, indicating rotation. The converse is also illustrated. (Adapted from White and Panjabi, 1978.)

to as coupled motion because each type of motion is always accompanied by another. Flexion-extension is coupled with transverse (horizontal) translation (Fig. 11–8), lateral flexion with rotation (Fig. 11–9), and rotation with axial (vertical) translation (Fig. 11–10).

SURFACE JOINT MOTION

The motion between the joint surfaces of two adjacent vertebrae may be analyzed by means of the instant center technique of Reuleaux, described in detail in Chapter 6. The method may be used to analyze surface motion of the cervical spine during flexion-extension and lateral flexion, but not rotation.

In a normal cervical spine the instant center of flexion-extension is located in the anterior part of the lower vertebra in each motion segment. Instant center analysis indicates that tangential motion (gliding) takes place between the facet joints as the cervical spine is flexed and extended (Fig. 11–11).

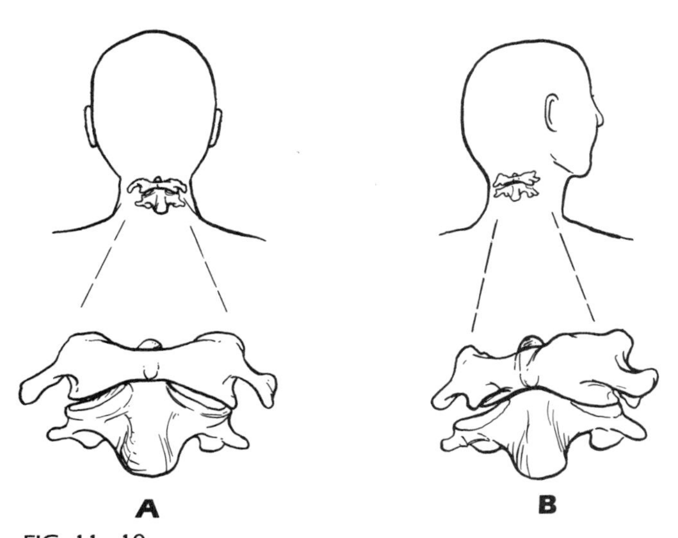

FIG. 11–10
Coupling of rotation and axial translation is depicted schematically. **A.** C1 and C2 are in the neutral position. **B.** "Telescoping" of C1 on C2 occurs as the head is rotated to the right. (Adapted from Fielding, 1957.)

The instant center of motion of the cervical spine may be displaced as a result of pathologic processes such as disc degeneration or ligament impairment. In such a case, instant center analysis may reveal, instead of gliding, distraction and jamming (compression) of the facet joint surfaces during flexion-extension (Fig. 11–12).

STABILITY OF THE CERVICAL SPINE

Spinal instability—loss of the spine's ability to limit excessive motion and displacement—has been addressed extensively by White and coworkers (1975) and by Panjabi and associates (1976). In White's group's investigation of cervical spine stability, loads simulating flexion and extension were imposed on cadaver motion segments from which the soft tissues and facets had been sequentially transected from posterior to anterior or vice versa (anterior and posterior components of the motion segment are shown in Fig. 11–3). The amount of rotation in the sagittal plane (flexion-extension) and the amount of horizontal (transverse) translation were calculated after each individual component was transected. The results may be summarized as follows:

1. It was generally found that a majority of the ligaments had to be transected before failure occurred. The motion segments did not demonstrate orderly incremental deformation with removal of each additional structure; small increments of change were followed without warning by sudden, abrupt failure.

2. Flexion was limited more effectively by the posterior ligaments than by the anterior ligaments.
3. The reverse was true for extension; the anterior ligaments provided more stability than did the posterior ligaments.
4. The facets played a significant role in the mechanics of the motion segment. Removal of the facets resulted in a decrease in angular displacement and an appreciable increase in horizontal displacement.

On the basis of their findings, White's group (1975) suggested that the adult cervical spine is unstable, or on the brink of instability, when any of the following conditions is present: all of the anterior or all of the posterior elements are destroyed or unable to function; more than 3.5 mm of transverse (horizontal) displacement of one vertebra in relation to an adjacent one is measured on lateral roentgenograms (resting or flexion-extension); more than an 11-degree difference in rotation from that of either adjacent vertebra is measured on a resting lateral or flexion-extension roentgenogram.

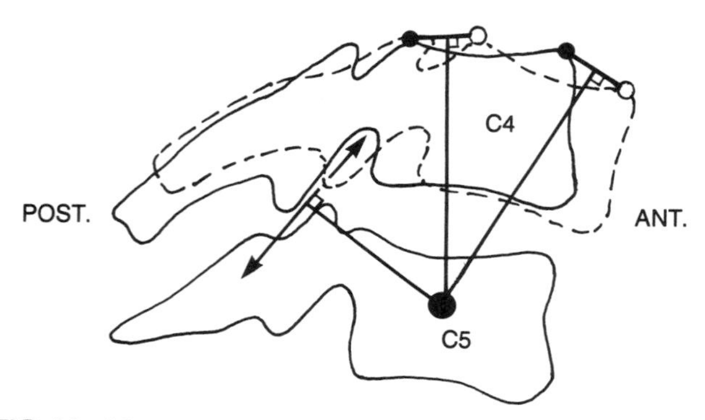

FIG. 11–11
Analysis of surface motion of the facet joints of the C4–C5 motion segment during flexion-extension. The schematic drawing represents superimposed roentgenograms of the motion segment in the neutral position and in slight flexion. The upper vertebra (C4) is considered to be the moving body, and the subjacent vertebra (C5) is the base vertebra. Two points have been identified and marked on the moving body in the neutral position (solid outline of C4), and the same two points have been marked on the second roentgenogram with the motion segment slightly flexed (dashed outline of C4). Lines connecting the two sets of points have been drawn, and their perpendicular bisectors have been added. The intersection of the perpendicular bisectors identifies the instant center of motion (large solid dot) for the degree of flexion under study. The perpendicular bisector (arrowed line) of a line drawn from the center of motion to the contact point of the facet joint surfaces indicates tangential motion, or gliding.

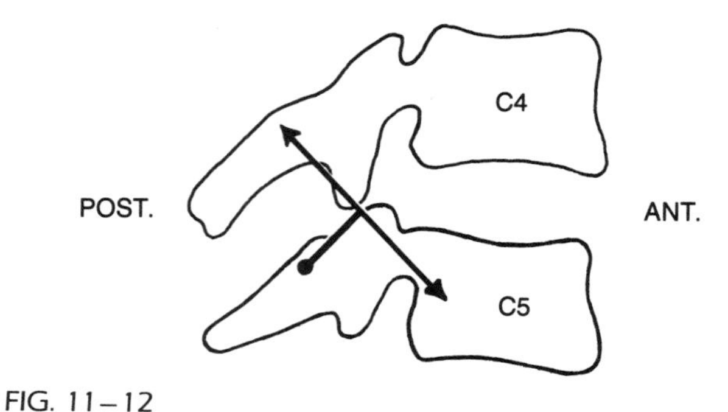

FIG. 11–12
Schematic drawing representative of a roentgenogram of the C4–C5 motion segment of a patient injured in a rear-end auto collision. The instant center of flexion-extension at this level (represented by the large solid dot) has been displaced from the anterior to the posterior part of C5 as a result of the injury process, which impaired the ligaments (compare Fig. 11–11). The analysis of surface motion shows compression and distraction of the facet joints with flexion and extension.

Such guidelines for evaluating clinical stability of the cervical spine may have specific applications in given cases. For example, if in examining a patient with a flexion injury a physician recognizes that the posterior ligament complexes may not be intact, precautions can be taken to prevent flexion of the neck during evaluation and treatment. If a surgical procedure is indicated, the approach should preserve the integrity of the anterior ligaments and remaining posterior ligaments lest damage to these structures render the spine completely unstable in both flexion and extension.

KINETICS

Loads on the cervical spine are produced mainly by the weight of the head, the activity of the surrounding muscles, the inherent tension of adjacent ligaments, and the application of external loads. Although few investigations have been conducted in vivo, existing studies confirm the obvious fact that the physiologic loads on this region are generally lower than those on the thoracic and lumbar spine. Still, in some parts of the world traditional practices of carrying objects on the head impose high loads on the cervical spine (Fig. 11–13).

As in the lumbar and thoracic spine, the static loads on the cervical spine vary with the position of the head and body. These loads are minimal during well-supported reclining and remain relatively low during upright relaxed standing and sitting. Rotation and lateral flexion of the cervical spine increase the loads only moderately. A more significant rise in these loads occurs during extreme flexion and extension. In the following sections, the function of various elements of the cervical spine during loading is described and the static and dynamic loads in several situations are examined.

FIG. 11–13
Traditional methods of transporting loads on the head can impose large loads on the cervical spine.

THE ROLE OF VARIOUS COMPONENTS OF THE CERVICAL SPINE DURING LOADING

The ligaments, discs, and different elements of the vertebrae influence the load-bearing characteristics of the cervical spine. Each vertebrae has a surface of cortical bone and a core of cancellous bone. The core forms a column-and-brace system of trabeculae that supports the surface (Whitehouse and Dupon, 1971; White and Panjabi, 1978) (see Figs. 1–2 and 1–3). In persons under 40 years of age, the surface and the core support approximately equal loads, but in persons over 40 years the cortical surface carries up to 65% of the load on the vertebra (Rockoff et al., 1969).

Because the trabecular surfaces are more actively involved in mineral exchange, the cancellous bone is resorbed to a greater extent than is the cortical bone (Walker, 1977). Increased resorption in elderly persons decreases the strength of the cancellous core, thereby reducing the load it can sustain before failing as well as the proportion of the total load it can carry (see Fig. 1–49). Although resorption at the cervical level is less pronounced than that in the lower segments, it can be extensive enough to weaken the resistance of the cervical spine to trauma.

Distribution of loads throughout the vertebra has been the subject of extensive investigation. Results of studies in which different experimental constructs were used indicate that the facets bear a portion of the load (King et al., 1975; Miller et al., 1978) and that as the spine moves from flexion into extension this portion increases to 33% of the total load (King et al., 1975), reducing the load sustained by the disc.

The properties of the ligamentum flavum (ligamentum nuchae in the cervical spine) allow the spine to both flex and extend; after lengthening during extension this ligament can shorten during flexion without ballooning into the spinal canal and impinging on the cord. Intradiscal pressure measurements in the lumbar spine (Nachemson and Morris, 1964) have shown that even with the spine in a neutral position the ligamentum flavum applies compressive forces to the intervertebral disc by virtue of its inherent tension (i.e., it prestresses the disc). Although to date no such investigations have been made in the cervical spine, it is assumed that a similar pattern of prestress on the disc exists.

Few studies have examined the torsional properties of the cervical intervertebral discs, but the alternating orientation of the collagen fiber bundles would appear to play a role in controlling axial rotation (Fig. 11–14). Indeed, disc fiber orientation must be a factor in controlling rotation because the cervical facets (excluding the joint surfaces between C1 and the occipital bone) do little to resist this motion. By comparison, the rib cage in the thoracic spine and the facets in the lumbar spine significantly control axial rotation (Markolf, 1972).

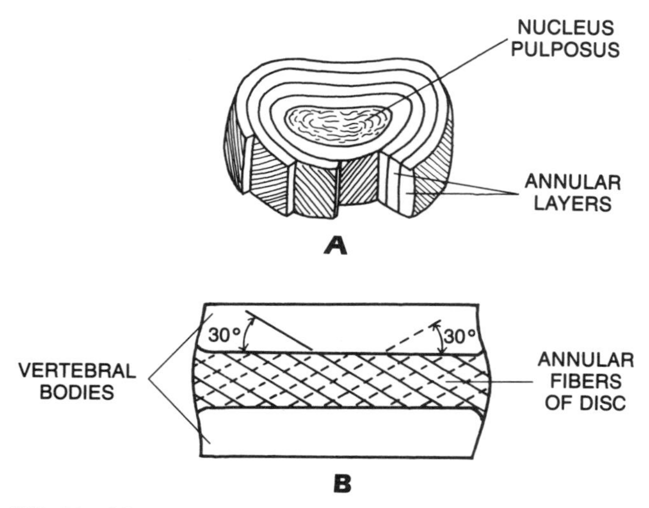

FIG. 11–14
Schematic drawings of an intervertebral disc showing the crisscross arrangement of its fibers. **A.** Concentric layers of the annulus fibrosus are depicted as cut away to show the alternating orientation of the collagen fibers. **B.** The layers of the annular fibers are oriented at a 30-degree angle to the vertebral body and at 120-degree angles to each other. (Adapted from White and Panjabi, 1978.)

STATIC LOADS ON THE CERVICAL SPINE WITH VARIOUS POSITIONS OF THE HEAD

Substantial loads on the cervical spine have been calculated during neck flexion, particularly in the lower cervical motion segments. Harms-Ringdahl (1986) calculated the bending moments generated around the axes of motion of the atlanto-occipital joint and the C7–T1 motion segment in seven subjects with the neck in five positions: full flexion, slight flexion, neutral, head upright with the chin tucked in, and full extension (Fig. 11–15).

The load on the junction between the occipital bone and C1 was lowest during extreme extension (ranging from an extension moment of 0.4 Nm to a flexion moment of 0.3 Nm). It was highest during extreme flexion (0.9 to 1.8 Nm) but showed only a slight increase over that produced when the neck was in the neutral position.

The load on the C7–T1 motion segment was low with the neck in the neutral position but became even lower when the head was held upright with the chin tucked in (ranging from an extension moment of 0.8 Nm to a flexion moment of 0.9 Nm). The load increased somewhat during extreme extension (ranging from 1.1 to 2.4 Nm) and substantially during slight flexion (reaching 3.0 to 6.2 Nm). The greatest loads were produced during extreme flexion, with moments ranging from 3.7 to 6.5 Nm.

Harms-Ringdahl (1986) used surface electrode electromyography to record the activity over the erector spinae muscles of the cervical spine with the neck in the five positions described above. Interestingly, the values obtained showed very low levels of muscle activity for all positions, even extreme flexion, during which the flexion moment on the C7–T1 motion segment increased more than threefold over the neutral position. The fact that the EMG levels over the neck extensors were low in this and other studies (Fountain et al., 1966; Takebe et al., 1974) suggests that the flexing moment is balanced by passive connective tissue structures such as the joint capsules and ligaments.

The values for the moments computed by Harms-Ringdahl (1986) are about 10% of the maximal values measured by Moroney and Schultz (1985) in 14 male subjects who resisted maximal and submaximal loads against the head while in an upright sitting position. The mean maximal voluntary moments were 10 Nm during axial rotation of the cervical spine, 12 to 14 Nm during flexion and lateral bending, and 30 Nm during extension. Calculation of the maximum (compressive) reaction forces on the C4–C5 motion segment ranged from 500 to 700 N during flexion, rotation, and lateral bending and rose to 1,100 N during extension. Anteroposterior and lateral shear forces reached 260 N and 110 N, respectively. Calculated moments and forces generally correlated well with mean measured myoelectric activities at eight sites around the perimeter of the neck at the C4 level.

STATIC LOADS ON THE CERVICAL SPINE DURING TRACTION TREATMENT

Cervical traction is commonly used to treat conditions causing neck pain. In this situation traction treatment is based on the principle of applying a tensile force to stretch the soft tissues of the posterior cervical spine and increase the size of the intervertebral foramina, thereby relieving compression of the nerve root.

Since normal neck flexion causes distraction of the spinous processes of the motion segment and enlargement of the intervertebral foramina (Fielding, 1957; Lysell, 1969) (see Fig. 11–8), several investigators have stressed the importance of applying traction with the neck flexed to enhance its effect (Cailliet, 1964; Colachis and Strohm, 1965; Crue and Todd, 1965). The magnitude of the flexing moment, or

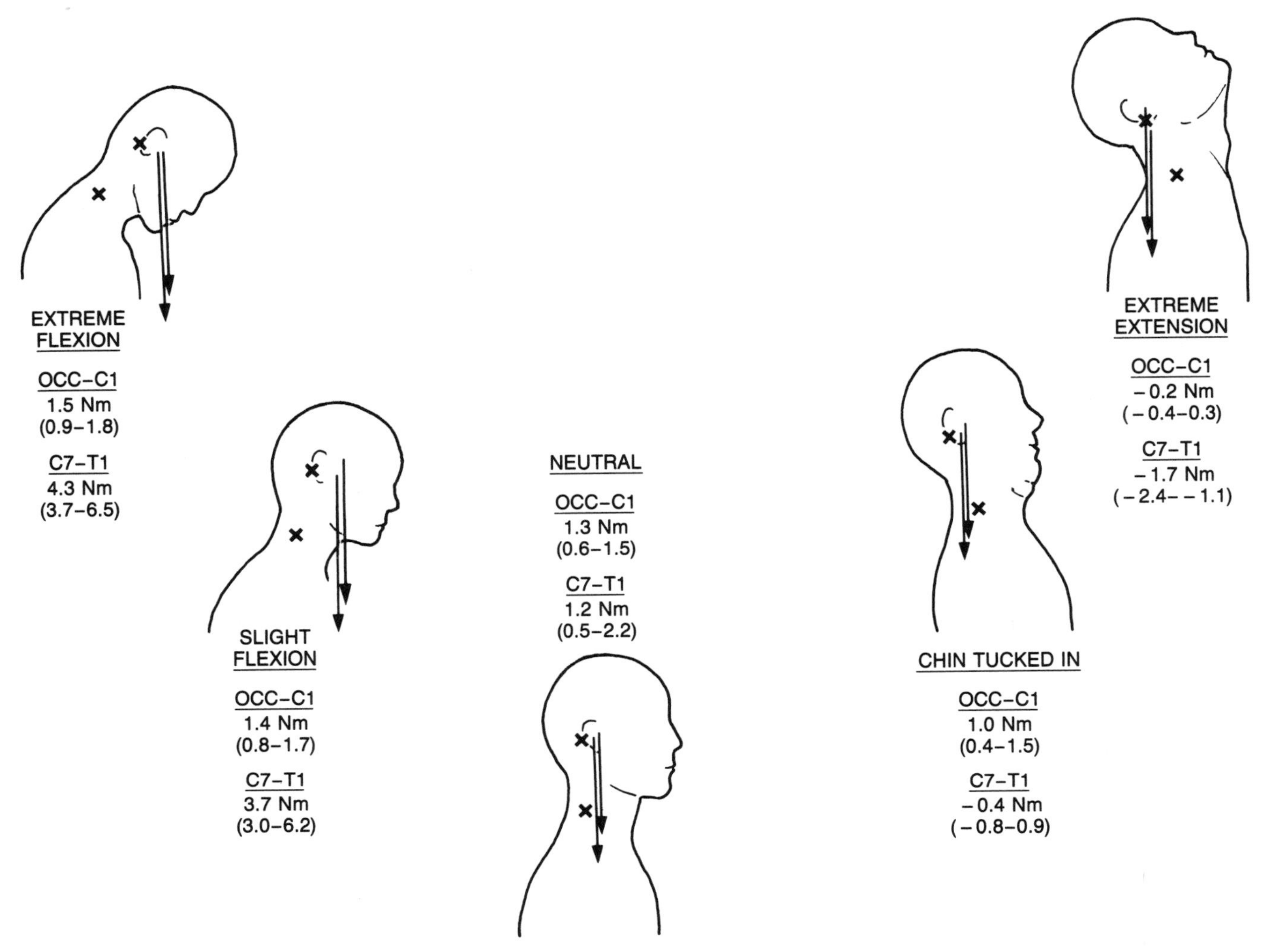

FIG. 11–15
Extension and flexion moments around the axes of motion of the atlanto-occipital (OCC-C1) joint and the C7–T1 motion segment (marked with X's) are presented for five positions of the head: extreme flexion, slight flexion, neutral, head upright with the chin tucked in, and extreme flexion. Values shown are the median and range for seven subjects; negative values indicate extension moments. The arrows represent the force vectors produced by the weight of the head. (Adapted from Harms-Ringdahl, 1986.)

torque, produced by such traction depends on the magnitude of the traction force and the length of its moment arms relative to the centers of motion in the atlanto-occipital joint and in the cervical motion segments (Moritz, 1975; Wiktorin and Nordin, 1986).

Some systems of cervical traction are adjustable, allowing considerable variation in the length of the chin strap and the direction of the traction force. Lengthening the chin strap and adjusting the direction of pull of the traction straps so that the point of application of the traction force falls dorsal to the center of rotation in the atlanto-occipital joint assures that the neck is flexed (Fig. 11–16A). On the contrary, a shortened chin strap and an adjustment of the traction straps that causes the point of application to fall ventral to the center of rotation pull the neck into extension, diminishing the effect of the traction treatment (Fig. 11–16B) (Wiktorin and Nordin, 1986).

DYNAMICS OF THE CERVICAL SPINE DURING EXCESSIVE LOADING

The structure of the head and neck, which is essentially that of a large mobile ball atop a slender pivot, makes the cervical spine and surrounding soft

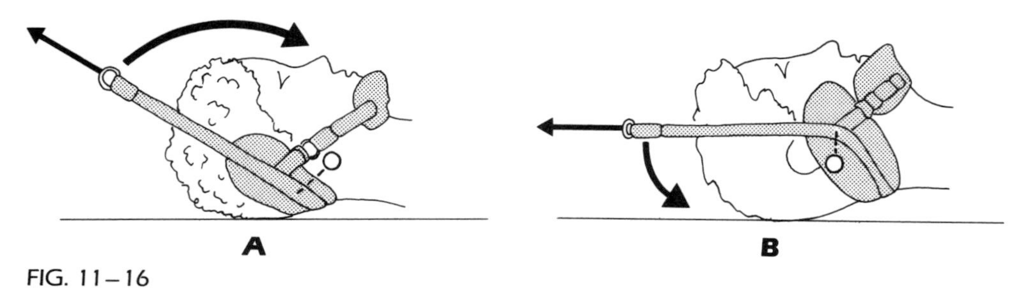

FIG. 11–16
Cervical traction with an adjustable system. **A.** The chin strap is lengthened and the direction of pull of the traction straps is such that the point of application of the traction force falls dorsal to the center of rotation in the atlanto-occipital joint (represented by the hollow circle). This adjustment produces a flexing moment, or torque, on the cervical spine, enhancing the effect of traction. The magnitude of the flexion moment on the atlanto-occipital joint depends on the magnitude of the traction force and the length of its moment arm (represented by the broken line). **B.** The chin strap is shortened and the adjustment of the traction straps causes the point of application of the traction force to fall ventral to the center of rotation. The cervical spine is pulled into extension and the effect of the traction treatment is diminished. (Adapted from Wiktorin and Nordin, 1986.)

tissues particularly vulnerable to dynamic injury. Excessive loading and motion of the relatively heavy head upon its "pivot" can easily create stresses and strains that exceed the strength of the stabilizing structures. The mechanisms responsible for excessive loading of the spine have received considerable attention because of the importance of preventing spine trauma and its grave consequences.

Flexion-extension injuries are the most common injuries of the cervical spine. These injuries often occur in youth as a consequence of diving into shallow water and even more frequently in persons of all ages as a result of vehicle collisions. The latter is the well-known whiplash injury.

The flexion-extension injury involves an impact that forces the head into flexion, causing disruption of the posterior ligaments. Depending on the exact orientation of the force, several types of injury may result (White and Panjabi, 1978; Gehweiler et al., 1980). If the force is mild and the head sustains purely forward flexion, the posterior ligaments may only be sprained. With a higher force these ligaments may rupture completely, allowing the facets of the upper vertebra of the motion segment to override those of the lower one and lock anteriorly when the spine recoils. The result is bilateral facet dislocation. Hyperflexion injuries involve an element of anterior compression that frequently results in wedging of the vertebral body. Alternatively, the inferior lip of the superior vertebral body may break off (a teardrop fracture), or if the force is severe, a comminuted fracture may result.

Excessive motion at the C1–C2 articulation can pose particular dangers to the cervical spine. At this level the spinal canal is encroached upon by the normal horizontal (transverse) translation that accompanies flexion-extension (see Fig. 11–8). During normal axial rotation the size of the canal is significantly decreased because the oval shape of the central foramen of C1 does not coincide with that of C2. Moreover, as the transverse process of C1 swings along its arc of rotation, the vertebral artery is stretched with it (Fielding, 1957). Should axial rotation be excessive, symptoms of vertebral artery insufficiency may occur, including dizziness, nausea, and transient ischemic attacks (Jernigan and Gardner, 1971). Because axial rotation takes place over the length of the cervical spine (about 50% at C1–C2 and the rest throughout C2–C7), stretching of the vertebral artery has less serious consequences at lower levels of the cervical spine than at C1–C2 unless complete dislocation occurs.

SUMMARY

1. The cervical spine is composed of two atypical vertebrae, C1 (the atlas) and C2 (the axis), and five typical vertebrae, C3–C7. The atlanto-occipital joint is also a functional part of the cervical spine.
2. The facets of C3–C7 are oriented at a 45-degree angle to the transverse plane, guiding motion but not limiting it.

3. The cruciform ligament is essential to the stability of the C1–C2 articulation, preventing posterior translation of the dens within the ring of C1.
4. The cervical spine is the most mobile region of the spine. Its total range of motion includes about 145 degrees of flexion-extension, about 180 degrees of axial rotation, and about 90 degrees of lateral flexion. The C1–C2 motion segment contributes 50% of the axial rotation.
5. Tangential motion (gliding) occurs between the facet joints during flexion and extension of a normal cervical spine.
6. The posterior ligaments provide more stability during flexion than do the anterior ligaments; the reverse is true during extension.
7. Loads on the cervical spine are relatively low during upright relaxed standing and sitting, and they increase only moderately during rotation and lateral flexion. A more significant rise in these loads occurs at the extremes of flexion-extension, particularly in the lower cervical motion segments.
8. Applying traction with the neck flexed enhances the effect of this treatment, since flexion causes distraction of the spinous processes of the motion segment and enlargement of the intervertebral foramina.
9. The most common injuries of the cervical spine are flexion-extension injuries, such as the whiplash injury sustained during a vehicle collision.

REFERENCES

Breasted, J. H.: The Edwin Smith Surgical Papyrus. Chicago, University of Chicago Press, 1930.

Cailliet, R.: Neck and Arm Pain. Philadelphia, F. A. Davis, 1964.

Colachis, S. C., and Strohm, B. R.: A study of tractive forces and angle of pull on vertebral interspaces in the cervical spine. Arch. Phys. Med., *46*:820, 1965.

Crue, B. L., and Todd, E. M.: The importance of flexion in cervical halter traction. Bull. Los Angeles Neurol. Soc., *30*:95, 1965.

Duff, C.: A Handbook on Hanging. London, The Bodley Head, 1938.

Fielding, J. W.: Cineroentgenography of the normal cervical spine. J. Bone Joint Surg., *39A*:1280, 1957.

Fountain, F. P., Minear, W. L., and Allison, R. D.: Function of longus colli and longissimus cervicis muscles in man. Arch. Phys. Med., *47*:665, 1966.

Gehweiler, J. A., Osborne, R. L., and Becker, R. F.: The Radiology of Vertebral Trauma. Philadelphia, W. B. Saunders, 1980.

Gray, H.: Anatomy of the Human Body. 30th American Ed. Edited by C. D. Clemente. Philadelphia, Lea & Febiger, 1985.

Harms-Ringdahl, K.: On assessment of shoulder exercise and load-elicited pain in the cervical spine. Biomechanical analysis of load—EMG—methodological studies of pain provoked by extreme position. Thesis, Karolinska Institute, University of Stockholm, 1986.

Hirsch, C., Schajowicz, F., and Galante, J.: Structural changes in the cervical spine: A study on autopsy specimens in different age groups. Acta Orthop. Scand., Suppl. *109*, 1967.

Jernigan, W., and Gardner, W.: Carotid artery injuries due to closed cervical trauma. Trauma, *11*:429, 1971.

King, A. I., Prasad, P., and Ewing, C. L.: Mechanism of spinal injury due to caudocephalad acceleration. Orthop. Clin. North Am., *6*:19, 1975.

Lysell, E.: Motion in the cervical spine. An experimental study on autopsy specimens. Acta Orthop. Scand., Suppl. *123*, 1969.

Markolf, K. L.: Deformation of the thoracolumbar intervertebral joint in response to external loads: A biomechanical study using autopsy material. J. Bone Joint Surg., *54A*:511, 1972.

Miller, M. D., et al.: Significant new observations on cervical spine trauma. Am. J. Roentgenol., *130*:659, 1978.

Moritz, U.: Traktionsbehandlingens mekaniska effekt på ryggraden. Sjukgymnasten, *12*:19, 1975.

Moroney, S. P., and Schultz, A. B.: Analysis and measurement of loads on the neck. Trans. Orthop. Res. Soc., *10*:329, 1985.

Nachemson, A., and Morris, J. M.: In vivo measurements of intradiscal pressure. Discometry, a method for the determination of pressure in the lower lumbar discs. J. Bone Joint Surg., *46A*:1077, 1964.

Panjabi, M. M., Hausfeld, J., and White, A. A.: Mechanical properties of the human thoracic spine as shown by three-dimensional load-displacement curves. J. Bone Joint Surg., *58A*:642, 1976.

Pope, O.: Decision at Trafalgar. Philadelphia, J. B. Lippincott, 1959.

Rockoff, S. D., Sweet, E., and Bleustein, J.: The relative contribution of trabecular and cortical bone to the strength of human lumbar vertebrae. Calcif. Tissue Res., *3*:163, 1969.

Takebe, K., Vitti, M., and Basmajian, J. V.: The functions of semispinalis capitis and splenius capitis muscles: An electromyographic study. Anat. Rec., *179*:477, 1974.

Walker, P. S.: Human Joints and Their Artificial Replacements. Springfield, Charles C Thomas, 1977, pp. 110–166.

White, A. A., and Panjabi, M. M.: Clinical Biomechanics of the Spine. Philadelphia, J. B. Lippincott, 1978.

White, A. A., et al.: Biomechanical analysis of clinical stability in the cervical spine. Clin. Orthop., *109*:85, 1975.

Whitehouse, W. J., and Dupon, E. D.: The scanning electron microscope studies of trabecular bone from a human vertebral body. J. Anat., *108*:481, 1971.

Wiktorin, C. v. H., and Nordin, M.: Introduction to Problem Solving in Biomechanics. Philadelphia, Lea & Febiger, 1986, pp. 145–170.

SUGGESTED READING

Aufdermaur, M.: Spinal injuries in juveniles. J. Bone Joint Surg., *56B*:513–519, 1974.

Beatson, T. R.: Fractures and dislocation of the cervical spine. J. Bone Joint Surg., *45B*:21–35, 1963.

Belytschko, T. B., Andriacchi, T., Schultz, A., and Galante, J.: Analog studies of forces in human spine: Computational techniques. J. Biomech., *6*:361–371, 1973.

Berkin, C. R., and Hirson, C.: Hyperextension injury of the neck with paraplegia. J. Bone Joint Surg., *36B*:57–61, 1954.

Buckwalter, J. A., Cooper, R. R., and Maynard, J. A.: Elastic fibers in human intervertebral discs. J. Bone Joint Surg., *58A*:73–76, 1976.

Cailliet, R.: Neck and Arm Pain. Philadelphia, F. A. Davis, 1964.

Colachis, S. C., and Strohm, B. R.: A study of tractive forces and angle of pull on vertebral interspaces in the cervical spine. Arch. Phys. Med., *46*:820–830, 1965.

Crue, B. L., and Todd, E. M.: The importance of flexion in cervical halter traction. Bull. Los Angeles Neurol. Soc., *30*:95–98, 1965.

Fielding, J. W.: Cineroentgenography of the normal cervical spine. J. Bone Joint Surg., *39A*:1280–1288, 1957.

Forsyth, H. F.: Extension injuries of the cervical spine. J. Bone Joint Surg., *46A*:1792–1797, 1964.

Fountain, F. P., Minear, W. L., and Allison, R. D.: Function of longus colli and longissimus cervicis muscles in man. Arch. Phys. Med., *47*:665–669, 1966.

Friedenberg, Z. B., and Miller, W. T.: Degenerative disc disease of the cervical spine. J. Bone Joint Surg., *45A*:1171–1178, 1963.

Gehweiler, J. A., Osborne, R. L., and Becker, R. F.: The Radiology of Vertebral Trauma. Philadelphia, W. B. Saunders, 1980.

Gehweiler, J. A., et al.: Cervical spine trauma: The common combined conditions. Radiology, *130*:77–86, 1979.

Harms-Ringdahl, K.: On assessment of shoulder exercise and load-elicited pain in the cervical spine. Biomechanical analysis of load—EMG—methodological studies of pain provoked by extreme position. Thesis, Karolinska Institute, University of Stockholm, 1986.

Hirsch, C., Schajowicz, F., and Galante, J.: Structural changes in the cervical spine: A study on autopsy specimens in different age groups. Acta Orthop. Scand., Suppl. *109*:1–77, 1967.

Jernigan, W. R., and Gardner, W. C.: Carotid artery injuries due to closed cervical trauma. Trauma, *11*:429–435, 1971.

King, A. I., Prasad, P., and Ewing, C. L.: Mechanism of spinal injury due to caudocephalad acceleration. Orthop. Clin. North Am., *6*:19–31, 1975.

Lysell, E.: Motion in the cervical spine. An experimental study on autopsy specimens. Acta Orthop. Scand., Suppl. *123*:1–62, 1969.

Marar, B. C.: Hyperextension injuries of the spine. J. Bone Joint Surg., *56A*:1655–1662, 1974.

Markolf, K. L.: Deformation of the thoracolumbar intervertebral joint in response to external loads: A biomechanical study using autopsy material. J. Bone Joint Surg., *54A*:511–533, 1972.

Miller, M. D., et al.: Significant new observations on cervical spine trauma. Am. J. Roentgenol., *130*:659–663, 1978.

Moritz, U.: Traktionsbehandlingens mekaniska effekt på ryggraden. Sjukgymnasten, *12*:19–22, 1975.

Moroney, S. P., and Schultz, A. B.: Analysis and measurement of loads on the neck. Trans. Orthop. Res. Soc., *10*:329, 1985.

Nachemson, A. L., and Evans, J. H.: Some mechanical properties of the third lumbar interlaminar ligament (ligamentum flavum). J. Biomech., *1*:211–220, 1968.

Nachemson, A., and Morris, J. M.: In vivo measurements of intradiscal pressure. Discometry, a method for the determination of pressure in the lower lumbar discs. J. Bone Joint Surg., *46A*:1077–1092, 1964.

Panjabi, M. M., Hausfeld, J., and White, A. A.: Mechanical properties of the human thoracic spine as shown by three-dimensional load-displacement curves. J. Bone Joint Surg., *58A*:642–652, 1976.

Roaf, R.: A study of the mechanics of spinal injuries. J. Bone Joint Surg., *42B*:810–823, 1960.

Rockoff, S. D., Sweet, E., and Bleustein, J.: The relative contribution of trabecular and cortical bone to the strength of human lumbar vertebrae. Calcif. Tissue Res., *3*:163–175, 1969.

Rothman, R. H., and Simeone, F. A.: Cervical disc disease. *In* The Spine. Philadelphia, W. B. Saunders, 1982, pp. 440–499.

Schmorl, G., and Junghanns, H.: The Human Spine in Health and Disease. New York, Grune & Stratton, 1971.

Takebe, K., Vitti, M., and Basmajian, J. V.: The functions of semispinalis capitis and splenius capitis muscles: An electromyographic study. Anat. Rec., *179*:477–480, 1974.

Walker, P. S.: Human Joints and Their Artificial Replacements. Springfield, Charles C Thomas, 1977, pp. 110–166.

White, A. A., and Panjabi, M. M.: Clinical Biomechanics of the Spine. Philadelphia, J. B. Lippincott, 1978.

White, A. A., Johnson, R. M., Panjabi, M. M., and Southwick, W. O.: Biomechanical analysis of clinical stability in the cervical spine. Clin. Orthop., *109*:85–96, 1975.

Whitehouse, W. J., and Dupon, E. D.: The scanning electron microscope studies of trabecular bone from a human vertebral body. J. Anat., *108*:481–496, 1971.

Wiktorin, C. v. H., and Nordin, M.: Introduction to Problem Solving in Biomechanics. Philadelphia, Lea & Febiger, 1986, pp. 145–170.

12

BIOMECHANICS OF THE SHOULDER

Joseph D. Zuckerman
Frederick A. Matsen III

The shoulder represents the first link in a mechanical chain of levers that extends from the shoulder to the fingertips. It is defined in a broad sense as the group of structures connecting the arm to the thorax. The main components of the shoulder are listed in Table 12–1. A comparison of these components with those of other articulations reveals that the shoulder is by far the most intricate joint complex in the body. The combined and coordinated movements of four distinct articulations—glenohumeral, acromioclavicular, sternoclavicular, and scapulothoracic—allow the arm to be positioned in space for efficient function (Fig. 12–1). The result is a range of motion that easily exceeds that of any other joint: the humerus can be moved through a space exceeding a hemisphere.

This chapter describes the anatomy of the various parts of the shoulder complex and shows how their structure allows for efficient biomechanical function. Because of the wide range of motion of the shoulder, its large number of diverse components, and considerable variability among individuals in the size and shape of these components, a complete quantitative biomechanical formulation for the shoulder joint is very difficult; however, investigators have generated estimates for mechanisms and forces by simplifying the actual situation. Kinematics and kinetics of the shoulder joint are presented in the context of these limitations.

KINEMATICS

The four articulations of the shoulder joint complex, acting in concert, provide the nearly global range of motion found in the shoulder, the sum of which is greater than the motion available at any single articulation. Movement of the spine can further extend the positions available to the humerus. In the following section the types and ranges of motion for the shoulder complex as a whole are discussed, and in subsequent sections the manner in which motion is achieved in each articulation is described.

RANGE OF MOTION OF THE SHOULDER COMPLEX

Shoulder elevation is defined as movement of the humerus away from the side of the thorax in any plane. It is measured in degrees from the vertical. Different types of shoulder elevation are possible depending on the plane of motion chosen.

Forward flexion is shoulder elevation in the sagittal plane (Fig. 12–2A), while abduction is elevation in the frontal plane (Fig. 12–2B). The normal range of forward flexion is about 180 degrees (AAOS, 1965). This range decreases with age (Germain and Blair, 1983; Murray et al., 1985), but the decrease has been shown to be significantly smaller in physically active

TABLE 12–1

MAIN COMPONENTS OF THE SHOULDER

BONES	JOINTS	LIGAMENTS	MUSCLES
Scapula	*Synovial*	Glenohumeral (capsular)	*Scapulohumeral and claviculohumeral*
	Glenohumeral	Superior	Superficial group
Clavicle	Acromioclavicular	Medial	Deltoid
Humerus	Sternoclavicular	Inferior	Pectoralis major (clavicular head)
Sternum	*Bone-muscle-bone articulation*	Coracohumeral	Deep group
	Scapulothoracic		Rotator cuff muscles
		Coracoacromial	Subscapularis
			Supraspinatus
		Acromioclavicular (capsular)	Infraspinatus
		Coracoclavicular	Teres minor
		Conoid	Other
		Trapezoid	Teres major
		Costoclavicular	*Scapuloradial*
			Biceps (long and short heads)
		Sternoclavicular (capsular)	*Scapuloulnar*
		Interclavicular	Triceps (long head)
			Thoracohumeral
			Latissimus dorsi
			Pectoralis major (sternocostal head)
			Thoracoscapular
			Serratus anterior
			Pectoralis minor
			Trapezius
			Levator scapulae
			Rhomboids
			Thoracoclavicular
			Subclavius

persons (Germain and Blair, 1983). The range of abduction is about 180 degrees (AAOS, 1965).

Forward elevation in the plane of the scapula (Fig. 12–3) has been discussed as the most functional form of elevation (Johnston, 1937; Saha, 1950) because in this plane the inferior part of the glenohumeral joint capsule is not twisted and the deltoid and supraspinatus muscles are optimally aligned for elevation of the arm. (In the shoulder literature, elevation in the plane of the scapula is sometimes referred to as "abduction in the plane of the scapula.") Since the scapula is oriented at an angle of approximately 30 to 45 degrees anterior to the frontal plane, elevation in the plane of the scapula is midway between forward flexion and abduction. Maximum elevation in this plane has been measured in men by Freedman and Munro (1966) and in women by Doody and coworkers (1970). Tabulation of their results according to the percentage of normal subjects able to attain a given degree of elevation indicates that the range is somewhat greater for the women than for the men: 28% of women and 4% of men exceeded 180 degrees (Table 12–2).

Rotation about the long axis of the humerus is another functionally important shoulder motion. Both internal and external rotation can be performed with the humerus in varying degrees of elevation. When the humerus is at the side of the thorax and the elbow is flexed 90 degrees, internal rotation moves the forearm closer to the body and external rotation moves it farther away (Fig. 12–4). With the arm in this position, internal rotation is restricted by the arm's contact with the body, whereas external rotation is limited by the soft tissues that surround the shoulder joint complex. When the humerus is abducted 90 degrees and the elbow is flexed 90 degrees, internal rotation moves the hand downward while external rotation moves it upward (Fig. 12–4B). The range of internal and external rotation varies with the degree of arm elevation, but in general each may be accomplished to about 90 degrees, yielding a maximum total range of 180 degrees.

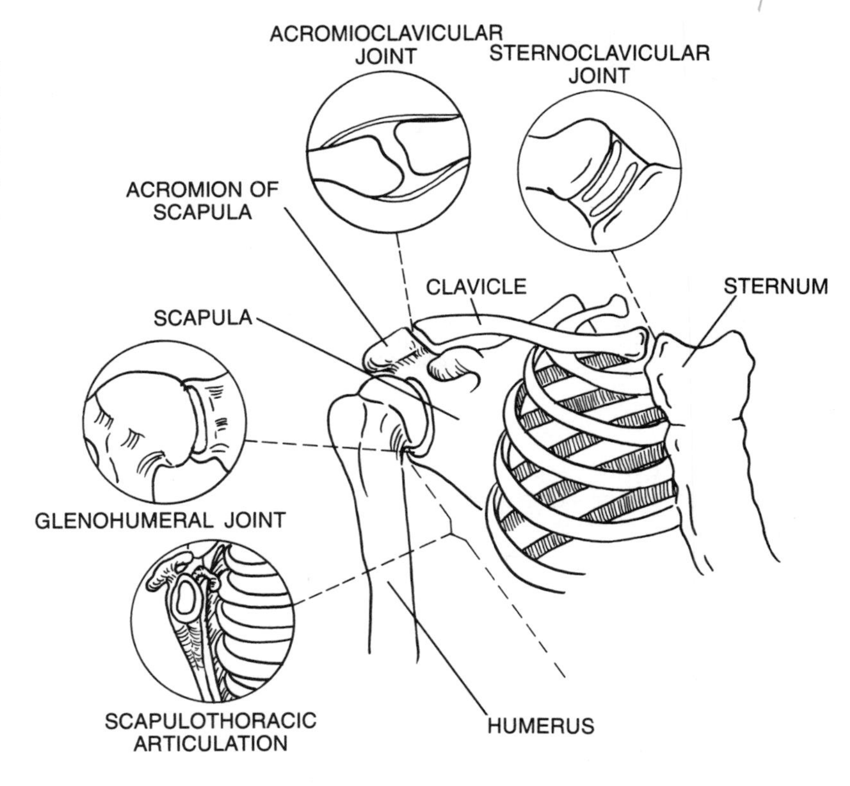

FIG. 12–1
Schematic depiction of the bony structures of the shoulder and their four articulations. The circular insets show front views of the three synovial joints—sternoclavicular, acromioclavicular, and glenohumeral—and a lateral view of the scapulothoracic joint, a bone-muscle-bone articulation. (Adapted from DePalma, 1983.)

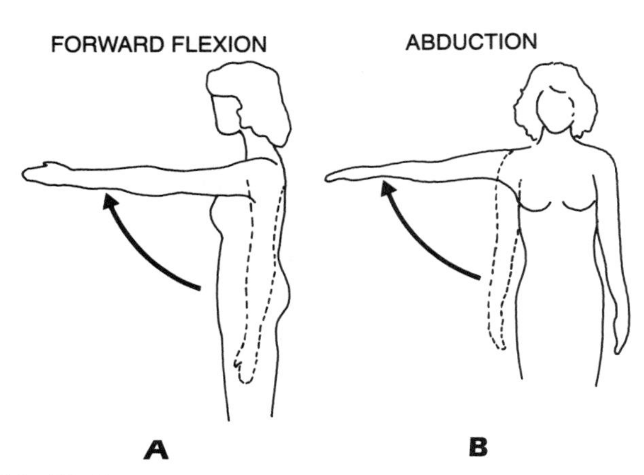

FIG. 12–2
A. Forward flexion. The humerus is in the sagittal plane. **B.** Abduction. The humerus is in the frontal plane.

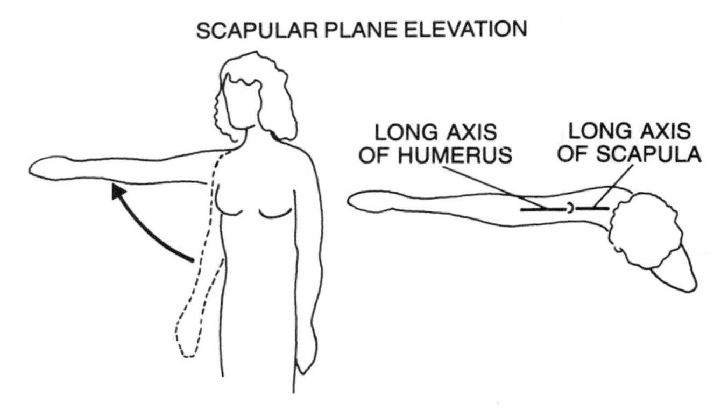

FIG. 12–3
Elevation in the scapular plane, which is midway between forward flexion and abduction. The humerus is in the plane of the scapula.

TABLE 12–2

MAXIMUM ELEVATION OF THE SHOULDER IN THE SCAPULAR PLANE

	>180 DEGREES (% OF SUBJECTS)	171–180 DEGREES (% OF SUBJECTS)	161–170 DEGREES (% OF SUBJECTS)	<161 DEGREES (% OF SUBJECTS)
Men*	4	33	46	17
Women†	28	60	12	0

*Data from Freedman and Munro, 1966. N = 61.
†Data from Doody et al., 1970. N = 25.

EXTERNAL AND INTERNAL ROTATION

HUMERUS AT THE SIDE

HUMERUS IN 90° OF ABDUCTION

A B

FIG. 12–4
Rotation around the long axis of the humerus. **A.** External and internal rotation with the humerus at the side. Internal rotation is shown with the arm behind the back, which is a functionally important form of this motion. **B.** External and internal rotation with the humerus in 90 degrees of abduction.

Several other shoulder motions are possible. Backward elevation, or extension in the sagittal plane, is possible to approximately 60 degrees (DePalma, 1973) (Fig. 12–5A). Adduction, or depression of the arm, is the action of bringing the humerus closer to the side of the thorax and is normally limited by contact with the body (Fig. 12–5B). Adduction with the arm moving in front of the body beyond the midline in an upward plane is possible to about 75 degrees (AAOS, 1965). Horizontal flexion is defined as forward movement of the arm in a horizontal (transverse) plane (Fig. 12–5C). Measured from a starting position of 90 degrees of abduction, the normal range of horizontal flexion is approximately 135 degrees. Movement in the opposite direction from the same starting point, horizontal extension, has a normal range of approximately 45 degrees (DePalma, 1973) (Fig. 12–5D). Thus, the shoulder is capable of about 180 degrees of motion in the horizontal plane.

MOTION AT THE FOUR SHOULDER ARTICULATIONS

Simultaneous synchronous movements of the four shoulder articulations provide the motion required for full function, such as that needed for throwing a javelin or performing gymnastics on the rings. This full shoulder function involves considerably more motion than that required for many daily activities. If the cervical spine, forearm, wrist, and hands are normal, one can feed oneself while keeping the whole shoulder complex immobile with the humerus at the side. Persons with an arthrodesis of the glenohumeral joint can reach the face and mouth

BACKWARD EXTENSION

ADDUCTION

A B

HORIZONTAL FLEXION

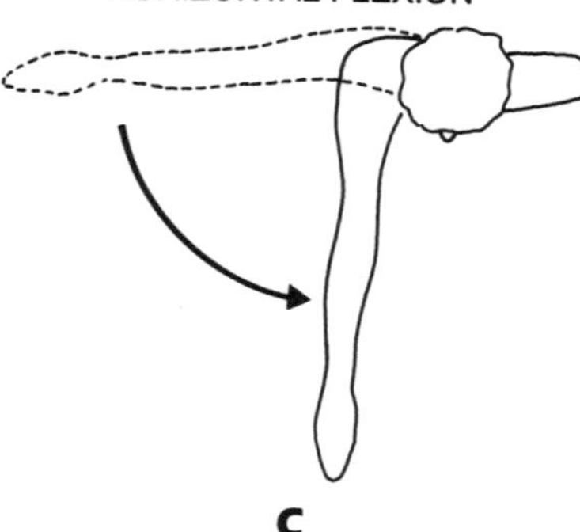

C

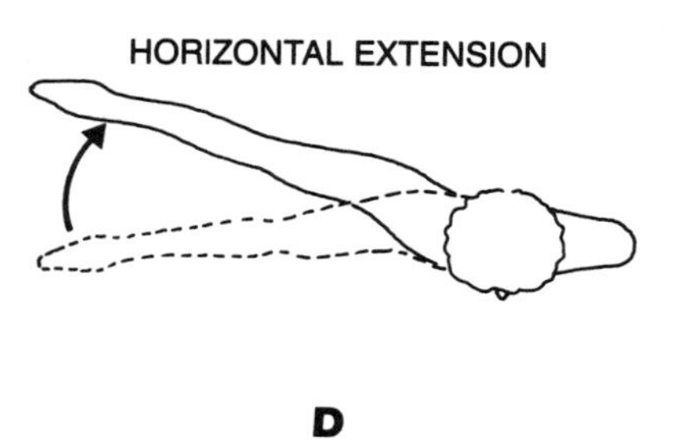

D

FIG. 12–5
A. Backward extension. The humerus is in the sagittal plane. **B.** Adduction, or depression, which can take place in any plane. Here the humerus is in the frontal plane. **C.** Horizontal flexion. **D.** Horizontal extension.

(Rowe, 1974). A patient with a pin or screw through the clavicle and coracoid process, which immobilizes the acromioclavicular joint, has a functional range of shoulder motion approximating full elevation (Rockwood and Green, 1975). Thus, although normal motion at all four shoulder joints is required for full shoulder function, other joints are capable of considerable compensatory motion when function at one or more of the four is limited.

The Glenohumeral Joint

The glenohumeral joint, a synovial articulation between the humeral head and the glenoid fossa of the scapula, consists of a nearly hemispheric convex humeral articular surface and a bone and soft-tissue socket (Fig. 12–6). A minimally constrained ball-and-socket joint, it allows much greater freedom of motion than does a more rigid ball-and-socket joint

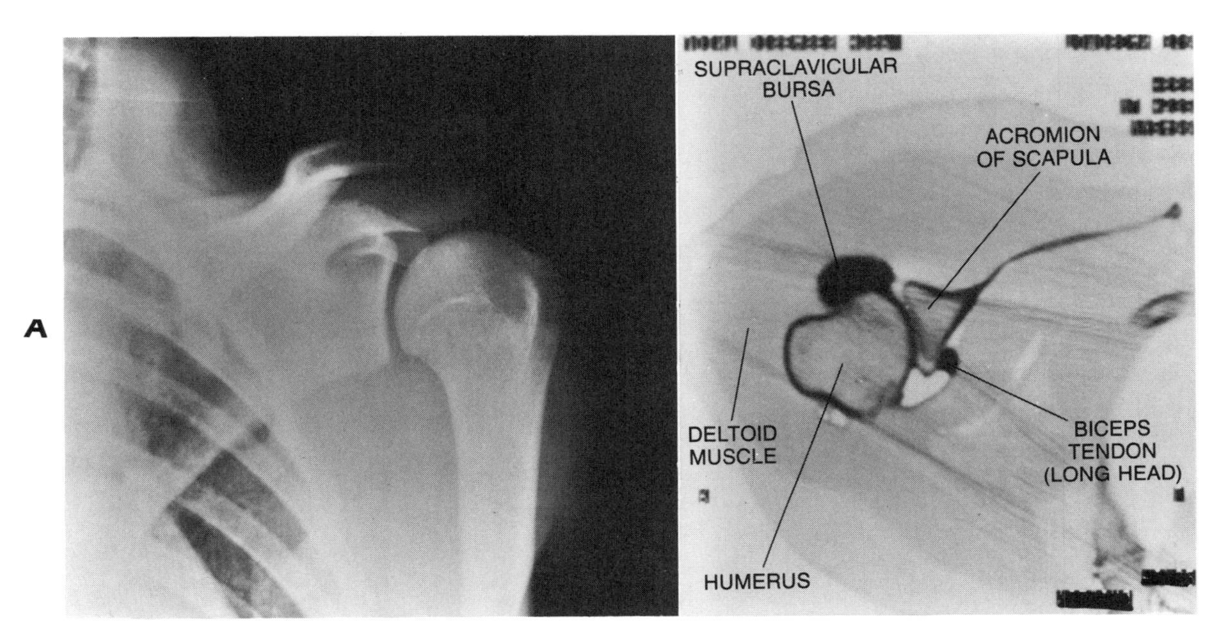

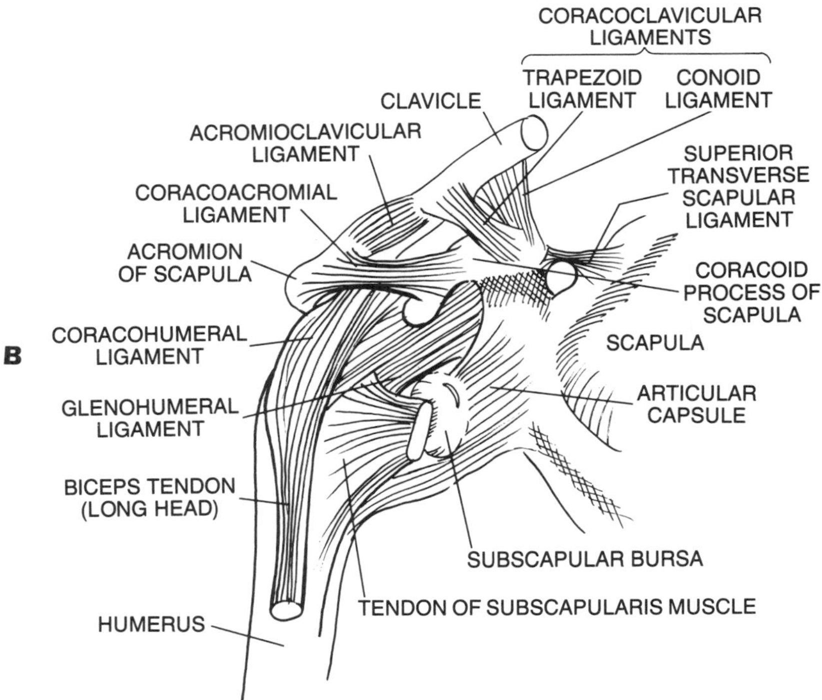

FIG. 12–6
A. Anteroposterior roentgenogram of the left shoulder complex with the humerus externally rotated (left). Computed tomogram of the glenohumeral joint (transverse cut) (right). (Courtesy of Alex Norman, M.D.) **B.** Schematic drawing of the glenohumeral and acromioclavicular joints showing major ligaments and the subscapular bursa. The subacromial bursa is hidden from view by the coracoacromial ligament.

such as the hip, thus contributing significantly to the large range of motion of the shoulder as a whole.

The diameter of the humeral articular surface (the ball) ranges from 37 to 55 mm (Maki and Gruen, 1976). The humeral head makes an angle of about 135 degrees with the humeral shaft and is retroverted about 32 degrees with respect to the axis of flexion of the elbow (see Fig. 13–2). The glenoid fossa (the socket) consists of a small, pear-shaped, cartilage-covered bony depression that measures about 41 mm longitudinally and about 25 mm in the transverse direction (Maki and Gruen, 1976). The surface area of the glenoid fossa is only one third to one fourth that of the humeral head (Kent, 1971). The longitudinal diameter of the glenoid fossa is about 75%, and its transverse diameter approximately 60%, of that of the humeral head (Saha, 1971). Saha (1971) found that 75% of 50 normal subjects had a posteriorly tilted (retrotilted) glenoid face, the tilt averaging 7.4 degrees. The perimeter of the glenoid fossa is lined by the fibrocartilaginous reflections of the joint capsule, the glenohumeral ligaments, and the tendon of the long head of the biceps. These reflections are known collectively as the glenoid labrum.

Although the glenoid fossa is deepened to a certain extent by the labrum, its shallowness allows significant freedom of movement of the humeral head on the glenoid surface. The small contribution of the glenoid bone to the shoulder socket allows a wide range of motion by minimizing bone-to-bone contact; the contact area between the two articulating surfaces is rather small (see Fig. 12–6A).

While the anatomic configuration of the glenohumeral joint allows significant motion, it makes the joint more susceptible to instability, which can be anterior, posterior, inferior, and/or superior. Dislocation and recurrent subluxation of the joint are common.

Unlike rigid ball-and-socket joints, which possess inherent bony stability, the glenohumeral joint relies primarily on soft tissue structures for its stability (see Fig. 12–6B). The glenohumeral joint capsule is reinforced anteriorly by the three glenohumeral ligaments (the superior, middle, and inferior), which may appear only as capsular thickenings, and superiorly by the coracohumeral ligament, which runs from the base of the coracoid process to the proximal end of the bicipital groove (Basmajian and Bazant, 1959). Together the coracoacromial ligament and acromion form an arch that prevents excessive upward displacement of the humeral head (Basmajian, 1969). The subacromial bursa facilitates smooth passage of the humeral head, with its overlying rotator cuff, beneath this arch. The tendons of four muscles—the subscapularis, supraspinatus, infraspinatus, and teres minor—blend with the glenohumeral joint capsule to form the rotator cuff, which provides dynamic restraints to anterior, posterior, and inferior displacement. These tendons insert on the lesser and greater tuberosities of the proximal humerus.

Several factors are important for the stability of the glenohumeral joint:

1. *Adequate size of the glenoid fossa.* Saha (1971) found that if the longitudinal diameter of the glenoid fossa was less than 75% and the transverse diameter less than 57% of that of the humeral head, the glenoid was relatively hypoplastic and the joint was more likely to be unstable.
2. *Posterior tilt of the glenoid fossa.* Saha (1971) found anteriorly tilted glenoid fossae in 80% of 21 unstable shoulders, while the incidence of this finding in 50 normal shoulders was 27%.
3. *Humeral head retroversion* (Saha, 1971).
4. *Intact capsule and glenoid labrum* (Reeves, 1968; Ovesen and Nielsen, 1985). Reeves (1968) found that young patients with anterior shoulder instability are likely to have a detached labrum, whereas older patients with this condition are likely to have a stretched capsule.
5. *Function of the muscles that control the anteroposterior position of the humeral head* (subscapularis, infraspinatus, and upper part of teres minor) (Saha, 1971).

The flexible glenohumeral joint displays the surface motion typical of a ball-and-socket joint, in which three types of surface motion may take place in any given plane—rotation, rolling, and translation (gliding). During rotation, the contact point on the socket remains constant while that on the ball changes as it rotates in the socket (Fig. 12–7A). During rolling, the contact point on each joint surface changes by an equal amount (Fig. 12–7B). During translation, or gliding, the contact point on the ball remains constant while that on the socket changes (Fig. 12–7C).

Surface motion at the glenohumeral joint is primarily rotational, but some combination of gliding and rolling also takes place. When the motion is not purely rotational, the humeral ball displaces with respect to the glenoid (see Figs. 12–7B, C). Poppen and Walker (1976) measured humeral ball excursion on the glenoid face during shoulder elevation in the plane of the scapula in 12 normal subjects. From zero to 30 degrees of elevation, and often from 30 to 60 degrees, the humeral ball moved upward on the glenoid fossa by approximately 3 mm, indicating that rolling and/or gliding had taken place. With each

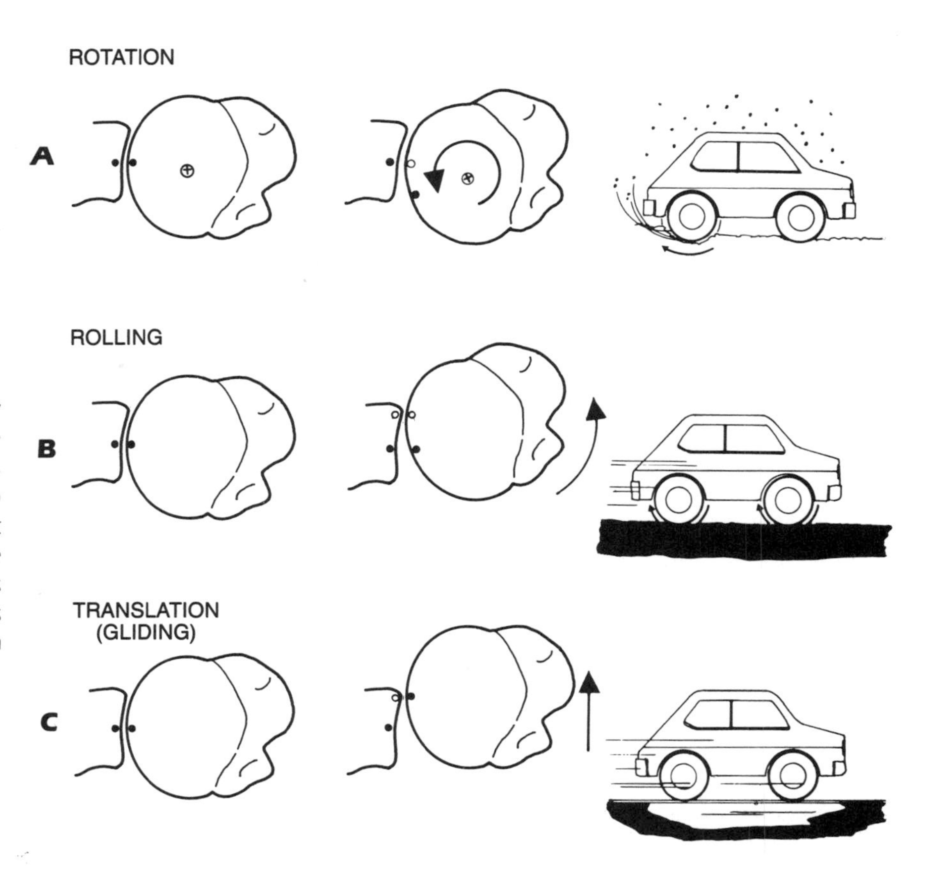

FIG. 12–7
Surface motion at the glenohumeral joint (top view). The three types of motion that may take place—rotation, rolling, and translation (gliding)—are depicted. **A.** During rotation the contact point on the glenoid surface remains constant while the contact point on the surface of the humeral head changes as the humerus rotates in the glenoid (analogous to the rear tire of a stuck automobile spinning in the snow). Original contact points are indicated by the solid circles; the new contact point is indicated by the hollow circle. **B.** During rolling the contact point on each of the joint surfaces changes by an equal amount (analogous to an automobile tire rolling along with perfect traction). **C.** During translation (gliding) the contact point on the humeral head remains the same while that of the glenoid changes (analogous to the tire of an automobile with locked brakes skidding on ice).

additional 30-degree interval of elevation, the humeral head moved only 1±0.5 mm up or down, indicating almost pure rotation. The upward excursion of the humeral head observed in the early stages of arm elevation is probably due to sagging of the humeral head in the dependent position before motion is initiated.

Poppen and Walker's (1976) analysis of the instant centers of rotation of the glenohumeral joint in the 12 normal subjects during scapular plane elevation also confirmed that the surface joint motion was mainly rotational. The instant centers lay quite close to each other and to the geometric center of the humeral ball (an average of 6±2 mm from the center of the ball). Measurements in 15 patients with shoulder lesions showed distinctly greater humeral head excursion and instant center values in seven patients who had a previous glenohumeral dislocation, a rotator cuff tear, or significant shoulder pain associated with a previous injury.

The Acromioclavicular Joint

The acromioclavicular joint is a small synovial articulation between the distal clavicle and the proximal acromion of the scapula (see Fig. 12–6B). The joint is surrounded by a dense, fibrous capsule, which includes the superior and inferior acromioclavicular ligaments. Joint stability is provided mainly by the two parts of the coracoclavicular ligament, the conoid and the trapezoid, which suspend the scapula from the clavicle. These ligaments permit the scapula to move on the clavicle about three axes:

1. The conoid ligament, which runs from the coracoid process of the scapula to the apex of the posterior curve of the clavicle, serves as a longitudinal (vertical) axis for scapular rotation (scapular protraction and retraction) (axis 1, Fig. 12–8).
2. The trapezoid ligament, located lateral to the conoid ligament, is a quadrilateral structure that runs from the coracoid process of the scapula to the clavicle, attaching broadly over the area from the conoid tubercle to the acromioclavicular joint. This ligament acts as a hinge for scapular motion about a transverse (horizontal) axis in the frontal plane (axis 2, Fig. 12–8).
3. Scapular motion is also possible through the acromioclavicular joint itself. Inman and associates (1944) have proposed that when the scapula rotates posteriorly relative to the clavicle (axis 2), a relative lengthening of the coracoclavicular ligament results that allows the scapula to rotate about a transverse axis that passes through the acromi-

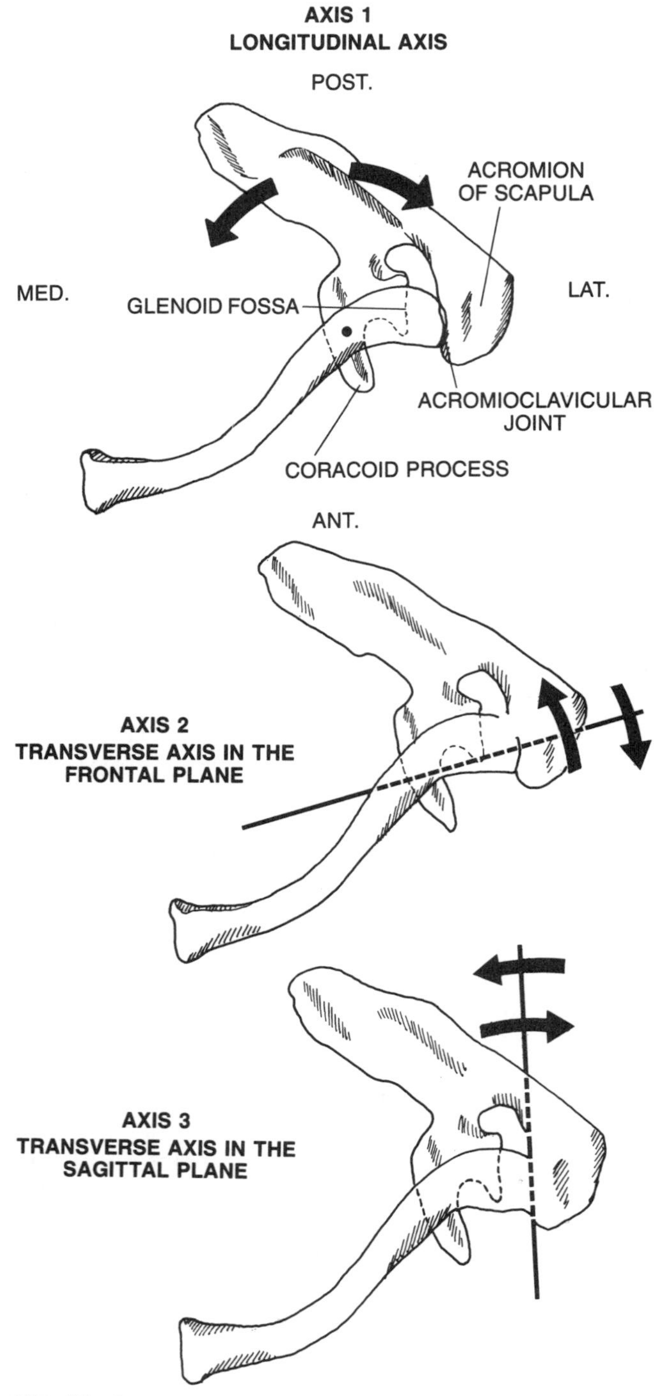

FIG. 12–8
Top view of the clavicle and scapula showing the axes of motion at the acromioclavicular joint. The conoid and trapezoid ligaments are hidden beneath the clavicle (see Fig. 12–6B). Axis 1: Longitudinal (vertical) axis (solid dot) for scapular rotation (protraction and retraction), which takes place through the conoid ligament. Axis 2: Transverse (horizontal) axis in the frontal plane for scapular rotation, which takes place through the trapezoid ligament. Axis 3: Transverse axis in the sagittal plane for scapular rotation through the acromioclavicular joint itself.

oclavicular joint in the sagittal plane (axis 3, Fig. 12–8).

The acromioclavicular joint usually has a meniscus that divides it into functional units. Rotation through the conoid ligament (axis 1) takes place between the acromion and the meniscus, and hinging on the trapezoid ligament (axis 2) takes place between the meniscus and the clavicle (Last, 1972).

The range of motion between the clavicle and scapula has been studied in cadaver preparations. Dempster (1965) showed a 30-degree range of rotation about the conoid ligament (axis 1), a 60-degree arc for hinging on the trapezoid ligament (axis 2), and a 30-degree arc for hinging about the transverse axis in the sagittal plane through the acromioclavicular joint (axis 3). The ranges of motion observed in cadaver preparations are obviously much greater than those in living subjects, whose scapuloclavicular motion is restricted by the thorax and muscle attachments. Inman and coworkers (1944) found that during shoulder abduction and forward flexion the total range of clavicular elevation at the acromioclavicular joint is 20 degrees, occurring primarily in the first 30 and the last 45 degrees of arm elevation.

The Sternoclavicular Joint

The sternoclavicular joint is the synovial articulation between the manubrium of the sternum and the proximal clavicle (Fig. 12–9). A fibrocartilaginous disc, or meniscus, similar to that in the acromioclavicular joint, is interposed between the two bony surfaces. The principal stabilizing structure of this joint, the costoclavicular ligament, securely attaches the clavicle to the first rib (Dempster, 1965; Last, 1972) and controls the motion between the relatively flat joint surfaces of the clavicle and manubrium, acting as a fulcrum for the significant gliding motion that takes place at this joint during shoulder motion (Last, 1972).

The meniscus, attached to the clavicle superiorly and cartilage of the first rib inferiorly, divides the sternoclavicular joint into two functional units for gliding (Dempster, 1965). Anteroposterior gliding (clavicular protraction and retraction) occurs between the sternum and the meniscus (Fig. 12–10A), while superoinferior gliding (clavicular elevation and depression) takes place between the clavicle and the meniscus (Fig. 12–10B). The clavicle also rotates about its long axis (Fig. 12–10C).

The range for various sternoclavicular joint motions has been observed by several investigators. Inman and associates (1944) noted a range of about 40 degrees during arm elevation in both the frontal and

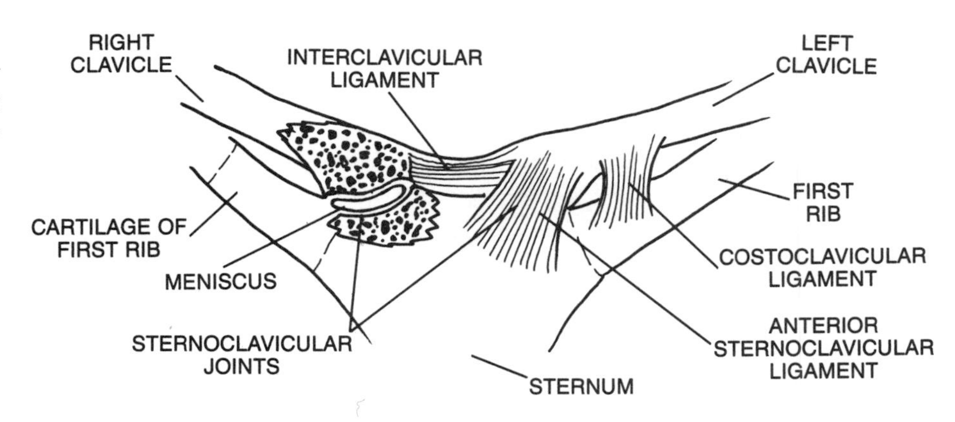

FIG. 12–9
The sternoclavicular joints (anterior view). The right joint is shown without ligaments to reveal the articular structures.

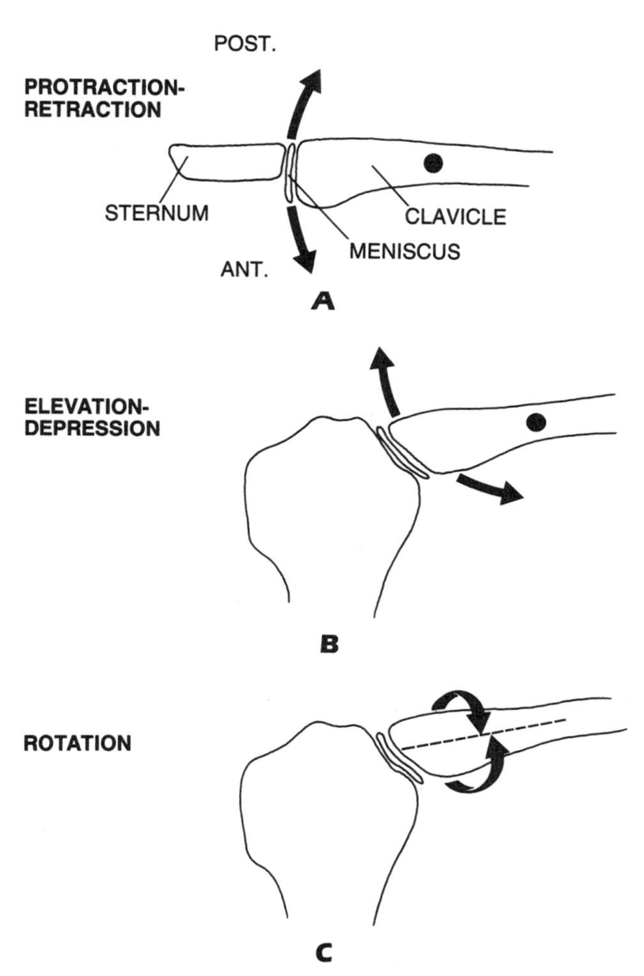

FIG. 12–10
Motion at the sternoclavicular joint. **A.** Top view showing clavicular protraction and retraction (anteroposterior gliding) in the transverse plane around a longitudinal axis (solid dot) through the costoclavicular ligament, not shown. Motion takes place between the sternum and the meniscus. **B.** Anterior view showing clavicular elevation and depression (superoinferior gliding) in the frontal plane around a sagittal axis (solid dot) through the costoclavicular ligament, not shown. Motion occurs between the clavicle and the meniscus. **C.** Anterior view depicting clavicular rotation around the longitudinal axis of the clavicle.

sagittal planes; 4 degrees of clavicular elevation occurred for each 10 degrees of arm elevation through the first 90 degrees, and beyond 90 degrees clavicular motion at this joint was almost negligible. Rotation about the long axis of the clavicle was approximately 40 degrees (Kent, 1971; Inman et al., 1944; Last, 1972). Since much of the shoulder girdle is palpable in normal subjects, readers can confirm most of these motions by feeling their own shoulders.

Motion at the sternoclavicular joint is reciprocal with motion at the acromioclavicular joint during clavicular protraction and retraction and elevation and depression, but not during rotation. For example, depression of the medial end of the clavicle takes place in conjunction with elevation of the lateral clavicle; also, when the lateral clavicle is protracted, the medial end retracts (Last, 1972). Rotation occurs in the same direction at both ends of the clavicle, however. These findings are not surprising when one considers that the clavicle is a bony strut whose ends form the articulation at both joints.

The Scapulothoracic Articulation

Except for its attachment through the acromioclavicular and sternoclavicular joints, the scapula is without bony or ligamentous connection to the thorax. While this seemingly unstable situation literally has the shoulder "hanging by the clavicle," it does allow for a wide range of scapular motion. These motions include protraction, retraction, elevation, depression, and rotation about a variable axis in the anteroposterior direction.

Scapular motion is enhanced by the scapulothoracic articulation, a bone-muscle-bone articulation between the scapula and the thoracic wall (Fig. 12–11). The broad anterior surface of the scapula is separated from the chest wall by the serratus anterior and the subscapularis muscles, which glide on each other during the various motions of the scapula. The serratus anterior holds the scapula in close apposition

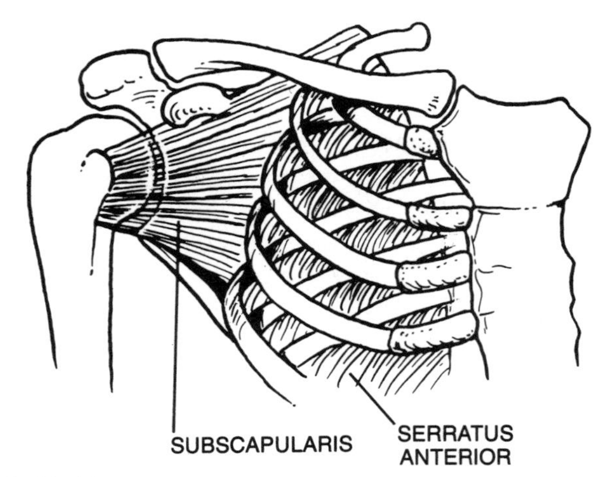

FIG. 12–11
Anterior view of the scapulothoracic articulation, a bone-muscle-bone articulation between the scapula and thorax. During scapular motion the subscapularis muscle, which attaches broadly to the costal surface of the scapula, glides on the serratus anterior muscle, which originates on the first eight ribs and inserts into the costal surface of the scapula along the length of its vertebral border.

to the chest wall throughout large ranges of scapular motion and prevents scapular winging. Hence, although the scapulothoracic articulation is not a joint in the truest sense, it contributes to the wide range of motion of the scapula, which greatly enhances the mobility of the entire shoulder complex.

Several investigators have attempted to relate glenohumeral and scapulothoracic motion during arm elevation in various planes (Inman et al., 1944; Freedman and Munro, 1966; Doody et al., 1970; Saha, 1973; Poppen and Walker, 1976). In their basic study, Inman's group (1944) examined arm elevation in the frontal and sagittal planes (abduction and forward flexion) and found that about two thirds of the motion (approximately 120 degrees) took place at the glenohumeral joint and one third (approximately 60 degrees) at the scapulothoracic articulation. During the first 30 to 60 degrees of arm elevation, scapular motion was highly irregular and appeared to depend on the position of the scapula at rest. Thereafter, the 2 to 1 ratio of glenohumeral to scapulothoracic motion remained quite constant. The investigators stated that the 60 degrees of scapular motion at the scapulothoracic articulation is possible only because an equal amount of motion takes place at the clavicular joints—20 degrees at the acromioclavicular joint and 40 degrees at the sternoclavicular joint. Moreover, clavicular rotation of about 40 degrees permits the motion at these two joints. Without rotation of the clavicle, arm elevation would be limited to about 120 degrees.

The 2 to 1 ratio Inman's group (1944) found during arm elevation in the frontal and sagittal planes is at variance with the ratios noted in more recent investigations in which glenohumeral motion and scapulothoracic motion were compared during arm elevation in the scapular plane. Freedman and Munro (1966) found a 1.35 to 1 (3 to 2) ratio for the range from zero to 135 degrees, with an increase of glenohumeral motion in the final stage of arm elevation. Doody and coworkers (1970) found a mean ratio of 1.74 to 1 from zero to approximately 180 degrees and noted that adding a weight in the hand tended to increase the contribution of scapulothoracic motion in the early stages of arm elevation. Poppen and Walker (1976) observed a 4.3 to 1 ratio during the first 30 degrees of elevation and a 1.25 to 1 (5 to 4) ratio from 30 to 180 degrees. Saha (1961) found a ratio

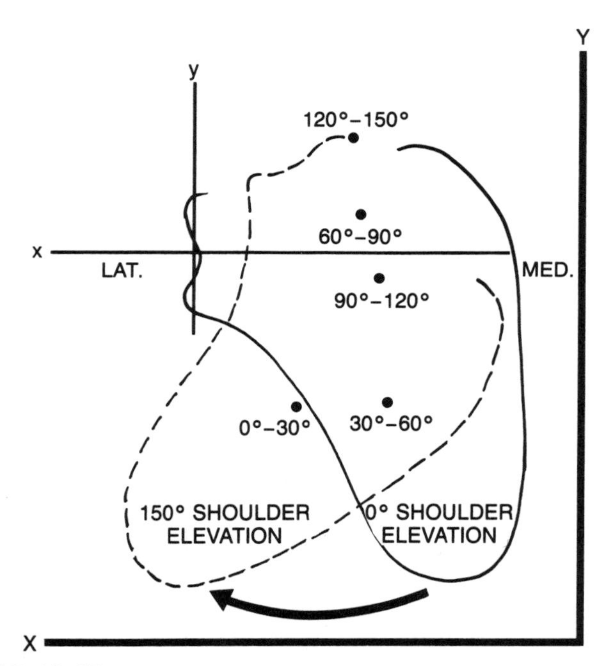

FIG. 12–12
Rotation of the scapula on the thorax in the scapular plane. Instant centers of rotation (solid dots) are shown for each 30-degree interval of motion during shoulder elevation in the scapular plane from zero to 150 degrees. The x and y axes are fixed in the scapula, whereas the X and Y axes are fixed in the thorax. From zero to 30 degrees the scapula rotated about its lower midportion; from 60 degrees onward rotation took place about the glenoid area, resulting in a medial and upward displacement of the glenoid face and a large lateral displacement of the inferior tip of the scapula. (Adapted from Poppen and Walker, 1976.)

of 2.3 to 1 for the range from 30 to 135 degrees of elevation. Differences in the ratios among the investigations may be due to the different measurement techniques used, to the different planes of arm elevation measured, and to anatomic variations among individuals, particularly in the initial position of the scapula.

Poppen and Walker (1976) analyzed the instant centers of rotation of the scapula in relation to fixed axes in the thorax during arm elevation in the scapular plane (Fig. 12–12) and found that as the arm moves into maximum elevation the glenoid face shifts medially, then tilts upward, and finally moves upward. Also, by plotting the course of the tips of the acromion and the coracoid process in sequential roentgenograms, they observed that during arm elevation the scapula twists about a transverse axis (Fig. 12–13). The mean amount of scapular twisting at maximum arm elevation was 40 degrees. This twisting is essentially external rotation of the scapula; the superior tip of the scapula moves away from the thorax while the inferior tip moves into it. The rotation is significant when considered with external rotation of the humerus, which often occurs beyond 90 degrees of arm elevation. It is apparent that the scapula and humerus move synchronously and that, to some extent, the amount of scapular rotation depends on the extent of humeral rotation.

The Contribution of the Spine

Although motion of the spine is usually not considered in a discussion of shoulder motion, it should be mentioned that the spine may play a significant role in orienting the arm with respect to the body's center of mass. A familiar example is extension of the range of overhead reach when the spine is tilted away from the reaching shoulder and the ribs are elevated on the reaching side (Fig. 12–14). The importance of spine motion during the act of throwing also has been demonstrated (Atwater, 1977). At the instant of release (or the instant of contact in the case of racket sports) the arm makes an angle of approximately 90 degrees with the trunk. Whether this is an overhead or side-arm motion is determined by the position of the trunk and not by the position of the glenohumeral or scapulothoracic joint.

KINETICS

The involvement of a great many muscles complicates exact calculation of the loads acting on the shoulder joint complex, but estimates have been made with the use of simplifying assumptions. In the following sections the actions of the various muscles

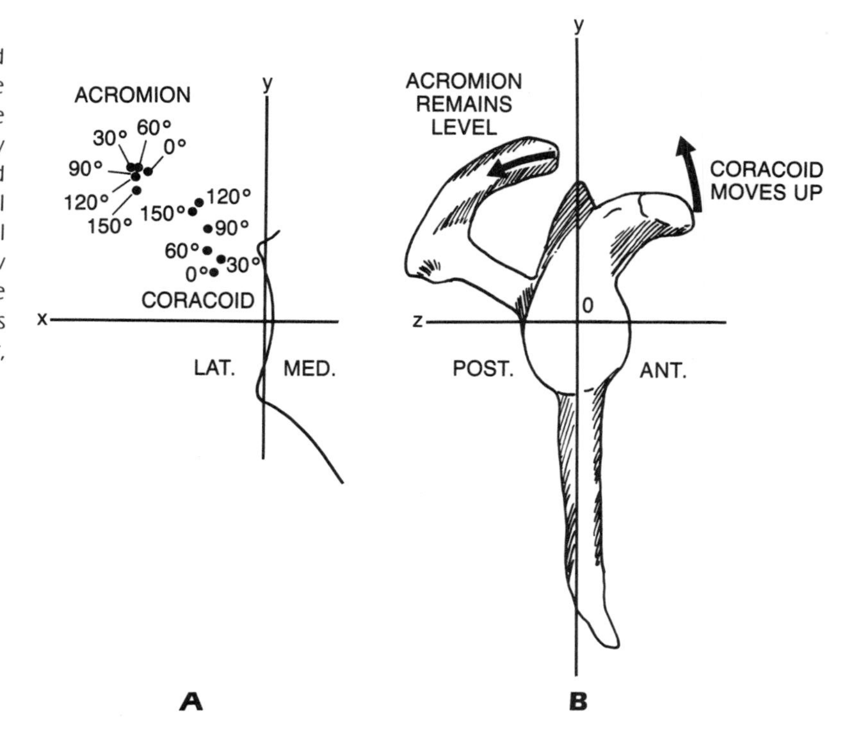

FIG. 12–13
A. A plot of the tips of the acromion and coracoid process on roentgenograms taken at successive intervals of arm elevation in the scapular plane shows upward movement of the coracoid and only a slight shift in the acromion relative to the glenoid face. This finding demonstrates twisting, or external rotation, of the scapula about the x axis. **B.** A lateral view of the scapula during this motion would show the coracoid process moving upward while the acromion remains on the same horizontal plane as the glenoid. (Adapted from Poppen and Walker, 1976.)

FIG. 12–14
The range of overhead reach is extended when the spine is tilted away from the reaching shoulder and the ribs are elevated on the reaching side.

affecting the shoulder are described and the loads on the glenohumeral joint are discussed. Of the four shoulder articulations, the glenohumeral joint bears the greatest loads, which are estimated to approach body weight. A final section covers the relationship of arm position to fatigue of the shoulder musculature.

SHOULDER MUSCLES AND THEIR ACTIONS

The seventeen muscles involved in shoulder motion are listed in Table 12–1 and illustrated in Figure 12–15. The inherent complexity of shoulder kinetics is due to the large number of muscles about the shoulder and also to the fact that the muscle actions have three unusual aspects:

1. Since the glenohumeral joint lacks rigid stability, a muscle exerting an effect on the humerus must act in concert with other muscles to avoid producing a dislocating force on the joint. (Compare the elbow joint, which can be stably extended by the triceps without other muscle contraction.)
2. The existence of multiple linkages in the shoulder (clavicle, scapula, and humerus) gives rise to the interesting situation in which a single muscle may span several joints, exerting an effect on each. For example, the latissimus dorsi, which originates from the chest wall and attaches to the humerus, spans the scapulothoracic, sternoclavicular, acromioclavicular, and glenohumeral joints.
3. The extensive range of shoulder motion causes muscle function to vary depending on the position of the arm in space. For example, the long head of the biceps acts as an accessory shoulder abductor when the humerus is externally rotated but not when it is internally rotated (Basmajian and Latif, 1957).

The complexity of the muscle actions involved in shoulder motion is demonstrated in the work of Inman and associates (1944) and of DeLuca and Forrest (1973), who used electromyography to study muscle activity during shoulder abduction under various loading conditions. During shoulder elevation without resistance in the frontal plane (Inman et al., 1944; DeLuca and Forrest, 1973) and in the sagittal plane (Inman et al., 1944), significant electromyographic activity was recorded in the deltoid, clavicular head of the pectoralis major, supraspinatus, infraspinatus, subscapularis, teres minor, upper and middle trapezius, serratus anterior, and rhomboids. When abduction was performed against resistance, the teres major also made a significant contribution.

A number of methods have been used to study the forces generated by the shoulder musculature during different motions (Inman et al., 1944; Colachis et al., 1969; Colachis and Strohm, 1971; Ekholm et al., 1978; Sigholm et al., 1984; Ito, 1980; Celli et al., 1985). Inman's group (1944), using the triceps brachii, were able to determine a direct relationship between the tension developed in the muscle and its action potential amplitude recorded electromyographically. They then used this relationship to estimate the force generated by the various shoulder muscles during forward flexion. Their results indicated that the subscapularis, teres minor, infraspinatus, supraspinatus, clavicular portion of the pectoralis major, and deltoid were active throughout the range of shoulder flexion. Although the exact force values could not be determined, it was evident that each of these muscles made a significant contribution throughout the range of forward flexion.

Colachis and coworkers (1969) and Colachis and Strohm (1971) studied the strength of various shoul-

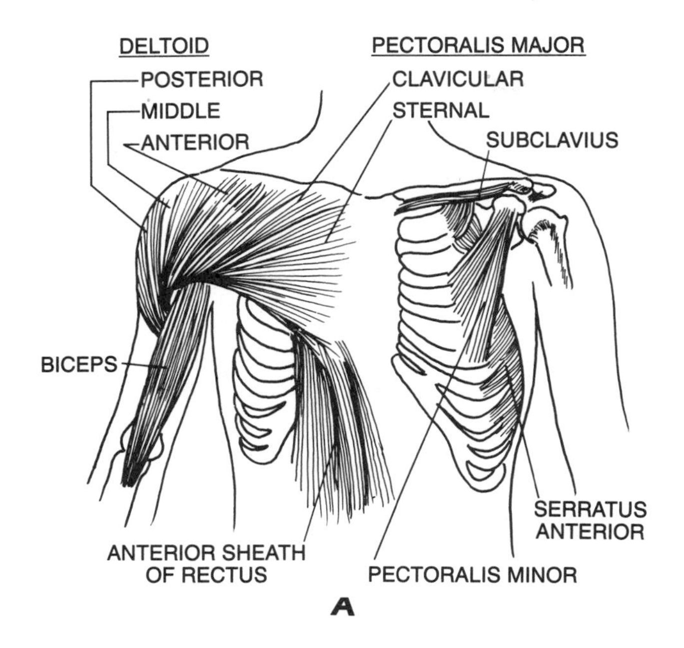

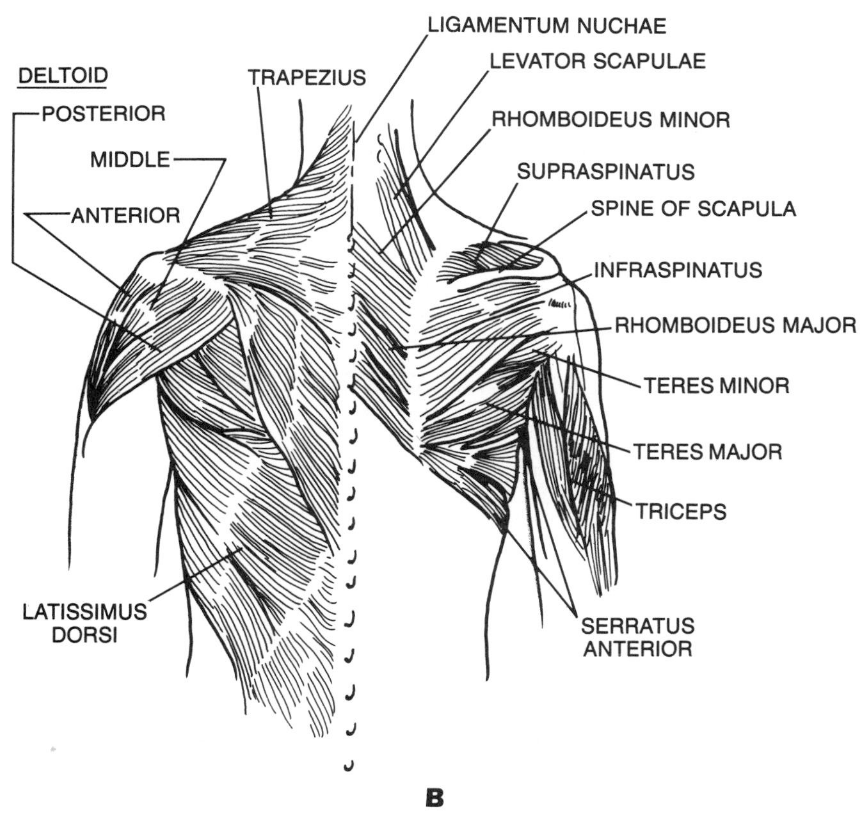

FIG. 12–15
Schematic drawings of the musculature of the shoulder joint complex. **A.** Anterior view showing the superficial muscles (left shoulder) and the deep muscles beneath the deltoid and pectoralis muscles (right shoulder). **B.** Posterior view showing the superficial muscles (left shoulder) and underlying muscles (right shoulder).

der motions in different positions before and after administration of suprascapular and axillary nerve blocks. A suprascapular nerve block, which eliminated the active contribution of the supraspinatus and infraspinatus muscles, reduced the force of elevation in the scapular plane by 35% at zero degrees of elevation and by 60% at 60 degrees. Above 60 degrees the loss of force became less apparent: only a 30% reduction was noted at 150 degrees. The force of external rotation was reduced by 50%. An axillary nerve block, which eliminated the active contribution of the deltoid muscle, reduced the force of elevation (measured in the frontal, scapular, and sagittal planes) by 35% at zero degrees of elevation and by

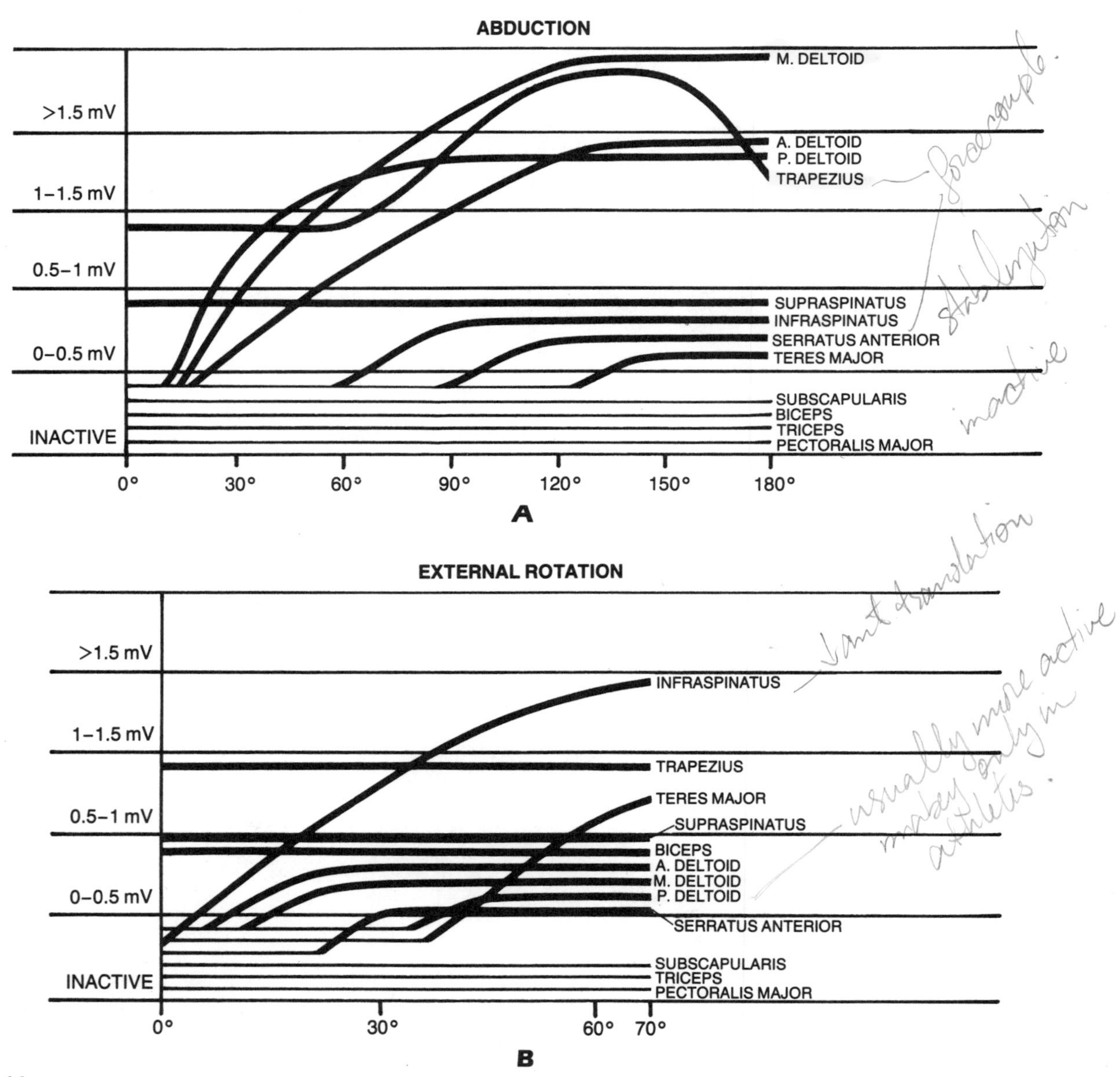

FIG. 12–16
Diagrammatic representation of the EMG activity of individual muscles (average of 20 observations) in a normal shoulder during abduction and external rotation. (Adapted from Celli et al., 1985.) **A.** During abduction the muscles that show rising curves ("movers") are the deltoid and upper trapezius. The muscles that show a steady tracing ("stabilizers") are the supraspinatus, infraspinatus, serratus anterior, and teres major. **B.** External rotation is activated by the infraspinatus and teres major (rising curves after the first 30 degrees). The muscles that show steady tracings are the supraspinatus, upper trapezius, biceps, deltoid, and serratus anterior.

approximately 60 to 80% at 150 degrees. The force of external rotation was reduced by 45%, and the force of horizontal arm motion was diminished by 60%.

Celli and coworkers (1985) studied muscles innervated by the axillary, suprascapular, and subscapular nerves with a different technique. Rather than blocking the nerve and measuring the muscular deficit produced, they stimulated the nerve and documented the shoulder motion that resulted. Stimulation of the axillary nerve (deltoid, teres minor) resulted in only 40 degrees of abduction. Suprascapular nerve stimulation (supraspinatus, infraspinatus) resulted in abduction to 90 degrees and external rotation to 45 degrees. Subscapular nerve stimulation (subscapularis) resulted in internal rotation of 25 degrees and flexion of 20 degrees. Stimulation of two of these nerves simultaneously produced partial, but never full, shoulder motion. In addition, patients with isolated paralysis of the supraspinatus and infraspinatus muscles could achieve no more than 45 degrees of active abduction, thereby demonstrating the importance of these muscles for shoulder elevation.

Ekholm and coworkers (1978) attempted to identify the muscles primarily involved in various shoulder movements. Electromyographic activity during a "maximum" isometric contraction was compared with the activity recorded during specific exercise movements for subjects using a resisted weight-and-pulley circuit. Forward flexion–abduction–external rotation intensively activated the infraspinatus and all three portions of the deltoid; during forward flexion–adduction–external rotation the anterior and middle portions of the deltoid and the infraspinatus were significantly activated. Backward extension–abduction–internal rotation vigorously activated the posterior portion of the deltoid, whereas backward extension–adduction–internal rotation strongly activated the sternocostal portion of the pectoralis major. Thus, although EMG activity occurred in many shoulder muscles during these combination movements, for each motion a few specific muscles showed the highest activation level.

Improved electromyographic techniques have recently revealed additional information about muscle activity during elevation of the arm. Celli and associates (1985) used electromyographic recording in combination with selective nerve stimulation to define three types of muscle groups about the shoulder: (1) the muscles that show increased electromyographic activity, which are responsible for the shoulder motion (deltoid and upper trapezius during abduction, infraspinatus during external rotation); (2) the muscles that show constant EMG activity, which are responsible for maintaining stability as the shoulder is moved (supraspinatus during abduction and external rotation); and (3) the inactive muscles. The EMG activity of these different muscle groups during abduction and external rotation is illustrated in Figure 12–16.

In a similar study of 10 subjects, Ito (1980) examined elevation versus depression in the scapular plane and in forward flexion. The motion of depression always registered less EMG activity than did elevation but showed the same pattern (Fig. 12–17). Up to 140 degrees of both scapular elevation and forward flexion, EMG activity showed an almost linear increase with an increase in arm elevation.

Sigholm and associates (1984) studied the effect of hand tool weight and arm position on shoulder muscle load with the use of electromyographic techniques. They found that when either a 1-kg or a 2-kg weight was held in the hand, EMG signals increased in all six muscles tested. The degree of elevation of the arm was found to correlate more closely with EMG activity than did the amount of load in the hand. In other words, the degree of upper arm elevation was the most important determinant of shoulder muscle loads. A load in the hand was found to affect the stabilizing muscles (particularly the infraspinatus, and to a lesser extent the supraspinatus and upper trapezius) more than the elevating muscles (the three portions of the deltoid). Sigholm and associates also determined that upper arm rotation and elbow flexion had little effect on shoulder muscle load. These results may be important in determining what positions best enable workers to avoid injuries.

Though, in general, the action of any muscle may be inferred from a knowledge of its origins and insertions, this is not the case in the shoulder. For example, when the arm is at the side, contraction of the fibers of the middle portion of the deltoid lifts the humerus along its axis but does not produce the motion of elevation because the line of action of the middle deltoid fibers is essentially parallel to the long axis of the humerus. However, if the action of the middle portion of the deltoid is "coupled" with other active or passive stabilizing forces, elevation can occur. In this case, the oblique rotator cuff muscles (infraspinatus, subscapularis, and teres minor) act to stabilize the humeral head on the glenoid, thereby providing a fixed fulcrum and allowing elevation to occur (Fig. 12–18). The importance of the stabilizing effect of the rotator cuff muscles is evident in patients

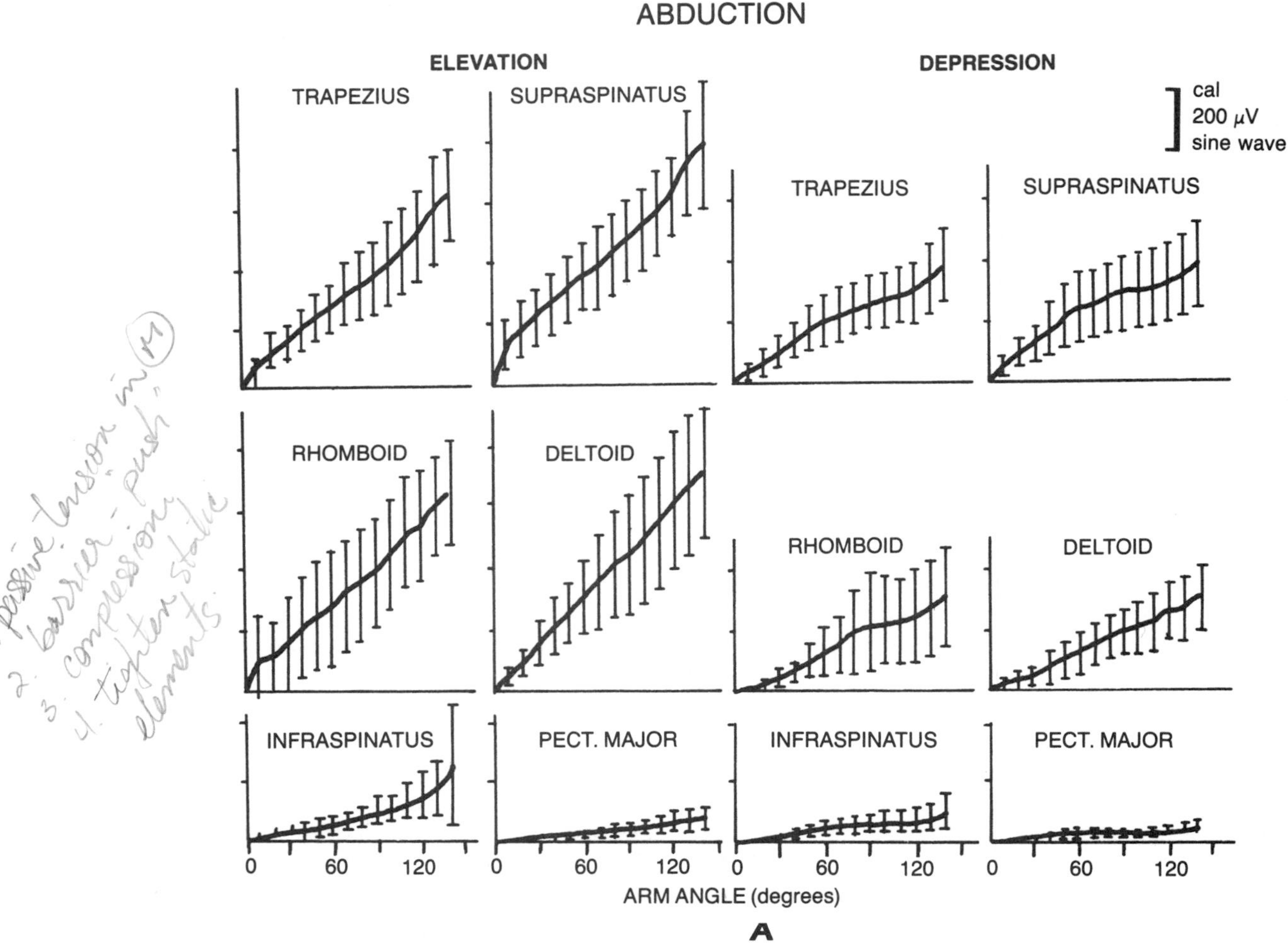

FIG. 12–17
Average linear envelopes of electromyograms obtained from 10 subjects during abduction **(A)** and forward flexion **(B)**. Each muscle showed less activity during depression than during elevation, but the EMG patterns were similar. The bars indicate standard deviation. (Adapted from Ito, 1980.)

with rotator cuff tears who lose their ability to elevate the arm despite intact deltoid muscles.

The rotator cuff muscles are unique in another way. They are oriented so that their contraction may not only result in a desired motion but may also resist displacement of the humeral head (Fig. 12–19). In this way they act as dynamic stabilizers of the glenohumeral joint. For example, contraction of the subscapularis compresses the humeral head against the glenoid, internally rotates the head of the humerus, and by pressing against the anterior aspect of the humeral head tends to displace it posteriorly. This is one way in which a strong subscapularis can prevent anterior subluxation of the shoulder during normal shoulder activity. In addition, this pushing action may be one means by which the supraspinatus prevents upward subluxation of the head of the humerus during strong contraction of the deltoid with the arm adducted.

LOADS AT THE GLENOHUMERAL JOINT

The glenohumeral joint is the most important component of the shoulder complex and makes the greatest contribution to total shoulder motion. When the shoulder complex is loaded, each articulation is subjected to increased stress; however, the glenohumeral joint receives the largest portion of the load because of its size relative to that of the acromioclavicular, sternoclavicular, and scapulothoracic articula-

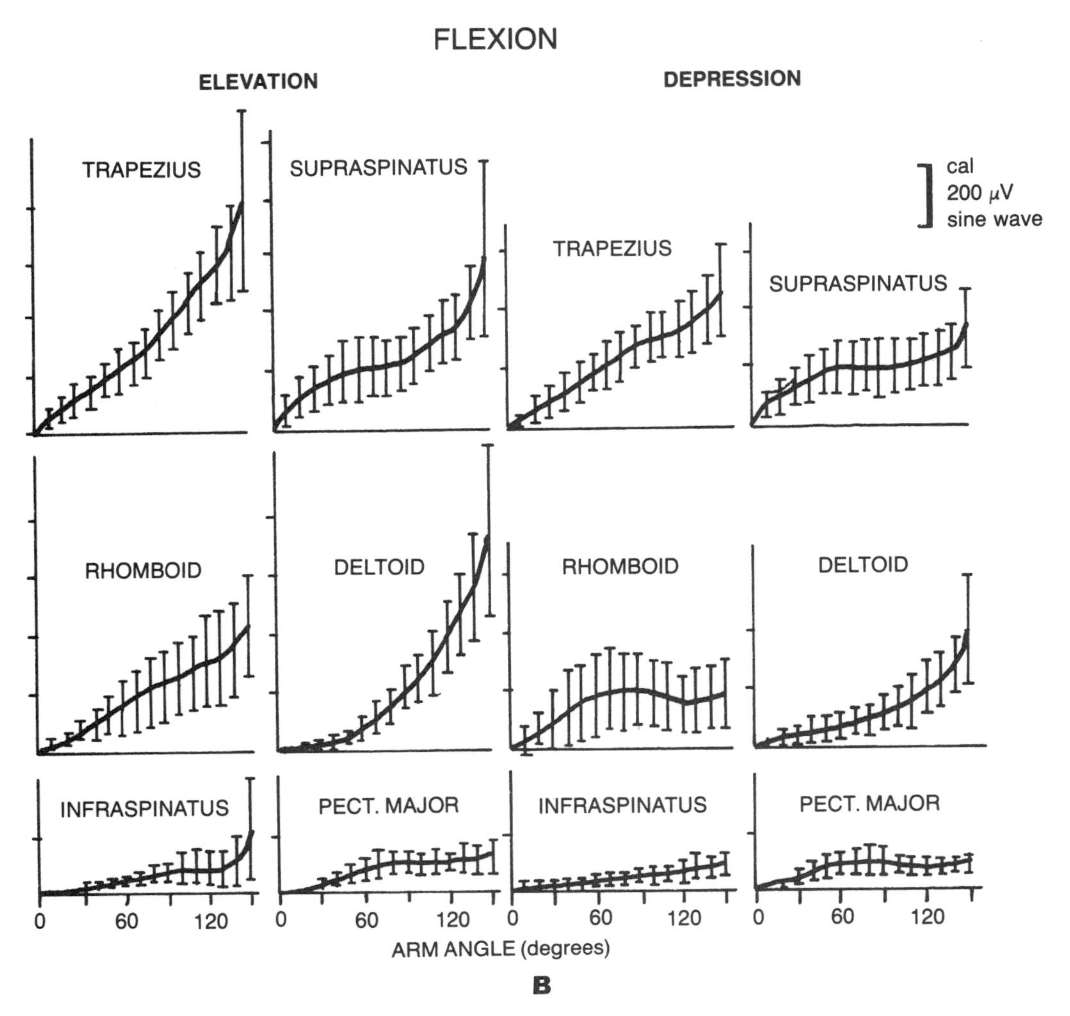

FIG. 12–17 *Continued*

tions. In this section the static and dynamic aspects of loads at the glenohumeral joint are discussed.

Statics

Recent studies (Ivey et al., 1985; Murray et al., 1985) have determined strength values for shoulder muscles during resisted motions. Ivey and coworkers (1985) used a Cybex II isokinetic dynamometer to measure torque forces for specific shoulder motions. This machine is a variable-speed, accommodating resistance-control device that allows dynamic torsional forces to be recorded at set velocities throughout a range of motion. They found that internal rotation strength exceeded external rotation strength by a ratio of 3 to 2, extension exceeded flexion strength by a ratio of 5 to 4, and adduction exceeded abduction strength by a ratio of 2 to 1. Overall comparison of strength values showed adduction to have the highest value, followed by extension, flexion, abduction, internal rotation, and external rotation. They found the males' strength to be greater than that of the females, but the difference decreased significantly when the results were normalized for lean body mass and exercise habit. There was no statistically significant difference between values for dominant and nondominant shoulders, although there was a consistent pattern of greater strength in the dominant shoulder.

Murray and coworkers (1985) measured maximum isometric shoulder muscle strength in male and female subjects of different ages (25 to 36 years and 55 to 66 years). They found that strength values for women were 45 to 66% of those for men; further, the older group had strength values that were 66 to 93%

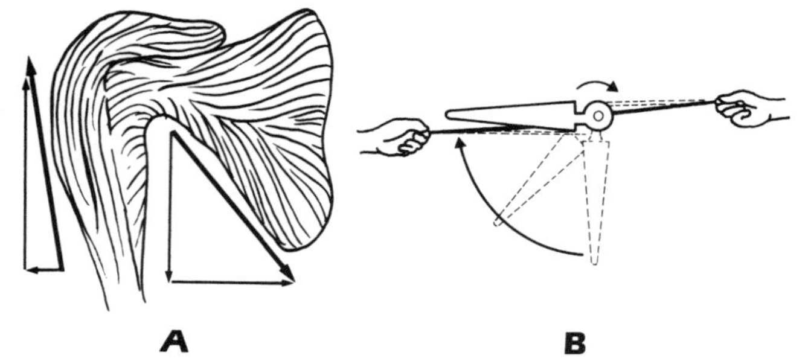

FIG. 12–18
The deltoid and the oblique rotator cuff muscles (infraspinatus, subscapularis, and teres minor) combine to produce elevation of the upper extremity by means of a force couple (two forces equal in magnitude but opposite in direction). With the arm at the side **(A)**, the directional force of the deltoid is upward and outward with respect to the humerus while the force of the oblique rotator cuff muscles is downward and inward. These two directional forces can be resolved into their respective vertical and horizontal components. The horizontal force of the deltoid acting below the center of rotation of the glenohumeral joint is opposite in direction to the horizontal force of the oblique rotators, which is applied above the center of rotation. These forces acting in opposite directions on either side of the center of rotation produce a powerful force couple, as illustrated by the arm signal **(B)**. The vertical forces offset each other, thereby stabilizing the humeral head on the glenoid and allowing elevation to take place. (Adapted from Lucas, 1973.)

of the younger group's values. They also observed that the strength of dominant and nondominant shoulders did not differ significantly.

Although fairly straightforward techniques can be utilized to measure muscle strength with different

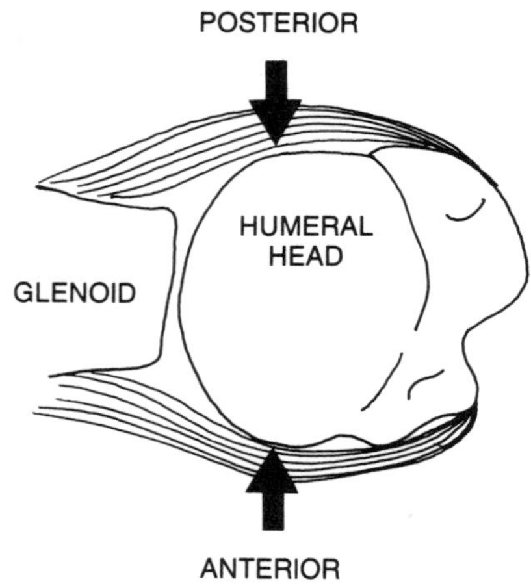

FIG. 12–19
Top view of the glenohumeral joint showing the muscles of the rotator cuff, which are oriented so that their tendons and muscle masses may push on the head of the humerus, thereby stabilizing the joint.

shoulder motions, calculating the reaction force on the glenohumeral joint remains a challenging task for two principal reasons: a large number of muscles are involved and the force contribution of each varies with differing loads, planes of shoulder elevation, and degrees of elevation. Nonetheless, by making some necessary simplifying assumptions, several investigators have obtained estimates of these forces. These estimations support the concept that the glenohumeral joint is a major load-bearing joint.

Inman and coworkers (1944) analyzed the forces produced during shoulder abduction (frontal plane) using a three-force system that included the force produced by the weight of the extremity, the deltoid muscle force, and the force required to oppose the other two combined forces. This last force was resolved into two components: the glenohumeral joint reaction force and the resultant force produced by the rotator cuff musculature. They estimated that the deltoid muscle force at 90 degrees of shoulder abduction was eight times the weight of the extremity (which they assumed to be 9% of body weight), or approximately 70% of body weight. The glenohumeral joint reaction force at 90 degrees of shoulder abduction was estimated to attain a maximum of 10 times the weight of the extremity, or 90% of body weight. The resultant force produced by the rotator cuff muscles (supraspinatus, infraspinatus, teres minor, and subscapularis) was estimated to be 9.6 times the weight of the extremity and reached a maximum at 60 degrees of abduction.

More recently, Poppen and Walker (1978) studied forces required for isometric arm elevation in the plane of the scapula. Their work was based on the assumption that the force in a muscle is proportional to its area multiplied by the integrated electromyographic signal. Considering all muscles to be active, they calculated that the resultant glenohumeral joint reaction force reached a maximum of 89% of body weight at 90 degrees of scapular plane elevation and that the shear force component on the face of the glenoid reached a maximum of 42% of body weight at 60 degrees of elevation.

Poppen and Walker (1978) also performed static analysis of the forces acting on the glenohumeral joint with the arm abducted (frontal plane). Making further simplifying assumptions, they analyzed the glenohumeral joint reaction force with the arm at 90 degrees of abduction and the elbow extended; only the deltoid muscle was considered to be active. A free body diagram of this static loading situation is shown in Figure 12–20A.

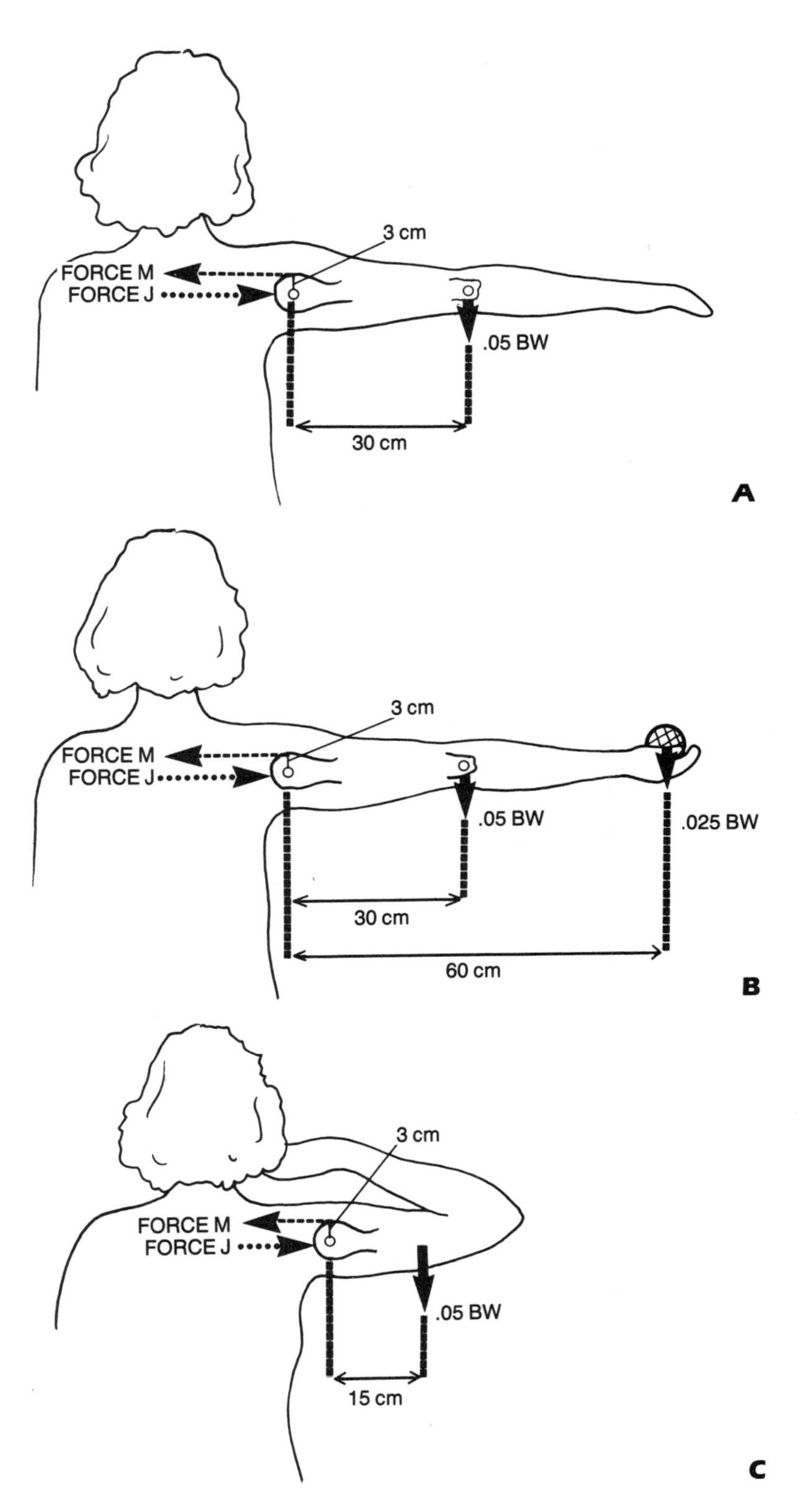

FIG. 12–20
Estimates of the reaction force on the glenohumeral joint are obtained with the use of simplifying assumptions (Poppen and Walker, 1978). **A.** In this example the arm is in 90 degrees of abduction, and it is assumed that only the deltoid muscle is active. The force produced through the tendon of the deltoid muscle (M) acts at a distance of 3 cm from the center of rotation of the joint (indicated by the hollow circle). The force produced by the weight of the arm is estimated to be 0.05 times body weight (BW) and acts at a distance of 30 cm from the center of rotation. The reaction force on the glenohumeral joint (J) may be calculated with the use of the equilibrium equation that states that for a body to be in moment equilibrium the sum of the moments must equal zero. In this example the moments acting clockwise are considered to be positive and counterclockwise moments are considered to be negative.

$$\Sigma M = 0$$
$$(30 \text{ cm} \times .05 \text{ BW}) - (M \times 3 \text{ cm}) = 0$$
$$M = \frac{30 \text{ cm} \times .05 \text{ BW}}{3 \text{ cm}}.$$

M is approximately one half body weight. Since M and J are almost parallel but opposite, they form a force couple and are of equal magnitude; thus, the joint reaction force is also approximately one half body weight. **B.** Similar calculations can be made to determine the value for M when a weight equal to 0.025 times body weight is held in the hand with the arm in 90 degrees of abduction.

$$\Sigma M = 0$$
$$(30 \text{ cm} \times .05 \text{ BW}) + (60 \text{ cm} \times .025 \text{ BW}) - (M \times 3 \text{ cm}) = 0$$
$$M = \frac{(30 \text{ cm} \times .05 \text{ BW}) + (60 \text{ cm} \times .025 \text{ BW})}{3 \text{ cm}}.$$

Once again, M and J are essentially equal and opposite, forming a force couple. Thus, the joint reaction force is approximately equal to body weight. **C.** The arm is held in the same position, but the elbow is now maximally flexed. Elbow flexion moves the center of gravity of the arm medially, shortening the lever arm of the gravitational force to 15 cm. The reaction force on the glenohumeral joint (J) is calculated with the use of the same equilibrium equation.

$$\Sigma M = 0$$
$$(15 \text{ cm} \times .05 \text{ BW}) - (M \times 3 \text{ cm}) = 0$$
$$M = \frac{15 \text{ cm} \times .05 \text{ BW}}{3 \text{ cm}}.$$

M is approximately one fourth body weight, as is the joint reaction force.

The force through the tendon of the deltoid muscle (M) acts parallel to the long axis of the arm. The distance between the center of rotation of the shoulder and the line of application of the muscle force (the lever arm of M) is approximately 3 cm. The weight of the arm produces a gravitational force of 0.05 times body weight, and its lever arm is 30 cm. The reaction force on the glenohumeral joint (J) is calculated with the equilibrium equation that states that for a body to be in moment equilibrium the sum of the moments must equal zero (see Fig. 12–20A). The muscle force required to keep the arm in this position (M) is calculated to be 0.5 times body weight. Since M and J are essentially parallel but opposite in direction, they form a force couple and are of equal magnitude. Therefore, the joint reaction force (J) is also 0.5 times body weight (10 times the weight of the extremity). It must be borne in mind that this is a minimum value, since only the force of the deltoid is considered and an unweighted extremity is used.

Similar calculations can be made for the arm at 90 degrees of abduction with the elbow extended and a 2-kg weight in the hand (Fig. 12–20B). If the subject's body weight is 80 kg, then the weight in the hand represents 2.5% of body weight. In this situation the addition of a relatively light weight to the outstretched arm doubles the glenohumeral joint reaction force, which is equal to body weight.

A further application of this simplified model may yield some clinically useful information. In this case the glenohumeral joint reaction force is calculated with the arm in the same position of abduction but with the elbow maximally flexed (Fig. 12–20C). The static analysis is the same as that for the first example, except that elbow flexion moves the center of the arm mass medially, thereby shortening the lever arm for the gravitational force from 30 to 15 cm. The reaction force on the glenohumeral joint (J) is calculated with the same equilibrium equations utilized in the first example and is found to equal 25% of body weight. Thus, compared with keeping the arm straight, flexing the elbow maximally during shoulder abduction reduces the muscle force and the glenohumeral joint reaction force by about 50%.

Dynamics

Dynamic aspects of shoulder muscle function have been investigated in the context of muscle fatigue and its relation to injuries in the workplace. Hagberg (1981) studied the relation of shoulder elevation to muscle fatigue as evidenced by electromyographic monitoring. Two positions were tested, 90 degrees of forward flexion and 90 degrees of abduction. In most subjects fatigue was evident within a few minutes in the upper trapezius and supraspinatus in both positions. This finding correlates well with epidemiologic studies of workers working with arms elevated. Supraspinatus tendinitis, as well as neck pain in the vicinity of the upper trapezius, is reported to be common (Herberts and Kadefors, 1976; Bjelle et al., 1979).

Herberts and associates (1980) expanded this study approach and analyzed the effect of shoulder and elbow position on muscle fatigue, also using electromyographic criteria. They found that localized muscle fatigue was present in all muscles studied (anterior and middle deltoid, supraspinatus, infraspinatus, upper trapezius) for overhead and shoulder-level work (and in some cases for waist-level work as well). Specifically, during overhead work, a lower fatigue value was seen in the anterior deltoid with the shoulder in 45 or 90 degrees of abduction than in zero degrees of abduction; fatigue in the supraspinatus muscle was significantly lower at 45 degrees of abduction than at zero or 90 degrees. During shoulder-level work, the upper trapezius fatigued more easily at 90 degrees than at 45 degrees of abduction. During waist-level work, moderate abduction did not produce significant localized muscle fatigue. Herberts and coworkers also found that, in general, the infraspinatus had the highest localized fatigue levels of all muscles tested.

Besides providing important insight into the dynamic function of the shoulder musculature, these fatigue data can be utilized to make recommendations on upper extremity positioning to workers who perform manual labor, recommendations that may have important implications for the incidence of occupational shoulder disorders.

SUMMARY

1. The shoulder complex consists of four distinct articulations: the glenohumeral joint, the acromioclavicular joint, the sternoclavicular joint, and the scapulothoracic articulation. The extensive range of motion of the shoulder (exceeding a hemisphere) is the result of synchronous, simultaneous contributions from each.
2. The glenohumeral joint, with its minimally constrained ball-and-socket configuration, contributes greatly to the wide range of motion of the

shoulder complex. Because it lacks inherent bony stability, this joint relies on static and dynamic soft tissue stabilizers including the joint capsule, the glenoid labrum, and the rotator cuff muscles. Stability depends on a glenoid of adequate size, a posteriorly titled glenoid fossa, a retroverted humeral head, an intact capsule and an intact glenoid labrum, and an intact and functioning rotator cuff.

3. Surface motion in the glenohumeral joint consists mainly of rotation but may also include some rolling and gliding (translation).
4. Shoulder motion, particularly elevation, is governed by the action of force couples. An important example is the interaction of the deltoid and the oblique rotator cuff muscles (subscapularis, infraspinatus, and teres minor) in producing shoulder elevation.
5. Although several simplifying assumptions must be made, it can be estimated that the reaction force on the glenohumeral joint approaches body weight when the arm is abducted 90 degrees; thus, this joint must be considered a major load-bearing joint.
6. Data from recent studies of shoulder musculature fatigue during overhead reach may be important for recommending optimal upper extremity positioning for workers engaged in manual labor.

REFERENCES

American Academy of Orthopaedic Surgeons: Joint Motion. Method of Measuring and Recording. Chicago, AAOS, 1965. Reprinted by the British Orthopaedic Association, 1966.

Atwater, A. E.: Biomechanics of throwing: Correction of common misconceptions. Paper presented at Joint Meeting of the National College Physical Education Association for Men and the National Association for Physical Education of College Women, Orlando, Florida, January 6–9, 1977.

Basmajian, J. V.: Recent advances in the functional anatomy of the upper limb. Am. J. Phys. Med., *48*:165, 1969.

Basmajian, J. V., and Bazant, F. J.: Factors preventing downward dislocation of the adducted shoulder joint. An electromyographic and morphological study. J. Bone Joint Surg., *41A*:1182, 1959.

Basmajian, J. V., and Latif, A.: Integrated actions and functions of the chef flexors of the elbow. A detailed electromyographic analysis. J. Bone Joint Surg., *39A*:1106, 1957.

Bjelle, A., Hagberg, M., and Michaelsson, G.: Clinical and ergonomic factors in prolonged shoulder pain among industrial workers. Scand. J. Work Environ. Health, *5*:205, 1979.

Celli, L., Balli, A., de Luise, G., and Rovesta, C.: Some new aspects of the functional anatomy of the shoulder. Ital. J. Orthop. Traumatol., *11*:83, 1985.

Colachis, S. C., and Strohm, B. R.: Effects of suprascapular and axillary nerve blocks on muscle force in upper extremity. Arch. Phys. Med. Rehabil., *52*:22, 1971.

Colachis, S. C., Strohm, B. R., and Brechner, V. L.: Effects of axillary nerve block on muscle force in the upper extremity. Arch. Phys. Med. Rehabil., *50*:647, 1969.

DeLuca, C. J., and Forrest, W. J.: Force analysis of individual muscles acting simultaneously on the shoulder joint during isometric abduction. J. Biomech., *6*:385, 1973.

Dempster, W. T.: Mechanisms of shoulder movement. Arch. Phys. Med. Rehabil., *46*:49, 1965.

DePalma, A. F.: Biomechanics of the shoulder. *In* Surgery of the Shoulder. 3rd Ed. Philadelphia, J. B. Lippincott, 1983, pp. 65–85.

Doody, S. G., Freedman, L., and Waterland, J. C.: Shoulder movements during abduction in the scapular plane. Arch. Phys. Med. Rehabil., *51*:595, 1970.

Ekholm, J., Arborelius, U. P., Hillered, L., and Örtqvist, Å.: Shoulder muscle EMG and resisting moment during diagonal exercise movements resisted by weight-and-pulley circuit. Scand. J. Rehab. Med., *10*:179, 1978.

Freedman, L., and Munro, R. R.: Abduction of the arm in the scapular plane: Scapular and glenohumeral movements. A roentgenographic study. J. Bone Joint Surg., *48A*:1503, 1966.

Germain, N. W., and Blair, S. N.: Variability of shoulder flexion with age, activity and sex. Am. Corr. Ther. J., *37*:156, 1983.

Hagberg, M.: Electromyographic signs of shoulder muscular fatigue in two elevated arm positions. Am. J. Phys. Med., *60*:111, 1981.

Herberts, P., and Kadefors, R.: A study of painful shoulder in welders. Acta Orthop. Scand., *47*:381, 1976.

Herberts, P., Kadefors, R., and Broman, H.: Arm positioning in manual tasks. An electromyographic study of localized muscle fatigue. Ergonomics, *23*:655, 1980.

Inman, V. T., Saunders, J. B. deC. M., and Abbott, L. C.: Observations on the function of the shoulder joint. J. Bone Joint Surg., *26A*:1, 1944.

Ito, N.: Electromyographic study of shoulder joint. J. Jpn. Orthop. Assoc., *54*:53, 1980.

Ivey, F. M., Jr., Calhoun, J. H., Rusche, K., and Bierschenk, J.: Isokinetic testing of shoulder strength: Normal values. Arch. Phys. Med. Rehabil., *66*:384, 1985.

Johnston, T. B.: The movements of the shoulder joint. A plea for the use of the "plane of the scapula" as the plane of reference for movements occurring at the humero-scapular joint. Br. J. Surg., *25*:252, 1937.

Kent, B. E.: Functional anatomy of the shoulder complex. A review. Phys. Ther., *51*:867, 1971.

Last, R. J.: Anatomy. Regional and Applied. Section 2. The Upper Limb. Edinburgh, Churchill Livingstone, 1972, pp. 79–111.

Lucas, D. B.: Biomechanics of the shoulder joint. Arch. Surg., *107*:425, 1973.

Maki, S., and Gruen, T.: Anthropometric study of the glenohumeral joint. Trans. Orthop. Res. Soc., *1*:173, 1976.

Murray, M. P., Gore, D. R., Gardner, G. M., and Mollinger, L. A.: Shoulder motion and muscle strength of normal men and women in two age groups. Clin. Orthop., *192*:268, 1985.

Poppen, N. K., and Walker, P. S.: Normal and abnormal motion of the shoulder. J. Bone Joint Surg., *58A*:195, 1976.

Poppen, N. K., and Walker, P. S.: Forces at the glenohumeral joint in abduction. Clin. Orthop., *135*:165, 1978.

Reeves, B.: Experiments on the tensile strength of the anterior capsular structures of the shoulder in man. J. Bone Joint Surg., *50B*:858, 1968.

Rockwood, C. A.: Acromioclavicular dislocation. *In* Fractures. Vol. 1. Edited by C. A. Rockwood and D. P. Green. Philadelphia, J. B. Lippincott, 1975, pp. 721–756.

Rowe, C. R.: Re-evaluation of the position of the arm in arthrodesis of the shoulder in the adult. J. Bone Joint Surg., *56A*:913, 1974.

Saha, A. K.: Mechanism of shoulder movements and a plea for the recognition of "zero position" of glenohumeral joint. Indian J. Surg., *12*:153, 1950.

Saha, A. K.: Dynamic stability of the glenohumeral joint. Acta Orthop. Scand., *42*:491, 1971.

Saha, A. K.: Mechanics of elevation of glenohumeral joint. Its application in rehabilitation of flail shoulder in upper brachial plexus injuries and poliomyelitis and in replacement of the upper humerus by prosthesis. Acta Orthop. Scand., *44*:668, 1973.

Sigholm, G., Herberts, P., Almström, C., and Kadefors, R.: Electromyographic analysis of shoulder muscle load. J. Orthop. Res., *1*:379, 1984.

SUGGESTED READING

Abbott, L. C., and Lucas, D. B.: The function of the clavicle. Its surgical significance. Ann. Surg., *140*:583–597, 1954.

American Academy of Orthopaedic Surgeons: Joint Motion. Method of Measuring and Recording. Chicago, AAOS, 1965. Reprinted by the British Orthopaedic Association, 1966.

Atwater, A. E.: Biomechanics of throwing: Correction of common misconceptions. Paper presented at Joint Meeting of the National College Physical Education Association for Men and the National Association for Physical Education of College Women, Orlando, Florida, January 6–9, 1977.

Basmajian, J. V.: Recent advances in the functional anatomy of the upper limb. Am. J. Phys. Med., *48*:165–177, 1969.

Basmajian, J. V., and Bazant, F. J.: Factors preventing downward dislocation of the adducted shoulder joint. An electromyographic and morphological study. J. Bone Joint Surg., *41A*:1182–1186, 1959.

Basmajian, J. V., and Latif, A.: Integrated actions and functions of the chief flexors of the elbow. A detailed electromyographic analysis. J. Bone Joint Surg., *39A*:1106–1118, 1957.

Bearn, J. G.: Direct observations on the function of the capsule of the sternoclavicular joint in clavicular support. J. Anat., *101*:105–170, 1967.

Bjelle, A., Hagberg, M., and Michaelsson, G.: Clinical and ergonomic factors in prolonged shoulder pain among industrial workers. Scand. J. Work Environ. Health, *5*:205–210, 1979.

Broome, H. L., and Basmajian, J. V.: The function of the teres major muscle: An electromyographic study. Anat. Rec., *170*:309–310, 1971.

Celli, L., Balli, A., de Luise, G., and Rovesta, C.: Some new aspects of the functional anatomy of the shoulder. Ital. J. Orthop. Traumatol., *11*:83–91, 1985.

Chinn, C. J., Priest, J. D., and Kent, B. E.: Upper extremity range of motion, grip, strength, and girth in highly skilled tennis players. Phys. Ther., *54*:474–483, 1974.

Colachis, S. C., and Strohm, B. R.: Effects of suprascapular and axillary nerve blocks on muscle force in upper extremity. Arch. Phys. Med. Rehabil., *52*:22–29, 1971.

Colachis, S. C., Strohm, B. R., and Brechner, V. L.: Effects of axillary nerve block on muscle force in the upper extremity. Arch. Phys. Med. Rehabil., *50*:647–654, 1969.

DeLuca, C. J., and Forrest, W. J.: Force analysis of individual muscles acting simultaneously on the shoulder joint during isometric abduction. J. Biomech., *6*:385–393, 1973.

Dempster, W. T.: Mechanisms of shoulder movement. Arch. Phys. Med. Rehabil., *46*:49–70, 1965.

DePalma, A. F.: Surgery of the Shoulder. 3rd Ed. Philadelphia, J. B. Lippincott, 1983.

Doody, S. G., Freedman, L., and Waterland, J. C.: Shoulder movements during abduction in the scapular plane. Arch. Phys. Med. Rehabil., *51*:595–604, 1970.

Dvir, Z., and Berme, N.: The shoulder complex in elevation of the arm: A mechanism approach. J. Biomech., *11*:219–255, 1978.

Ekholm, J., Arborelius, U. P., Hillered, L., and Örtqvist, Å.: Shoulder muscle EMG and resisting moment during diagonal exercise movements resisted by weight-and-pulley circuit. Scand. J. Rehab. Med., *10*:179–185, 1978.

Freedman, L., and Munro, R. R.: Abduction of the arm in the scapular plane: Scapular and glenohumeral movements. A roentgenographic study. J. Bone Joint Surg., *48A*:1503–1510, 1966.

Furlani, J.: Electromyographic study of the m. biceps brachii in movements at the glenohumeral joint. Acta Anat., *96*:270–284, 1976.

Germain, N. W., and Blair, S. N.: Variability of shoulder flexion with age, activity and sex. Am. Corr. Ther. J., *37*:156–160, 1983.

Hagberg, M.: Electromyographic signs of shoulder muscular fatigue in two elevated arm positions. Am. J. Phys. Med., *60*:111–121, 1981.

Hagberg, M.: Work load and fatigue in repetitive arm elevations. Ergonomics, *24*:543–555, 1981.

Herberts, P., and Kadefors, R.: A study of painful shoulder in welders. Acta Orthop. Scand., *47*:381–387, 1976.

Herberts, P., Kadefors, R., and Broman, H.: Arm positioning in manual tasks. An electromyographic study of localized muscle fatigue. Ergonomics, *23*:655–665, 1980.

Inman, V. T., Saunders, J. B. deC. M., and Abbott, L. C.: Observations on the function of the shoulder joint. J. Bone Joint Surg., *26A*:1–30, 1944.

Ito, N.: Electromyographic study of shoulder joint. J. Jpn. Orthop. Assoc., *54*:53–64, 1980.

Ivey, F. M., Jr., Calhoun, J. H., Rusche, K., and Bierschenk, J.: Isokinetic testing of shoulder strength: Normal values. Arch. Phys. Med. Rehabil., *66*:384–386, 1985.

Jensen, R. K., and Bellow, D. G.: Upper extremity contraction moments and their relationship to swimming training. J. Biomech., *9*:219–225, 1976.

Johnston, T. B.: The movements of the shoulder joint. A plea for the use of the "plane of the scapula" as the plane of reference for movements occurring at the humero-scapular joint. Br. J. Surg., *25*:252–260, 1937.

Jonsson, B., Olofsson, B. M., and Steffner, L. Ch.: Function of the teres major, latissimus dorsi and pectoralis major muscles. A preliminary study. Acta Morphol. Neerl. Scand., *9*:275–280, 1971/1972.

Kent, B. E.: Functional anatomy of the shoulder complex. A review. Phys. Ther., *51*:867–888, 1971.

Last, R. J.: Anatomy. Regional and Applied. Section 2. The Upper Limb. Edinburgh, Churchill Livingstone, 1972, pp. 79–111.

Lucas, D. B.: Biomechanics of the shoulder joint. Arch. Surg., *107*:425–432, 1973.

Maki, S., and Gruen, T.: Anthropometric study of the glenohumeral joint. Trans. Orthop. Res. Soc., *1*:173, 1976.

McMillan, J.: Therapeutic exercise for shoulder disabilities. J. Am. Phys. Ther. Assoc., *46*:1052–1067, 1966.

Moseley, H. F.: The clavicle: Its anatomy and function. Clin. Orthop., *58*:17–27, 1968.

Murray, M. P., Gore, D. R., Gardner, G. M., and Mollinger, L. A.: Shoulder motion and muscle strength of normal men and women in two age groups. Clin. Orthop., *192*:268–273, 1985.

Ovesen, J., and Nielsen, S.: Stability of the shoulder joint: Cadaver study of stabilizing structures. Acta Orthop. Scand., *56*:149–151, 1985.

Poppen, N. K., and Walker, P. S.: Normal and abnormal motion of the shoulder. J. Bone Joint Surg., *58A*:195–201, 1976.

Poppen, N. K., and Walker, P. S.: Forces at the glenohumeral joint in abduction. Clin. Orthop., *135*:165–170, 1978.

Reeves, B.: Experiments on the tensile strength of the anterior capsular structures of the shoulder in man. J. Bone Joint Surg., *50B*:858–865, 1968.

Reid, D.: The shoulder girdle: Its function as a unit in abduction. Physiotherapy, *55*:57–59, 1969.

Rockwood, C. A.: Acromioclavicular dislocation. *In* Fractures. Vol. 1. Edited by C. A. Rockwood and D. P. Green. Philadelphia, J. B. Lippincott, 1975, pp. 721–756.

Rothman, R. H., Marvel, J. P., and Heppenstall, R. B.: Anatomic considerations in the glenohumeral joint. Orthop. Clin. North Am., *6*:341–352, 1975.

Rowe, C. R.: Re-evaluation of the position of the arm in arthrodesis of the shoulder in the adult. J. Bone Joint Surg., *56A*:913–922, 1974.

Saha, A. K.: Mechanism of shoulder movements and a plea for the recognition of "zero position" of glenohumeral joint. Indian J. Surg., *12*:153–165, 1950.

Saha, A. K.: Dynamic stability of the glenohumeral joint. Acta Orthop. Scand., *42*:491–505, 1971.

Saha, A. K.: Mechanics of elevation of glenohumeral joint. Its application in rehabilitation of flail shoulder in upper brachial plexus injuries and poliomyelitis and in replacement of the upper humerus by prosthesis. Acta Orthop. Scand., *44*:668–678, 1973.

Shelvin, M. G., Lehmann, J. F., and Lucci, J. A.: Electromyographic study of the function of some muscles crossing the glenohumeral joint. Arch. Phys. Med. Rehabil., *50*:264–270, 1969.

Sigholm, G., Herberts, P., Almström, C., and Kadefors, R.: Electromyographic analysis of shoulder muscle load. J. Orthop. Res., *1*:379–386, 1984.

Singleton, M. C.: Functional anatomy of the shoulder. J. Am. Phys. Ther. Assoc., *46*:1043–1051, 1966.

Walker, P. S., and Poppen, N. K.: Biomechanics of the shoulder joint during abduction in the plane of the scapula. Bull. Hosp. Joint Dis., *38*:107–111, 1977.

Wiktorin, C. v. H., and Nordin, M.: Introduction to Problem Solving in Biomechanics. Philadelphia, Lea & Febiger, 1986, pp. 49–57.

13

BIOMECHANICS OF THE ELBOW

Joseph D. Zuckerman
Frederick A. Matsen III

The elbow is the anatomic area that joins the arm with the forearm. The elbow-forearm complex represents the second link in a mechanical chain of levers that begins at the shoulder and ends at the fingertips. The shoulder, as the first link, functions to permit the hand to be positioned anywhere within an imaginary sphere that represents the full excursion of shoulder motion. Elbow motion allows the height and length of the upper extremity to be adjusted, whereas forearm rotation allows the hand to be placed in the most effective position for function.

The bony structures of the elbow are the distal end of the humerus and the proximal ends of the radius and ulna (Fig. 13–1). The distal end of the humerus is formed by the hyperboloid trochlea medially and the convex capitellum laterally. The proximal end of the ulna is marked by the prominent olecranon process posteriorly and the coronoid process anteriorly, with the concave trochlear fossa lying between them; the radial notch is found on the lateral surface. The proximal end of the radius has a cup-shaped articular surface with an elevated rim called the radial head.

The elbow joint complex is essentially three separate synovial articulations (see Fig. 13–1). The humeroulnar joint is the articulation between the trochlea of the distal humerus and the reciprocally shaped trochlear fossa of the proximal ulna. The humeroradial joint is formed by the articulation between the capitellum of the distal humerus and the head of the radius. The proximal radioulnar joint is formed by the head of the radius and the radial notch of the proximal ulna.

In this chapter the role of the three elbow articulations and surrounding structures in providing motion and stability to the joint complex is discussed. Also, estimates of the joint reaction forces acting on the elbow in common static and dynamic situations are presented.

KINEMATICS

The elbow joint complex allows two degrees of freedom in motion: flexion-extension and pronation-supination (Steindler, 1970). The humeroulnar and humeroradial articulations, by allowing flexion and extension, make the elbow a ginglymoid, or hinged, joint. The proximal radioulnar articulation allows only forearm rotation (pronation and supination) and is classified as a trochoid joint. Therefore, the elbow is a composite trochoginglymoid joint. The humeroradial joint is anatomically a ball-and-socket joint much like the glenohumeral joint of the shoulder. It does not allow the global motion one would expect from this configuration, however, because the close association of the humeroulnar and radioulnar joints restricts its motion to two axes.

FLEXION AND EXTENSION

Elbow flexion and extension are accomplished through the humeroulnar and humeroradial articulations. The range of flexion and extension can be pre-

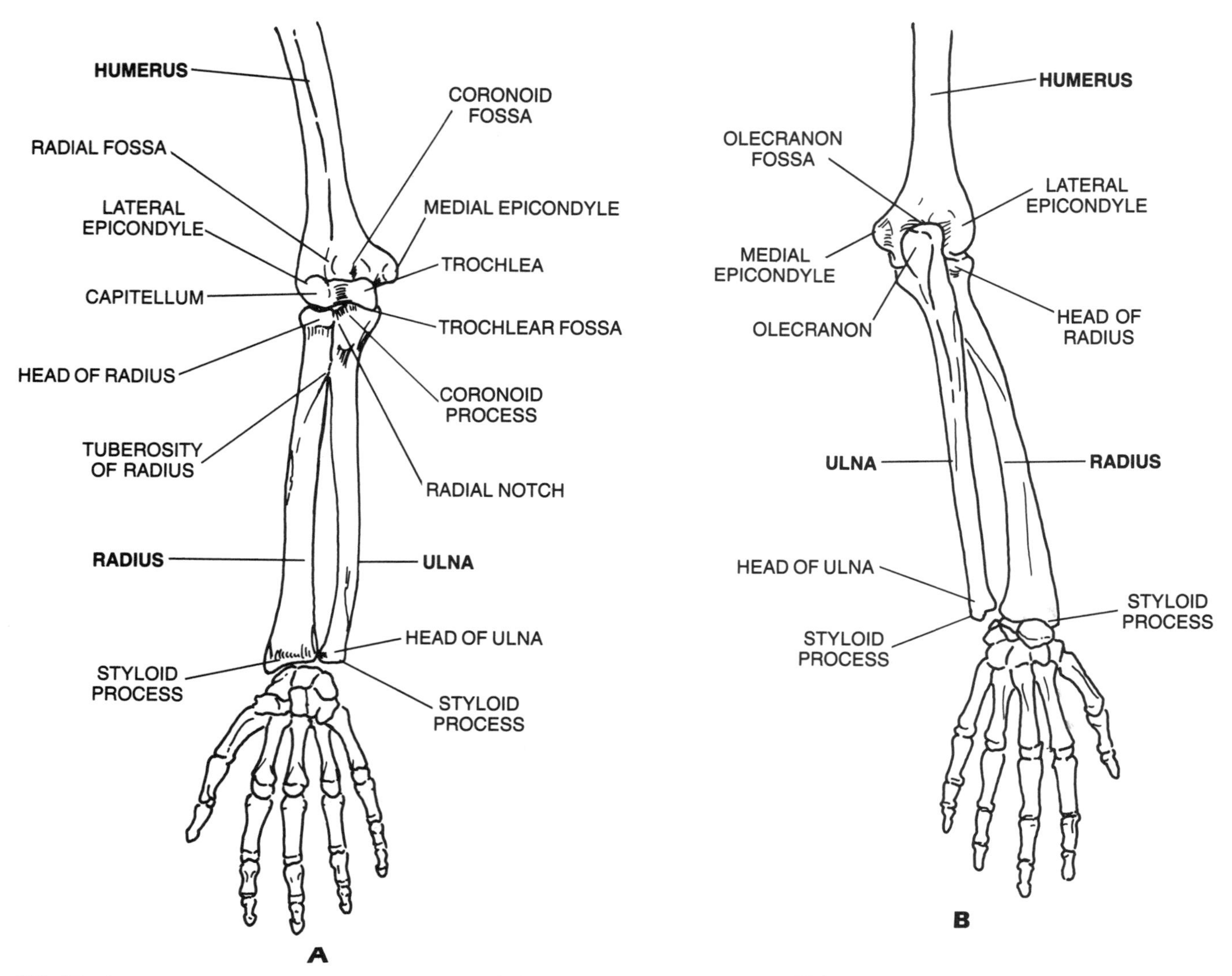

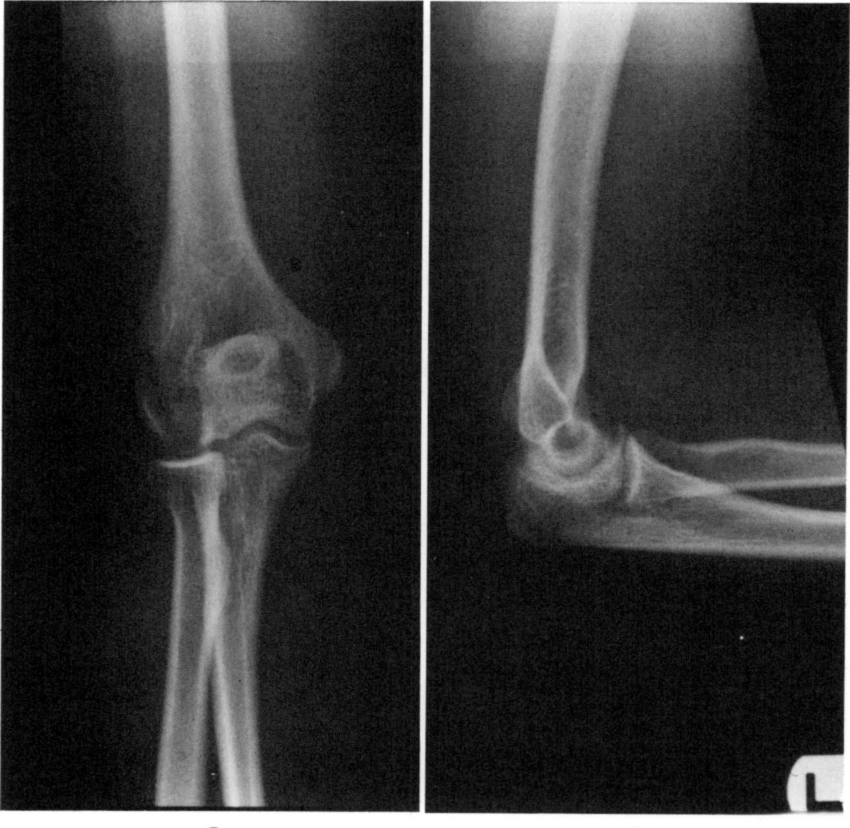

FIG. 13–1
The bony structures of the elbow. **A.** Anterior view of the right limb showing the three articulations of the elbow joint complex: the humeroulnar, the humeroradial, and the proximal radioulnar joints. **B.** Posterior view of the right limb. **C.** Anteroposterior roentgenogram of the left elbow in extension. (Courtesy of Steven Lubin, M.D.) **D.** Lateral roentgenogram of the left elbow in semiflexion. (Courtesy of Steven Lubin, M.D.)

dicted from the angular characteristics of the involved bony components (i.e., the portion of an arc subtended by the articular surface). The angular value of the articular surface of the trochlea of the humerus is 330 degrees, while that of the trochlear fossa of the ulna is 190 degrees. The difference is 140 degrees, which is the range of flexion-extension of the elbow. Similarly, 140 degrees is the difference between the angular value of the articular surface of the capitellum (180 degrees) and that of the proximal radial head (40 degrees). Indeed, studies conducted to determine the normal flexion-extension arc of the elbow joint have shown the range to be from 140 to 146 degrees (Morrey et al., 1981).

The axis of flexion and extension has been studied by several investigators. Morrey and Chao (1976) found that the axis passes through the middle of the trochlea, bisecting the angle formed by the longitudinal axes of the humerus and the ulna. They also noted that the instant centers of flexion and extension vary within 2 to 3 mm of this axis. London (1981), using different experimental techniques, found that the axis of flexion-extension of the elbow passes through the center of the concentric arcs outlined by the bottom of the trochlear sulcus and the periphery of the capitellum (Fig. 13–2). London also noted that the surface joint motion during flexion-extension is principally a gliding one. In the final 5 to 10 degrees of both flexion and extension, gliding changes to rolling. Rolling occurs at the end of flexion, as the coronoid process of the ulna comes into contact with the floor of the humeral coronoid fossa, and in the final stage of extension, as the olecranon of the ulna is received by the floor of the humeral olecranon fossa.

With the elbow fully extended and the forearm fully supinated, the longitudinal axes of the humerus and ulna normally intercept at a valgus angle referred to as the carrying angle (Fig. 13–3). In adults this angle is usually 10 to 15 degrees, and normally it is greater, on average, in women.

The valgus angulation is present because the trochlea extends farther distally than does the capitellum. In addition, the trochlea is asymmetric, with its outer lip extending farther distally than does the inner lip. Morrey and Chao (1976) found that, as a result of this asymmetry, flexing the elbow from zero to 120 degrees changed the carrying angle from 11 degrees of valgus to 6 degrees of varus; however, this finding has been disputed by London (1981), who

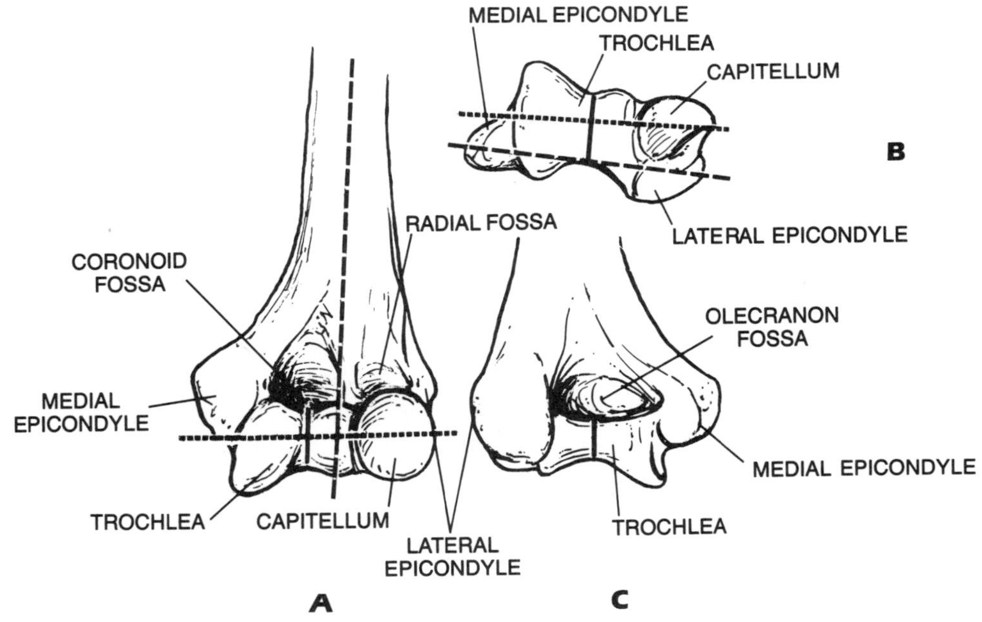

FIG. 13–2
Left humerus from the front **(A)**, from below **(B)**, and from behind **(C)**. As the ulna flexes about the humerus, the path it takes is determined by the deep sulcus in the trochlea (solid lines in **A**, **B**, and **C**). The plane defined by the sulcus is neither parallel to the long axis of the humerus (dashed line in **A**) nor perpendicular to the plane of the epicondyles (dashed line in **B**). The axis of elbow flexion-extension (dotted line in **A** and **B**) passes through the center of arcs formed by the trochlear sulcus and the capitellum and is not coincident with the line through the epicondyles (dashed line in **B**). (Adapted from London, 1981.)

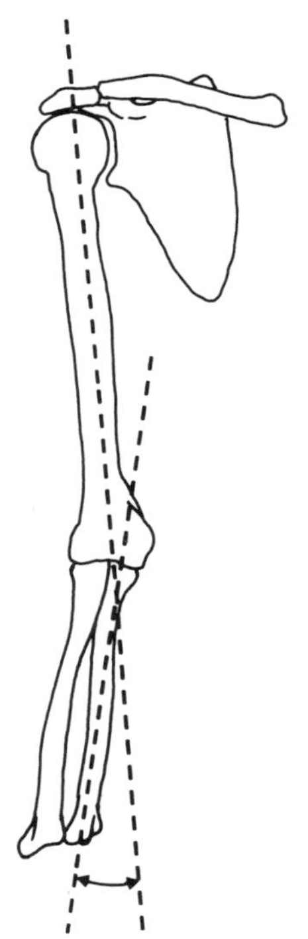

FIG. 13–3
The carrying angle of the elbow, formed by the interception of the long axes of the humerus and ulna (represented by dashed lines) with the elbow fully extended and the forearm supinated (anterior view of the right limb). Valgus angulation normally ranges from 10 to 15 degrees.

determined that the carrying angle changed less than 1 degree as the elbow moved from full extension to full flexion. London felt that the change in carrying angle observed by Morrey and Chao was a function of the positions in which the elbow was studied and did not represent the true kinematic characteristics of the joint.

In an attempt to understand the different measurements reported, An and coworkers (1984) carried out a rigorous theoretical analysis of elbow kinematics based on available anatomic and geometric data on the skeleton. They described three different definitions of carrying angle and found that the change in carrying angle from full extension to full flexion depended on the specific definition utilized. Another important factor was the anthropometric variations found for the oblique angle between the trochlea and the humerus and the angle between the olecranon and the ulna, which they described as the "joint flexion angle." The discrepancy between Morrey and Chao's (1976) and London's (1981) data could be attributed to differing definitions of carrying angle as well as anthropometric variations in the specimens used. An's group (1984) concluded that the dynamic change of carrying angle that occurs with elbow flexion and extension probably has little clinical significance.

PRONATION AND SUPINATION

Pronation and supination are motions in which the forearm rotates about a longitudinal axis passing through the center of the radial head and the distal ulnar articular surface. Therefore, the axis is oblique to the longitudinal axis of the radius and ulna. This rotation involves primarily the humeroradial joint and the proximal radioulnar joint. In pronation the palm faces posteriorly if the elbow is extended and down if the elbow is flexed 90 degrees. In supination the palm faces anteriorly if the elbow is extended and upward if the elbow is flexed 90 degrees.

Carret and associates (1976) studied in detail the instant centers of rotation at the proximal and distal radioulnar joints with the forearm in varying degrees of pronation and supination. They found that the proximal instant center varied with differences in the curvature of the radial head among individuals. Chao and Morrey (1978) investigated the effect of pronation and supination on the position of the ulna. They found no significant axial rotation or valgus deviation of the ulna during forearm pronation and supination when the elbow was fully extended.

Studies of the range of forearm pronation and supination in normal subjects have produced differing results. Steindler (1970) found the range to be from 120 to 140 degrees. Wagner (1977) found that pronation averaged 71 degrees, while supination averaged 88 degrees. The American Academy of Orthopaedic Surgeons (1965) reported average pronation to be about 70 degrees and average supination to be about 85 degrees. Morrey and coworkers (1981), using a triaxial electrogoniometer on normal subjects, measured average pronation at 68 degrees and average supination at 74 degrees.

FUNCTIONAL MOTION OF THE ELBOW

Most activities of daily living can be performed within a much more limited range of elbow motion than the 140- to 146-degree range of flexion-extension and the 142-degree range of pronation and supination noted by Morrey's group (1981). Using triaxial electrogoniometry, Chao and associates (1980) and Morrey and coworkers (1981) measured the elbow

TABLE 13–1

ELBOW MOTION REQUIRED FOR CARRYING OUT SELECTED DAILY ACTIVITIES

ACTIVITY	MEAN FLEXION (DEGREES)			MEAN ROTATION (DEGREES)		
	MINIMUM	MAXIMUM	ARC	PRONATION	SUPINATION*	ARC
Pouring from a pitcher	36	58	22	43	22	65
Cutting with a knife	89	107	18	42	−27	15
Putting fork to mouth	85	128	43	10	52	62
Using a telephone	43	136	93	41	23	64
Reading a newspaper	78	104	26	49	−7	42
Rising from a chair	20	95	75	34	−10	24
Opening a door	24	57	33	35	23	58
Putting glass to mouth	45	130	85	10	13	23

(Data from Morrey et al., 1981. N = 33.)
*Negative value indicates pronation.

range of motion required for common daily activities such as pouring from a pitcher, using a telephone, and rising from a chair. The specific ranges of flexion-extension and pronation-supination for these and other activities are summarized in Table 13–1. Most activities of daily living can be accomplished with 100 degrees of forearm rotation (an arc from 50 degrees of pronation to 50 degrees of supination) and 100 degrees of elbow flexion (an arc from 30 to 130 degrees). Patients with limited elbow motion can obtain compensatory motion by means of shoulder abduction and rotation, trunk flexion and rotation, and movements of the head (Chao et al., 1980; Ishizuki, 1979).

ELBOW JOINT STABILITY

Unlike the shoulder, the elbow joint complex possesses significant inherent stability because of the interlocking configuration of the articulating surfaces. The main contributor to bony stability is the articulation between the trochlea of the humerus and the trochlear fossa of the ulna (Schwab et al., 1980). The coronoid process provides an important block to posterior displacement as the elbow flexes. The humeroradial articulation provides some resistance to valgus stress across the elbow and inhibits posterior dislocation at 90 degrees of flexion or more. The proximal radioulnar articulation does not provide significant elbow stability but simply allows forearm pronation and supination to take place.

Soft tissue stability is provided by the ligamentous structures that surround the elbow. The most important is the medial collateral ligament, which is the major stabilizer against valgus stress. It originates from the inferior surface of the medial epicondyle of the distal humerus and inserts along the medial edge of the olecranon from the coronoid process anteriorly to the midportion of the olecranon. It is composed of two portions: the anterior oblique ligament, which is tight in extension, and the posterior oblique ligament, which is tight in flexion. The anterior oblique ligament is the more important stabilizing component (Schwab et al., 1980).

The elbow does not possess a true lateral collateral ligament. The anatomic structure designated as such originates on the lateral epicondyle of the distal humerus and attaches to the annular ligament but does not attach directly to bone. Therefore, this lateral stabilizing structure is able to resist only minimal tensile forces. Pauly and associates (1967) showed that the anconeus, a muscle located on the lateral aspect of the elbow, provides additional stability against varus stress. The lack of strong lateral stabilizing structures does not pose a significant problem, because valgus stability is much more important functionally than varus stability. If one considers typical elbow motions involving high forces (e.g., throwing, using a heavy tool, or falling on an outstretched arm), it can be appreciated that the primary tensile stresses are sustained on the medial side. The valgus carrying angle of the elbow probably contributes to this situation because it produces increased tensile stresses medially and increased compressive stresses laterally. The annular ligament surrounds the neck of the radius and allows pronation and supination while simultaneously preventing radial head displacement.

An additional soft tissue stabilizer is the interosseous membrane, which binds together the radial and ulnar shafts. These fibers run obliquely downward and medially (from the radius to the ulna). The interosseous ligament prevents separation or migration (longitudinal shifts) of the radius and ulna with respect to each other.

KINETICS

Elbow motion—flexion, extension, supination, and pronation—is the result of the actions of the muscles surrounding the elbow (Fig. 13–4). Each specific motion generally results from the action of more than one muscle; furthermore, some muscles participate in producing more than one specific motion.

Most of the muscles involved in elbow function and stability originate on the humerus and insert on either the radius or ulna (see Fig. 13–4). Humeroradial muscles include the biceps, brachioradialis, and pronator teres. Humeroulnar muscles include the brachialis, triceps, and anconeus. A radioulnar group includes the supinator and the pronator quadratus. Two additional muscles of importance for elbow motion insert on the wrist and hand. The extensor carpi radialis muscles originate on the distal humerus and insert dorsally on the hand; the flexor carpi radialis originates on the proximal ulna and inserts volarly on the wrist.

MUSCLE ACTIONS DURING FLEXION

Using electromyographic data, Basmajian and Latif (1957) determined that the brachialis, which arises from the anterior aspect of the humerus and inserts on the anterior aspect of the proximal ulna, was the primary flexor of the elbow. They found that it functioned as a flexor regardless of forearm position and referred to it as the "workhorse of the elbow joint." The biceps, which arises via a long head tendon from the supraglenoid tubercle and a short head tendon from the coracoid process of the scapula and inserts on the bicipital tuberosity of the radius, was found to be active in flexion only when the forearm was supinated or in the neutral position. The brachioradialis, which originates from the lateral two thirds of the distal humerus and inserts on the distal lateral aspect of the radius near the radial styloid, was found to be active during rapid flexion movements of the forearm or when a weight was lifted during a slow flexion movement.

An and associates (1981) analyzed the muscles about the elbow using a serial cross-sectional analysis and a special dissection technique to measure muscle volume. From the data they were able to determine the work capacity of each muscle. They determined that the brachialis, biceps, brachioradialis, and extensor carpi radialis were the major flexors of the elbow, the brachialis possessing the greatest work capacity.

Other authors (Steindler, 1970; Pauly et al., 1967; Larson, 1969) analyzed the flexors of the elbow using various techniques and reported findings at variance with those of Basmajian and Latif (1957) and of An's group (1981). Most of these studies draw their conclusions from electromyographic analysis. In interpreting them, it must be kept in mind that the quantitative electromyographic method has not always been found to correlate well with maximum muscle force (Currier, 1972).

Larson (1969) measured the isometric elbow flexor force that normal subjects could generate with the elbow in 65 degrees of flexion. This force, which averaged 420 N, was maximal when the forearm was supinated or in the neutral position and lowest when it was pronated.

MUSCLE ACTIONS DURING EXTENSION

The primary extensor of the elbow, the triceps (Basmajian, 1969), is composed of three separate heads. The long head originates from the inferior aspect of the glenoid of the scapula, and the medial and lateral heads originate from the posterior aspect of the humerus. The three heads form one tendinous insertion into the olecranon process of the ulna. Basmajian (1969) determined that the medial head is the primary extensor and that the lateral and long heads act in reserve. This auxiliary function is particularly true for the long head of the triceps because it originates on the glenoid of the scapula.

The anconeus muscle, which arises from the posterolateral aspect of the distal humerus and inserts on the posterolateral aspect of the proximal ulna, is also active in extension. In a detailed electromyographic study, Pauly and coworkers (1967) concluded that the anconeus was active in initiating and maintaining elbow extension and in stabilizing the elbow during upper extremity motion, particularly against varus forces. For example, they recorded

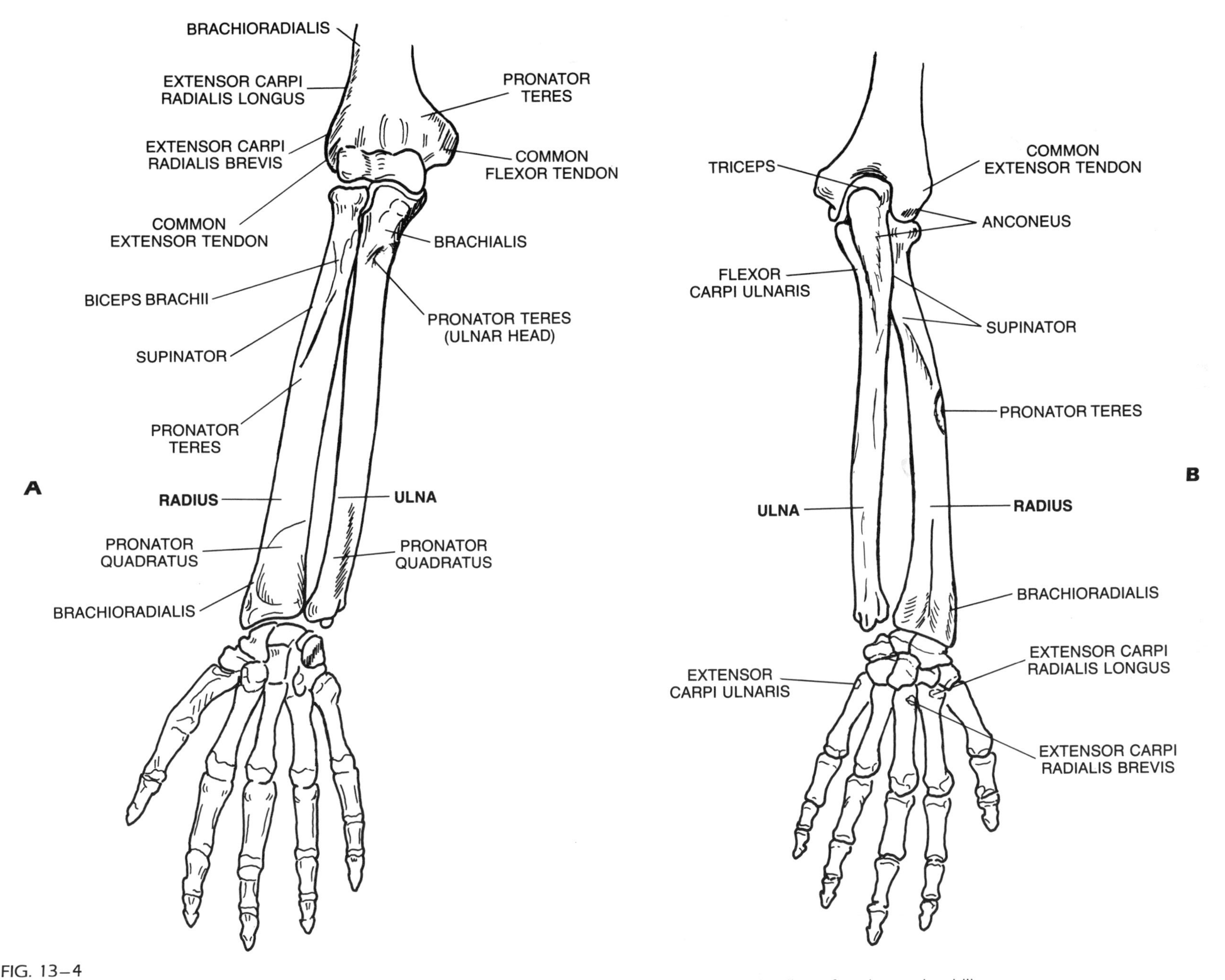

FIG. 13–4
Schematic drawing of the bones of the forearm showing attachment sites of major muscles involved in elbow function and stability.
A. Anterior (volar) view of the right limb. **B.** Posterior (dorsal) view of the right limb.

active contraction of the anconeus during forced finger flexion and extension. Other muscles around the elbow, such as the biceps, the brachioradialis, and the triceps, also participated in this type of stabilization, even though they were not required for the primary action. Apparently then, the elbow acquires additional stability by using mutually antagonistic muscles to increase the compressive load across the joint. An et al. (1981) recognized the importance of the triceps and anconeus as extensors but determined that the flexor carpi ulnaris was active in extension as well. In addition, they determined that the triceps had the largest work capacity of all elbow extensors.

Little and Lehmkuhl (1966) measured the elbow extension force generated in 60 young women 17 to 21 years of age with the elbow in three different positions of flexion. The extension force ranged from 80 to 110 N, depending on the position of the arm. Currier (1972), using a cable tensiometer, measured the maximum isometric extension force in 41 male subjects with the elbow in different amounts of flexion. The maximum tension of 220 N developed when the arm was in 90 degrees of flexion.

MUSCLE ACTIONS DURING PRONATION

Muscles involved in pronation include the pronator quadratus and pronator teres. The pronator quadratus takes its origin from the volar aspect of the distal ulna and inserts on the distal and lateral aspect of the supinated radius. The pronator teres is more proximally located, arising from the medial epicondyle of the humerus and inserting on the lateral aspect of the supinated radius at approximately the midpoint.

Basmajian (1969), using electromyographic analysis, showed that the pronator quadratus is the primary pronator of the forearm regardless of the position of the forearm or the amount of flexion of the elbow. He found that the pronator teres acted as a secondary pronator whenever rapid pronation was required or during pronation against resistance. In an earlier study (Basmajian and Travill, 1961), he determined that the amount of elbow flexion had no influence on pronator teres activity. This finding is not what would be anticipated. One would expect the pronator teres to function better with the elbow flexed because in full extension the lever arm for this muscle is shortest. Steindler (1970) suggested that other muscles such as the flexor carpi radialis may serve as accessory pronators, but Basmajian (1969) determined that the flexor carpi radialis, brachioradialis, and extensor carpi ulnaris do not function in pronation.

MUSCLE ACTIONS DURING SUPINATION

Muscles involved in supination of the forearm include the supinator and the biceps. The supinator arises from the lateral epicondyle of the humerus and the proximal lateral aspect of the ulna, and it inserts into the anterior aspect of the supinated proximal radius. The origin and insertion of the biceps are described above, in the section on muscle actions during elbow flexion.

The supinator muscle is the primary supinator of the forearm. Slow, unresisted supination is brought about by the independent action of the supinator regardless of the position of the elbow (Basmajian and Travill, 1961), but rapid unresisted supination with the elbow flexed, or any supination against resistance regardless of the position of the elbow, is assisted by the biceps. Since the biceps also functions as an elbow flexor whenever it functions as a supinator, the action of the elbow extensors (triceps and anconeus) is required to cancel its flexor action. Basmajian (1969) determined that the brachioradialis, which had long been believed to be a supinator of the forearm, was actually not active in supination.

JOINT FORCES AT THE ELBOW

Because a large number of muscles participate in producing flexion and extension of the elbow, a few simplifying assumptions must be made in order for the joint reaction force at the elbow to be estimated in certain static and dynamic situations. In the following static example, the simplified free body technique for coplanar forces is used to calculate the joint reaction force at the elbow during flexion with and without an object in the hand (Fig. 13–5). The three main coplanar forces acting on the elbow are analyzed, and the equilibrium equations for moments and forces are applied. The elbow is flexed 90 degrees; it is assumed that the predominant elbow flexors are the brachialis and biceps and that the force produced through the tendons of these muscles (M) acts perpendicular to the longitudinal axis of the forearm. The distance between the center of rotation of the elbow joint and the point of insertion of the tendons of these muscles (the lever arm of M) is approximately 5 cm. The mass of the forearm (2 kg) produces a gravitational force (W)

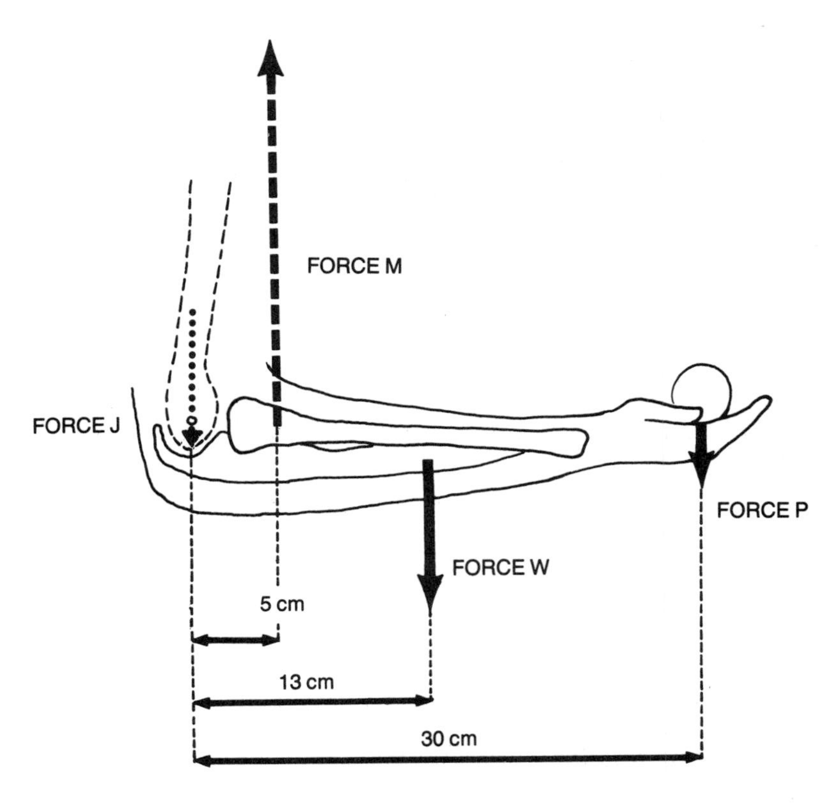

FIG. 13–5
The reaction force on the elbow joint during elbow flexion with and without an object in the hand can be calculated by means of the simplified free body technique for coplanar forces and the equilibrium equations that state that the sum of the moments and the sum of the forces acting on the elbow joint must be zero. The primary elbow flexors are assumed to be the biceps and the brachialis muscles. The force produced through the tendons of these muscles (M) acts at a distance of 5 cm from the center of rotation of the joint (indicated by the hollow circle). The force produced by the weight of the forearm (W), taken to be 20 N, acts at a distance of 13 cm from the center of rotation. The force produced by any weight held in the hand (P) acts at a distance of 30 cm from the center of rotation.

Case A. No object is held in the hand. M is calculated with the equilibrium equation for moments. Clockwise moments are considered to be positive, whereas counterclockwise moments are considered to be negative.

$$\Sigma M = 0$$
$$(13\ \text{cm} \times W) + (30\ \text{cm} \times P) - (5\ \text{cm} \times M) = 0.$$
$$\text{If } W = 20\ \text{N and } P = 0,$$
$$M = \frac{13\ \text{cm} \times 20\ \text{N}}{5\ \text{cm}}.$$

M is calculated to be 52 N.

J, the reaction force on the trochlear fossa of the ulna, can now be calculated by means of the equilibrium equation for forces. Gravitational forces are negative; forces in the opposite direction are positive.

$$\Sigma F = 0$$
$$M - J - W - P = 0$$
$$J = 52\ \text{N} - 20\ \text{N} - 0\ \text{N}.$$

J is found to be 32 N.

Case B. An object of 1 kg is held in the hand, producing a force of 10 N (P).

$$\Sigma M = 0.$$
$$\text{If } W = 20\ \text{N and } P = 10\ \text{N},$$
$$(13\ \text{cm} \times 20\ \text{N}) + (30\ \text{cm} \times 10\ \text{N}) - (5\ \text{cm} \times M) = 0$$
$$M = \frac{260\ \text{Ncm} + 300\ \text{Ncm}}{5\ \text{cm}}.$$

M is found to be 112 N.

The joint reaction force can now be calculated.

$$\Sigma F = 0$$
$$M - W - P - J = 0$$
$$J = M - W - P$$
$$J = 112\ \text{N} - 20\ \text{N} - 10\ \text{N}.$$

J is found to be 82 N. Thus, in this example a 1-kg object held in the hand with the elbow flexed 90 degrees increases the joint reaction force by 50 N.

equal to 20 N. The lever arm of W, the distance from the center of rotation of the elbow to the midpoint of the forearm, is 13 cm. The force produced by any weight held in the hand (P) acts at a distance of 30 cm from the center of rotation of the elbow joint.

The muscle force required to keep the elbow in the flexed position (M) is calculated with the equilibrium equation for moments. The equilibrium equation for forces is then used to calculate the joint reaction force on the trochlear fossa (J). When no object is held in the hand (case A, Fig. 13–5), the muscle force is calculated to be 52 N and the joint reaction force, 32 N. By contrast, when a 1-kg weight is held in the hand (case B, Fig. 13–5), producing a gravitational force (P) of 10 N at a distance of 30 cm from the center of elbow rotation, the required muscle force (M) rises to 112 N and the joint reaction force more than doubles, reaching 82 N. Thus, small loads applied to the hand dramatically increase the elbow joint reaction force.

An estimation of the joint reaction force can also be made for the elbow during extension. In the following example, the elbow is held in 90 degrees of flexion with the forearm positioned over the head and parallel to the ground (Fig. 13–6). In this position, action of the elbow extensors is required to offset the gravitational force on the forearm that tend to increase elbow flexion. Therefore, the joint reaction force for extension can be determined without having the elbow extended.

It is assumed that the triceps is the predominant extensor and that the force through the tendon of this muscle acts perpendicular to the longitudinal axis of the forearm. Therefore, the three main coplanar forces acting on the elbow include the force produced by the weight of the arm (W), the tensile force exerted through the tendon of the triceps muscle (M), and the joint reaction force on the trochlear fossa of the ulna (J). The distance between the center of rotation of the elbow and the point of insertion of the tendon of the triceps muscle (the lever arm of M) is approximately 3 cm.

M and J are calculated with equilibrium equations. The joint reaction force for the elbow in extension is 107 N, compared with 32 N in flexion. This more than threefold increase can be explained by the fact that the lever arm for the elbow extensor force is shorter than that for the flexor force—3 cm as opposed to 5 cm. Thus, a greater muscle force (87 N as opposed to 52 N) is required for the forearm to be maintained in the extended position, and the joint reaction force is higher as a result.

These simplified calculations provide an estimation of the joint reaction force during elbow positioning and lifting of a light object. The joint reaction force at the elbow during other common activities has been estimated by Nicol (1977), who used three-dimensional biomechanical analysis in three healthy male subjects as well as in cadavers specimens. During dressing and eating activities the compressive loads at the elbow (the joint reaction forces) were found to be 300 N. Supporting oneself with the arms while rising from a chair resulted in a joint reaction force of 1,700 N on the medial aspect of the elbow and 800 N on the lateral aspect. Pulling a table generated a joint reaction force of 1,900 N. These forces can be placed in proper perspective if one considers that the gravitational force generated by a 70-kg man is 700 N. Therefore, it can be seen that these common activities can give rise to elbow joint loads of more than twice body weight. These data challenge the simplified view that the elbow is not a load-bearing joint.

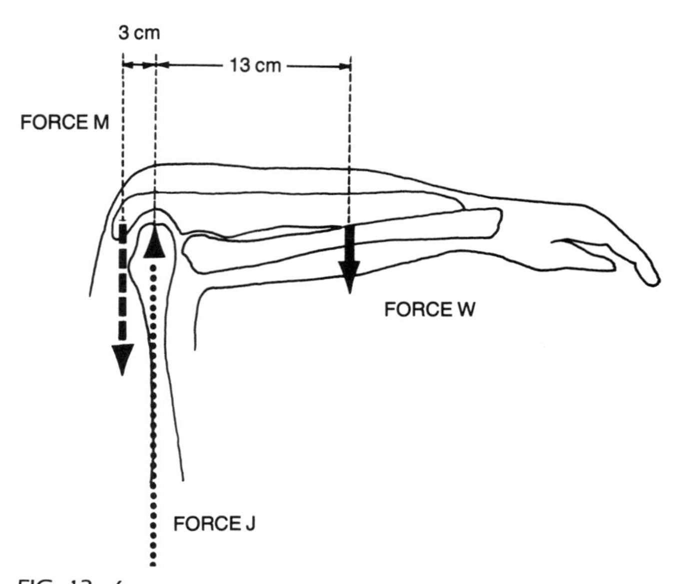

FIG. 13–6
The joint reaction force during elbow extension can be calculated by means of the same method:

$$\Sigma M = 0$$
$$(13\text{ cm} \times W) - (3\text{ cm} \times M) = 0.$$
If $W = 20$ N,
$$M = \frac{13\text{ cm} \times 20\text{ N}}{3\text{ cm}}.$$

M is found to be 87 N.

$$\Sigma F = 0$$
$$J - M - W = 0$$
$$J = M + W$$
$$J = 87\text{ N} + 20\text{ N}.$$

J is found to be 107 N. Thus, in this example the joint reaction force during elbow extension is 75 N greater than during elbow flexion.

SUMMARY

1. The elbow joint complex is essentially three articulations in one: the humeroulnar, the humeroradial, and the proximal radioulnar.
2. This joint complex provides two types of motion, flexion-extension and pronation-supination (forearm rotation).
3. Stability of the elbow joint is provided mainly by the ligamentous apparatus surrounding the joint and the interlocking of the distal humerus, proximal radius, and proximal ulna.
4. The relatively large number of muscles producing the various motions in the elbow complicates an exact force analysis for this joint complex. Estimates suggest that static loads approach—and dynamic loads exceed—body weight.

REFERENCES

American Academy of Orthopaedic Surgeons: Joint Motion. Method of Measuring and Recording. Chicago, AAOS, 1965. Reprinted by the British Orthopaedic Association, 1966.

An, K. N., Morrey, B. F., and Chao, E. Y. S.: Carrying angle of the human elbow joint. J. Orthop. Res., *1*:369, 1984.

An, K. N., et al.: Muscles across the elbow joint: A biomechanical analysis. J. Biomech., *14*:659, 1981.

Basmajian, J. V.: Recent advances in the functional anatomy of the upper limb. Am. J. Phys. Med., *48*:165, 1969.

Basmajian, J. V., and Latif, M. A.: Integrated actions and functions of the chief flexors of the elbow. J. Bone Joint Surg., *39A*:1106, 1957.

Basmajian, J. V., and Travill, A. A.: Electromyography of the pronator muscles of the forearm. Anat. Rec., *139*:45, 1961.

Carret, J.-P., Fischer, L. P., Gonon, G. P., and Dimnet, J.: Etude cinematique de la prosupination au niveau des articulations radiocubitales (radio ulnaris). Bull. Assoc. Anat., *60*:279, 1976.

Chao, E. Y., and Morrey, B. F.: Three-dimensional rotation of the elbow. J. Biomech., *11*:57, 1978.

Chao, E. Y., An, K. N., Ashew, L. J., and Morrey, B. F.: Electrogoniometer for the measurement of human elbow joint rotation. J. Biomech. Eng., *102*:301, 1980.

Currier, D. P.: Maximal isometric tension of the elbow extensors at varied positions. Part 2. Assessment of extensor components by quantitative electromyography. Phys. Ther., *52*:1265, 1972.

Ishizuki, M.: Functional anatomy of the elbow joint and three-dimensional quantitative motion analysis of the elbow joint. J. Jpn. Orthop. Assoc., *53*:989, 1979.

Larson, R. F.: Forearm positioning on maximal elbow-flexor force. Phys. Ther., *49*:748, 1969.

Little, A. D., and Lehmkuhl, D.: Elbow extension force. Measured in three test positions. J. Am. Phys. Ther. Assoc., *46*:7, 1966.

London, J. T.: Kinematics of the elbow. J. Bone Joint Surg., *63A*:529, 1981.

Morrey, B. F., and Chao, E. Y. S.: Passive motion of the elbow joint. A biomechanical analysis. J. Bone Joint Surg., *58A*:501, 1976.

Morrey, B. F., Askew, L. J., An, K. N., and Chao, E. Y.: A biomechanical study of normal elbow motion. J. Bone Joint Surg., *63A*:872, 1981.

Nicol, A. C., Berme, N., and Paul, J. P.: A biomechanical analysis of elbow joint function. *In* Institution of Mechanical Engineers Conference Publications 1977–5, Joint Replacement in the Upper Limb, pp. 45–51. Conference sponsored by the Medical Engineering Section of the Institution of Mechanical Engineers and the British Orthopaedic Association, London, April 18–20, 1977.

Pauly, J. E., Rushing, J. L., and Scheving, L. E.: An electromyographic study of some muscles crossing the elbow joint. Anat. Rec., *159*:47, 1967.

Schwab, G. H., Bennett, J. B., Woods, G. W., and Tullos, H. S.: Biomechanics of elbow stability: The role of the medial collateral ligament. Clin. Orthop., *146*:42, 1980.

Steindler, A.: Kinesiology of the Human Body under Normal and Pathological Conditions. Springfield, Charles C Thomas, 1970.

Wagner, C.: Determination of the rotary flexibility of the elbow joint. Eur. J. Appl. Physiol., *37*:47, 1977.

SUGGESTED READING

American Academy of Orthopaedic Surgeons: Joint Motion. Method of Measuring and Recording. Chicago, AAOS, 1965. Reprinted by the British Orthopaedic Association, 1966.

Amis, A. A., Dowson, D., and Wright, V.: Analysis of elbow forces due to high speed forearm movements. J. Biomech., *13*:825–831, 1980.

Amis, A. A., Dowson, D., and Wright, V.: Elbow joint force predictions for some strenuous isometric actions. J. Biomech., *13*:765–775, 1980.

Amis, A. A., Miller, J. H., Dowson, D., and Wright, V.: Biomechanical aspects of the elbow: Joint forces related to prosthesis design. Eng. Med., *10*:65–68, 1981.

An, K. N., and Morrey, B. F.: Biomechanics of the elbow. *In* The Elbow and Its Disorders. Edited by B. F. Morrey. Philadelphia, W. B. Saunders, 1985, pp. 43–61.

An, K. N., Morrey, B. F., and Chao, E. Y. S.: Angle of the human elbow joint. J. Orthop. Res., *1*:369–378, 1984.

An, K. N., et al.: Muscles across the elbow joint: A biomechanical analysis. J. Biomech., *14*:659–669, 1981.

Anderson, G. B. J., and Schultz, A. B.: Transmissions of moments across the elbow joint and the lumbar spine. J. Biomech., *12*:747–755, 1979.

Basmajian, J. V.: Recent advances in the functional anatomy of the upper limb. Am. J. Phys. Med., *48*:165–177, 1969.

Basmajian, J. V., and Latif, M. A.: Integrated actions and functions of the chief flexors of the elbow. J. Bone Joint Surg., *39A*:1106–1118, 1957.

Basmajian, J. V., and Travill, A. A.: Electromyography of the pronator muscles of the forearm. Anat. Rec., *139*:45–49, 1961.

Carret, J.-P., Fischer, L. P., Gonon, G. P., and Dimnet, J.: Etude cinematique de la prosupination au niveau des articulations radiocubitales (radio ulnaris). Bull. Assoc. Anat., *60*:279–295, 1976.

Chao, E. Y., and Morrey, B. F.: Three-dimensional rotation of the elbow. J. Biomech., *11*:57–73, 1978.

Chao, E. Y., An, K. N., Ashew, L. J., and Morrey, B. F.: Electrogoniometer for the measurement of human elbow joint rotation. J. Biomech. Eng., *102*:301–310, 1980.

Currier, D. P.: Maximal isometric tension of the elbow extensors at varied positions. Part 2. Assessment of extensor components by quantitative electromyography. Phys. Ther., *52*:1265–1276, 1972.

Dehaven, K. E., and Evarts, C. M.: Throwing injuries of the elbow in athletes. Orthop. Clin. North Am., *4*:801–808, 1973.

Goel, V. K., Singh, D., and Bylani, V.: Contact areas in human elbow joints. J. Biomech. Eng., *104*:169–175, 1982.

Hang, Y.-S., et al.: Biomechanical study of the pitching elbow. Int. Orthop., *3*:217–223, 1979.

Ishizuki, M.: Functional anatomy of the elbow joint and three-dimensional quantitative motion analysis of the elbow joint. J. Jpn. Orthop. Assoc., *53*:989–996, 1979.

Larson, R. F.: Forearm positioning on maximal elbow-flexor force. Phys. Ther., *49*:748–756, 1969.

Le Bozec, S., Maton, B., and Crockaert, J. C.: The synergy of elbow extensor muscles during static work in man. Eur. J. Appl. Physiol., *43*:57–68, 1980.

Le Bozec, S., Maton, B., and Crockaert, J. C.: The synergy of elbow extensor muscles during dynamic work in man. I. Elbow extension. Eur. J. Appl. Physiol., *44*:255–269, 1980.

Le Bozec, S., Maton, B., and Crockaert, J. C.: The synergy of elbow extensor muscles during dynamic work in man. II. Braking of elbow flexion. Eur. J. Appl. Physiol., *44*:271–278, 1980.

Little, A. D., and Lehmkuhl, D.: Elbow extension force. Measured in three test positions. J. Am. Phys. Ther. Assoc., *46*:7–17, 1966.

London, J. T.: Kinematics of the elbow. J. Bone Joint Surg., *63A*:529–535, 1981.

Morrey, B. F., and Chao, E. Y. S.: Passive motion of the elbow joint. A biomechanical analysis. J. Bone Joint Surg., *58A*:501–508, 1976.

Morrey, B. F., Askew, L. J., An, K. N., and Chao, E. Y.: A biomechanical study of normal elbow motion. J. Bone Joint Surg., *63A*:872–877, 1981.

Nicol, A. C., Berme, N., and Paul, J. P.: A biomechanical analysis of elbow joint function. *In* Institution of Mechanical Engineers Conference Publications 1977–5, Joint Replacement in the Upper Limb, pp. 45–51. Conference sponsored by the Medical Engineering Section of the Institution of Mechanical Engineers and the British Orthopaedic Association, London, April 18–20, 1977.

Pauly, J. E., Rushing, J. L., and Scheving, L. E.: An electromyographic study of some muscles crossing the elbow joint. Anat. Rec., *159*:47–54, 1967.

Petrofsky, J. S., and Phillips, C. A.: The effect of elbow angle on the isometric strength and endurance of the elbow flexors in men and women. J. Human Ergol., *9*:125–131, 1980.

Ray, R. D., Johnson, R. J., and Jameson, R. M.: Rotation of the forearm. An experimental study of pronation and supination. J. Bone Joint Surg., *33A*:993–996, 1951.

Schwab, G. H., Bennett, J. B., Woods, G. W., and Tullos, H. S.: Biomechanics of elbow stability: The role of the medial collateral ligament. Clin. Orthop., *146*:42, 1980.

Steindler, A.: Kinesiology of the Human Body under Normal and Pathological Conditions. Springfield, Charles C Thomas, 1970.

Wagner, C.: Determination of the rotary flexibility of the elbow joint. Eur. J. Appl. Physiol., *37*:47–59, 1977.

Wiktorin, C. v. H., and Nordin, M.: Introduction to Problem Solving in Biomechanics. Philadelphia, Lea & Febiger, 1986, pp. 36–48.

Youm, Y., et al.: Biomechanical analyses of forearm pronation-supination and elbow flexion-extension. J. Biomech., *12*:245–255, 1979.

14

BIOMECHANICS OF THE WRIST

Steven Stuchin

The wrist, or carpus, is the collection of bones and soft tissue structures that connects the hand to the forearm. This joint complex is capable of a substantial arc of motion that augments hand and finger function, yet it possesses a considerable degree of stability. The wrist functions kinematically by allowing for changes in the location and orientation of the hand relative to the forearm, and kinetically by transmitting loads from hand to forearm and vice versa.

Although the function of all joints of the upper extremity is to position the hand in space so that it can perform the activities of daily living, the wrist appears to be the key to hand function. Stability of the wrist is essential for proper functioning of the digital flexor and extensor muscles, and wrist position affects the ability of the fingers to flex and extend maximally and to grasp effectively during prehension.

ANATOMY OF THE WRIST

The wrist joint complex consists of the multiple articulations of the eight carpal bones with the distal radius, the structures within the ulnocarpal space, the metacarpals, and each other (Fig. 14–1). The soft tissue structures surrounding the carpal bones include the tendons that cross the carpus or attach to it and the ligamentous structures that connect the carpal bones to each other and to the bony elements of the hand and forearm.

THE WRIST BONES

Conventionally, the eight carpal bones are divided into two rows, proximal and distal. The bones of the distal row—the trapezium, trapezoid, capitate, and hamate—constitute a relatively immobile transverse unit that articulates with the metacarpals to form the carpometacarpal joints. All four bones in the distal row fit tightly against each other and are held together by stout interosseous ligaments. The more mobile proximal row, composed of the lunate and the triquetrum, articulates with the radius to form the radiocarpal joint. The scaphoid spans both rows anatomically and functionally. The eighth carpal bone, the pisiform, functions as a sesamoid bone that enhances the mechanical advantage of the wrist's most powerful motor, the flexor carpi ulnaris, and forms its own small joint with the triquetrum. Between the proximal and distal rows of carpal bones is the midcarpal joint, and between adjacent bones of these rows are the intercarpal joints (see Fig. 14–1). The palmar surface of the carpus as a whole is concave, constituting the floor and walls of the carpal tunnel (Fig. 14–2).

The distal radius and the ulnar carpal bones (lunate and triquetrum) articulate with the distal ulna through a ligamentous and cartilaginous structure, the ulnocarpal complex. The components of this complex are illustrated in Figure 14–3.

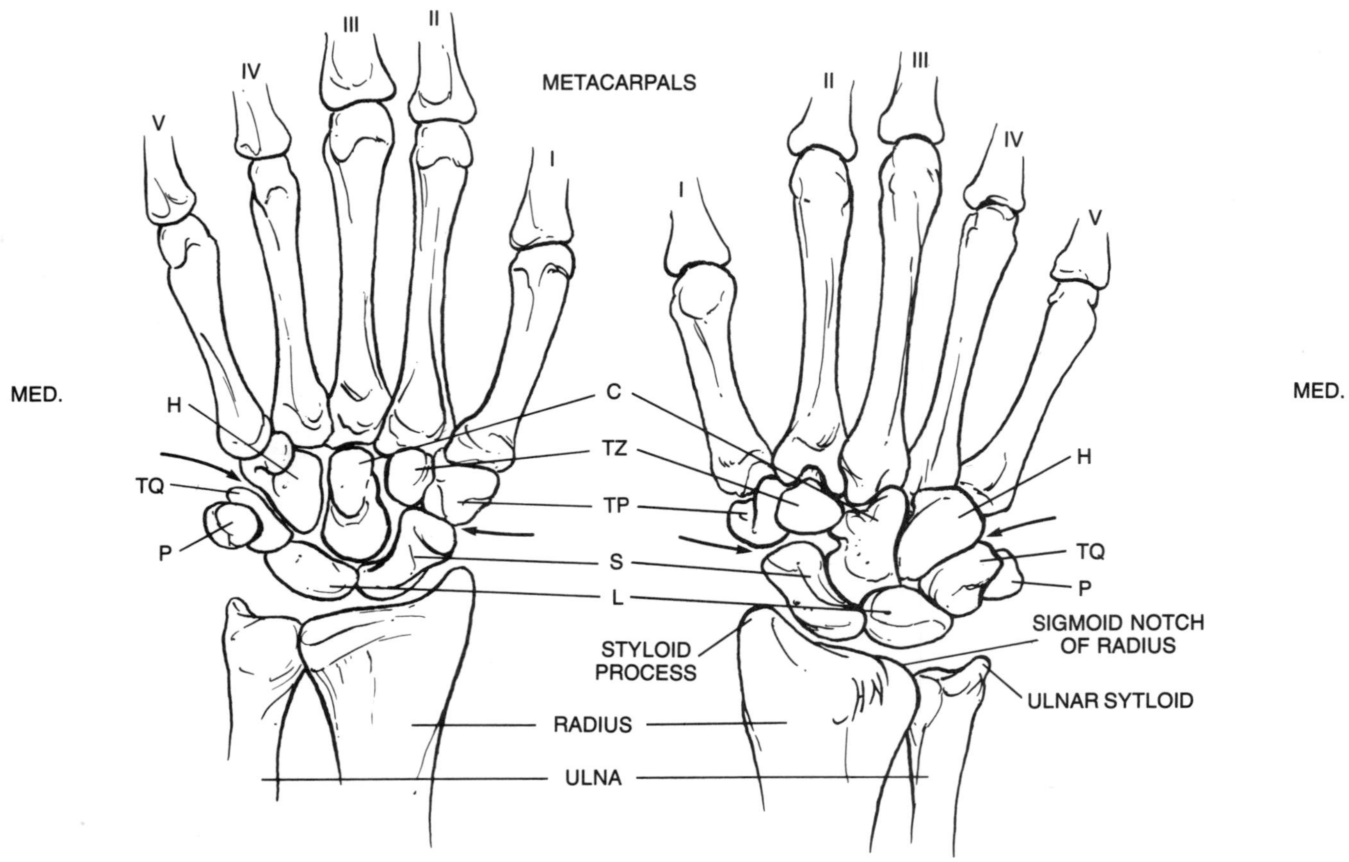

FIG. 14–1
Schematic drawings of the wrist joint complex showing the eight carpal bones and their articulations with the distal radius, the metacarpal bones of the hand, and each other. Palmar view (left) and dorsal view (right) of the right hand. H, hamate; C, capitate; TZ, trapezoid; TP, trapezium; TQ, triquetrum; P, pisiform; L, lunate; S, scaphoid. The arrows indicate the line of the midcarpal joint. (Adapted from Taleisnik, 1985.)

THE WRIST LIGAMENTS

In most joints the function of the ligaments is limited to restricting joint motion and supporting joint integrity. By contrast, the ligaments of the wrist are capable of inducing bony displacements and of transmitting precise loads at a distance (Taleisnik, 1985). The palmar ligaments (Fig. 14–4A) are thick and strong, whereas the dorsal ligaments (Fig. 14–4B) are much thinner and fewer in number (Taleisnik, 1976, 1985). It may be that a strong palmar system is necessary to stabilize against extension and that less strength is needed dorsally to stabilize against flexion.

The highly developed, complex ligament system of the wrist can be divided into extrinsic and intrinsic components (Table 14–1). The extrinsic ligaments run from radius to carpus and from carpus to metacarpals. The intrinsic ligaments originate and insert on the carpus.

The palmar extrinsic system consists of the radial collateral ligament, the palmar radiocarpal ligaments, and components of the ulnocarpal complex. The radial collateral ligament is actually more palmar than lateral. It should be viewed as the most lateral of all palmar radiocarpal fascicles rather than as a collateral ligament, since the function of a true collateral ligament is not possible in a joint like the wrist.

The palmar radiocarpal ligaments are arranged in superficial and deep layers. In the superficial layer most fibers assume a V shape, providing restraint and support. The deep ligaments are three strong fascicles named according to their points of origin and insertion: the radioscaphocapitate (or radiocapitate) ligament, which supports the waist of the scaphoid; the radiolunate ligament, which supports the lunate;

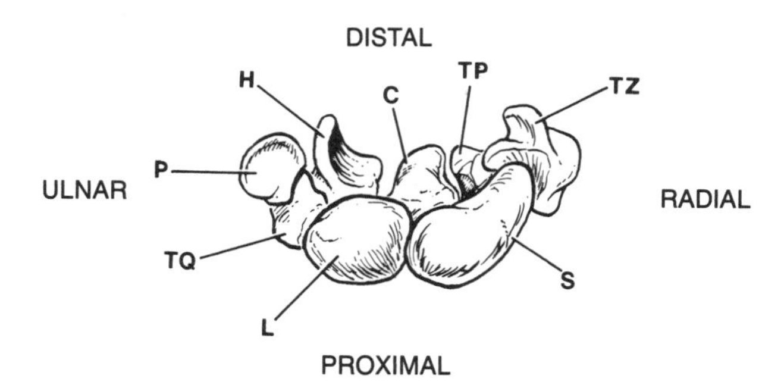

FIG. 14–2
Longitudinal view of the right hand from proximal to distal showing the palmar surface of the bones. This concave surface constitutes the floor and walls of the carpal tunnel, through which the median nerve and the flexor tendons pass. The carpal tunnel is bordered laterally by the prominent tubercle of the trapezium and medially by the hook of the hamate. The motor branch of the ulnar nerve (not shown) winds around the base of the hook before entering the deep palmar compartment. S, scaphoid; L, lunate; TQ, triquetrum; P, pisiform; H, hamate; C, capitate; TP, trapezium; TZ, trapezoid. (Adapted from Taleisnik, 1985.)

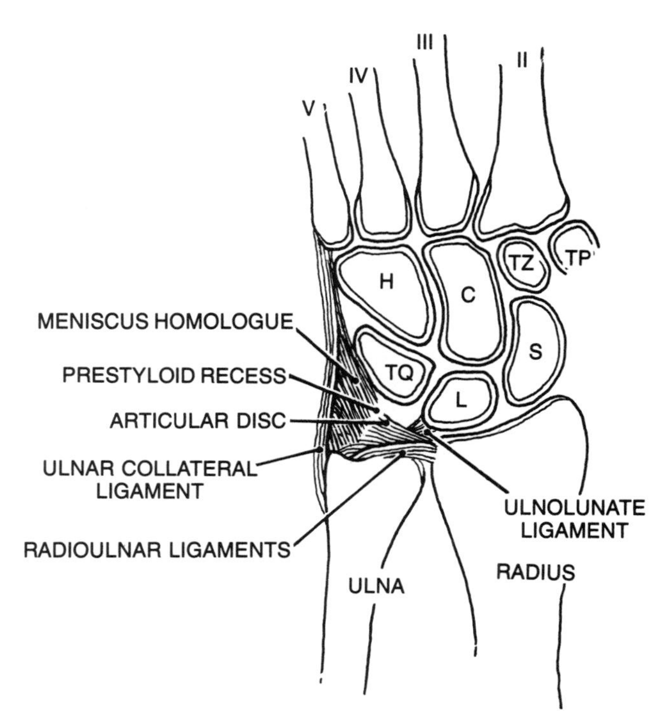

FIG. 14–3
Longitudinal section (frontal plane) of the right wrist and hand viewed from the palmar side. The components of the ulnocarpal complex are visible between the distal ulna and the lunate and triquetrum. S, scaphoid; L, lunate; TQ, triquetrum (the pisiform is not shown); H, hamate; C, capitate; TZ, trapezoid; TP, trapezium. (Adapted from Palmer and Werner, 1981.)

and the radioscapholunate ligament, which connects the scapholunate articulation with the palmar portion of the distal radius. This ligament acts as a checkrein for scaphoid flexion and extension.

The components of the ulnocarpal complex are the meniscus homologue (radiotriquetral ligament), the triangular fibrocartilage (articular disc), the ulnolunate ligament, the ulnar collateral ligament, and the poorly distinguishable dorsal and palmar radioulnar ligaments (see Fig. 14–3). The meniscus homologue and the triangular fibrocartilage have a strong common origin from the dorsoulnar corner (sigmoid notch) of the radius. From there the meniscus swings toward the palm and around the ulnar border of the wrist to insert firmly into the triquetrum, while the triangular fibrocartilage extends horizontally to insert into the base of the ulnar styloid. Between the meniscus and the triangular fibrocartilage there is often a triangular area, the prestyloid recess, which is filled with synovium. Dorsally the ulnocarpal complex has a weak attachment to the carpus except where some of its fibers join the flexor carpi ulnaris sheath dorsolaterally. The ulnolunate ligament connects the palmar border of the triangular fibrocartilage with the lunate. The ulnar collateral ligament arises from the ulnar styloid and extends distally to the base of the fifth metacarpal.

The dorsal extrinsic system is composed of the dorsal radiocarpal ligament. Originating from the rim of the radius, its three fascicles insert firmly into the lunate, triquetrum, and scaphoid.

The intrinsic ligaments can be grouped into three categories (short, long, and intermediate) according to their length and the relative intercarpal movement they allow. Overall, the palmar intrinsic ligaments are more robust (thicker and stronger) than the dorsal ones.

The three short intrinsic ligaments—palmar, dorsal, and interosseous—are stout, unyielding fibers that bind the adjacent carpal bones tightly. These strong ligaments convert all four bones of the distal row into essentially one functional unit (Taleisnik, 1976; Weber, 1984). Three intermediate intrinsic ligaments are located between the lunate and triquetrum, the scaphoid and lunate, and the scaphoid and trapezium.

Of the two long intrinsic ligaments—dorsal intercarpal and palmar intercarpal—the palmar is the more important. Also called the deltoid, or V, ligament, it stabilizes the capitate, since it attaches to its neck and fans out proximally to insert into the

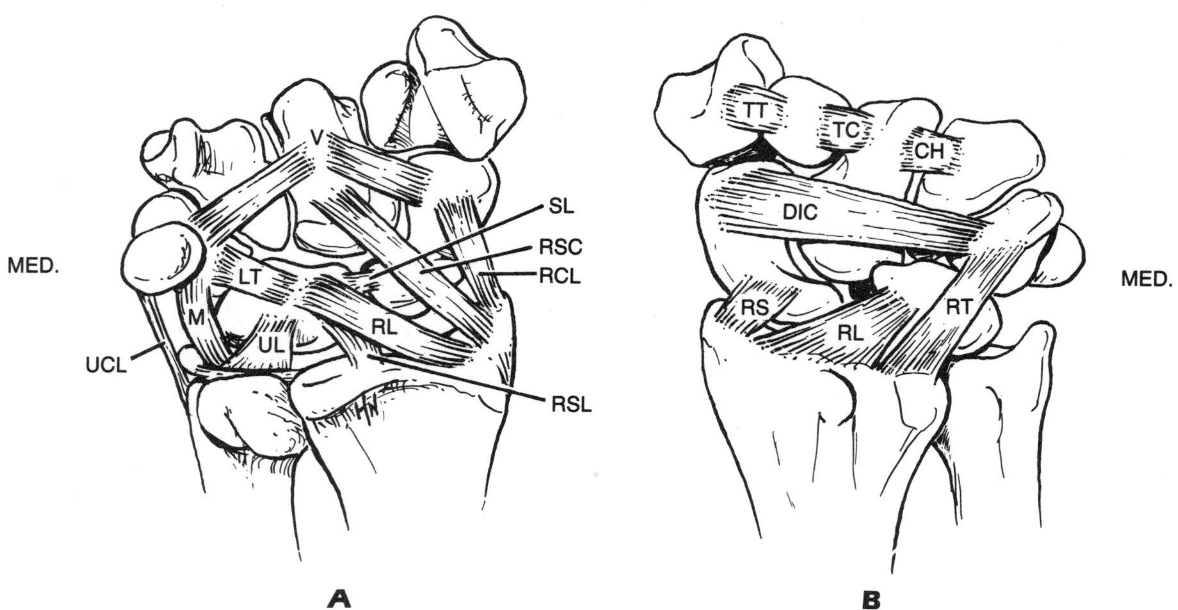

FIG. 14–4
The ligaments of the wrist. (Adapted from Taleisnik, 1985.) **A.** The palmar wrist ligaments (right hand). Extrinsic ligaments: RSC, radioscaphocapitate ligament; RCL, radial collateral ligament; RL, radiolunate ligament; RSL, radioscapholunate ligament; UL, ulnolunate ligament; M, meniscus homologue (radiotriquetral ligament); UCL, ulnar collateral ligament. The superficial palmar radiocarpal ligament and the triangular fibrocartilage are not shown. Intrinsic ligaments: SL, scapholunate ligament; LT, lunotriquetral ligament; V, palmar intercarpal (deltoid, or V) ligament. The short palmar intrinsics are not shown. **B.** The dorsal wrist ligaments of the right hand. Extrinsic ligaments: radiotriquetral (RT), radiolunate (RL), and radioscaphoid (RS) fascicles of the dorsal radiocarpal ligament. Intrinsic ligaments: dorsal intercarpal (DIC); trapeziotrapezoid (TT), trapeziocapitate (TC), and capitohamate (CH) fascicles of the short intrinsic ligaments. The scaphotrapezium ligament is not shown.

TABLE 14–1

LIGAMENTS OF THE WRIST

EXTRINSIC LIGAMENTS	INTRINSIC LIGAMENTS
Proximal (radiocarpal)	Short
Radial collateral	Palmar
Palmar radiocarpal	Dorsal
Superficial	Interosseous
Deep	Intermediate
Radioscaphocapitate (radiocapitate)	Lunotriquetral
Radiolunate	Scapholunate
Radioscapholunate	Scaphotrapezium
Ulnocarpal complex	Long
Meniscus homologue (radiotriquetral)	Palmar intercarpal (V, deltoid)
Triangular fibrocartilage (articular disc)	Dorsal intercarpal
Ulnolunate ligament	
Ulnar collateral ligament	
Dorsal radiocarpal	
Distal (carpometacarpal)	

(Modified from Taleisnik, 1985.)

scaphoid and triquetrum. The dorsal intercarpal ligament originates from the triquetrum and courses laterally and obliquely to insert on the scaphoid and trapezium (Jeanne and Mouchet, 1919; Testut and Latarjet, 1951).

THE WRIST MUSCLES AND TENDONS

The wrist joint complex is surrounded at its periphery by the 10 wrist tendons, whose muscles and their actions are listed in Table 14–2. The three flexors and three extensors are the motors of the wrist, controlling radial and ulnar deviation as well as wrist flexion and extension. Four additional muscles control pronation and supination of the forearm. Eight of the muscles originate from the forearm, and two, the brachialis and extensor carpi radialis longus, originate above the elbow. Except for the flexor carpi ulnaris tendon, which attaches to the pisiform, all of the wrist muscle tendons traverse the carpal bones to

TABLE 14–2

MUSCLES OF THE WRIST AND THEIR ACTIONS

MUSCLE	ACTION
Flexors	
Flexor carpi ulnaris	Flexion of wrist; ulnar deviation of hand
Flexor carpi radialis	Flexion of wrist; radial deviation of hand
Palmaris longus	Tension of the palmar fascia
Extensors	
Extensor carpi radialis longus and brevis	Extension of wrist; radial deviation of hand
Extensor carpi ulnaris and brevis	Extension of wrist; ulnar deviation of hand
Pronators-Supinators	
Pronator teres	Forearm pronation
Pronator quadratus	Forearm pronation
Supinator	Forearm supination
Brachioradialis	Pronation or supination, depending on position of forearm

(Modified from Strickland, 1987.)

insert on the metacarpals; hence, dynamic stability at the wrist is limited.

Each wrist tendon has a substantial amplitude of excursion. The extensor carpi radialis brevis and longus each have a maximal excursion of about 37 mm. The flexor carpi radialis's excursion is approximately 40 mm, and that of the flexor carpi ulnaris is about 33 mm. The pronator teres excursion is approximately 50 mm (Boyes, 1970). Impairment of the excursion of any of these tendons owing to adhesions after trauma or surgery can seriously limit wrist motion.

KINEMATICS

Motions at the wrist joint complex are exceedingly complicated. With the use of techniques such as sonic digitizers (Andrews and Youm, 1979; Youm and Yoon, 1979; Berger et al., 1982; Brumbaugh et al., 1982), stereoscopic photography (Erdman et al., 1979; Peterson and Erdman, 1983), six-degree-of-freedom instrumented spatial linkages (Sommer and Miller, 1976), and roentgen-stereophotogrammetry (de Lange et al., 1985), these movements are slowly being revealed, but our understanding of just how they take place is far from complete.

RANGE OF MOTION

The articulations of the wrist joint complex allow motion in two planes: flexion-extension (palmar flexion and dorsiflexion) in the sagittal plane and radial-ulnar deviation (abduction-adduction) in the frontal plane. Combinations of these motions are also possible, the greatest range of wrist motion taking place from radial deviation and extension to ulnar deviation and palmar flexion.

The spherical shape of the proximal pole of the capitate suggests that the capitate-scaphoid-lunate articulation may act as a ball-and-socket joint capable of axial rotation. Although small amounts of axial rotation are possible and may exist in some individual wrists at this articulation, from a practical standpoint such rotation does not occur through the carpal complex (Volz, 1976; Youm et al., 1978). Axial rotation of the hand, expressed as pronation and supination, results instead from motion arising at the proximal and distal radial joints and is dependent on their normal alignment (Volz et al., 1980).

Flexion-Extension

The normal wrist range is 85 to 90 degrees of flexion and 75 to 80 degrees of extension, but it can vary widely among individuals. In a radiographic study of 55 normal wrists, Sarrafian and coworkers (1977) observed an average total arc of carpal flexion-extension of 121 degrees, with ranges varying from 84 to 169 degrees. The average arc of flexion was 66 degrees, with ranges from 38 to 102 degrees; extension averaged 55 degrees and ranged from 31 to 70 degrees. Owing to a slight palmar tilt of the distal radial plates, flexion exceeds extension by an average of 10 degrees.

Investigators have also found various values for the contribution of the proximal and distal carpal rows to the total arc of flexion and extension (Fick, 1901; Wright, 1935/1936; Horwitz, 1940; Bunnell, 1956; von Lanz and Wachsmuth, 1959; Kaplan, 1965; Kapandji, 1968; Fisk, 1970; Sarrafian et al., 1977). Sarrafian and coworkers (1977) noted that approximately 60% of flexion occurs at the midcarpal joint and 40% in the radiocarpal joint, while about 67% of extension takes place at the radiocarpal joint and 33% at the midcarpal joint (Fig. 14–5). They noted a significant variation in this pattern, however: 27% of the wrists showed greater radiocarpal than midcarpal flexion and 14% revealed a greater arc of midcarpal than of radiocarpal extension.

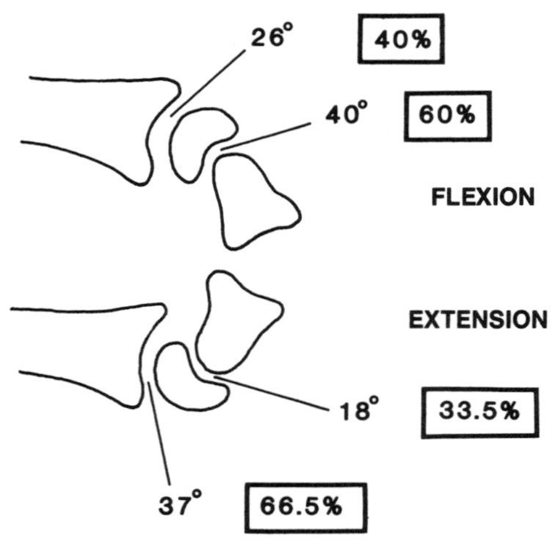

FIG. 14–5
About 60% of wrist flexion (top) occurs at the midcarpal joint, whereas approximately two thirds of wrist extension (bottom) arises at the radiocarpal joint. (Adapted from Sarrafian et al., 1977.)

Radial-Ulnar Deviation

The total arc of radial-ulnar deviation is approximately 50 degrees, 15 to 20 degrees radially and 35 to 37 degrees ulnarly (Youm et al., 1978; Volz et al., 1980). While the elements of flexion-extension can be easily apportioned between the radiocarpal and midcarpal joints, analysis of radial-ulnar deviation is more complex. With radial deviation of the hand, the proximal carpal row moves ulnarly while the distal carpal row is displaced radially. It can be appreciated that if radial-ulnar deviation occurred in a single (frontal) plane, the distal row would swing radially during radial deviation only to jam the scaphoid against the radial styloid, precluding motion. Hence, deviation is rather more complex than this description suggests.

During radial deviation, a shift takes place in the distal pole of the scaphoid, provoked by the encroachment of the scaphoid and trapezoid on the radial styloid process (Fig. 14–6A). The distal pole of the scaphoid rotates toward the palm. This scaphoid motion is transmitted across the proximal row through the scapholunate ligament. Thus, in radial deviation the scaphoid flexes and the proximal row in turn is brought into some flexion. This change in scaphoid position is reversed as the hand is brought into ulnar deviation (Fig. 14–6C). The scaphoid and, by virtue of its ligamentous attachments, the proximal row come into extension.

The function of the triquetrum is similar to, but opposite from, that of the scaphoid. With ulnar deviation, the triquetrum glides distally on the hamate and extends, bringing the lunate into the extended position.

A double-V system formed by the palmar intercarpal ligament and the radiolunate and ulnolunate ligaments helps to support radial-ulnar deviation (Fig. 14–7). The apex of the proximal V is at the lunate and that of the distal V is at the capitate. In ulnar deviation the medial arm of the proximal V, the ulnolunate ligament, becomes somewhat transverse and inhibits radial displacement of the lunate, while the lateral arm, the radiolunate ligament, orients longitudinally and limits lunate extension. The V configuration is now an L. The distal V also becomes an L, but in the opposite direction. The lateral intrinsic ligamentous fibers connecting the scaphoid and capitate become somewhat transverse to check the central ulnar translation of the capitate during this motion. The medial fibers from triquetrum to capitate shift longitudinally and control capitate flexion. In radial deviation the opposite configurations apply (Taleisnik, 1985).

Forearm Pronation-Supination

The motions of forearm pronation and supination through the distal radioulnar joint, although not part of wrist motion proper, play an intricate part in the function of the wrist and the positioning of the hand in space. Pronation-supination, or axial rotation, of up to 150 degrees occurs at the distal radioulnar joint, with the distal radius and its fixed distal member, the hand, rotating about the ulnar head (Slater and Darcus, 1953; Rose-Innes, 1960; Palmer and Werner, 1984). Modest lateral movement of the ulnar head of up to 9 degrees in the direction opposite that of the distal radius has been demonstrated during this motion (Ray et al., 1951). The ulnar head glides in the sigmoid notch of the radius from a dorsal distal position to a palmar proximal position as the forearm moves from full pronation into full supination (Bunnell, 1956; Rose-Innes, 1960; Vesely, 1967; Palmer and Werner, 1984).

FUNCTIONAL WRIST MOTION

Because the joints proximal to the wrist may provide compensatory motion, even a considerable loss of wrist motion may not interfere significantly with activities of daily living. An electrogoniometric study of the range of wrist flexion-extension required for accomplishing 14 activities showed that an arc of 45 degrees (10 degrees of flexion to 35 degrees of extension) was sufficient for performing most of them (Brumfield and Champoux, 1984). Seven activities of

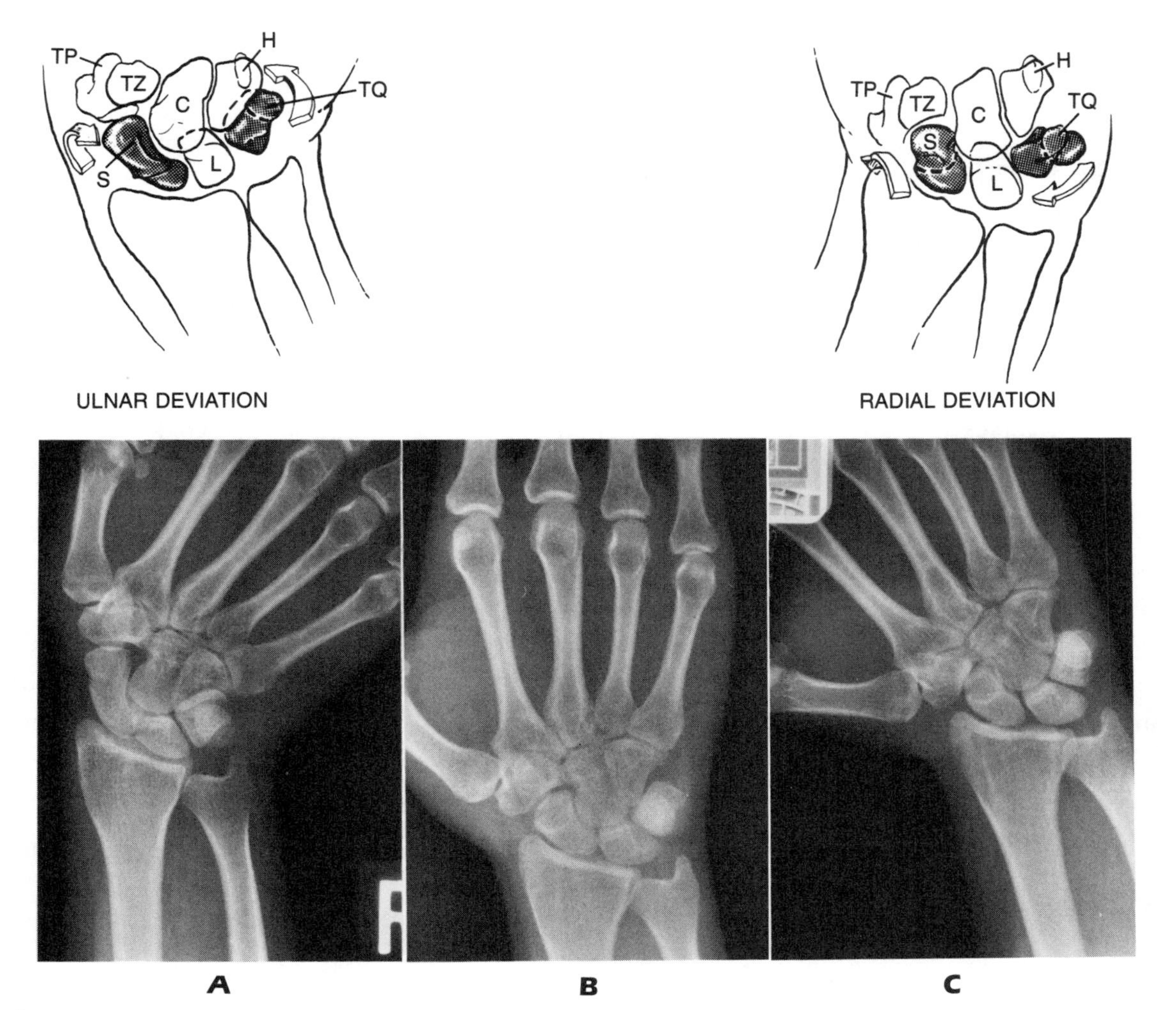

FIG. 14–6
Roentgenograms of the right wrist and hand (dorsal view) showing the position of the carpal bones in ulnar deviation **(A)**, in the neutral position **(B)**, and in radial deviation **(C)**. Arrows in the schematic drawings above roentgenograms **A** and **C** indicate general movement of the bones of the proximal row with wrist motion. In radial deviation the bones of the proximal row are flexed toward the palm. The scaphoid appears foreshortened, the lunate appears triangular, and the triquetrum is proximal in relation to the hamate. In ulnar deviation the bones of the proximal row are extended. The scaphoid appears elongated, the shape of the lunate appears trapezoidal, and the triquetrum is distal in relation to the hamate. TP, trapezium; TZ, trapezoid; C, capitate; H, hamate; TQ, triquetrum; L, lunate; S, scaphoid. (Roentgenograms courtesy of Alex Norman, M.D.; drawings adapted from Taleisnik, 1985.)

personal care that require placing the hand at various locations on the body were accomplished within a range of 10 degrees of flexion to 15 degrees of extension, and most were performed with the wrist slightly flexed (Table 14–3). Other necessary activities requiring an arc of wrist motion, such as eating, drinking, using a telephone, and reading, were accomplished by motion of 5 degrees of flexion to 35 degrees of extension (Table 14–4). Nearly all of these continuous tasks required only extension. Rising from a chair employed the greatest arc of motion, nearly 63 degrees.

Volz and coworkers (1980) also found that loss of wrist mobility did not seriously impede performance of the activities of daily living. Volunteers with wrists immobilized in four different positions were asked to rate their performance on 10 activities, and performance averages were then computed for each position of immobilization. The results disclosed the least compromise of hand function with wrists immobi-

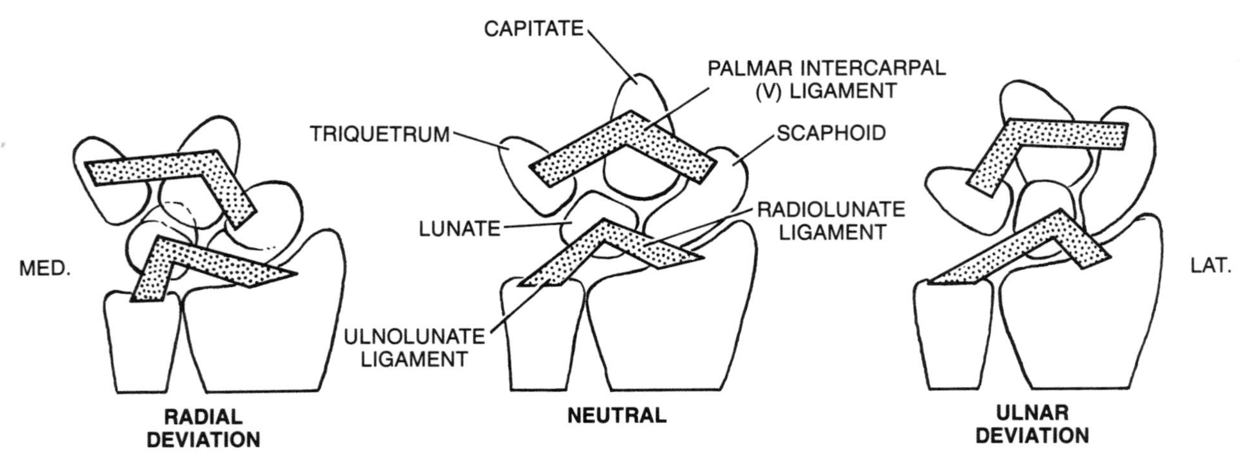

FIG. 14–7
Diagrammatic representation of the changes in alignment of the double-V system formed by the ulnolunate and radiolunate ligaments and the palmar intercarpal (V, or deltoid) ligament with the wrist in radial deviation, the neutral position, and ulnar deviation (palmar view of right hand). (Adapted from Taleisnik, 1985.)

lized in 15 degrees of extension (88% of normal performance) and the greatest disability with the wrists placed in 20 degrees of ulnar deviation (71% of normal function) (Table 14–5).

SURFACE JOINT MOTION

The multiplicity of wrist articulations and the complexity of joint motion make it difficult to calculate the instant center of motion for either flexion-extension or radial-ulnar deviation. Studies have placed the instant center of radial and ulnar deviation of the hand relative to the forearm in various positions in the capitate, including the head (proximal capitate) (von Bonin, 1929; Wright, 1935/1936; MacConaill, 1941; Volz, 1976; Volz et al., 1980; Brumbaugh et al., 1983), body (distal capitate) (Landsmeer, 1961), and neck (Linscheid and Dobyns, 1971). Using a modification of the Reuleaux method, Youm and coworkers (1978) identified the instant center for radial-ulnar deviation in the proximal fourth of the capitate.

Findings for flexion-extension of the hand relative to the forearm are also inconsistent. MacConaill (1941) and Volz (1976) stated that there is a single axis of rotation that remains in the head of the capitate. Mayfield and associates (1976) located the axis of flexion-extension at the junction of the radioscapho-capitate and capitotriquetral ligaments. Youm and

TABLE 14–3

POSITION OF THE WRIST DURING ACTIVITIES OF PERSONAL CARE AND HYGIENE

HAND POSITION	WRIST EXTENSION, MEAN AND STANDARD DEVIATION* (DEGREES)
Head (occiput)	12.7± 9.9
Head (vertex)	−2.3±12.5
Shirt (neck)	−4.6± 8.5
Shirt (chest)	−18.9± 8.9
Shirt (waist)	−15.6± 8.3
Sacrum	−0.6± 9.8
Shoe	14.2±10.6

*Negative value indicates flexion.
(Data from Brumfield and Champoux, 1984. N = 19; 12 men and 7 women, age range 25 to 60 years [mean 33 years].)

TABLE 14–4

AMOUNT OF WRIST MOTION REQUIRED FOR PERFORMANCE OF SELECTED DAILY ACTIVITIES

ACTIVITY	MEAN EXTENSION* (DEGREES) MINIMUM	MAXIMUM	ARC
Lift glass to mouth	11.2	24.0	12.8
Pour from pitcher	8.7	29.7	21.0
Cut with knife	−3.5	20.2	23.7
Lift fork to mouth	9.3	36.5	27.2
Use telephone	−0.1	42.6	42.7
Read newspaper	1.7	34.9	33.2
Rise from chair	0.6	63.4	62.8

*Negative value indicates flexion.
(Data from Brumfield and Champoux, 1984. N = 19; 12 men and 7 women, age range 25 to 60 years [mean 33 years].)

TABLE 14–5

WRIST FUNCTION DURING IMMOBILIZATION

POSITION OF WRIST IMMOBILIZATION	PERFORMANCE AVERAGE* (%)
15° extension	88
Neutral	81
15° flexion	76
20° ulnar deviation	71

*Tasks evaluated were (1) opening and closing doors; (2) writing; (3) eating; (4) handling buttons; (5) handling zippers; (6) driving a car; (7) tying shoes; (8) taking care of personal hygiene (hair, teeth); (9) preparing food (handling jar lids and can openers, grasping and pouring); and (10) toileting. Subjects then rated their performance on a scale of 1 to 4 (1 = tasks not possible to perform; 2 = tasks performed with great difficulty with two hands; 3 = tasks requiring assistance from the other hand; 4 = tasks performed independently). A gross score was obtained, and performance averages were computed. (Data from Volz et al., 1980.)

coworkers (1978) identified the instant center for flexion-extension in the proximal capitate near the lunate, somewhat more proximal than their location for the instant center of radial-ulnar deviation.

WRIST STABILITY

The existence of both a proximal and a midcarpal joint in the wrist creates a double-hinged system that provides inherent stability. According to the laws governing a bimuscular, biarticular chain, this construction is subject to zigzag collapse under compressive load (Landsmeer, 1976) (see Fig. 15–18). As virtually no muscles insert on the carpus to provide dynamic stability, the compressive forces of the long flexors and extensors should be expected to cause the carpus to buckle at the proximal and midcarpal joints. Intricate ligamentous constraints and the precise opposition of multifaceted articular surfaces produce stability.

A longitudinal sagittal section of the wrist reveals that both the scaphoid and the lunate are wedge-shaped, the palmar aspect of both bones being wider than the dorsal (Kauer, 1980). Since compression tends to squeeze a wedge to its narrowest portion, both the lunate and the scaphoid would tend to rotate into extension. This arrangement has an advantage over a symmetrical biarticular system because instability is focused in only one direction and can be countered by a single force applied in the opposite direction (Kauer, 1980; Kauer and Landsmeer, 1981).

As both scaphoid and lunate tend to be forced into extension, stabilization forces must be directed primarily toward flexion. It is here that the contribution of the scaphoid spanning both distal and proximal carpal rows can be appreciated. The scaphoid's natural tendency to extend is stabilized at the midcarpal level; the trapezium and trapezoid articulate with the dorsal aspect of the scaphoid, pushing its distal pole down into flexion. Hence, the scaphoid counteracts the extension tendency of the lunate, lending some stability to the biarticular carpal complex during flexion and extension (Fig. 14–8).

In an investigation of the tensile properties of the carpal ligaments, Mayfield and associates (1979) showed that the radioscaphocapitate ligament failed in tension with less force (170 N) than did the meniscus homologue (radiotriquetral ligament) (210 N) or the dorsal radiocarpal ligament (240 N). The radioscaphoid ligament was the weakest, failing in tension with a force of 54 N. The meniscus homologue was stiffer than the radioscaphocapitate ligament (57% and 74% elongation before failure, respectively). The radial collateral ligament was the weakest of the collateral ligaments, failing in tension at 70 N. These findings suggest that the weakest link between the carpus and the forearm is through the radioscaphocapitate and radial collateral ligaments, both of which are on the radial side of the wrist.

Dynamic stabilization of the wrist joint complex during muscle activity of the fingers and thumb requires a fine balance of extrinsic and intrinsic forces. The arrangement of digital and wrist extensor and flexor systems around the wrist axis makes for antagonist groupings of motor forces. The extensor digitorum communis and extensor indicis proprius pair against the flexor carpi radialis and flexor pollicis longus. The extensor carpi ulnaris works against the extensor pollicis brevis, and the abductor pollicis longus and extensor carpi radialis longus pair against the flexor carpi ulnaris and the flexor digitorum pollicis (Steindler, 1955).

The contribution of the extensor carpi ulnaris–extensor pollicis brevis–abductor pollicis longus axis

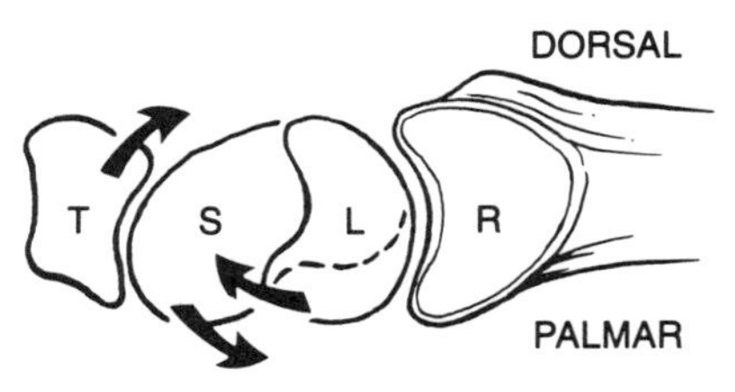

FIG. 14–8
Schematic drawing of the trapezoid (T), scaphoid (S), lunate (L), and radius (R) in a sagittal view. The tendency of the wedge-shaped lunate (palmar pole larger than the dorsal pole) to rotate into extension is counteracted by the scaphoid, which provides a palmar-flexing force induced by the trapezium and trapezoid. (Adapted from Taleisnik, 1985.)

was assessed electromyographically during wrist flexion (Kauer, 1979, 1980). In addition to demonstrating their expected muscle actions, these muscles were found to function as a dynamic "adjustable collateral system" that acts as a true collateral support, the extensor carpi ulnaris for the ulnar side of the wrist and the extensor pollicis brevis and abductor pollicis longus for the radial side.

INTERACTION OF WRIST AND HAND MOTION

Wrist motion is essential for augmenting the fine motor control of the fingers and hand. Positioning the wrist in the direction opposite that of the fingers alters the functional length of the digital tendons so that maximal finger movement can be attained. Wrist extension is synergistic to finger flexion and increases the length of the finger flexor muscles, allowing increased flexion with stretch (Fig. 14–9A) (Tubiana, 1984). Conversely, some flexion of the wrist puts tension on the long extensors, causing the fingers to open automatically and aiding full finger extension (Fig. 14–9B).

The synergistic movements of the wrist extensors and the more powerful digital flexors are facilitated by the architecture of the wrist. The digital flexor tendons cross the wrist within the depths of the carpal arch and are held close to the axis of wrist flexion-extension, affecting wrist position minimally. By contrast, the extrinsic wrist flexors and extensors are positioned widely about the periphery to provide maximal moment arms for positioning the wrist.

As the wrist changes its position and the functional lengths of the digital flexor tendons are altered, the resultant forces in the fingers vary, affecting the ability to grip. Volz and associates (1980) evaluated electromyographically the relationship of grip strength and wrist position. Grip strengths of 67, 134, 201, and 268 N were analyzed with the wrist in five positions: 40 and 20 degrees of flexion, neutral, and 20 and 40 degrees of extension. They found that grip strength was maximal at about 20 degrees of wrist extension and least at 40 degrees of wrist flexion. With the wrist in 40 degrees of extension and in the neutral position, grip strength was slightly less than the maximal values.

Studies by Hazelton and coworkers (1975) of the influence of wrist position on the force produced at the middle and distal phalanges revealed that the greatest force was generated with the wrist in ulnar deviation, the next greatest in extension, and the least in palmar flexion. Taken together, Volz and associates' (1980) and Hazelton and coworkers' (1975) results suggest that for grip to be effective and have maximal force, the wrist must be stable and must be in slight extension and ulnar deviation. Linscheid's (1986) arthrotomographic finding—that the contact area of the radiocarpal joint surfaces is greatest with the wrist in this position—further supports this concept.

The position of the wrist also changes the position of the thumb and fingers, thus affecting the ability to grip. When the wrist is flexed with the hand relaxed, the pulp of the thumb reaches only the level of the distal interphalangeal joint of the index finger; with the wrist extended, the pulps of the thumb and index finger are passively in contact, creating an optimal situation for gripping or pinching (Fig. 14–10).

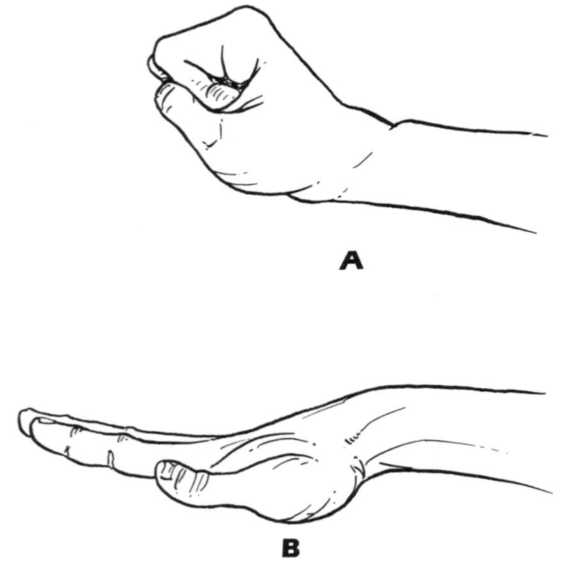

FIG. 14–9
Role of wrist position in finger function. **A.** Slight extension of the wrist allows the flexor muscles to attain maximal functional length, permitting full flexion. **B.** Slight flexion of the wrist places tension on the digital extensor tendons, automatically opening the hand and aiding full finger extension.

KINETICS

The main kinetic function of the carpus is to transmit compressive loads from the hand to the forearm and vice versa. A study of this function is complicated by the observation that the degree of joint conformity differs in the midcarpal and radiocarpal interfaces. In the midcarpal articulation the opposing joint surfaces conform rather precisely, whereas the surfaces of the proximal carpal–radial complex display much less conformity (Volz et al., 1980).

Volz's group (1980) analyzed the pattern of contact between the proximal carpal row and the distal radial-ulnar surfaces by means of transfer prints obtained when the convex surfaces of the scaphoid, lunate, and trapezium were coated with graphite solution and the joint complex was subjected to compressive loads in zero extension. With small loads the initial contact area was between the scaphoid, lunate, and the distal radial plate, but with increasing loads the contact area extended to the fibrocartilaginous surface overlying the distal ulna. Removing the triangular fibrocartilage diminished the contact area between the lunate and the distal radial-ulnar surface, thus increasing the stress per unit area between these structures.

Volz's group (1980) concluded that compressive loads are directed across the carpus along a vector force pattern that passes through the head of the capitate to the scapholunate junction and then to the distal radial-ulnar triangular fibrocartilage surfaces. They suggested that any alteration in the alignment of the structures of the proximal and distal carpal rows might provoke an increase in stress in localized areas, which would then accelerate articular cartilage wear.

Palmer and Werner (1984) also studied the pattern of load transmission across the wrist joint and confirmed the importance of the triangular fibrocartilage and surrounding structures in cushioning compressive loads. Axial loading of 16 upper extremity specimens (22.5 N applied through the wrist motors) showed that 82% of the load was borne by the distal radius and 18% by the distal ulna. Removal of the ulnocarpal complex (termed the "triangular fibrocartilage complex" by Palmer and Werner) decreased the loads borne by the distal radius and the distal ulna by about 12% each.

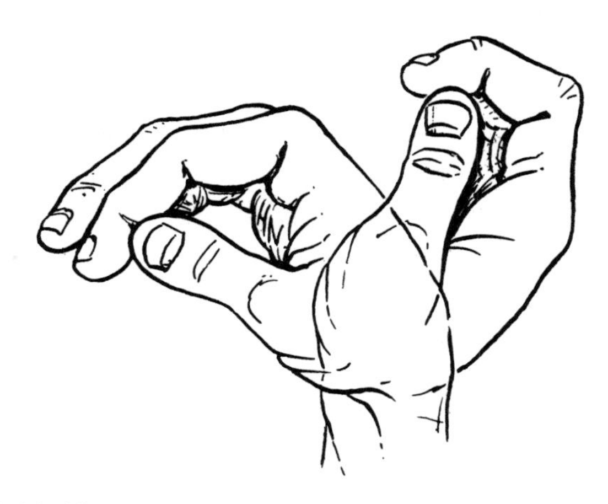

FIG. 14–10
When the wrist is flexed, the tip of the thumb is level with the distal interphalangeal joint of the index finger. With the wrist in extension the pulps of the thumb and index finger come passively into contact. (Adapted from Tubiana, 1984.)

An analysis of muscle forces acting around the instant center of motion of the wrist carried out by Volz and associates (1980) revealed that the flexor carpi ulnaris was the most powerful of all wrist motors, tending to place the wrist in a position of flexion and ulnar deviation. These investigators noted further that the summation of all muscle forces crossing the carpus tends to place the wrist in this position.

SUMMARY

1. The wrist is a complicated joint complex consisting of the multiple articulations of the eight carpal bones with the distal radius, the structures within the ulnocarpal space, the metacarpals, and each other. The carpal bones are conventionally divided into a proximal and a distal row.
2. Motions at the wrist include flexion-extension and radial-ulnar deviation. Stability during radial-ulnar deviation is provided by a double-V system formed by the palmar intrinsic ligament and the radiolunate and ulnolunate ligaments.
3. Functional wrist motion requires a flexion-extension arc of only about 65 degrees; the greater range possible in the wrist stabilizes the hand and allows its optimal positioning in space.
4. The proximal and distal carpal rows form a bimuscular, biarticular chain that is subject to zigzag collapse under compression. Stability is provided by precise opposition of the articular surfaces and intricate intrinsic and extrinsic ligament constraints.
5. The extensor carpi ulnaris, extensor pollicis brevis, and abductor pollicis longus act as a dynamic collateral system for the wrist.
6. Wrist position affects the ability of the fingers to flex and extend maximally and to grasp effectively.
7. The ulnocarpal complex plays a significant role in cushioning compressive loads across the wrist joint.
8. The flexor carpi ulnaris is the most powerful wrist motor and tends to place the wrist in a position of flexion and ulnar deviation.

REFERENCES

Andrews, J. G., and Youm, Y.: A biomechanical investigation of wrist kinematics. J. Biomech., *12*:83, 1979.

Berger, R. A., Crowninshield, R. D., and Flatt, A. E.: The three-dimensional rotational behaviors of the carpal bones. Clin. Orthop., *167*:303, 1982.

Boyes, J. H. (ed.): Bunnell's Surgery of the Hand. 5th Ed. Philadelphia, J. B. Lippincott, 1970.

Brumbaugh, R. B., Crowninshield, R. D., Blair, W. F., and Andrews, J. G.: An in vivo study of normal wrist kinematics. J. Biochem. Eng., *104*:176, 1982.

Brumfield, R. H., and Champoux, J. A.: A biomechanical study of normal functional wrist motion. Clin. Orthop., *187*:23, 1984.

Bunnell, S.: Surgery of the Hand. 3rd Ed. Philadelphia, J. B. Lippincott, 1956.

Erdman, A. G., et al.: Kinematic and kinetic analysis of the human wrist by stereoscopic instrumentation. J. Biomech. Eng., *101*:124, 1979.

Fick, R.: Ergebnisse einer Untersuchung der Handbewegungen mit X-Strahlen. Verh. Anat. Ges., *15*:175, 1901.

Fick, R.: Handbuch der Anatomie and Mechanik der Gelenke. Vol. 3. Jena, G. Karger, 1911.

Fisk, G. R.: Carpal instability and the fractured scaphoid, Hunterian lecture. Ann. R. Coll. Surg. Engl., *46*:63, 1970.

Hazelton, F. T., Smidt, G. L., Flatt, A. E., and Stephens, R. L.: The influence of wrist position on the force produced by the finger flexors. J. Biomech., *8*:301, 1975.

Horwitz, T.: An anatomic and roentgenologic study of the wrist joint. Surgery, *7*:773, 1940.

Jeanne, L. A., and Mouchet, A.: Les lésions traumatiques fermees du poignet. 28th Congrès Français de Chirurgie, 1919. Cited in Taleisnik, J.: The Wrist. New York, Churchill Livingstone, 1985, p. 12.

Kapandji, I. A.: Physiologie Articulaire. Vol. 1. Paris, Lib. Maloine S. A., 1968, p. 138.

Kaplan, E. B.: Functional and Surgical Anatomy of the Hand. 2nd Ed. Philadelphia, J. B. Lippincott, 1965.

Kauer, J. M. G.: The collateral ligament function in the wrist joint. Acta Morphol. Neer. Scand., *17*:252, 1979.

Kauer, J. M. G.: Functional anatomy of the wrist. Clin. Orthop., *149*:9, 1980.

Kauer, J. M. G.: The mechanism of the carpal joint. Clin. Orthop., *202*:16, 1986.

Kauer, J. M. G., and Landsmeer, J. M. F.: Functional anatomy of the wrist. *In* The Hand. Vol. 1. Edited by R. Tubiana. Philadelphia, W. B. Saunders, 1981.

Landsmeer, J. M. F.: Studies in the anatomy of articulation. I. The equilibrium of the "intercalated" bone. Acta Morphol. Neerl. Scand., *3*:287, 1961.

Landsmeer, J. M. F.: Atlas of Anatomy of the Hand. Edinburgh, Churchill Livingstone, 1976.

Lange, A. de, Kauer, J. M. G., and Huiskes, R.: Kinematic behavior of the human wrist joint: A roentgen-stereophotogrammetric analysis. J. Orthop. Res., *3*:56, 1985.

Linscheid, R. L.: Kinematic considerations of the wrist. Clin. Orthop., *202*:27, 1986.

Linscheid, R. L., and Dobyns, J. H.: Rheumatoid arthritis of the wrist. Orthop. Clin. North Am., *2*:649, 1971.

MacConaill, M. A.: The mechanical anatomy of the carpus and its bearings on some surgical problems. J. Anat., *75*:166, 1941.

Mayfield, J. K., Johnson, R. P., and Kilcoyne, R. F.: The ligaments of the human wrist and their functional significance. Anat. Rec., *186*:417, 1976.

Mayfield, J. K., et al.: Biomechanical properties of human carpal ligaments. Orthop. Trans., *3*:143, 1979.

Palmer, A. K., and Werner, F. W.: The triangular fibrocartilage complex of the wrist—anatomy and function. J. Hand Surg., *6*:153, 1981.

Palmer, A. K., and Werner, F. W.: Biomechanics of the distal radioulnar joint. Clin. Orthop., *187*:26, 1984.

Peterson, S. W., and Erdman, A. G.: Analysis and display of human wrist motion. *In* Biostereometrics, '82. Edited by R. E. Herron. Proc. SPIE 361, 1983, pp. 257–261.

Ray, R. D., Johnson, R. J., and Jameson, R. M.: Rotation of the forearm. An experimental study of pronation and supination. J. Bone Joint Surg., *33A*:993, 1951.

Rose-Innes, A. P.: Anterior dislocation of the ulna at the inferior radio-ulnar joint. J. Bone Joint Surg., *42B*:515, 1960.

Sarrafian, S. K., Melamed, J. L., and Goshgarian, G. M.: Study of wrist motion in flexion and extension. Clin. Orthop., *126*:153, 1977.

Slater, N., and Darcus, H. D.: The amplitude of forearm and humeral rotation. J. Anat., *87*:407, 1953.

Sommer, H. G., III, and Miller, N. R.: A technique for kinematic modeling of anatomical joints. J. Biomech. Eng., *102*:311, 1980.

Steindler, A.: Kinesiology of the Human Body. Springfield, Charles C Thomas, 1955, p. 534.

Strickland, J. W.: Anatomy and kinesiology of the hand. *In* Hand Splinting. Principles and Methods. 2nd Ed. Edited by E. E. Fess and C. A. Philips. St. Louis, C. V. Mosby, 1987, pp. 3–41.

Taleisnik, J.: The ligaments of the wrist. J. Hand. Surg., *1*:110, 1976.

Taleisnik, J.: The Wrist. New York, Churchill Livingstone, 1985.

Testut, L., and Latarjet, A.: Tratado de anatomia humana. Vol. 1. 9th Ed. Buenos Aires, Salvat Editores, 1951.

Tubiana, R.: Architecture and functions of the hand. *In* Examination of the Hand and Upper Limb. Edited by R. Tubiana, J.-M. Thomine, and E. Mackin. Philadelphia, W. B. Saunders, 1984, pp. 1–97.

Vesely, D. G.: The distal radioulnar joint. Clin. Orthop., *51*:75, 1967.

Volz, R. G.: The development of a total wrist joint. Clin. Orthop., *116*:209, 1976.

Volz, R. G., Lieb, M., and Benjamin, J.: Biomechanics of the wrist. Clin. Orthop., *149*:112, 1980.

von Bonin, G.: A note on the kinematics of the wrist-joint. J. Anat., *63*:259, 1929.

von Lanz, T., and Wachsmuth, W.: Praktische Anatomie. Berlin, Springer-Verlag, 1959, p. 236.

Weber, E. R.: Concepts governing the rotational shift of the intercalated segment of the carpus. Orthop. Clin. North Am., *15*:193, 1984.

Wright, R. D.: A detailed study of the movement of the wrist joint. J. Anat., *70*:137, 1935/1936.

Youm, Y., and Yoon, Y. S.: Analytical development in investigation of wrist kinematics. J. Biomech., *12*:613, 1979.

Youm, Y., et al.: Kinematics of the wrist. I. An experimental study of radial-ulnar deviation and flexion-extension. J. Bone Joint Surg., *60A*:423, 1978.

SUGGESTED READING

Andrews, J. G., and Youm, Y.: A biomechanical investigation of wrist kinematics. J. Biomech., *12*:83–93, 1979.

Berger, R. A., Crowninshield, R. D., and Flatt, A. E.: The three-dimensional rotational behaviors of the carpal bones. Clin. Orthop., *167*:303–310, 1982.

Boyes, J. H. (ed.): Bunnell's Surgery of the Hand. 5th Ed. Philadelphia, J. B. Lippincott, 1970.

Brumbaugh, R. B., Crowninshield, R. D., Blair, W. F., and Andrews, J. G.: An in vivo study of normal wrist kinematics. J. Biochem. Eng., *104*:176–181, 1982.

Brumfield, R. H., and Champoux, J. A.: A biomechanical study of normal functional wrist motion. Clin. Orthop., *187*:23–25, 1984.

Erdman, A. G., et al.: Kinematic and kinetic analysis of the human wrist by stereoscopic instrumentation. J. Biomech. Eng., *101*:124–133, 1979.

Fisk, G. R.: Carpal instability and the fractured scaphoid, Hunterian lecture. Ann. R. Coll. Surg. Engl., *46*:63, 1970.

Hazelton, F. T., Smidt, G. L., Flatt, A. E., and Stephens, R. L.: The influence of wrist position on the force produced by the finger flexors. J. Biomech., *8*:301–306, 1975.

Horwitz, T.: An anatomic and roentgenologic study of the wrist joint. Surgery, *7*:773, 1940.

Kapandji, I. A.: Physiologie Articulaire. Vol. 1. Paris, Lib. Maloine S. A., 1968, p. 138.

Kaplan, E. B.: Functional and Surgical Anatomy of the Hand. 2nd Ed. Philadelphia, J. B. Lippincott, 1965.

Kauer, J. M. G.: The interdependence of carpal articulation chains. Acta Anat., *88*:481–501, 1974.

Kauer, J. M. G.: The collateral ligament function in the wrist joint. Acta Morphol. Neerl. Scand., *17*:252, 1979.

Kauer, J. M. G.: Functional anatomy of the wrist. Clin. Orthop., *149*:9–20, 1980.

Kauer, J. M. G.: The mechanism of the carpal joint. Clin. Orthop., *202*:16–26, 1986.

Kauer, J. M. G., and Landsmeer, J. M. F.: Functional anatomy of the wrist. *In* The Hand. Vol. 1. Edited by R. Tubiana. Philadelphia, W. B. Saunders, 1981.

Landsmeer, J. M. F.: Studies in the anatomy of articulation. I. The equilibrium of the "intercalated" bone. Acta Morphol. Neerl. Scand., *3*:287–321, 1961.

Landsmeer, J. M. F.: Atlas of Anatomy of the Hand. Edinburgh, Churchill Livingstone, 1976.

Lange, A. de, Kauer, J. M. G., and Huiskes, R.: Kinematic behavior of the human wrist joint: A roentgen-stereophotogrammetric analysis. J. Orthop. Res., *3*:56–64, 1985.

Linscheid, R. L.: Kinematic considerations of the wrist. Clin. Orthop., *202*:27–39, 1986.

Linscheid, R. L., and Dobyns, J. H.: Rheumatoid arthritis of the wrist. Orthop. Clin. North Am., *2*:649–665, 1971.

Linscheid, R. L., Dobyns, J. H., Beabout, J. W., and Bryan, R. S.: Traumatic instability of the wrist. Diagnosis, classification, and pathomechanics. J. Bone Joint Surg., *54A*:1612–1632, 1972.

MacConaill, M. A.: The mechanical anatomy of the carpus and its bearings on some surgical problems. J. Anat., *75*:166–175, 1941.

Mayfield, J. D.: Wrist ligamentous anatomy and pathogenesis of carpal instability. Orthop. Clin. North Am., *15*:209–216, 1984.

Mayfield, J. K., Johnson, R. P., and Kilcoyne, R. F.: The ligaments of the human wrist and their functional significance. Anat. Rec., *186*:417–428, 1976.

Mayfield, J. K., et al.: Biomechanical properties of human carpal ligaments. Orthop. Trans., *3*:143, 1979.

Palmer, A. K., and Werner, F. W.: The triangular fibrocartilage complex of the wrist—anatomy and function. J. Hand Surg., *6*:153–162, 1981.

Palmer, A. K., and Werner, F. W.: Biomechanics of the distal radioulnar joint. Clin. Orthop., *187*:26–35, 1984.

Ray, R. D., Johnson, R. J., and Jameson, R. M.: Rotation of the forearm. An experimental study of pronation and supination. J. Bone Joint Surg., *33A*:993, 1951.

Rose-Innes, A. P.: Anterior dislocation of the ulna at the inferior radio-ulnar joint. J. Bone Joint Surg., *42B*:515, 1960.

Sarrafian, S. K., Melamed, J. L., and Goshgarian, G. M.: Study of wrist motion in flexion and extension. Clin. Orthop., *126*:153–159, 1977.

Slater, N., and Darcus, H. D.: The amplitude of forearm and humeral rotation. J. Anat., *87*:407, 1953.

Sommer, H. G., III, and Miller, N. R.: A technique for kinematic modeling of anatomical joints. J. Biomech. Eng., *102*:311–317, 1980.

Strickland, J. W.: Anatomy and kinesiology of the hand. *In* Hand Splinting. Principles and Methods. 2nd Ed. Edited by E. E. Fess and C. A. Philips. St. Louis, C. V. Mosby, 1987, pp. 3–41.

Taleisnik, J.: The ligaments of the wrist. J. Hand Surg., *1*:110–118, 1976.

Taleisnik, J.: Post-traumatic carpal instability. Clin. Orthop., *149*:73–82, 1980.

Taleisnik, J.: The Wrist. New York, Churchill Livingstone, 1985.

Testut, L., and Latarjet, A.: Tratado de anatomia humana. Vol. 1. 9th Ed. Buenos Aires, Salvat Editores, 1951.

Tubiana, R.: Architecture and functions of the hand. *In* Examination of the Hand and Upper Limb. Edited by R. Tubiana, J.-M. Thomine, and E. Mackin. Philadelphia, W. B. Saunders, 1984, pp. 1–97.

Vesely, D. G.: The distal radioulnar joint. Clin. Orthop., *51*:75, 1967.

Vinh, T. S., and Kuhlmann, J. N.: Mesure des pressions intra-articulaires du poignet. Acta Orthop. Belgica, *48*:576–588, 1982.

Volz, R. G.: The development of a total wrist joint. Clin. Orthop., *116*:209–214, 1976.

Volz, R. G.: Total wrist arthroplasty: A clinical and biomechanical analysis. *In* AAOS Symposium on Total Joint Replacements of the Upper Extremity. Edited by A. E. Inglis. St. Louis, C. V. Mosby, 1982, pp. 273–288.

Volz, R. G., Lieb, M., and Benjamin, J.: Biomechanics of the wrist. Clin. Orthop., *149*:112–117, 1980.

Weber, E. R.: Concepts governing the rotational shift of the intercalated segment of the carpus. Orthop. Clin. North Am., *15*:193–207, 1984.

Wright, R. D.: A detailed study of the movement of the wrist joint. J. Anat., *70*:137–143, 1935/1936.

Youm, Y., and Flatt, A. E.: Kinematics of the wrist. Clin. Orthop., *149*:21–32, 1980.

Youm, Y., and Yoon, Y. S.: Analytical development in investigation of wrist kinematics. J. Biomech., *12*:613–621, 1979.

Youm, Y., McMurtry, R. Y., Flatt, A. E., and Gillespie, T. E.: Kinematics of the wrist. I. An experimental study of radial-ulnar deviation and flexion-extension. J. Bone Joint Surg., *60A*:423–431, 1978.

15

BIOMECHANICS OF THE HAND

Fadi J. Bejjani
Johan M. F. Landsmeer

The hand is the final link in the mechanical chain of levers that begins at the shoulder. The mobility of the shoulder, the elbow, and the wrist, all operating in different planes, allows the hand to move within a large volume of space and to reach all parts of the body with relative ease. The hand itself is remarkably mobile and malleable. It is capable of a great variety of functions, from grasping objects of various shapes (prehension), to tactile exploration, to emphasizing an idea being expressed. The unique arrangement and mobility of the 19 bones and 14 joints of the hand provide the structural foundation for the hand's extraordinary functional adaptability.

Fingers and thumb are the elementary components of the hand (Fig. 15–1). Because each digital unit extends into the middle of the hand, the term "finger ray" is used to indicate the entire chain, composed of one metacarpal and three phalanges (two in the thumb). The finger rays are numbered from the radial to the ulnar side: I (thumb), II (index finger), III (middle finger), IV (ring finger), and V (little finger).

Each finger ray articulates proximally with a particular carpal bone in a carpometacarpal (CMC) joint. The next joint in each ray—the metacarpophalangeal (MCP) joint—links the metacarpal bone to the proximal phalanx. Between the phalanges of the fingers a proximal (PIP) and a distal (DIP) interphalangeal joint are found; the thumb has only one interphalangeal (IP) joint. The thenar eminence at the palmar side of the first metacarpal is formed by the intrinsic muscles of the thumb. Its ulnar counterpart, the hypothenar eminence, is created by muscles of the little finger and an overlying fat pad.

The bones of the hand are arranged in three arches (Fig. 15–2), two transverse and one longitudinal (Tubiana, 1969; Flatt, 1974). The proximal transverse arch, with the capitate as its keystone, lies at the level of the distal carpus and is relatively fixed. The distal transverse arch, with the head of the third metacarpal as its center, passes through all the metacarpal heads and is more mobile. The two transverse arches are connected by the longitudinal arch, composed of the four digital rays and the proximal carpus. The second and third metcarpal bones form the central pillar of this arch (Flatt, 1974). Although the extrinsic flexor and extensor muscles are largely responsible for changing the shape of the working hand, the intrinsic muscles of the hand are primarily responsible for maintaining the configuration of the three arches. Collapse in the arch system resulting from bone injury, rheumatic disease, or paralysis of the intrinsic muscles can contribute to severe disability and deformity.

As the fingers flex during grasp, their longitudinal arches fold in a strict mathematical pattern known as the equiangular spiral, or the logarithmic spiral (Littler, 1973) (Fig. 15–3). Basic to the construction of this spiral is a series of isosceles triangles with an apical angle of 36 degrees. This natural biological pattern is found in the whorls seen in flowers, is exemplified in the shape of an egg, and is perfectly formed in the accretive shell of the chambered nautilus. The equiangular curve described by the

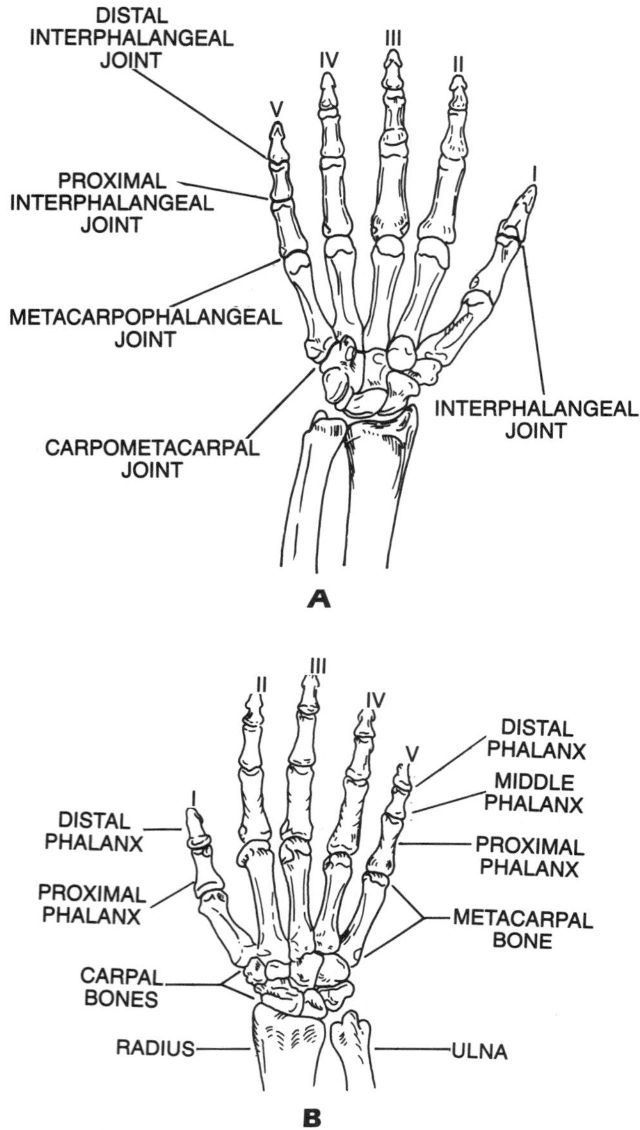

FIG. 15–1
Schematic drawings of the skeleton of the hand. The finger rays are numbered from the medial to the lateral side. **A.** Anterior (palmar) view of the right hand. The joints are labeled. **B.** Posterior (dorsal) view of the right hand. The bones are labeled.

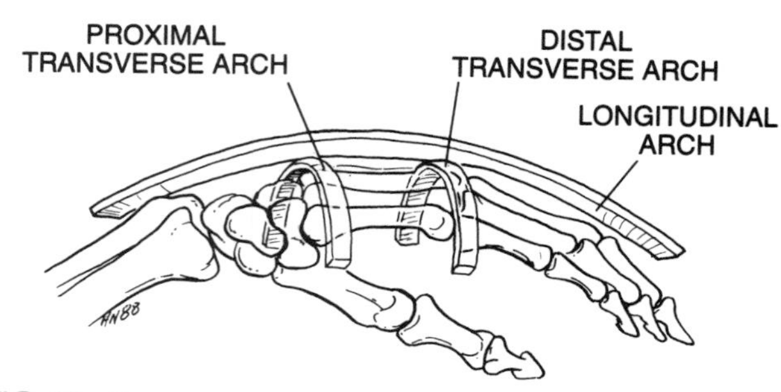

FIG. 15–2
The three skeletal arches of the hand (mediolateral view). The relatively fixed proximal transverse arch passes through the distal carpus, while the more mobile distal transverse arch passes through the metacarpal heads. The longitudinal arch is composed of the four finger rays and the proximal carpus. (Adapted from Strickland, 1987.)

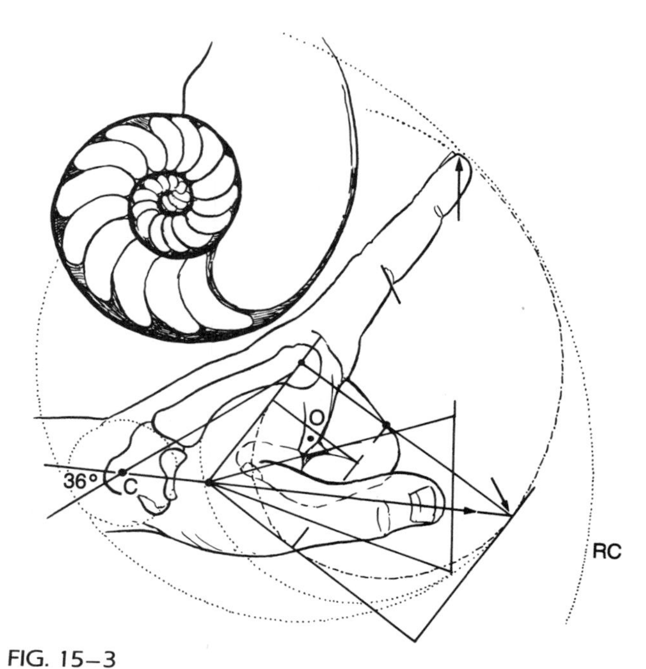

FIG. 15–3
The sweep of the index finger from full extension to full flexion executes a curve congruent with an equiangular spiral as found in the perfectly formed accretive shell of the chambered nautilus. The construction of this spiral is a series of isosceles triangles, each with an apical angle of 36 degrees. This curve allows for the hand's limitless adaptability in grasping. O, polar axis of the equiangular spiral; RC, radiocarpal arc; C, head of the capitate, the center of radiocarpal movement. (Adapted from Littler, 1973.)

fingertips as they close is determined by the ratios of the average interarticular lengths of the metacarpals (7.1 cm), proximal phalanges (4.6 cm), middle phalanges (2.8 cm), and distal phalanges (1.8 cm), which approximate a Fibonacci sequence. In this sequence, discovered by Fibonacci in 1202, each number is the sum of the previous two; the ratio of any two consecutive numbers is constant and approximates 1.618 (Hoggart, 1969), which is the ratio characteristic of all equiangular spirals. This explains how the hand, in spite of its segmental length inequalities, grows without ever changing its shape (Thompson, 1963).

The importance of the sensory function of the hand must not be overlooked. Because of the interplay between sensibility and motor function, the hand is a unique instrument of active touch, the only sense under voluntary control (Bell, 1833).

KINEMATICS

The hand is an extremely mobile organ that can coordinate an infinite variety of movements in relation to each of its components. The blending of hand and wrist movements enables the hand to mold itself to the shape of an object being palpated or grasped. The great mobility of the hand is due to the articular contours, the position of the bones in relation to one another, and the actions of an intricate system of muscles.

RANGE OF MOTION

The varying shapes of the CMC, MCP, and IP joints of the fingers are responsible for the differences in degrees of freedom at these joints. The unique orientation of the thumb, the large web space, and the special configuration of the thumb CMC joint afford this digit great mobility and versatility.

The Fingers

The second and third metacarpals are linked to the trapezoid and capitate and to each other by tight-fitting joints that are basically immobile (Fig. 15–4). As a result, these metacarpal and carpal bones constitute the "immobile unit" of the hand. The

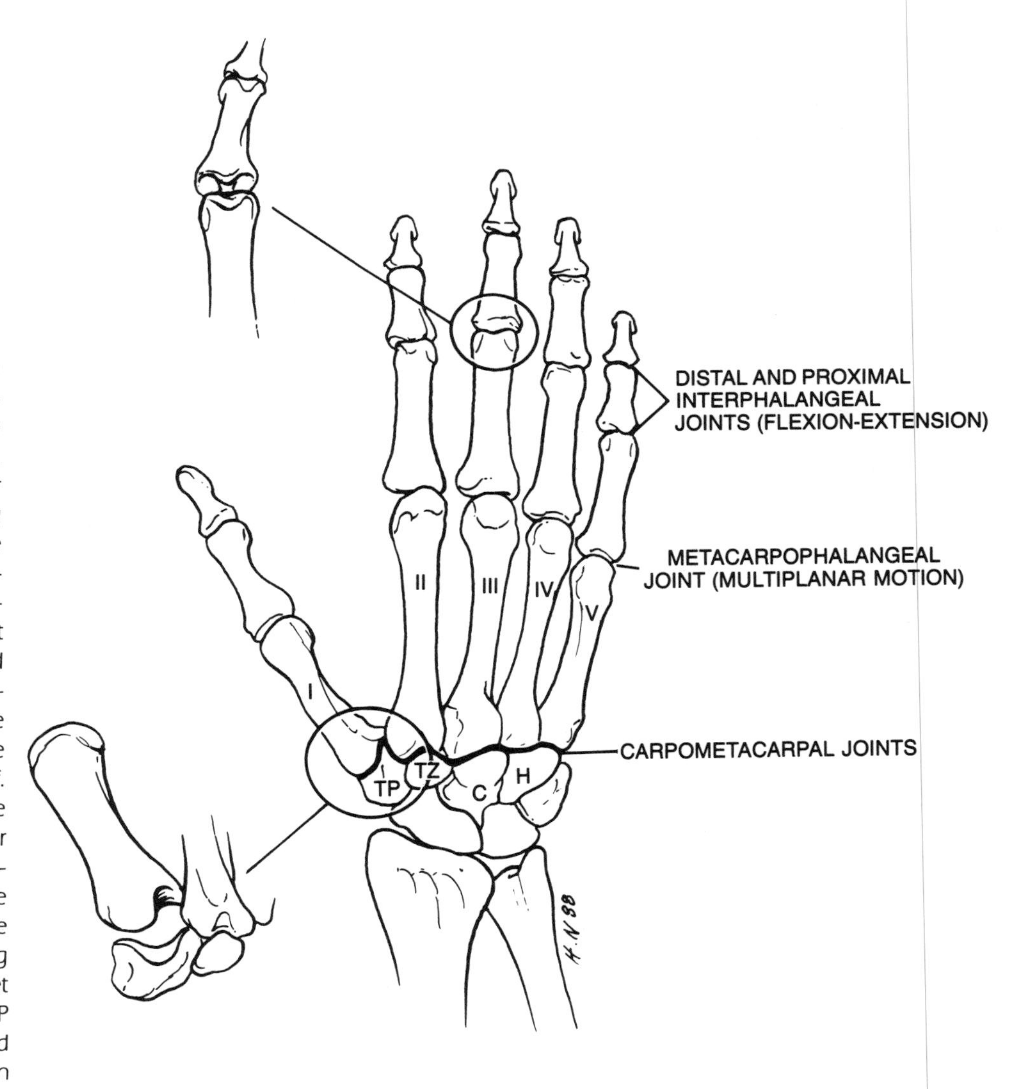

FIG. 15–4
Schematic representations of the joints of the finger rays (dorsal view of the right hand). The CMC joint between the first metacarpal and trapezium (TP) is composed of two saddle-shaped surfaces, the convexity of one fitting tightly into the concavity of the other (inset shows enlargement). This arrangement allows for the positioning of the thumb in a wide arc of motion. The tight-fitting joints that link the second and third metacarpals with the trapezoid (TZ) and capitate (C), respectively, and with each other are basically immobile, rendering these four bones the "immobile unit" of the hand. The joints between the fourth and fifth metacarpals and the hamate (H) permit a modest amount of flexion and extension. The unicondylar configuration of the MCP joints of the four fingers allows motion in three planes and combinations thereof. By contrast, the tongue-and-groove articular contours of the bicondylar hinge joints between the phalanges limit motion to one plane (flex-ion-extension) and contribute to these joints' stability in resisting shear and rotatory forces (inset shows enlargement of a typical IP joint in an oblique view). (Adapted from Strickland, 1987, and Van Zwieten, 1980.)

articulations of the fourth and fifth metacarpals with the hamate permit a modest amount of motion: 10 to 15 degrees of flexion-extension at the fourth CMC joint and 20 to 30 degrees at the fifth. Limited palmar displacement, or descent, of these metacarpals may thus take place. This motion allows cupping of the hand and is essential for gripping, as will be described below.

The MCP joints of the four fingers are unicondylar diarthrodial joints (see Figs. 15–4 and 15–10), allowing motion in three planes: flexion-extension (sagittal plane), abduction-adduction (frontal plane), and a bit of pronation-supination (transverse plane), which is coupled with abduction-adduction (Hagert, 1981).

The range of MCP flexion from the zero position is approximately 90 degrees (Fig. 15–5A), but this value differs among the fingers, the little finger demonstrating the most flexion (about 95 degrees) and the index finger, about 70 degrees (Batmanabane and Malathi, 1985). Extension beyond the zero position varies considerably among human populations and also among individuals, depending on joint laxity.

The proximal and distal IP joints of the four digits are bicondylar hinge joints as a result of the tongue-and-groove fit of their articular surfaces (see Figs. 15–4 and 15–10). These surfaces are closely congruent throughout the range of flexion-extension, which is the only motion possible in these joints. Flexion is measured from the zero position with the finger in the plane of the hand. The largest range of flexion, 110 degrees or more, occurs in the PIP joint (Fig. 15–5B). Flexion of about 90 degrees takes place in the DIP joint (Fig. 15–5C). Extension beyond the zero position, termed hyperextension, is a regular feature of the DIP and PIP joints, although it depends largely on ligament laxity, especially in the PIP joint.

The Thumb

At the CMC level, the base of the thumb metacarpal forms a saddle joint with the trapezium (see Fig. 15–4). This configuration allows the thumb metacarpal a wide range of motion through a conical space extending from the plane of the hand in a palmar direction. Motion of metacarpal I should be described in degrees of abduction from metacarpal II, thereby defining the plane in which this motion is carried out with respect to the plane of the hand. The terms "flexion" and "extension" with respect to the thumb should be reserved for motions of the MCP and IP joints.

Functionally, the most important motion of the thumb is opposition, in which abduction coupled with rotation at the CMC joint moves the thumb toward the tip of the little finger; flexion at the MCP

FLEXION OF THE FINGER

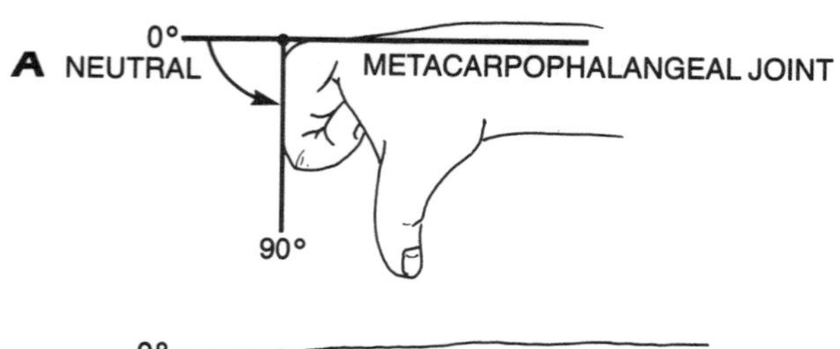

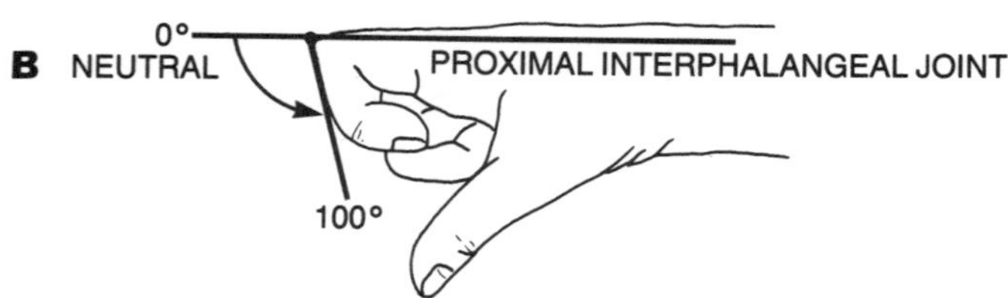

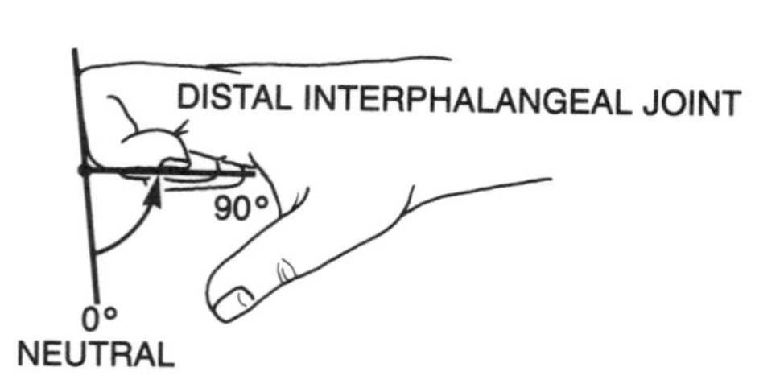

FIG. 15–5
Flexion of the three joints of the finger, beginning with the neutral position in which the extended fingers are in the plane of the dorsal hand and wrist. **A.** Flexion of the MCP joint, averaging 70 to 90 degrees. **B.** Flexion at the PIP joint, averaging 100 degrees or more. **C.** Flexion at the DIP joint, averaging 90 degrees. (Adapted from AAOS, 1965.)

and IP joints then brings the thumb closer to the fingertips (Fig. 15–6).

The MCP joint of the thumb resembles those of the fingers. The range of flexion from the zero position varies considerably among individuals, from as little as 30 degrees to as much as 90 degrees; extension from the zero position is approximately 15 degrees (Batmanabane and Malathi, 1985).

STABILITY AND CONTROL OF THE HAND

A number of anatomic features contribute to the stability and control of the various articulations of the hand. The coordinated actions of the extrinsic and intrinsic muscles of the hand permit control of the finger rays; a dorsal tendinous complex known as the extensor assembly contributes to the control and stability of the IP joints; and a well-developed flexor tendon sheath pulley system facilitates smooth and stable flexion of these joints. The bony and ligamentous asymmetry of the MCP joints lends the hand its functional versatility. The IP joints gain their stability from the shape of their articular contours and from special ligamentous restraints.

All of the digital articulations have one essential feature in common: they are designed to function in flexion. Each joint has firm collateral ligaments bilaterally and a thick anterior capsule reinforced by a fibrocartilaginous structure known as the palmar plate. By comparison, the dorsal capsule is thin and lax. The palmar tendinous apparatus, composed of the two flexor tendons, is much stronger than the dorsal extensor assembly, and even the skin is thicker on the palmar side.

The Extrinsic and Intrinsic Muscles of the Fingers

The finger rays are controlled by extrinsic and intrinsic muscles (Table 15–1). The extrinsic muscles originate in the arm and forearm. The intrinsic muscles are entirely confined to the hand (Fig. 15–7; see also Fig. 15–19B). Although the contribution of each system is distinctly different, the coordinated functioning of intrinsic and extrinsic muscle systems is essential for the satisfactory performance of the hand in a wide range of tasks.

The independent operation of each finger is restrained to some extent by junctions between the extensor tendons (juncturae tendinum) in the middle of the hand (see Fig. 15–11). The middle, ring, and little fingers are further restricted in their independent functioning because their flexor profundus tendons emerge from the same muscle. The index finger is capable of greater functional independence

OPPOSITION OF THE THUMB

COMPOSITE OF THREE MOTIONS

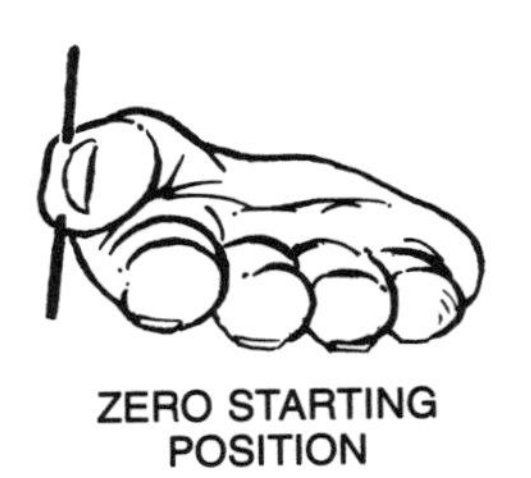

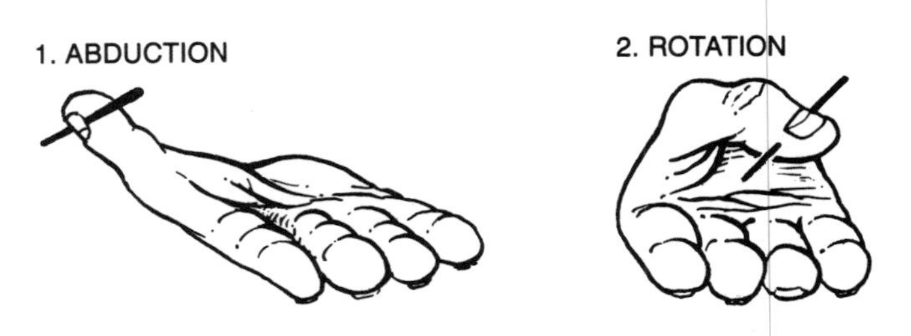

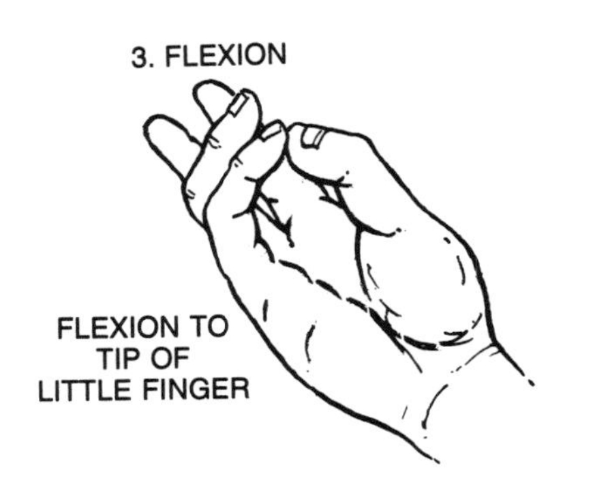

FIG. 15–6
Opposition of the thumb, which begins with the extended thumb in line with the index finger, is the combined motion of abduction and rotation of the CMC joint. Flexion in the MCP and IP joints then brings the tip of the thumb closer to the little finger. Palmar displacement, or descent, of the fourth and fifth metacarpals and flexion in the MCP and PIP joints of the little finger result in tip-to-tip contact between the thumb and little finger. (Adapted from AAOS, 1965.)

because its flexor profundus tendon emerges from a fairly individualized muscle belly (Fahrer, 1971).

The Digital Extensor Assembly

The long extensor tendons are flat structures that emerge from their synovial sheaths at the dorsal side of the carpus and run over the MCP joint; they are held in this position by the sagittal bands. At the dorsum of the proximal phalanx, these extensor tendons and parts of the interossei interweave so as to form a tendinous complex, the extensor assembly (also known as the extensor mechanism), which extends over both IP joints (Fig. 15–8).

Trifurcation of the long extensor tendon and fanning of interosseous fibers result in the formation of one medial and two lateral bands. The middle band (or central slip) runs dorsally over the trochlea of the proximal phalanx and inserts into the base of the middle phalanx. The two lateral bands run at the shoulders of the PIP joint. These bands pursue their way distally and merge over the dorsum of the middle phalanx, forming the terminal tendon, which inserts into the dorsal tubercle of the distal phalanx. This terminal tendon is linked to the proximal phalanx by means of the oblique retinacular ligaments. These ligaments originate from the proximal phalanx and run laterally around the PIP joint, just palmar to the center of motion of this joint in the extended position, to join the terminal tendon.

Illustrating the action of the extensor assembly in coupling PIP and DIP joint motion, Landsmeer (1949) described the "release of the distal phalanx" (Fig. 15–9). If a finger is flexed at the PIP joint only, the whole trifurcated extensor assembly is pulled distally, following the central slip. This slip alone is taut because the distal pull occurs at the middle phalanx; the lateral bands remain slack but are allowed to shift distally over the same distance. Only part of the slack of the lateral bands is required for flexion of the PIP joint, because these bands run closer to the center of motion of this joint than does the central slip. Therefore, some of the slack will remain, allowing passive or active flexion of the distal phalanx but no active extension. The "released" distal phalanx is the functional basis for the coupled flexion and extension of the DIP and PIP joints.

Conversely, if the DIP is actively flexed, the entire extensor assembly is displaced distally. This relaxes the central slip and simultaneously increases the tension in the oblique retinacular ligaments, a tension that creates a flexion force at the PIP joint. Since the central slip is already unloaded, flexion of this joint is then unavoidable. The release of the distal phalanx is fundamental for pulp-to-pulp pinch. It also allows, through intermittent contraction of the flexor profundus, a change from pulp-to-pulp to tip-to-tip pinch, a

TABLE 15–1

MUSCLES OF THE HAND AND THEIR ACTIONS

MUSCLE	ACTION
Extrinsic Muscles	
Flexors	
Flexor digitorum superficialis	Flexion of PIP and MCP joints
Flexor digitorum profundus	Flexion of DIP, PIP, and MCP joints
Flexor pollicis longus	Flexion of IP and MCP joints of thumb
Extensors	
Extensor pollicis longus	Extension of IP and MCP joints of thumb; secondary adduction of the thumb
Extensor pollicis brevis	Extension of MCP joint of thumb
Abductor pollicis longus	Abduction of thumb
Extensor indicis proprius	Extension of index finger
Extensor digitorum communis	Extension of fingers
Extensor digiti quinti proprius	Extension of fingers
Intrinsic Muscles	
Interossei (all)	Flexion of MCP joints and extension of PIP and DIP joints
Dorsal interossei	Spread of index and ring fingers away from long finger
Palmar interossei	Adduction of index, ring, and little fingers toward long finger
Lumbricals	With extensor, extension of PIP and DIP joints
Thenar muscles	
Abductor pollicis brevis	Abduction of thumb
Flexor pollicis brevis	Flexion and rotation of thumb
Opponens pollicis	Rotation of first metacarpal toward palm
Hypothenar muscles	
Abductor digiti quinti	Abduction of little finger (flexion of proximal phalanx, extension of PIP and DIP joints)
Flexor digiti quinti brevis	Flexion of proximal phalanx of little finger and forward rotation of fifth metacarpal
Adductor pollicis	Adduction of thumb

(Modified from Strickland, 1987.)

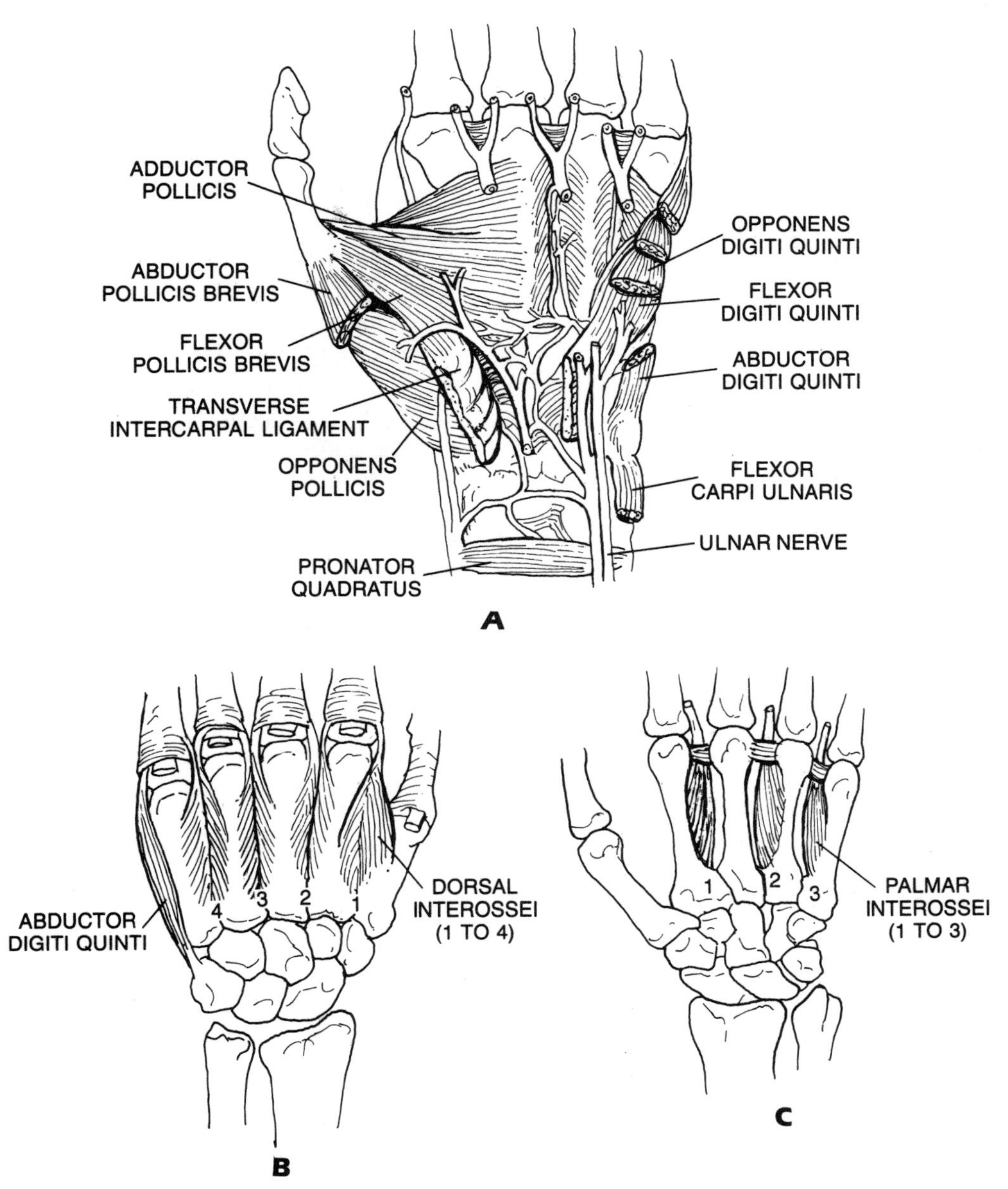

FIG. 15–7
The intrinsic muscles of the hand. **A.** Palmar view of the left hand. **B.** Dorsal view of the left hand showing the four dorsal interossei and the abductor of the little finger. These muscles spread the fingers (i.e., move them away from the midline of the hand). **C.** Palmar view of the left hand showing the three palmar interossei. These muscles adduct the second, fourth, and fifth fingers, flex the MCP joint, and extend the PIP joint. (Adapted from Strickland, 1987, and Caillet, 1982.)

mechanism used in precision handling such as needlework and active tactile exploration.

Sarrafian and coworkers (1970) used strain gauges to measure the tension in different parts of the extensor mechanism during finger flexion and further elaborated on this phenomenon. They found an increase in the central slip tension beyond 60 degrees of PIP flexion; at 90 degrees of flexion there was total relaxation of the lateral bands.

The Metacarpophalangeal Joints

A unique feature of the MCP joint is its asymmetry, which is apparent both in the bony configu-

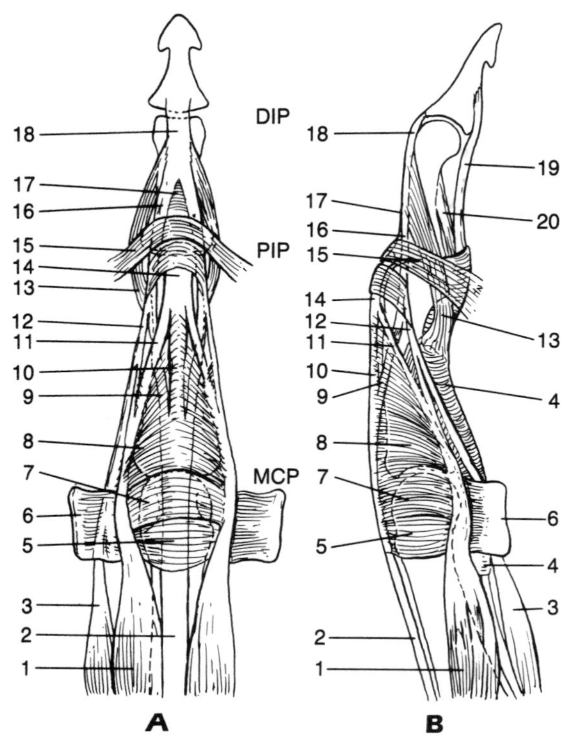

FIG. 15–8
Schematic drawings of the anatomy of the extensor assembly. MCP, metacarpophalangeal joint; PIP, proximal interphalangeal joint; DIP, distal interphalangeal joint. 1, Interosseous muscle; 2, long extensor (extensor communis) tendon; 3, lumbrical muscle; 4, flexor tendon fibrous sheath; 5, sagittal band; 6, intermetacarpal ligament; 7, transverse fibers of interosseous hood; 8, oblique fibers of hood; 9, lateral band of long extensor tendon; 10, medial band of long extensor tendon; 11, central band of interosseous tendon; 12, lateral band of interosseous tendon; 13, oblique retinacular ligament; 14, medial band of long extensor tendon in central slip; 15, transverse retinacular ligament; 16, lateral band of extensor tendon; 17, triangular ligament; 18, terminal tendon; 19, flexor profundus tendon; 20, flexor digitorum superficialis tendon. **A.** Dorsal view. Just proximal to the PIP joint the long extensor tendon within the central tendon slip trifurcates into one medial and two lateral bands. The medial band inserts into the base of the middle phalanx. The lateral bands merge over the dorsum of the middle phalanx to form the terminal tendon, which inserts on the distal phalanx. **B.** Sagittal view. The oblique retinacular ligaments, which originate from the proximal phalanx, run laterally around the PIP joint just palmar to the center of motion, then join the terminal tendon. (Adapted from Tubiana, 1984.)

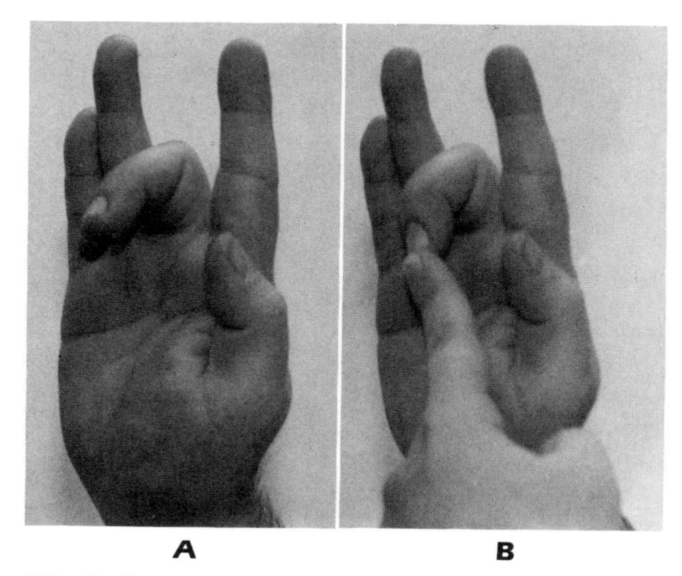

FIG. 15–9
Release mechanism of the distal phalanx. **A.** All fingers are extended, and the PIP joint of the middle finger is flexed. The DIP joint of this finger is totally out of control. **B.** The DIP joint is very loose and can be flexed or extended only passively.

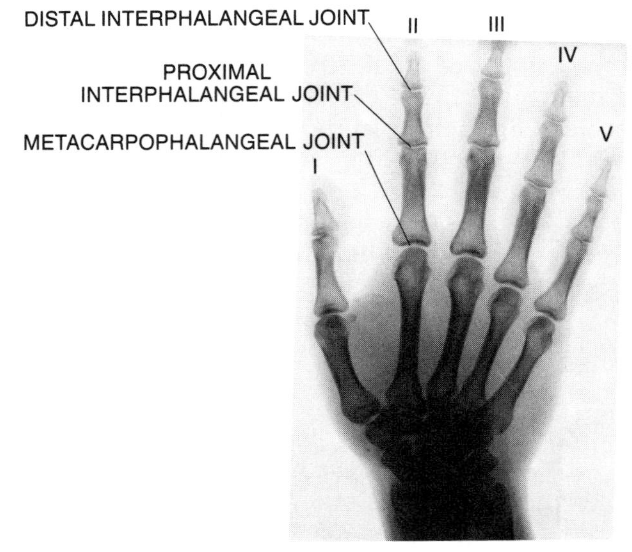

FIG. 15–10
Posteroanterior roentgenogram of the right hand and wrist revealing asymmetry in the configuration of the metacarpal heads. Also apparent is the disparity in the diameters of the proximal and distal surfaces of the PIP joints, the distal surfaces being considerably wider.

ration of the metacarpal head (Fig. 15–10) and in the location of the radial and ulnar collateral ligament attachments to it (Landsmeer, 1955). The collateral ligaments of the MCP joint extend obliquely forward from their proximal attachment at the dorsolateral aspect of the metacarpal head to their insertion on the palmolateral aspect of the base of the proximal phalanx. The bilateral asymmetry in the site of attachment of these ligaments manifests itself particularly in the asymmetric range of abduction-adduction in these joints. The asymmetric bilateral arrangement of the interossei also contributes to the overall asymmetry of the MCP joints (see Fig. 15–7).

Quantification of the length changes in the collateral ligaments during MCP joint motion was accomplished by Minami and associates (1984), who used biplanar roentgenographic techniques to analyze the lengths of the dorsal, middle, and palmar thirds of the radial and ulnar collateral ligaments of the index finger at various degrees of joint flexion. When the MCP joint was flexed from zero to 80 degrees, the dorsal portion of the ligaments lengthened 3 to 4 mm, the middle portion elongated slightly, and the palmar portion shortened 1 to 2 mm. When the MCP joint moved into hyperextension, the dorsal portion of the ligaments shortened 2 to 3 mm, the middle third shortened slightly, and the palmar third lengthened slightly. Thus, the dorsal portions of both collateral ligaments appear to provide the principal restraining force when the MCP joint is flexed, while the palmar portions provide a restraining force during extension. This study confirms the rationale for immobilizing the MCP joint in 50 to 70 degrees of flexion to prevent extension contracture.

Other studies by Minami and coworkers (1985) demonstrated that both the radial and the ulnar collateral ligaments play primary roles in stabilizing the MCP joint in the following modes of joint displacement: distal distraction, dorsopalmar dislocation, adduction-abduction, and axial rotation.

Just palmar to the radial and ulnar collateral ligaments are the accessory collateral ligaments, which originate from the metacarpal and insert into the thick palmar fibrocartilaginous plate. The palmar plate is firmly fixed to the phalangeal base and reinforces the joint capsule anteriorly. One of its functions is to limit hyperextension of the MCP joint. It also provides the structural base for the flexor tendon sheath pulley system. The transverse intermetacarpal ligament, which connects the palmar plates, gives additional stability to the MCP region (Fig. 15–11). The extensor tendons are linked to this transverse structure by the transverse laminae, which hold them in position on the dorsal side of the MCP joint.

The Proximal Interphalangeal Joints

The tongue-in-groove configuration of the PIP joint gives it stability, particularly in resisting shear and rotational forces (see Fig. 15–4). Ligamentous stability at this joint is ensured by two strong collateral ligaments and a three-sided supporting cradle produced by the junction of the palmar plate with the base of the middle phalanx and the accessory collateral ligament (Fig. 15–12). This confluence of ligaments is firmly anchored to the proximal phalanx and the flexor tendon sheath by proximal and lateral extensions of the palmar plate known as the checkrein ligaments (Eaton, 1971; Watson and Light, 1979).

The Digital Flexor Tendon Sheath Pulley System

Most tendons in the hand are restrained to some extent by sheaths and retinacula that keep them close to the skeletal plane so that they maintain a relatively constant moment arm rather than bowstringing across the joints. The pulley system of the flexor tendon sheath in the finger is the most highly developed of these restraints.

As they extend from their muscles, the digital flexor tendons pass through the carpal tunnel before fanning out toward their respective digits. The flexor superficialis tendon inserts on the middle phalanx and the flexor profundus, on the distal phalanx. In each digit these two tendons, surrounded by their synovial sheaths, are held against the phalanges by a fibrous sheath. At strategic locations along the sheath are five dense annular pulleys (designated as A1, A2, A3, A4, and A5) and three thinner cruciform pulleys (C1, C2, and C3) (Fig. 15–13). These pulleys allow for a smooth curve so that no sharp or angular bends exist in the course of the tendon. Hence, local points of high pressure (stress raisers) between tendon and sheath are minimized.

At the point where the A3 pulley traverses the PIP joint, the tension in the tendon generated by joint flexion tends either to pull the pulley away from its attachment to the bone or to pull the bone away from the joint. This is no problem in a normal, stable joint, but when the joint has already been rendered unstable, as in a patient with rheumatoid arthritis, problems of instability may arise and there may be a danger of severe subluxation.

To appreciate the magnitude of these subluxating forces and how they increase with increased flexion, let us consider two flexed positions of a PIP joint: 60 degrees and 90 degrees. At 60 degrees, the two limbs of the flexor tendon form an angle of 120 degrees (Fig. 15–14A). At that point, the tension in the restraining pulley must equal the tension in the

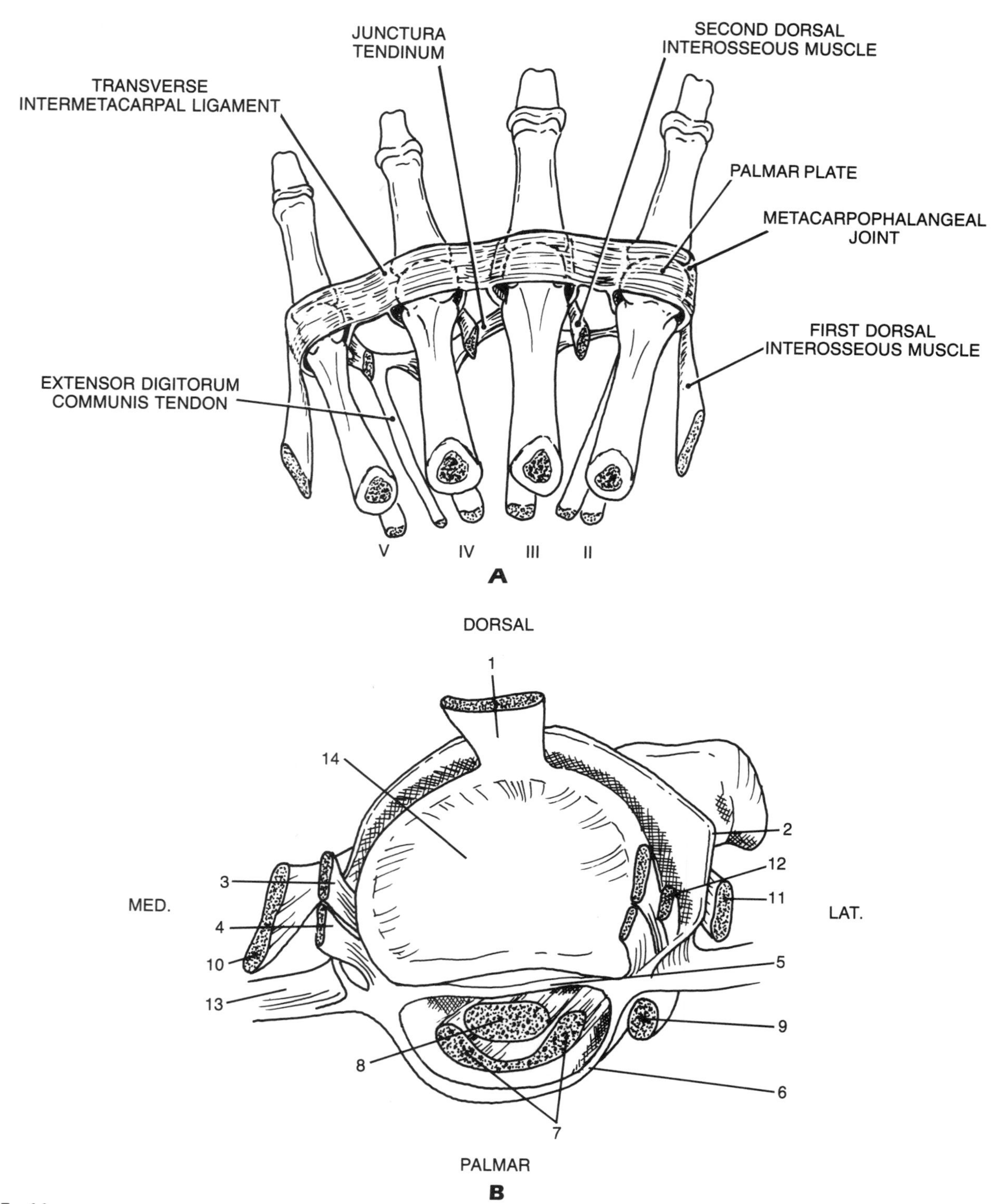

FIG. 15–11

A. Fibrous structures of the proximal transverse (MCP) arch (palmar view of the right hand). (Adapted from Tubiana, 1984.) **B.** Capsuloligamentous structures of the MCP joint (transverse view of the proximal phalangeal joint surface, middle finger, left hand). (Adapted from Zancolli, 1979.) 1, Extensor digitorum communis tendon; 2, sagittal band; 3, collateral ligament; 4, accessory collateral ligament; 5, palmar plate; 6, flexor tendon sheath; 7, flexor digitorum superficialis tendon; 8, flexor digitorum profundus tendon; 9, lumbrical muscle; 10, dorsal interosseous muscle; 11, dorsal interosseous muscle; 12, insertion of dorsal interosseous muscle into base of phalanx; 13, transverse intermetacarpal ligament; 14, articular surface of proximal phalanx.

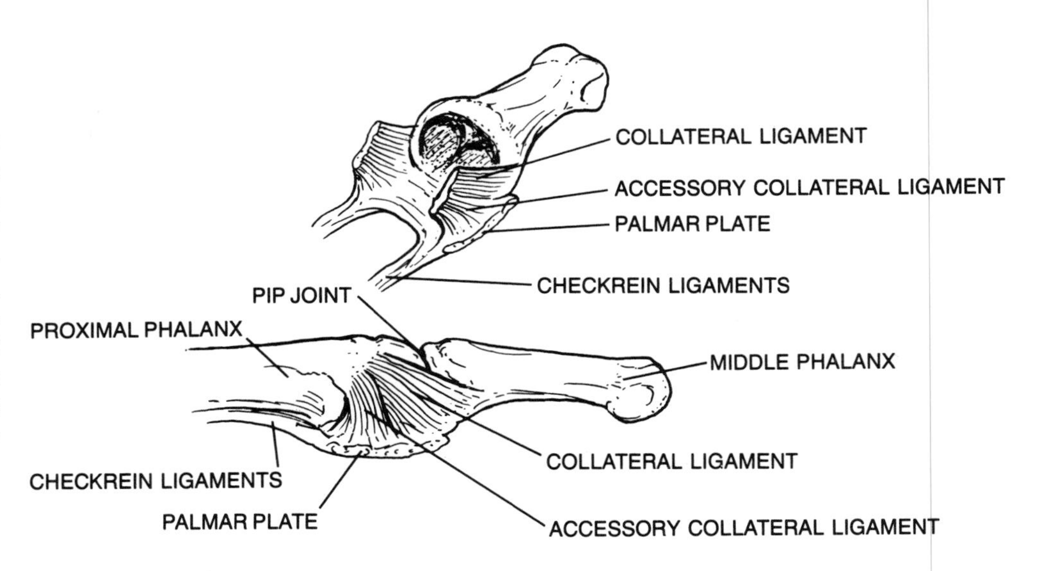

FIG. 15–12
Oblique (top) and mediolateral (bottom) views of the PIP joint. This joint gains stability from a strong, three-sided ligamentous support system produced by the collateral ligament, the accessory collateral ligament, and the palmar fibrocartilaginous plate, which is anchored to the proximal phalanx by proximal and lateral extensions known as the checkrein ligaments. (Adapted from Strickland, 1987, as modified from Eaton, 1971.)

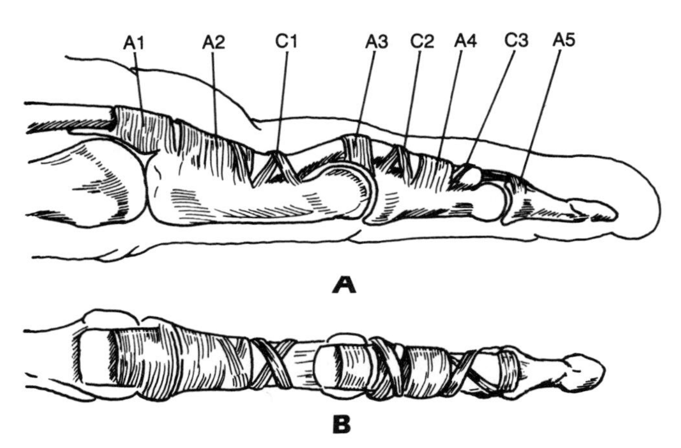

FIG. 15–13
Schematic drawings of the components of the digital flexor tendon sheath. The five sturdy annular pulleys (A1, A2, A3, A4, A5) are important in assuring efficient digital motion by keeping the tendons close to the phalanges. The three thin, pliable cruciate pulleys (C1, C2, and C3) allow flexibility of the sheath while maintaining its integrity. **A.** Mediolateral view. **B.** Palmar view of the sheath without its tendons. (Adapted from Doyle and Blythe, 1975, and Strickland, 1987.)

tendon for the system to be in equilibrium. At 90 degrees of flexion, however, the pulley must sustain 40% more tension than the tendon (Fig. 15–14B) (Brand, 1985).

TENDON EXCURSIONS

As the finger moves, each tendon slides a certain distance, which defines the "excursion of the tendon." Excursion takes place simultaneously in the flexor and extensor tendons during joint motion; the tendons of the agonist muscles displace in one direction, whereas the tendons of the antagonist muscles displace in the opposite direction to accommodate the motion. Knowledge of tendon excursions has applications in theoretical calculations of muscle forces, in splinting and rehabilitation of the hand after tendon repair, and in surgical procedures such as tendon transfers.

Use of basic rules of geometry allows the measurement of tendon excursion in relation to angular motion of a joint (Brand, 1975). When a lever rotates around an axis of an angle (θ), the distance (E) moved by every point on the lever is proportional to its own distance (r) from the axis ($E = r[\theta]$). In particular, when a lever rotates around an axis through an angle of one radian (about 60 degrees), every point on the lever moves through a distance equal to its own distance from the axis ($E = r$)—its moment arm (Fig. 15–15).

An and coworkers (1983b) developed a method for determining tendon excursions and moment arms that does not require knowledge of the exact center of rotation, which, for the finger joints, is still not known precisely. The instantaneous moment arm of a tendon in the plane of motion at a specific joint configuration is obtained from the slope of the plot of the tendon excursion versus joint rotational displacement. When this method is used the relationship between excursion and joint angle is not quite linear, except for the extensors at the MCP joints during flexion and extension. Elliot and McGrouther's (1986) study of extensor tendon excursion confirmed this observation.

In the digital tendons, the moment arms, and hence the excursions, are larger in the more proximal

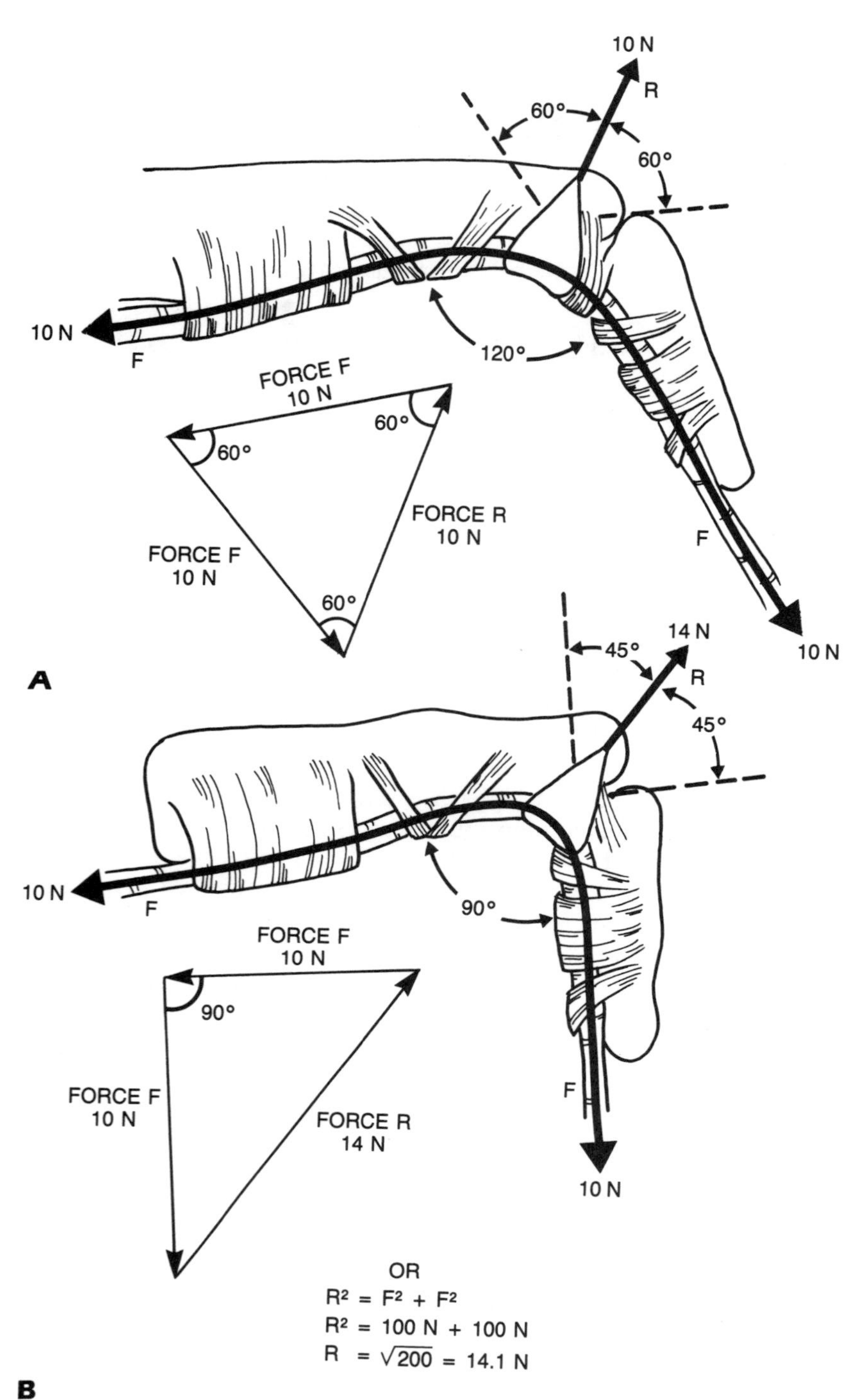

FIG. 15–14
Lateral view of the flexor tendon sheath pulley system at the PIP joint. **A.** The PIP joint is flexed 60 degrees. With the system in equilibrium, the resultant force (R) in the pulley system is equal to the vector sum of the two components of the tensile force (F) in the flexor tendon (i.e., 10 N). These three forces are presented graphically in an equiangular triangle of forces. **B.** The PIP joint is flexed 90 degrees. A triangle of forces shows that the resultant force R in the pulley system equals 14 N. Therefore, R equals 1.4 F. The value for R is also found by use of the pythagorean theorem, which states that in a right triangle the square of the hypotenuse equals the sum of the squares of the sides. (Adapted from Brand, 1985.)

joints (Fig. 15–16; see also Fig. 15–26A). The flexor superficialis tendons have a longer overall excursion than do the flexor profundus tendons. The excursion of the flexor tendons is larger than that of the extensor tendons, and the excursion of the extrinsic muscle tendons is larger in general than that of the intrinsic tendons (Table 15–2).

The values for tendon excursions given in Figure 15–16 and Table 15–2 are for normal tendons in close contact with the bone; the tendon moment arm and the joint axis thus remain essentially constant during joint motion. If either the location of the joint axis or the length of the moment arm changes during the movement, the measured tendon excursion per radian represents the average moment arm for that movement.

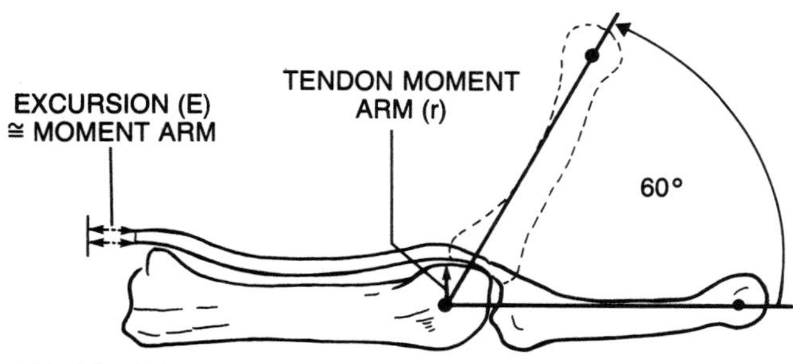

FIG. 15–15
When the metacarpophalangeal joint is flexed 60 degrees (approximately one radian), the tendon excursion (E) is equal to the tendon moment arm (r), which is the perpendicular distance between the tendon and the center of motion of the joint. (Adapted from Brand, 1985.)

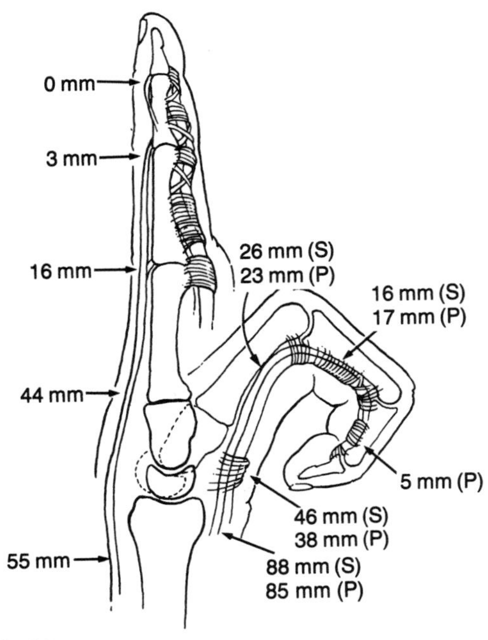

FIG. 15–16
Schematic drawing of a sagittal section through the finger showing the excursion of the flexor and extensor tendons at the DIP, PIP, MCP, and CMC joints. The numbers on the dorsum of the extended finger represent the excursion of the extensor tendons required at each level to bring all distal joints from full flexion to full extension. The numbers associated with the palmar aspect of the flexed finger represent the excursion for the superficialis (S) and profundus (P) tendons required at each level to bring the finger from full extension to full flexion. The measurements are in millimeters. (Adapted from Strickland, 1987, as modified from Verdan, 1979.)

Disruption of the pulley system is one case in which the length of the tendon moment arm changes during joint movement. In this instance the pulley may bowstring across one or more joints, increasing the tendon excursion requirement. Strickland (1983) found that biomechanical alteration of the finger flexor pulley system by excision of the distal half of the A2 pulley, the entirety of the C1, A3, and C2 pulleys, and the proximal half of the A4 pulley resulted in a considerable increase in the flexor profundus moment arm and in the tendon excursion requirement at the PIP joint (Fig. 15–17).

TABLE 15–2

APPROXIMATE TOTAL EXCURSIONS OF FINGER MUSCLE TENDONS

MUSCLE	TENDON EXCURSION (MM)
Interossei	30
Extensor pollicis brevis	30
Abductor pollicis longus	30
Lumbricals	40
Thenar muscles	40
Finger extensors	50–60
Finger flexors	60–70

(Data from Urbaniak, 1984.)

Extra excursion required at any one joint results in inadequate excursion and subsequent weakness in the more distal joints (Brand, 1975). When such weakness affects the PIP or DIP joints, a finger flexed to its limit may fail to touch the palm. Failure to touch the palm can be measured and used to assess the importance of a given pulley or group of pulleys. The use of this method by Doyle and Blythe (1975) revealed that the only pulleys required for normal flexor tendon function are the widest annular bands, A2 and A4, which traverse the proximal and middle phalanges, respectively (Table 15–3; see Fig. 15–13). Of these two critical pulleys, A2 was noted to be the more important, since the fingertip more closely approximated the palm with only A2 intact (12 to 15 mm) than with only A4 intact (20 to 25 mm).

BALANCE OF MUSCLE POWER IN THE FINGER

The motion of flexion-extension at the three finger joints does not result solely from the antagonistic action of the long flexors and extensors. The intrinsic muscles (interossei and lumbricals) act as moderators between these forces. This role can be explained by the use of the anatomic model system developed by Landsmeer (1955, 1961, 1963) and Spoor and Landsmeer (1976), in which the hand is viewed as a linked system of intercalated bony segments whose joints are spanned by ligaments,

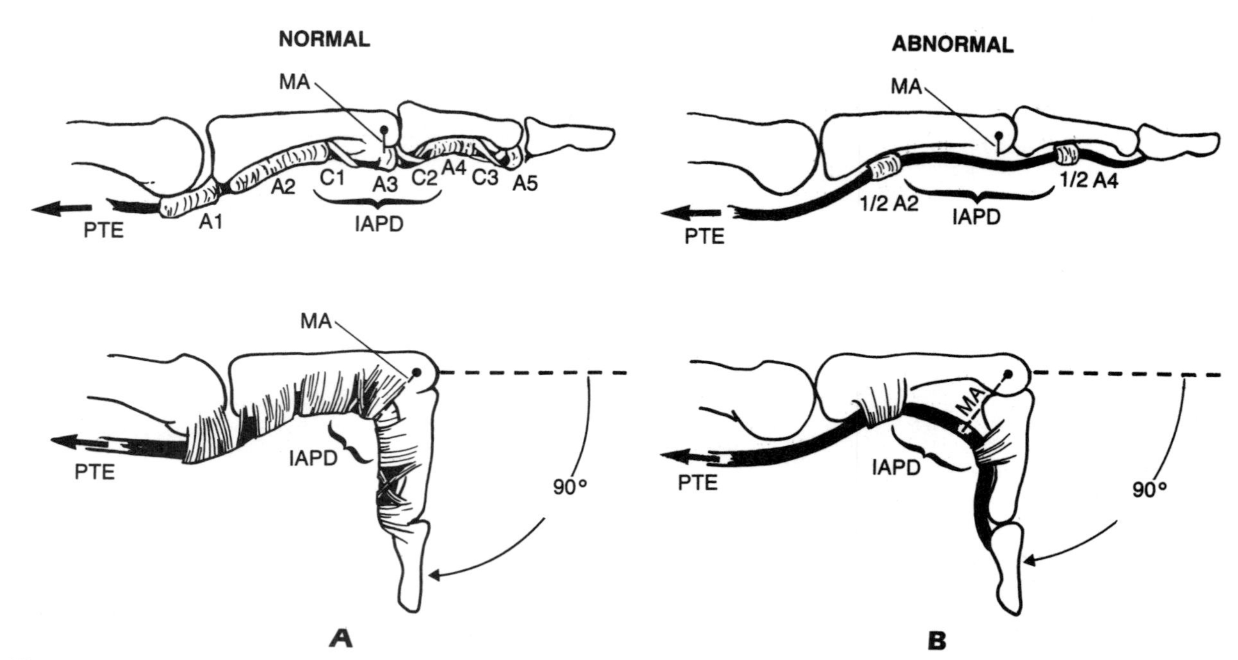

FIG. 15–17
Finger flexor tendon sheath pulley system with the PIP joint in extension and in 90 degrees of flexion. The five annular pulleys (A1 to A5) and the three cruciate pulleys (C1 to C3) are shown. IAPD indicates the intra-annular pulley distance between the A2 and A4 pulleys. **A.** In a normal finger with an intact pulley system, the moment arm of the flexor profundus tendon (MA) is small and the tendon excursion (PTE), which occurs within the intact osteofibrous canal, is minimal. **B.** Biomechanical alteration of the pulley system has been produced by excision of the distal half of the A2 pulley, all of the C1, A3, and C2 pulleys, and the proximal half of the A4 pulley. The flexor profundus tendon moment arm (MA) has increased dramatically, and a greater profundus tendon excursion is required to produce 90 degrees of flexion because of the bowstringing that results from the loss of pulley support. (Adapted from Strickland, 1983.)

TABLE 15–3

RELATIVE FUNCTIONAL IMPORTANCE OF DIGITAL FLEXOR TENDON PULLEYS DETERMINED BY SERIAL RESECTION AND RESULTANT FAILURE OF FINGERTIP TO TOUCH PALM

INTACT PULLEYS							FAILURE TO TOUCH PALM (MM)
A1	A2	C1	A3	C2	A4	*	0
*	A2	C1	A3	C2	A4	C3	0
*	A2	*	*	*	A4	C3	0
*	A2	*	*	*	A4	*	0
A1	A2	C1	A3	C2	*	*	2–5
A1	A2	C1	A3	*	*	*	5–8
A1	A2	C1	*	*	*	*	10–12
*	A2	*	*	*	*	*	12–15
*	*	C1	A3	C2	A4	C3	12–15
*	*	*	*	*	A4	*	20–25
*	*	*	*	*	*	*	25–30

NOTE: A1, A2, A3, and A4 indicate annular pulleys; C1, C2, and C3 indicate cruciform pulleys (see Fig. 15–17).
*Pulley resected.
(Data from Doyle and Blythe, 1975.)

tendons, and muscles. Most tendons in the hand span two or more joints, forming a biarticular or polyarticular system or chain that must be regarded as an anatomic unit. In its simplest form, a biarticular chain consists of two joints bridged by two tendons. The bone between the joints is called the intercalated bone (Fig. 15–18).

The following is given:

- Two hypothetical tendons, E and F, run on opposite sides of two adjacent joints, I and II.
- The lever arms of E about joints I and II are b and d, respectively, and a and c are the respective lever arms of F about joints I and II.
- γ is the angular change in joint I and φ, that in joint II.

The following convention will be adopted: proximal excursion will be considered negative and distal excursion positive.

Using the geometric rule mentioned in the section on tendon excursion and considering this system in

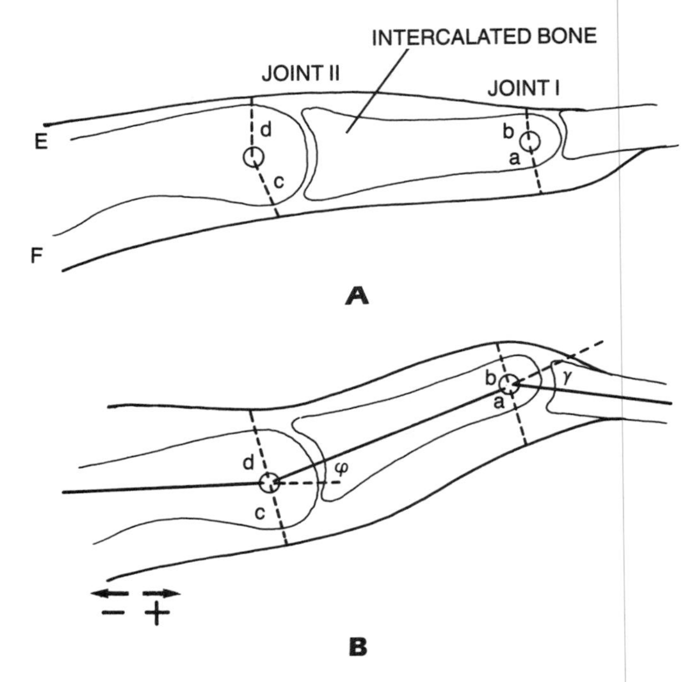

FIG. 15–18
Schematic representations of a biarticular chain consisting of two joints, I and II, bridged by two tendons, E and F. The middle phalanx is the intercalated bone. The moment arms of tendon F are represented by a and c and those of tendon E by b and d. The hollow circles represent the centers of rotation of the joints. Proximal excursion is considered negative and distal excursion positive. **A.** The biarticular chain is in the straight position. **B.** The chain is in the collapsed position. γ indicates the angular change of joint I; φ indicates the angular change of joint II.

equilibrium and the excursion of tendon E to equal 0, we have the following equation:

$$b\gamma = d\varphi. \quad (1)$$

The total excursion of tendon F can be written as follows:

$$a\gamma + c\varphi.$$

The resulting excursion of tendon F has to be in the negative (proximal) direction. Therefore, the following inequality holds:

$$a\gamma + c\varphi < 0.$$

Using equation (1), we can also write this inequality as:

$$a\gamma + \frac{cb}{d}\gamma < 0. \quad (2)$$

The only way that this inequality can be true is if the absolute value of the negative term is larger than the absolute value of the positive term.

It is obvious that this biarticular chain model system can be applied to any two adjacent joints. It is fundamental to the understanding of wrist kinematics, for example, and is important for elucidating finger kinematics. Let us suppose that the system in Figure 15–18 is the proximal part of a finger with joint II representing the MCP joint, joint I the PIP joint, F the flexor profundus tendon, and E the extensor tendon. The following is anatomically true:

$$a > b \text{ and } c > d. \text{ Thus, } ac > bd \text{ and } a > \frac{cb}{d}. \quad (3)$$

Therefore, in inequality (2), aγ would be the negative term and cφ the positive term, which means that the PIP joint will have to flex and the MCP joint will have to extend. In other words, if only the extensor muscle acts the proximal phalanx tends toward retroversion (MCP joint extension) and the middle phalanx toward anteversion, or IP joint flexion (Fig. 15–19A). The zigzag collapse of the polyarticular chain in this manner is termed the "intrinsic minus" position, indicating loss of the action of the intrinsic muscles. Clinically, this tendency, known as clawing, is a marked sign of intrinsic muscle palsy.

The anatomic location of the intrinsic muscles specifically enables them to stabilize the middle phalanx and to counteract the clawing tendency of the extrinsic muscles (Fig. 15–19B). The intrinsic tendons at the MCP joint lie palmar to the axis of motion of the joint, and their action tends to oppose that of the long extensor while reinforcing that of the long flexors. At the level of the IP joint, however, the intrinsic tendons have a dorsal position, and hence their action is synergistic with that of the extensor.

In a cadaver study simulating conditions in vivo, Ranney and coworkers (1987) showed that isolated concentric contraction of the lumbrical acting only against the passive resistance of the extrinsic muscles caused the finger to move from the claw position to a straight position and that further shortening brought the finger to the intrinsic plus position (MCP flexion and IP extension). In a clinical study of 10 subjects, Ketchum and associates (1978) showed that the

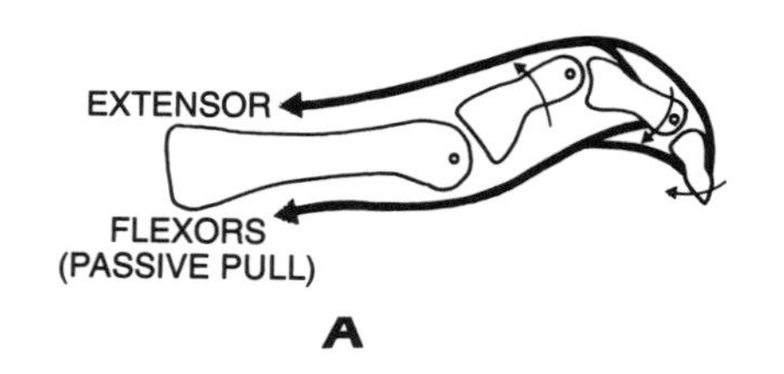

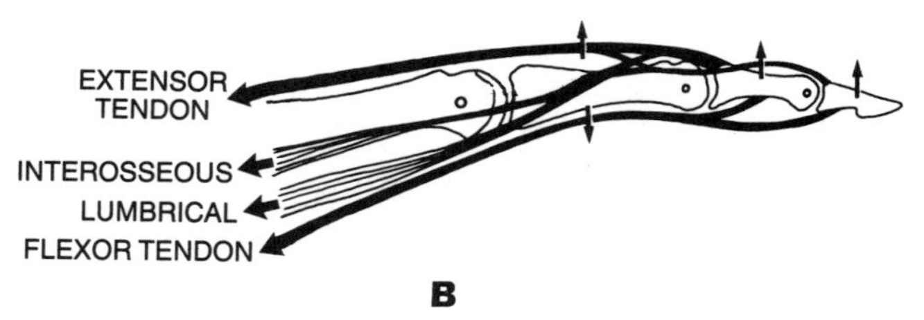

FIG. 15–19
Schematic representations of intrinsic-extrinsic muscle antagonism in a finger (lateral view). The hollow circles indicate the centers of motion of the DIP, PIP, and MCP joints. **A.** Isolated contraction of the extensor, without intrinsic muscle action, extends the MCP joint and flexes the IP joints (see arrows) because the flexor profundus and superficialis exert passive traction on the middle and distal phalanges. This tendency, known as the intrinsic minus position, or clawing, is a clinical sign of intrinsic muscle palsy. (Adapted from Caillet, 1982.) **B.** The anatomic location of the intrinsic muscles (the interossei and lumbrical) enables them to stabilize the middle and distal phalanges and to counteract the clawing tendency of the extrinsic muscles during finger extension. Contraction of the lumbrical diminishes the passive traction force of the flexors in the interphalangeal joints, thus allowing full extension of these joints. (Adapted from Tubiana, 1984.)

lumbrical to the index finger contributes 20% of the flexor force produced by all intrinsic muscles when the MCP joint is held in the flexed position while the IP joints are held in the extended position.

Electromyographic findings have confirmed that the lumbrical muscles endow the finger with a characteristic "counterclawing bias," which becomes apparent in the opening and closing of the hand (Landsmeer and Long, 1965; Thomas et al., 1968). The lumbricals are part of a mechanism that actively ensures an opening pattern of the hand in which IP joint extension does not lag behind MCP joint extension; in addition, this mechanism passively ensures a closing pattern of the hand in which IP joint flexion occurs simultaneously with, or lags behind, MCP joint flexion.

Stretching of the dorsal skin (Zancolli, 1979) and compression of the palmar pads during flexion also tend to produce a counterclawing effect. Brand (1985) showed that dorsal digital skin requires 12 mm of lengthening for 90 degrees of flexion. This requirement increases to 19 mm with a 5-mm thickness caused by soft tissue edema; the range of flexion is thus limited and joint stiffness may result.

Unlike the digital joints, those of the thumb chain do not appear to be linked, so dynamic stabilization of this chain is more complicated. Not only does the thumb have many supportive static stays to maintain stability, including the collateral ligaments, capsule, and palmar plate, but it also has eight active musculotendinous units (Fig. 15–20). Acting together, the flexor pollicis longus and the extensor pollicis longus produce a clawing type of posture characterized by strong flexion of the IP joint and extension of the CMC joint. The opponens pollicis counteracts this extrinsic effect by pulling the first metacarpal out of the plane of the hand; however, this situation results in another constrained position: flexion of the IP joint, extension of the MCP joint, and neutral positioning of the CMC joint. A release from this new constrained position can be achieved by the actions of the flexor pollicis brevis and the adductor pollicis, each of which produces a strong flexion moment about the MCP joint.

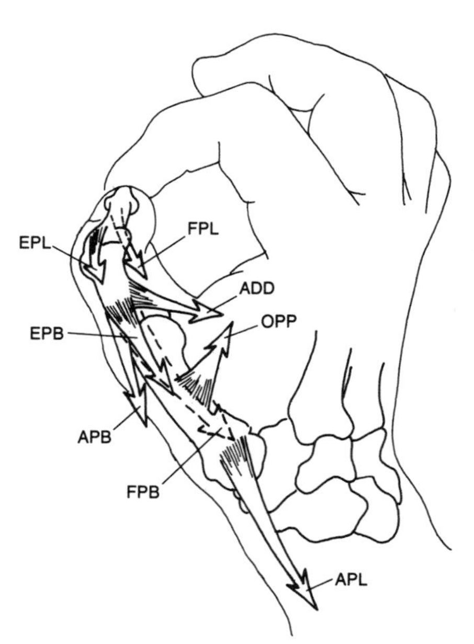

FIG. 15–20
Musculotendinous forces (arrows) acting on the thumb. EPL, extensor pollicis longus; EPB, extensor pollicis brevis; FPL, flexor pollicis longus; FPB, flexor pollicis brevis; APL, abductor pollicis longus; APB, abductor pollicis brevis; ADD, adductor pollicis; OPP, opponens pollicis. (Adapted from Cooney and Chao, 1977.)

KINETICS

The following sections present a discussion of the muscle forces acting on the fingers and thumb during free movement, the forces on the fingers and thumb during basic prehensile functions of the hand, the forces on the joints of the hand when an external load is applied to the fingertip, and the forces on the thumb IP joint during two methods of holding bricks.

MUSCLE FORCES IN THE HAND

The values most often cited for the strengths of the extrinsic muscles of the hand are those of Von Lanz and Wachsmuth (1970). Their values (Table 15–4) show that the strength of the finger flexors is over twice that of the extensors.

PATTERNS OF PREHENSILE HAND FUNCTION

Prehensile movements of the hand are those in which an object is seized and held partly or wholly within the compass of the hand. Such movements are used in a broad range of purposive activities involving handling of objects of all shapes and sizes. Efficient prehensile function depends on a multitude of factors, the most important of which are (1) mobility of the CMC joint of the thumb and, to a lesser extent, of the fourth and fifth MCP joints; (2) relative rigidity of the second and third CMC joints; (3) stability of the longitudinal finger and thumb arches; (4) balanced synergism and antagonism between the long extrinsic muscles and the intrinsic muscles; and (5) adequate sensory input from all areas of the hand. The precise relationships among the length, mobility, and position of each ray also play an essential role.

TABLE 15–4

STRENGTH VALUES OF THE EXTRINSIC MUSCLES OF THE HAND

MUSCLE	STRENGTH (NM)
Flexor pollicis longus	12
Extensor pollicis longus	1
Abductor pollicis longus	
As a wrist flexor	1
As a wrist abductor	4
Extensor pollicis brevis	1
Flexor digitorum superficialis	48
Flexor digitorum profundus communis	45
Extensor digitorum communis	17
Extensor indicis proprius	5

(Data from Von Lanz and Wachsmuth, 1970.)

Many attempts have been made to classify different patterns of prehensile hand function (McBride, 1942; Griffiths, 1943; Slocum and Pratt, 1946; Napier, 1956; Landsmeer, 1962). Napier (1956) distinguished two distinct patterns of prehensile movement in the normal hand: power grip and precision grip. He emphasized that the fundamental requisite to prehension, stability, can be met by either posture.

Power grip, or power grasp, is a forceful act performed with the finger flexed at all three joints so that the object is held between the finger and palm, with the thumb positioned on the palmar side of the object to force it securely into the palm (Fig. 15–21A). It is usually performed with the wrist deviated ulnarly and dorsiflexed slightly to augment the tension in the flexor tendons. Precision grip involves the manipulation of small objects between the thumb and the flexor aspects of the fingers in a finely controlled manner (Fig. 15–21B). The wrist position varies so as to increase the manipulative range. The fingers are generally in a semiflexed position, and the thumb is abducted and opposed. Certain prehensile activities involve both power and precision grips (Fig. 15–22).

As a refinement of Napier's classification, Landsmeer (1962) suggested that the precision grip be termed "precision handling," since it involves no forceful gripping of the object and is a dynamic process without a static phase. In both power grip and precision handling, full opposition of the thumb to the ring and little fingers is obtained via palmar displacement of the metacarpals of these fingers.

A variant of precision handling is the often used "dynamic tripod" (Capener, 1956), wherein the thumb, index finger, and middle finger have a dynamic action, working in close synergy for precision handling of the object, while the ring and little fingers are used largely for support and static control (Fig. 15–23). A further refinement is pinching a small object between the thumb and index finger. Such maneuvers are commonly classified as tip pinch, palmar pinch, lateral (or key) pinch, and pulp (or ulnar) pinch, depending on the parts of the phalanges brought to bear on the object being handled (Fig. 15–24).

An important distinction between power grip and precision handling is the fundamentally different position of the thumb in each posture. In the power grip the thumb is adducted; in precision handling it is

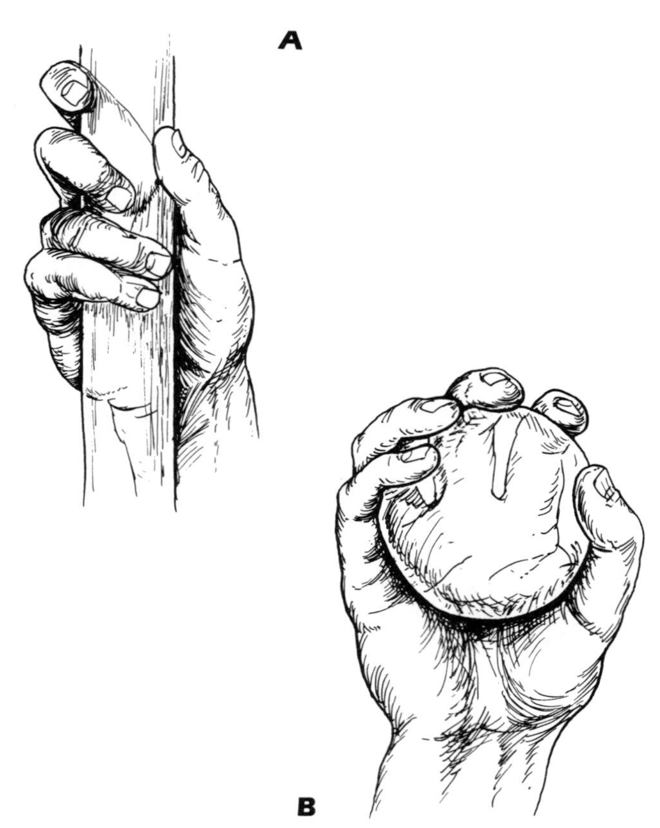

FIG. 15–21
The two fundamental patterns of prehensile hand function. (Adapted from Landsmeer, 1955.) **A.** A typical power grip. The adducted thumb forms a clamp with the partly flexed fingers and the palm. The palmar descent of metacarpals IV and V and additional flexion in their respective MCP joints enable these fingers to hold the object firmly against the palm. Counter-pressure is applied by the thumb, which lies approximately in the plane of the palm. The wrist is deviated ulnarly and dorsiflexed to increase the tension in the flexor tendons. Grip of an object along the oblique palmar axis (palmar groove), as shown here, involves a larger area of contact, and thus more control, than does grip along the transverse palmar axis. **B.** A typical precision maneuver. The object is pinched between the flexor aspects of the fingers and the thumb. The fingers are semiflexed, and the thumb is abducted and opposed. The wrist is dorsiflexed.

abducted (see Fig. 15–21). The relationship of the hand to the forearm also differs strikingly. In the power grip (see Fig. 15–21A), the hand is usually deviated ulnarly and the wrist is held approximately in a neutral position so that the long axis of the thumb coincides with that of the forearm. In this way pronation and supination can be transmitted from the forearm to the object. In precision handling (see Fig. 15–21B), the hand is generally held midway between radial and ulnar deviation, and the wrist is markedly

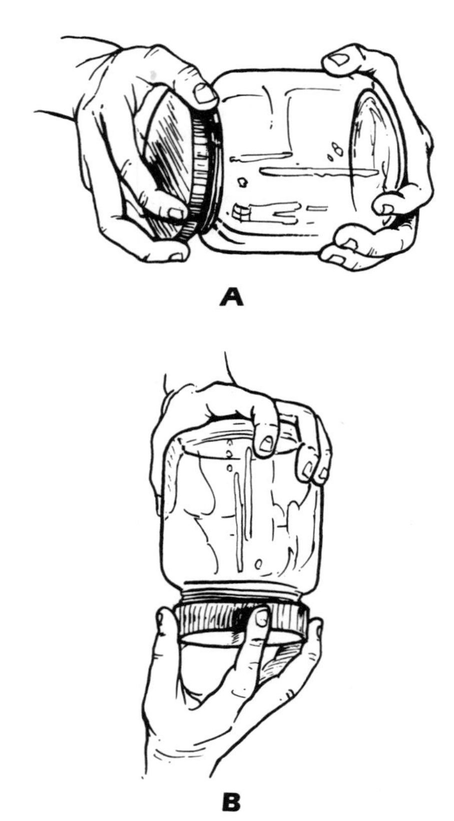

FIG. 15–22
The two fundamental patterns of hand function are used in unscrewing the lid of a tightly closed jar. (Adapted from Napier, 1956.) **A.** As the motion is begun, the right hand assumes a power grip posture. **B.** As the lid loosens, the hand assumes a precision posture to perform the final stages of unscrewing.

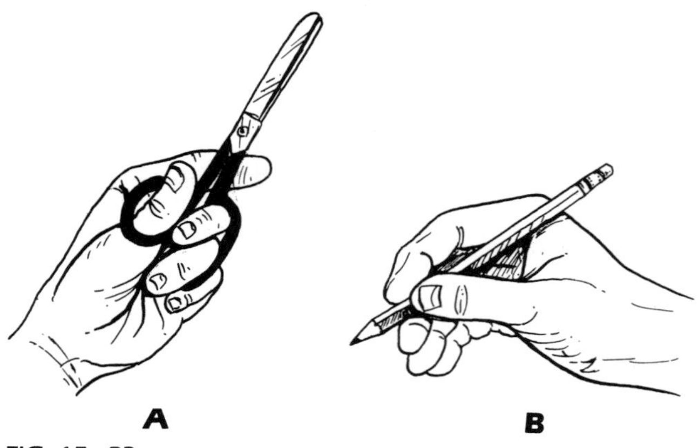

FIG. 15–23
The "dynamic tripod," a type of precision handling wherein the thumb, index finger, and middle finger work in close synergy for precision handling of the object while the ring finger and little finger offer support and static control. This functional configuration is illustrated by the use of scissors **(A)** and a pencil **(B).**

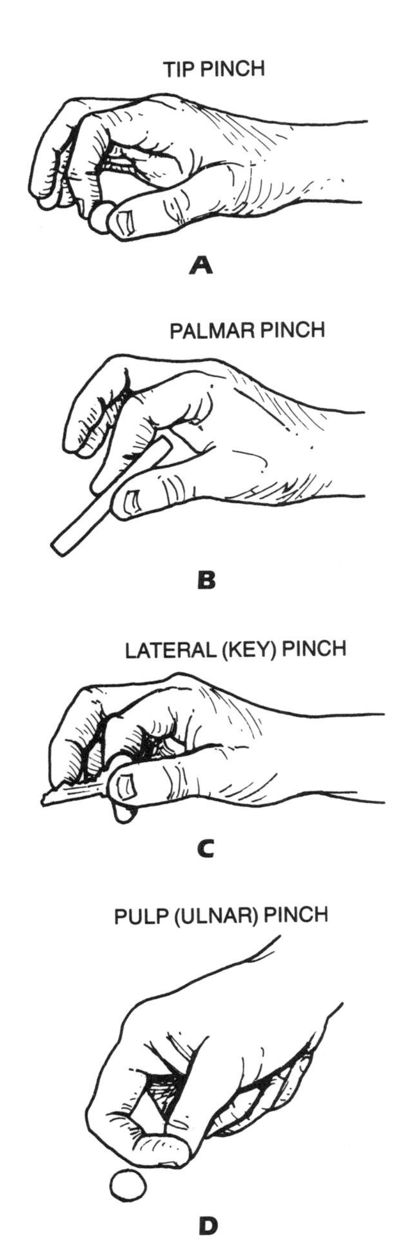

FIG. 15–24
Examples of precision handling in which small objects are pinched between the thumb and index finger. These grips are classified according to the parts of the phalanges brought to bear on the object handled. **A.** Tip pinch. **B.** Palmar pinch. **C.** Lateral (key) pinch. **D.** Pulp (ulnar) pinch.

dorsiflexed, so that the long axis of the thumb no longer aligns with that of the forearm.

Both power and precision play a role in nearly all power grips and precision postures, but the predominance of one factor over the other is responsible for the anatomic characteristics of each posture. The element of precision in the power grip complex is reflected in the posture of the thumb. When the demand for precision is minimal or absent, the thumb is wrapped over the dorsum of the middle phalanges of the digits and acts purely as a reinforcing mechanism. When an element of precision is required in what is predominately a power grip, such as the fencing grip (Fig. 15–25A), the thumb is adducted and aligned with the long axis of the cylinder so that, by means of small adjustments of posture, it can control the direction in which the force is being applied. At the other extreme of the power grip range is the coal-hammer grip (Fig. 15–25B), the crudest form of prehensile function, where the thumb is wholly occupied in reinforcing the clamping action of the digits. An example of this extreme in an empty hand is the bunched fist (Fig. 15–25C).

Rotating the thumb into an opposing position is a requirement of almost every hand function, whether

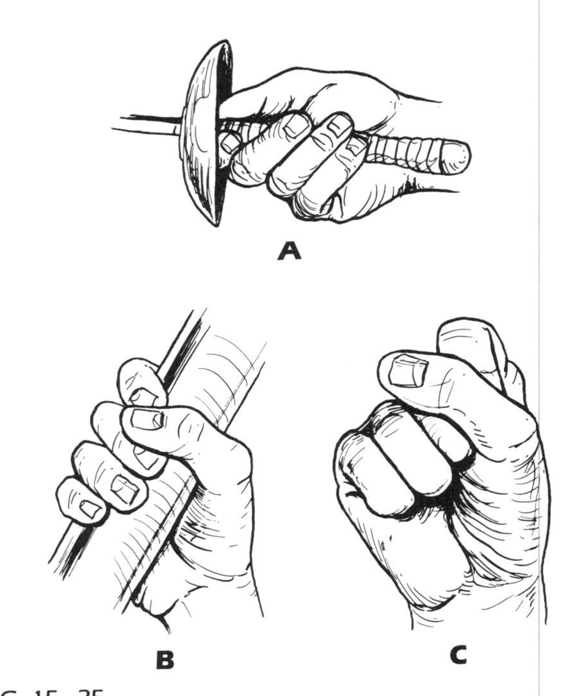

FIG. 15–25
The fencing grip **(A)** is a power grip posture in which the element of precision plays a large part. Rather than being wrapped over the dorsum of the digits, the thumb aligns with the long axis of the cylinder so that it can control the direction in which the force is being applied. In doing so, the thumb loses its effect as a powerful buttress on the lateral side of the hand. Hence, some of the power of grip is sacrificed in the interest of precision. The coal-hammer grip **(B)** and the bunched fist **(C)** are examples of a strong power grip with no element of precision. The ulnar deviation characteristic of a power grip is apparent in both cases. (Adapted from Napier, 1956.)

it be a strong grip or a delicate precision pinch. In some instances, however, the thumb may not be involved at all, as in the hook grip, in which the fingers are flexed so that their pads lie parallel and slightly away from the palm, forming a hook. This posture requires relatively little muscle activity to maintain and is used when precision requirements are minimal and when power must be exerted continuously for long periods. It may also be employed when only a fingertip hold can be obtained, as in raising a sash window. In a normal person this grip has limited potential and is rarely used; in a disabled person whose intrinsic hand muscles are paralyzed, however, it is virtually the only form of prehension possible (Napier, 1956).

TABLE 15–5

AVERAGE STRENGTH OF THE INDEX FINGER DURING ISOMETRIC HAND FUNCTIONS

HAND FUNCTION*	STRENGTH (N)
Tip pinch	24–95
Key pinch	37–106
Pulp pinch	30–83
Grasp	
Distal phalanx	38–109
Middle phalanx	7–38
Proximal phalanx	23–73

*For tip and pulp pinch, the forces were applied on tip and pulp of the distal phalanx, respectively. For key pinch, these forces were applied on the radial side of the middle phalanx. For grasp function, forces were applied at the middle of each phalanx.
(Data from An et al., 1985.)

FORCES IN THE HAND DURING PINCH AND GRIP

Several investigators have conducted three-dimensional analyses of the internal forces in the joints of the fingers and thumb during various isometric hand functions, namely power grip and various types of thumb-index pinch (Chao et al., 1976; Cooney and Chao, 1977; An et al., 1979, 1985; An et al., 1983a).

Elaborating on the three-dimensional analytic model developed by Chao and coworkers (1976), An's group (1985) made further studies of the forces in the index finger during pinch, grip, and the simulated daily activities of gripping a briefcase, holding a glass, and opening a big jar. The average values for strength of the index finger during grip and the various pinch positions, as these values can be measured, are listed in Table 15–5.

Using the optimization method, An's group (1985) calculated both muscle forces and constraint forces (such as joint compression forces). As assumptions must be made to obtain a solution for a basically indeterminate condition, their estimated values show a certain range.

From pinches to grasp to opening a big jar, the flexor profundus was required to build up a force that progressed from twice to more than five times the applied force. The flexor superficialis was particularly active in tip pinch and also in grasp, generating a force that was up to twice the external force. In the lateral key pinch, the flexor superficialis carried a minimal load but the long extensors and the two radial intrinsic muscles contributed large forces. The radial interosseous provided most of the balance necessary to prevent ulnar deviation at the MCP joint, creating a flexion moment at this joint that required counteraction by the long extensor. The intrinsic muscles appeared to produce more force during pinch than grasp to stabilize the MCP joint.

The compressive forces at the three finger joints for the various hand functions in units of applied force are shown in Table 15–6. These forces were least in the DIP joints and became progressively greater in the PIP and MCP joints. The greatest forces were sustained during lateral key pinch, particularly in the MCP joint because of the muscle forces acting to prevent ulnar deviation. The next greatest forces were sustained during the simulated activity of opening a big jar.

Studies of the internal forces in the thumb by Cooney and Chao (1977) showed that during pinch with an externally applied force of 10 N the joint compression force ranged from 24 to 36 N at the IP joint, from 46 to 66 N at the MCP joint, and from 60 to 134 N at the CMC joint. Extrinsic and intrinsic

TABLE 15–6

JOINT COMPRESSIVE FORCE (IN UNITS OF APPLIED FORCE) AT THE DIP, PIP, AND MCP JOINTS OF THE INDEX FINGER DURING ISOMETRIC HAND FUNCTIONS

HAND FUNCTION	DIP JOINT	PIP JOINT	MCP JOINT
Tip pinch	2.4–2.7	4.4–4.9	3.5–3.9
Key pinch	2.9–12.5	4.9–19.4	14.7–27.1
Pulp pinch	3.0–4.6	4.8–5.8	4.0–4.6
Grasp	2.8–3.4	4.5–5.3	3.2–3.7
Briefcase grip	0.0–0.0	1.7–1.9	1.0–1.3
Holding glass	2.5–2.9	4.3–4.4	4.0–4.1
Opening big jar	5.2–9.5	7.2–14.2	14.8–24.3

(Data from An et al., 1985.)

tendons of the thumb sustained tensile forces of 10 to 30 N during pinch when an external force of 10 N was applied. Forces of up to 500 N were produced in individual tendons during grip when an external force of 100 N was applied. In general, during normal pinch and grip, the extrinsic tendon forces were 4 to 5 times the applied external force, and the intrinsic tendon forces were 1.5 to 3 times this applied force.

Analysis of the forces produced during various hand functions has ergonomic applications. In a study assessing the relationship of work method to incidence of carpal tunnel syndrome, Armstrong and Chaffin (1979) found that 18 women with carpal tunnel syndrome used pinch hand positions more frequently during their work in production sewing than did a matched control group of 18 disease-free women (51.9% as opposed to 43.9% of the time); moreover, they tended to exert significantly greater forces while using the pinch position than did the controls. Lending support to these findings are Chao and coworkers' (1976) data showing that, for a given hand force, pinch can result in 20 to 50% more force in tendons adjacent to the median nerve than does grip.

EXTERNAL FORCES ON THE FINGER

When an external load is applied orthogonal to the fingertip, the moment arm of this load increases enormously from the most distal joint (DIP) to the most proximal (CMC). The moment arm of the main counterbalancing force, produced through the flexor profundus tendon, increases only moderately from the most distal to the most proximal joint (Fig. 15–26A), so muscles other than the flexor profundus must come into play to balance the extension forces of this externally applied load, as shown in the following example (Brand, 1985).

A load of 20 N is applied orthogonal to the tip of the index finger, producing extension moments of 0.4 Nm at the DIP joint, 1.1 Nm at the PIP joint, 2.1 Nm at the MCP joint, and 4.0 Nm at the CMC joint (Fig. 15–26B). Flexor profundus moment arms at the DIP, PIP, MCP, and CMC joints are assumed to be 0.5, 0.75, 1.0, and 1.25 cm, respectively.

If the profundus tendon were acting alone in this situation, it would have to produce a tensile force of 80 N at the DIP joint (0.4 Nm divided by its moment arm at this joint—0.005 m) and increasingly greater tensile forces at the more proximal joints (up to 320 N at the CMC joint [4.0 Nm/0.0125 m]) for equilibrium to be reached (i.e., for the sum of the moments to equal zero). Since the tension in a tendon is assumed to be the same along its entire length, this task cannot be accomplished by the flexor profundus tendon alone.

Although the flexor profundus tendon is capable of balancing the external extension moment (0.4 Nm) at the DIP joint with a tensile force of 80 N, it lacks a moment of 0.5 Nm needed to balance the extension moment at the PIP joint (1.1 Nm − [80 N × 0.0075 m]). This flexion moment lag at the PIP joint must be provided by the flexor superficialis, which will need to produce a tensile force of 66.6 N (0.5 Nm/0.0075 m) at this joint.

The superficialis, whose moment arms are assumed to be the same as those for the profundus, will also add to the flexion moments at the MCP and CMC joints: 0.667 Nm (66.6 N × 0.01 m) and 0.833 Nm (66.6 N × 0.0125 m), respectively, though the flexor superficialis would still lack a moment of 0.63 Nm needed to balance the extension moment at the MCP joint (2.1 Nm − [80 N × 0.01 m] − [66.6 N × 0.01 m]). This flexion moment at the MCP joint must be supplied by the intrinsic muscles. The flexion moment lag of 2.167 Nm at the CMC joint (4.0 Nm −[80 N × 0.0125 m] − [66.6 N × 0.0125 m]) would have to be supplied by one or more wrist flexors, such as the flexor carpi radialis, acting with a moment arm of 2.0 cm (106.6 N × 0.02 m) (Table 15–7).

It must be noted that the figures in this example are approximations and do not take into account the joint position or the actions of other muscles. They do indicate how the tensile force requirements increase progressively in the more proximal joints when the external load is distal. They also demonstrate the impossibility of a single tendon controlling a number of joints in a chain.

ANALYSIS OF A GRIP PATTERN

Some ergonomic applications of various grip patterns can be elucidated through the following comparison of two methods of holding bricks. The first method involves holding two bricks in one hand, between the thumb and the other fingers; the second involves using both hands flat to hold four bricks (Fig. 15–27A). Free body analysis of the forces on the thumb IP joint is performed for both methods (Fig. 15–27B).

In the free body diagram of the IP joint of one thumb shown in Figure 15–27B, let us assume the following:

- Force W, the external force, is applied orthogonally to the middle of the distal phalanx.

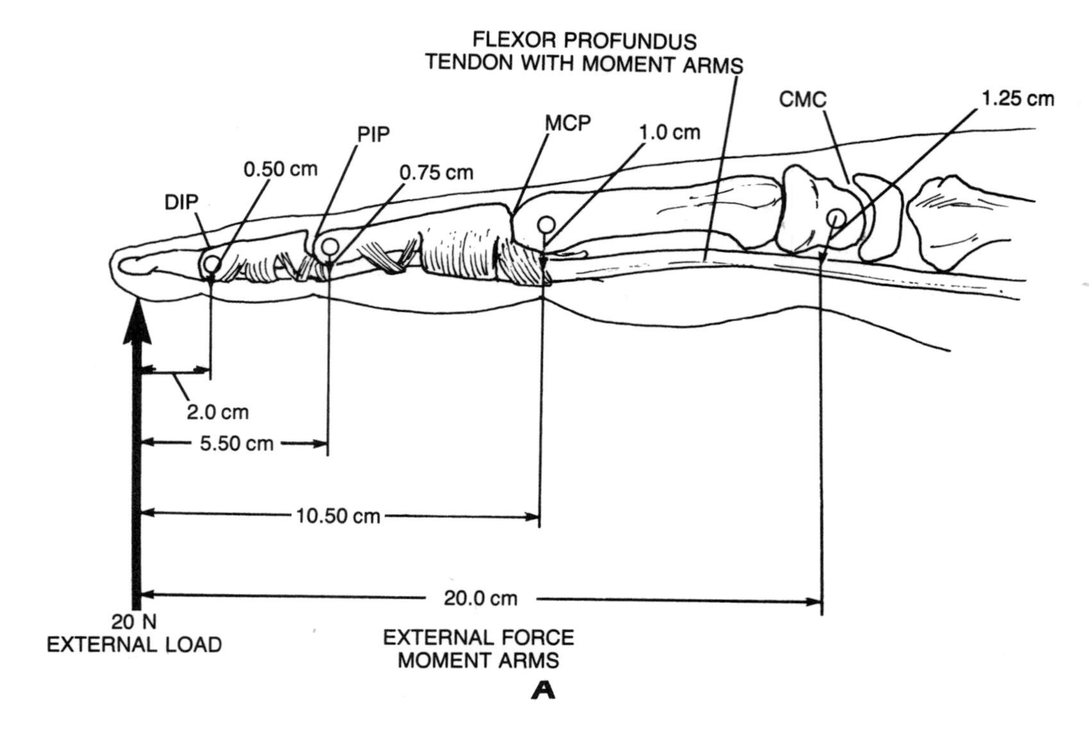

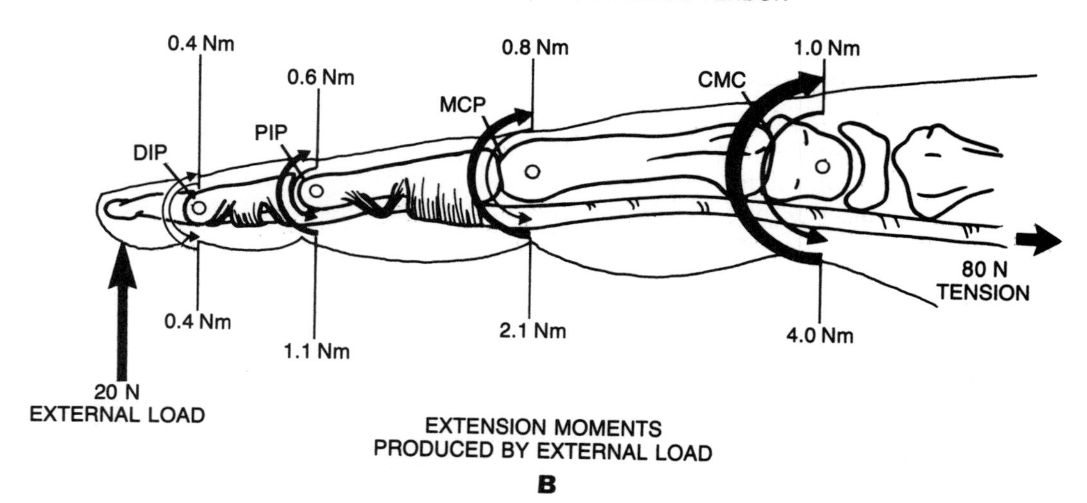

FIG. 15–26

An index finger in equilibrium with a 20-N load applied to the palmar surface of the fingertip. (Adapted from Brand, 1985.) **A.** The flexor profundus tendon is shown with hypothetical but realistic moment arms about the DIP, PIP, MCP, and CMC joints. The hollow circles indicate the centers of motion of these joints. Also shown are the moment arms at each joint for an external load of 20 N applied to the fingertip. Whereas the tendon moment arms increase moderately from distal to proximal, the moment arms of the external load increase enormously. **B.** The externally applied force of 20 N on the palmar fingertip creates extension moments on the finger joints (lower figures) that increase in magnitude from distal to proximal because of the considerable increase in the moment arms of this force. These extension moments are only partially counterbalanced by the flexion moments produced through the flexor profundus tendon (upper figures).

TABLE 15–7

MUSCLE TENSILE FORCES AND MOMENTS NEEDED AT THE DIP, PIP, MCP, AND CMC JOINTS TO SUPPORT AN EXTERNAL LOAD OF 20 NEWTONS APPLIED TO THE FINGERTIP

MUSCLE	TENSILE FORCE (N)	MOMENT OR TORQUE (NM)			
		DIP	PIP	MCP	CMC
Profundus	80.0	0.4	0.6	0.8	1.0
Superficialis	66.6	—	0.5	0.67	0.83
Intrinsics		—	—	0.63	—
Wrist flexors	106.6	—	—	—	2.17
Totals		0.4	1.1	2.1	4.0

(Data from Brand, 1985.)

Therefore, its lever arm (b) about the center of rotation of the IP joint is about 1 cm.

- The lever arm (a) of the flexor profundus muscle force (F), through its tendon, is 0.5 cm.
- The coefficient of friction between the brick and the skin is equal to 1 and is disregarded.
- Force W is equal to the weight of one or more bricks, depending on the method used.

For the thumb IP joint to be in complete equilibrium, the sum of the moments must equal zero and the sum of the forces must equal zero.

$$\Sigma \text{ moments} = 0$$
$$F \times 0.5 - W \times 1.0 = 0$$
$$F = 2\,W.$$

Since the point of application, direction, and magnitude of W and of F are known, J, the IP joint reaction force, can be found either graphically by constructing a triangle of forces (Fig. 15–27C) or mathematically:

$$\Sigma \text{ forces} = 0$$
$$J^2 = W^2 + F^2 = W^2 + (2\,W)^2 = 5\,W^2$$
$$J^2 = 5 \times W^2.$$

In the first method, the load is evenly distributed between the thumb as a unit and the rest of the

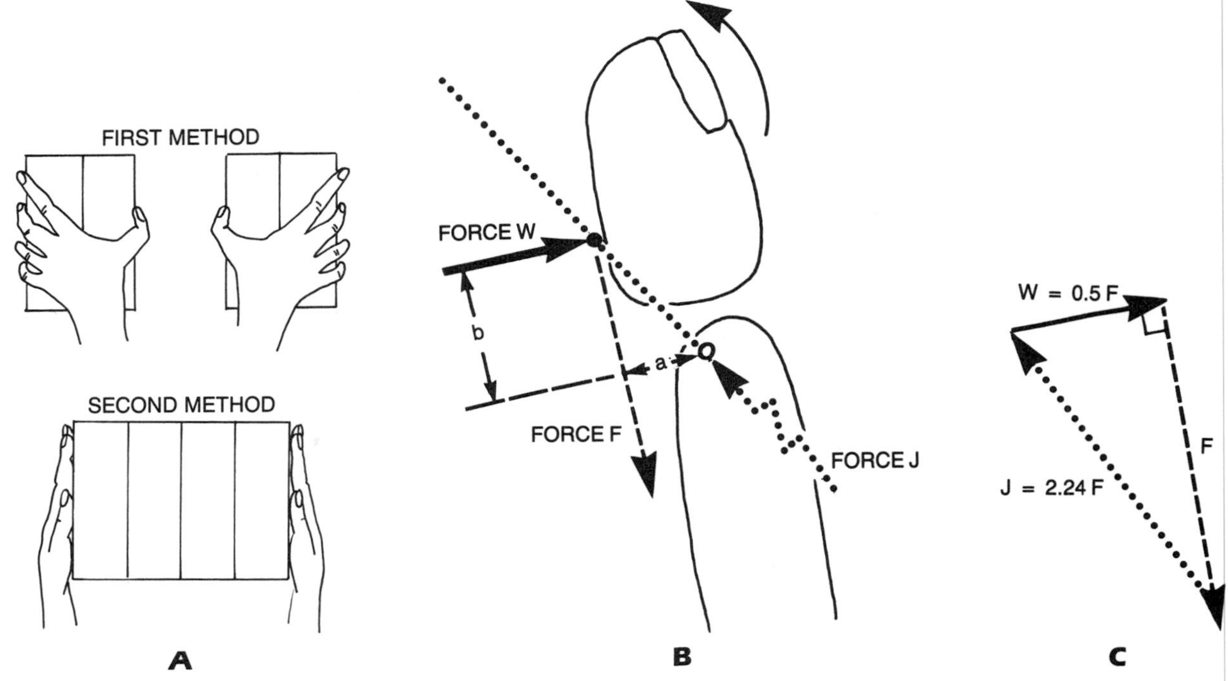

FIG. 15–27
A. Two methods of holding four bricks. The first method involves holding two bricks, clamped between thumb and fingers, in each hand. The second method involves holding all four bricks between both hands, with the hands flat. **B.** Free body diagram of a thumb IP joint for either method. W is the externally applied force, and its point of application is represented by the solid dot. The center of rotation of the IP joint is indicated by the hollow dot. The lever arm of force W is b. F is the flexor profundus muscle force through its tendon, and a is its lever arm. J is the joint reaction force. **C.** Triangle of forces acting on the thumb IP joint in static equilibrium.

fingers as a unit. Therefore, W = B (weight of one brick).

$$J^2 = 5 \times W^2$$
$$J = 2.24\ W = 2.24\ B$$
$$\text{and } F = 2\ B.$$

In the second method, the load is evenly distributed among all five fingers and between two hands. Therefore:

$$W = \frac{2\ B}{5}$$

$$J = 5 \times \frac{2\ B}{5} = 0.9\ B$$

$$\text{and } F = \frac{4\ B}{5} = 0.8\ B.$$

Hence, when the first method is used, both the IP joint reaction force and the flexor profundus muscle force are about 2.5 times greater than when the second method is used.

PATHOMECHANICS

Although a comprehensive presentation of pathomechanics of the hand is beyond the scope of this chapter, some commonly seen pathologic states

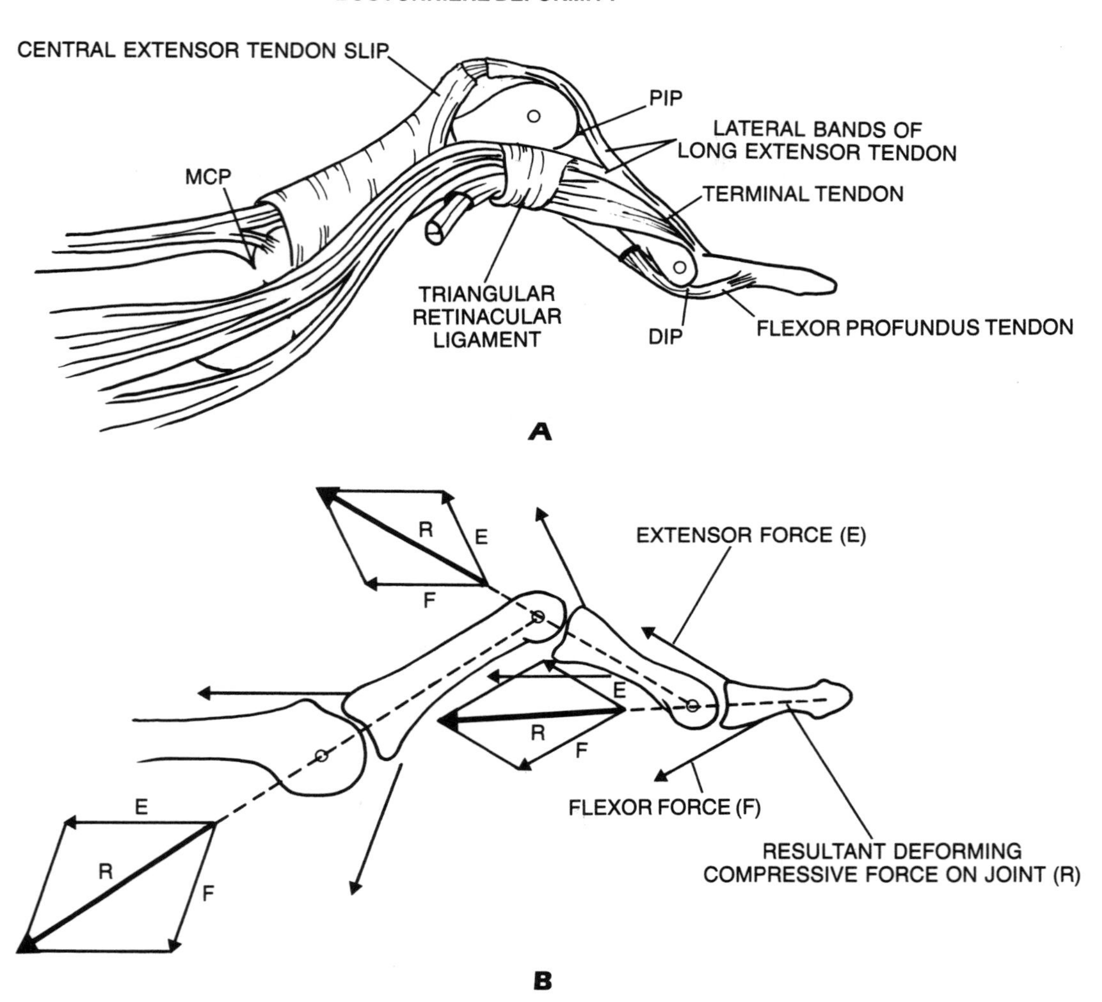

FIG. 15–28
A. Boutonnière deformity, a contracture consisting of MCP joint extension, PIP joint flexion, and DIP joint extension. **B.** Dynamic pathomechanics of a boutonnière deformity demonstrated by diagrams of the deforming forces from the PIP joint to the DIP joint to the MCP joint. The hollow circles indicate the centers of motion of these joints. R is the resultant compressive deforming force on the joint, F is the flexor force, and E is the extensor force. (Adapted from Swanson, 1973.)

are discussed below to exemplify the relationship of pathomechanics to normal biomechanics of the hand.

DEGENERATIVE JOINT DISEASE

The joints of the human finger are geometrically and kinematically similar, but the preferential degeneration of the DIP joint is well documented (Kellgren and Moore, 1952; Kellgren and Lawrence, 1957; Acheson et al., 1970; Radin et al., 1971; Ehrlich, 1975). Heberden's nodes, a classic sign of arthrosis in the hand, are also common in this joint (Kellgren and Moore, 1952). The great anatomic congruency of the DIP joint (see Fig. 15–10) results in steep pressure gradients around the perimeter of this joint, which may contribute to its propensity toward degeneration. Although the joint reaction forces increase in magnitude from the DIP to the MCP joint (Chao et al., 1976), the highest average contact pressures do occur in the DIP joint, primarily because it has the smallest contact area (Moran et al., 1985).

Moran and coworkers (1985) found that pressures in the DIP joint were higher during pinch than during grip, a finding that corroborates those of Hadler and associates (1978), who studied the DIP joints of 64 women textile workers who had performed the same repetitive tasks for at least 20 years. These tasks were broken down into three categories: burling, winding, and spinning. Winding involves power grip and considerable wrist motion, whereas burling and spinning require pinching with the index and middle fingers. Burlers and spinners had significantly higher degenerative joint disease scores for the DIP joints of the index and middle fingers than did winders.

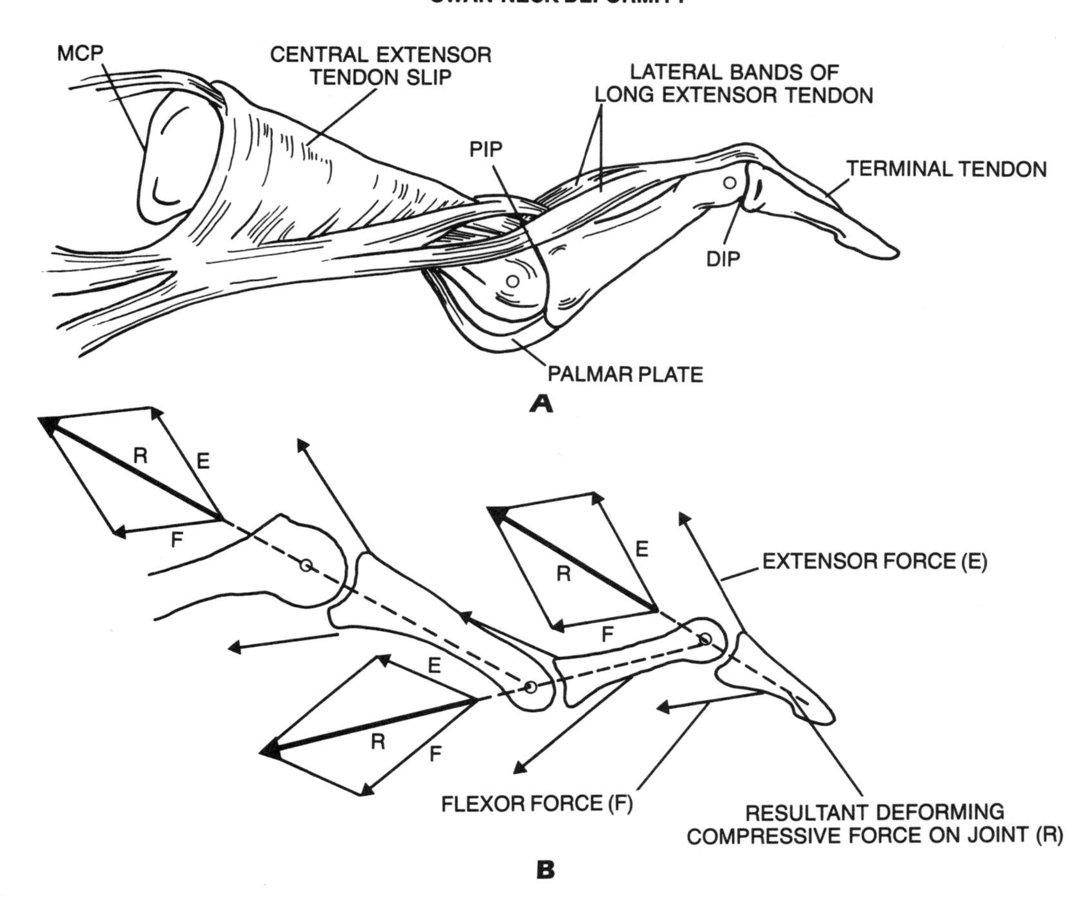

FIG. 15–29
A. Swan-neck deformity, a contracture consisting of MCP joint flexion, PIP joint extension, and DIP joint flexion. **B.** Dynamic pathomechanics of a swan-neck deformity, with diagrams of the deforming forces from the MCP joint to the PIP joint to the DIP joint. The hollow circles indicate the centers of motion of these joints. R is the resultant compressive deforming force on the joint, F is the flexor force, and E is the extensor force. (Adapted from Swanson, 1973.)

HAND DEFORMITIES CAUSED BY RHEUMATOID ARTHRITIS

Rheumatoid arthritis in the hand causes multiple deformities. Although various factors are responsible for the classic ulnar drift observed in this disease, two are especially important: the subluxating forces on the pulleys, and the tendency toward ulnar deviation during the power grip.

Two relatively common rheumatoid deformity patterns in the hand—the boutonnière deformity (15% of all cases) and the swan-neck deformity (28% of all cases)—involve a collapse deformity of the three-joint system of the fingers. This disturbance is characterized by hyperextension of one joint and reciprocal flexion of the two adjacent joints. This zigzag collapse in the normal flexion-extension pattern of the biarticular chain results from the loss of the balanced tendon mechanisms and ligament restrictions that normally prevent hyperextension (Swanson, 1973).

The boutonnière deformity (Fig. 15–28A, see p. 298) is a contracture consisting of MCP joint extension, PIP joint flexion, and DIP joint extension. This deformity begins with dorsal synovitis and capsular distension at the PIP joint (Fig. 15–28B), which creates lengthening of the central extensor slip and the triangular retinacular ligament at this level, and therefore a tendency toward dorsal subluxation of the proximal phalanx and palmar displacement of the lateral bands of the long extensor. This constraining situation increases the extensor pull on the distal phalanx. Extension of the MCP joint occurs later as a consequence of the resultant forces, with progressive joint destruction and fixed contracture.

The swan-neck deformity (Fig. 15–29A, see p. 299), a contracture consisting of MCP joint flexion, PIP joint extension, and DIP joint flexion, starts with flexor synovitis that increases the flexor pull on the MCP joint. Constant efforts to extend the finger against this pull lead to stretching of the collateral and retinacular ligaments and the palmar plate at the PIP joint (Fig. 15–29B). Dorsal displacement of the lateral bands of the long extensor, along with intrinsic tightening and MCP joint flexion, follows. This new constrained position of the lateral bands increases the pull of the long extensor tendon so that reciprocal flexion of the DIP joint occurs. Progressive disease leads to joint destruction and fixed contracture.

SUMMARY

1. The 19 bones of the hand are arranged in three arches: one longitudinal and two transverse. Derangement or collapse of the arch system as a result of bone injury, rheumatic disease, or paralysis of the intrinsic muscles of the hand can contribute to severe disability and deformity.
2. The hand is the principal instrument of touch. The combination of sensibility and motor function gives the hand its great importance as an organ of information and accomplishment.
3. The trapezoid, capitate, and second and third metacarpals, with their tight-fitting articulations, form the immobile unit of the hand. In their articulation with the hamate the fourth and fifth metacarpals are permitted a modest amount of palmar displacement, a motion essential for gripping.
4. Functionally, the most important motion of the thumb is opposition, in which abduction coupled with rotation at the CMC joint moves the thumb toward the tip of the little finger.
5. The finger rays are controlled by the coordinated action of the extrinsic and intrinsic muscle systems. The operation of each ray is not completely independent of its neighbors'.
6. The components of the extensor assembly, especially the oblique retinacular ligaments, account for the release of the distal phalanx and the coupling of PIP and DIP joint motion.
7. A unique feature of the MCP joints is their asymmetry, reflected in the bony configuration of the metacarpal heads, in the attachments of the collateral ligaments, and in the arrangement of the interossei.
8. The MCP joints are stabilized primarily by the radial and ulnar collateral ligaments and also by the transverse intermetacarpal ligament, which links the palmar plates to each other.
9. The flexor tendon sheath pulley system is essential for permitting the flexor tendons to maintain a relatively constant moment arm and for minimizing stress raisers between tendon and sheath. The second and fourth annular pulleys play a particularly important role in this respect.
10. The flexor superficialis tendon has a greater overall excursion than does the flexor profundus.

The excursion of the flexors is larger than that of the extensors, and the excursion of the extrinsic muscle tendons is generally greater than that of the intrinsic tendons.

11. Extra excursion required at any one joint owing to disruption of the pulley system results in inadequate excursion and subsequent weakness in the more distal joints.
12. The biarticular chain model system demonstrates that if only the extrinsic muscles act during finger extension the proximal phalanx tends toward retroversion (MCP joint extension) and the middle phalanx toward anteversion (IP joint flexion). This tendency is known clinically as clawing.
13. The intrinsic muscles, especially the lumbricals, provide a third force that endows the finger with a characteristic "counterclawing bias." This effect is further enhanced by stretching of the dorsal skin and compression of the palmar pads during flexion.
14. The strength of the finger flexors is over twice that of the extensors.
15. Efficient prehensile function depends on mobility of the thumb CMC joint and the fourth and fifth MCP joints, relative rigidity of the second and third CMC joints, balanced synergism-antagonism between the extrinsic and intrinsic muscles, and adequate sensory input. The relative lengths of the metacarpals and phalanges and the finger rays as a whole are also important.
16. The position of the thumb and the relationship between hand and forearm are the most important differences between power grip and precision handling.
17. Whereas the moment arm of an external load orthogonal to the fingertip increases enormously from the most distal to the most proximal joint, the moment arm of the main counterbalancing force, produced through the flexor profundus tendon, increases only moderately from distal to proximal. Hence, other muscles besides the flexor profundus must come into play to balance the extension forces of the externally applied load.
18. Although joint reaction forces increase in magnitude from the DIP to the MCP joint, the highest average contact pressures occur in the DIP joint, primarily because it has the smallest contact area. Pressures in the DIP joint are higher during pinch than during grip.
19. The boutonnière and swan-neck deformities involve a zigzag collapse of the three-joint system of the finger, characterized by hyperextension of one joint and reciprocal flexion of the two adjacent joints.

REFERENCES

Acheson, R. M., Chan, Y. K., and Clemett, A. R.: New Haven survey of joint diseases. XII: Distribution and symptoms of osteoarthrosis in the hands with reference to handedness. Ann. Rheum. Dis., *29*:275, 1970.

American Academy of Orthopaedic Surgeons: Joint Motion. Method of Measuring and Recording. Chicago, AAOS, 1965. Reprinted by the British Orthopaedic Association, 1966.

An, K. N., Chao, E. Y., Cooney, W. P., III, and Linscheid, R. L.: Normative model of human hand for biomechanical analysis. J. Biomech., *12*:775, 1979.

An, K. N., Chao, E. Y., Cooney, W. P., and Linscheid, R. L.: Forces in the normal and abnormal hand. J. Orthop. Res., *3*:202, 1985.

An, K. N., et al.: Determination of forces in extensor pollicis longus and flexor pollicis longus of the thumb. J. Appl. Physiol., *54*:714, 1983a.

An, K. N., et al.: Tendon excursion and moment arm of index finger muscles. J. Biomech., *16*:419, 1983b.

Armstrong, T. J., and Chaffin, D. B.: Carpal tunnel syndrome and selected personal attributes. J. Occup. Med., *21*:481, 1979.

Batmanabane, M., and Malathi, S.: Movements at the carpometacarpal and metacarpophalangeal joints of the hand and their effect on the dimensions of the articular ends of the metacarpal bones. Anat. Rec., *213*:102, 1985.

Bell, C.: The Hand: Its Mechanism and Vital Endowments as Evincing Design. Bridgewater Treatise. London, William Pickering, 1833.

Brand, P. W.: Clinical Mechanics of the Hand. St. Louis, C. V. Mosby, 1985, pp. 30–60.

Brand, P. W., Cranor, K. C., and Ellis, J. C.: Tendon and pulleys at the metacarpophalangeal joint of a finger. J. Bone Joint Surg., *57A*:779, 1975.

Caillet, R.: Hand Pain and Impairment. 3rd Ed. Philadelphia, F. A. Davis, 1982.

Capener, N.: The hand in surgery. J. Bone Joint Surg., *38B*:128, 1956.

Chao, E. Y., Opgrande, J. D., and Axmear, F. E.: Three-dimensional force analysis of finger joints in selected isometric hand functions. J. Biomech., *9*:387, 1976.

Cooney, W. P., and Chao, E. Y.: Biomechanical analysis of static forces in the thumb during hand function. J. Bone Joint Surg., *59A*:27, 1977.

Doyle, J. R., and Blythe, W.: The finger flexor tendon sheath and pulleys: Anatomy and reconstruction. *In* AAOS Symposium on Tendon Surgery in the Hand. St. Louis, C. V. Mosby, 1975, pp. 81–87.

Eaton, R. G.: Joint Injuries of the Hand. Springfield, Charles C Thomas, 1971.

Ehrlich, G. E.: Osteoarthritis beginning with inflammation. J.A.M.A., *232*:157, 1975.

Elliot, D., and McGrouther, D. A.: The excursions of the long extensor tendons of the hand. J. Hand Surg., *11B*:77, 1986.

Fahrer, M.: Considerations on functional anatomy of the flexor digitorum profundus. Ann. Chir., *25*:945, 1971.

Flatt, A. E.: The Care of the Rheumatoid Hand. St. Louis, C. V. Mosby, 1974, pp. 12–32.

Griffiths, H. E.: Treatment of the injured workman. Lancet, *I*:729, 1943.

Hadler, N. M., et al.: Hand structure and function in an industrial setting. Influence of three patterns of stereotyped, repetitive usage. Arthritis Rheum., *21*:210, 1978.

Hagert, C.-G.: Anatomical aspects on the design of metacarpophalangeal implants. Reconstr. Surg. Traumatol., *18*:92, 1981.

Hoggart, V. E., Jr.: Fibonacci and Lucas Numbers. Boston, Houghton Mifflin, 1969.

Kellgren, J. H., and Lawrence, J. S.: Radiological assessment of osteo-arthroses. Ann. Rheum. Dis., *16*:494, 1957.

Kellgren, J. H., and Moore, R.: Generalized osteoarthritis and Heberden's nodes. Br. Med. J., *1*:181, 1952.

Ketchum, L. D., Thompson, D., Pocock, G., and Wallingford, D.: A clinical study of forces generated by the intrinsic muscles of the index finger and the extrinsic flexor and extensor muscles of the hand. J. Hand Surg., *3*:571, 1978.

Landsmeer, J. M. F.: The anatomy of the dorsal aponeurosis of the human finger and its functional significance. Anat. Rec., *104*:31, 1949.

Landsmeer, J. M. F.: Anatomical and functional investigations on the articulation of the human fingers. Acta Anat., Suppl. *24*, 1955.

Landsmeer, J. M. F.: Studies in the anatomy of articulation. I. The equilibrium of the "intercalated" bone. Acta Morphol. Neerl.-Scand., *3*:287, 1961.

Landsmeer, J. M. F.: Power grip and precision handling. Ann. Rheum. Dis., *21*:164, 1962.

Landsmeer, J. M. F.: The coordination of finger-joint motions. J. Bone Joint Surg., *45A*:1654, 1963.

Landsmeer, J. M. F., and Long, C.: The mechanism of finger control, based on electromyograms and location analysis. Acta Anat., *60*:330, 1965.

Littler, J. W.: On the adaptability of man's hand (with reference to the equiangular curve). Hand, *5*:187, 1973.

McBride, E. D.: Disability Evaluation and Principles of Treatment of Compensable Injuries. 3rd Ed. Philadelphia, J. B. Lippincott, 1942, pp. 177–212.

Minami, A., et al.: Ligamentous structures of the metacarpophalangeal joint: A quantitative anatomic study. J. Orthop. Res., *1*:361, 1984.

Minami, A., et al.: Ligament stability of the metacarpophalangeal joint: A biomechanical study. J. Hand Surg., *10A*:255, 1985.

Moran, J. M., Hemann, J. H., and Greenwald, A. S.: Finger joint contact areas and pressures. J. Orthop. Res., *3*:49, 1985.

Napier, J. R.: The prehensile movements of the human hand. J. Bone Joint Surg., *38B*:902, 1956.

Radin, E. L., Parker, H. G., and Paul, I. L.: Pattern of degenerative arthritis. Lancet, *I*:377, 1971.

Ranney, D. A., Wells, R. P., and Dowling, J.: Lumbrical function: Interaction of lumbrical contraction with the elasticity of the extrinsic finger muscles and its effect on metacarpophalangeal equilibrium. J. Hand Surg., *12A*:566, 1987.

Sarrafian, S. K., et al.: Strain variation in the components of the extensor apparatus of the finger during flexion and extension: A biomechanical study. J. Bone Joint Surg., *52A*:980, 1970.

Slocum, D. B., and Pratt, D. R.: Disability evaluation for the hand. J. Bone Joint Surg., *28*:491, 1946.

Spoor, C., and Landsmeer, J. M. F.: Analysis of the zigzag movement of the human finger under influence of the extensor digitorum tendon and the deep flexor tendon. J. Biomech., *9*:561, 1976.

Strickland, J. W.: Management of acute flexor tendon injuries. Orthop. Clin. North Am., *14*:827, 1983.

Strickland, J. W.: Anatomy and kinesiology of the hand. In Hand Splinting. Principles and Methods. 2nd Ed. Edited by E. E. Fess and C. A. Philips. St. Louis, C. V. Mosby, 1987, pp. 3–41.

Swanson, A. B.: Flexible Implant Resection Arthroplasty in the Hand and Extremities. St. Louis, C. V. Mosby, 1973, pp. 71–86.

Thomas, D. H., Long, C., and Landsmeer, J. M. F.: Biomechanical considerations of lumbrical behavior in the human finger. J. Biomech., *1*:107, 1968.

Thompson, D'A.W.: On Growth and Form. Vol. II. 2nd Ed. Cambridge, Cambridge University Press, 1963.

Tubiana, R.: Anatomical and physiopathological features. *In* The Rheumatoid Hand. Edited by R. Tubiana with the collaboration of P. C. Archach et al. Group d'etude de la main. Paris, L'Expansion Scientifique Française, 1969, pp. 23–31.

Tubiana, R.: Architecture and functions of the hand. *In* Examination of the Hand and Upper Limb. Edited by R. Tubiana, J.-M. Thomine, and E. Mackin. Philadelphia, W. B. Saunders, 1984, pp. 1–97.

Urbaniak, J. R.: Tendon Transfers for Radial, Median and Ulnar Nerve Palsies. Orthopaedic Surgery Update Series. Vol. 3, Lesson 12, 1984.

Van Zwieten, K. J.: The extensor assembly of the finger in man and non-human primates: A morphological, functional, and comparative anatomical study. Thesis, Department of Anatomy and Embryology, University of Leiden, The Netherlands, 1980.

Verdan, C.: Introduction à la chirurgie des tendons. *In* Tendon Surgery of the Hand. Edited by C. Verdan in collaboration with J. H. Boyes. 1st English Ed. Edinburgh, Churchill Livingstone, 1979, pp. 9–11.

Von Lanz, T., and Wachsmuth, W.: Functional anatomy. *In* Bunnell's Surgery of the Hand. 5th Ed. Edited by J. H. Boyes. Philadelphia, J. B. Lippincott, 1970.

Watson, H. K., and Light, T. R.: Checkrein resection for flexion contracture of the middle joint. J. Hand Surg., *4*:67, 1979.

Zancolli, E.: Structural and Dynamic Bases of Hand Surgery. 2nd Ed. Philadelphia, J. B. Lippincott, 1979, pp. 3–63.

SUGGESTED READING

Acheson, R. M., Chan, Y. K., and Clemett, A. R.: New Haven survey of joint diseases. XII: Distribution and symptoms of osteoarthrosis in the hands with reference to handedness. Ann. Rheum. Dis., *29*:275–286, 1970.

American Academy of Orthopaedic Surgeons: Joint Motion. Method of Measuring and Recording. Chicago, AAOS, 1965. Reprinted by the British Orthopaedic Association, 1966.

An, K. N., Chao, E. Y., Cooney, W. P., III, and Linscheid,

R. L.: Normative model of human hand for biomechanical analysis. J. Biomech., *12*:775–788, 1979.

An, K. N., et al.: Forces in the normal and abnormal hand. J. Orthop. Res., *3*:202–211, 1985.

An, K. N., et al.: Determination of forces in extensor pollicis longus and flexor pollicis longus of the thumb. J. Appl. Physiol., *54*:714–719, 1983.

An, K. N., et al.: Tendon excursion and moment arm of index finger muscles. J. Biomech., *16*:419–425, 1983.

Armstrong, T. J., and Chaffin, D. B.: Carpal tunnel syndrome and selected personal attributes. J. Occup. Med., *21*:481–486, 1979.

Batmanabane, M., and Malathi, S.: Movements at the carpometacarpal and metacarpophalangeal joints of the hand and their effect on the dimensions of the articular ends of the metacarpal bones. Anat. Rec., *213*:102–110, 1985.

Bell, C.: The Hand: Its Mechanism and Vital Endowments as Evincing Design. Bridgewater Treatise. London, William Pickering, 1833.

Brand, P. W.: Clinical Mechanics of the Hand. St. Louis, C. V. Mosby, 1985, pp. 30–60.

Brand, P. W., Cranor, K. C., and Ellis, J. C.: Tendon and pulleys at the metacarpophalangeal joint of a finger. J. Bone Joint Surg., *57A*:779–784, 1975.

Caillet, R.: Hand Pain and Impairment. 3rd Ed. Philadelphia, F. A. Davis, 1982.

Capener, N.: The hand in surgery. J. Bone Joint Surg., *38B*:128–151, 1956.

Chao, E. Y., Opgrande, J. D., and Axmear, F. E.: Three-dimensional force analysis of finger joints in selected isometric hand functions. J. Biomech., *9*:387–396, 1976.

Cooney, W. P., and Chao, E. Y.: Biomechanical analysis of static forces in the thumb during hand function. J. Bone Joint Surg., *59A*:27–36, 1977.

Cooney, W. P., Lucca, M. J., Chao, E. Y., and Linscheid, R. L.: The kinesiology of the thumb trapeziometacarpal joint. J. Bone Joint Surg., *63A*:1371–1381, 1981.

Doyle, J. R., and Blythe, W.: The finger flexor tendon sheath and pulleys: Anatomy and reconstruction. *In* AAOS Symposium on Tendon Surgery in the Hand. St. Louis, C. V. Mosby, 1975, pp. 81–87.

Eaton, R. G.: Joint Injuries of the Hand. Springfield, Charles C Thomas, 1971.

Ehrlich, G. E.: Osteoarthritis beginning with inflammation. J.A.M.A., *232*:157–159, 1975.

Elliot, D., and McGrouther, D. A.: The excursions of the long extensor tendons of the hand. J. Hand Surg., *11B*:77–80, 1986.

Fahrer, M.: Considerations on functional anatomy of the flexor digitorum profundus. Ann. Chir., *25*:945–950, 1971.

Flatt, A. E.: The Care of the Rheumatoid Hand. St. Louis, C. V. Mosby, 1974, pp. 12–32.

Flatt, A. E.: The Care of Minor Hand Injuries. 4th Ed. St. Louis, C. V. Mosby, 1979, pp. 3–25.

Griffiths, H. E.: Treatment of the injured workman. Lancet, *I*:729, 1943.

Hadler, N. M., et al.: Hand structure and function in an industrial setting. Influence of three patterns of stereotyped, repetitive usage. Arthritis Rheum., *21*:210–220, 1978.

Hagert, C.-G.: Advances in hand surgery: Finger joint implants. Surg. Ann., *10*:253–275, 1978.

Hagert, C.-G.: Anatomical aspects on the design of metacarpophalangeal implants. Reconstr. Surg. Traumatol., *18*:92–110, 1981.

Hakstian, R. W., and Tubiana, R.: Ulnar deviation of the fingers. The role of joint structure and function. J. Bone Joint Surg., *49A*:299–316, 1967.

Hoggart, V. E., Jr.: Fibonacci and Lucas Numbers. Boston, Houghton Mifflin, 1969.

Kellgren, J. H., and Lawrence, J. S.: Radiological assessment of osteo-arthroses. Ann. Rheum. Dis., *16*:494–502, 1957.

Kellgren, J. H., and Moore, R.: Generalized osteoarthritis and Heberden's nodes. Br. Med. J., *1*:181–187, 1952.

Ketchum, L. D., Thompson, D., Pocock, G., and Wallingford, D.: A clinical study of forces generated by the intrinsic muscles of the index finger and the extrinsic flexor and extensor muscles of the hand. J. Hand Surg., *3*:571–578, 1978.

Kiefhaber, T. R., Stern, P. J., and Grood, E. S.: Lateral stability of the proximal interphalangeal joint. J. Hand Surg., *11A*:661–668, 1986.

Landsmeer, J. M. F.: The anatomy of the dorsal aponeurosis of the human finger and its functional significance. Anat. Rec., *104*:31–44, 1949.

Landsmeer, J. M. F.: Anatomical and functional investigations on the articulation of the human fingers. Acta Anat., Suppl. *24*:1–69, 1955.

Landsmeer, J. M. F.: Studies in the anatomy of articulation. I. The equilibrium of the "intercalated" bone. Acta Morphol. Neerl.-Scand., *3*:287–303, 1961.

Landsmeer, J. M. F.: Power grip and precision handling. Ann. Rheum. Dis., *21*:164–169, 1962.

Landsmeer, J. M. F.: The coordination of finger-joint motions. J. Bone Joint Surg., *45A*:1654–1662, 1963.

Landsmeer, J. M. F.: The human hand in phylogenetic perspective. Bull. Hosp. Joint Dis. Orthop. Inst., *44*:276–287, 1984.

Landsmeer, J. M. F., and Long, C.: The mechanism of finger control, based on electromyograms and location analysis. Acta Anat., *60*:330–347, 1965.

Linscheid, R. L., and Chao, E. Y. S.: Biomechanical assessment of finger function in prosthetic joint design. Orthop. Clin. North Am., *4*:317–330, 1973.

Littler, J. W.: On the adaptability of man's hand (with reference to the equiangular curve). Hand, *5*:187–191, 1973.

Littler, J. W.: The finger extensor system. Some approaches to the correction of its disabilities. Orthop. Clin. North Am., *17*:483–492, 1986.

Lundborg, G., Myrhage, R., and Rydevik, B.: The vascularization of human flexor tendons within the digital synovial region—structural and functional aspects. J. Hand Surg., *2*:417, 1977.

Mansat, M. F.: Volar aspect of the proximal interphalangeal joint. An anatomical study and pathological correlations. Bull. Hosp. Joint Dis. Orthop. Inst., *44*:309–317, 1984.

McBride, E. D.: Disability Evaluation and Principles of Treatment of Compensable Injuries. 3rd Ed. Philadelphia, J. B. Lippincott, 1942, pp. 177–212.

Minami, A., et al.: Ligamentous structures of the metacarpophalangeal joint: A quantitative anatomic study. J. Orthop. Res., *1*:361–368, 1984.

Minami, A., et al.: Ligament stability of the metacarpophalangeal joint: A biomechanical study. J. Hand Surg., *10A*:255–260, 1985.

Moran, J. M., Hemann, J. H., and Greenwald, A. S.: Finger joint contact areas and pressures. J. Orthop. Res., *3*:49–55, 1985.

Napier, J. R.: The form and function of the carpometacarpal joint of the thumb. J. Anat., *89*:362–369, 1955.

Napier, J. R.: The prehensile movements of the human hand. J. Bone Joint Surg., *38B*:902–913, 1956.

Radin, E. L., Parker, H. G., and Paul, I. L.: Pattern of degenerative arthritis. Lancet, *1*:377–379, 1971.

Ranney, D. A., Wells, R. P., and Dowling, J.: Lumbrical function: Interaction of lumbrical contraction with the elasticity of the extrinsic finger muscles and its effect on metacarpophalangeal equilibrium. J. Hand Surg., *12A*:566–575, 1987.

Sarrafian, S. K., et al.: Strain variation in the components of the extensor apparatus of the finger during flexion and extension: A biomechanical study. J. Bone Joint Surg., *52A*:980–990, 1970.

Slocum, D. B., and Pratt, D. R.: Disability evaluation for the hand. J. Bone Joint Surg., *28*:491–495, 1946.

Spoor, C., and Landsmeer, J. M. F.: Analysis of the zigzag movement of the human finger under influence of the extensor digitorum tendon and the deep flexor tendon. J. Biomech., *9*:561–566, 1976.

Strickland, J. W.: Management of acute flexor tendon injuries. Orthop. Clin. North Am., *14*:827–849, 1983.

Strickland, J. W.: Anatomy and kinesiology of the hand. *In* Hand Splinting. Principles and Methods. 2nd Ed. Edited by E. E. Fess and C. A. Philips. St. Louis, C. V. Mosby, 1987, pp. 3–41.

Swanson, A. B.: Flexible Implant Resection Arthroplasty in the Hand and Extremities. St. Louis, C. V. Mosby, 1973, pp. 71–86.

Thomas, D. H., Long, C., and Landsmeer, J. M. F.: Biomechanical considerations of lumbrical behavior in the human finger. J. Biomech., *1*:107–115, 1968.

Thompson, D'A.W.: On Growth and Form. Vol. II. 2nd Ed. Cambridge, Cambridge University Press, 1963.

Tubiana, R.: Anatomical and physiopathological features. *In* The Rheumatoid Hand. Edited by R. Tubiana with the collaboration of P. C. Archach et al. Group d'etude de la main. Paris, L'Expansion Scientifique Française, 1969, pp. 23–31.

Tubiana, R.: Architecture and functions of the hand. *In* Examination of the Hand and Upper Limb. Edited by R. Tubiana, J.-M. Thomine, and E. Mackin. Philadelphia, W. B. Saunders, 1984, pp. 1–97.

Verdan, C.: Introduction à la chirurgie des tendons. *In* Tendon Surgery of the Hand. Edited by C. Verdan in collaboration with J. H. Boyes. 1st English Ed. Edinburgh, Churchill Livingstone, 1979, pp. 9–11.

Von Lanz, T., and Wachsmuth, W.: Functional anatomy. *In* Bunnell's Surgery of the Hand. 5th Ed. Edited by J. H. Boyes. Philadelphia, J. B. Lippincott, 1970.

Watson, H. K., and Light, T. R.: Checkrein resection for flexion contracture of the middle joint. J. Hand Surg., *4*:67–71, 1979.

Youm, Y., Gillespie, T. E., Flatt, A. E., and Sprague, B. L.: Kinematic investigation of normal MCP joint. J. Biomech., *11*:109–118, 1978.

Zancolli, E.: Structural and Dynamic Bases of Hand Surgery. 2nd Ed. Philadelphia, J. B. Lippincott, 1979, pp. 3–63.

Zancolli, E. H.: Claw-hand caused by paralysis of the intrinsic muscle. J. Bone Joint Surg., *39A*:1076–1080, 1980.

GLOSSARY OF BIOMECHANICAL TERMS

Abduction—Motion away from the midline

Acceleration—The change in velocity of a body divided by the time over which change occurs

Adduction—Motion toward the midline

Agonistic muscles—Muscles that initiate and carry out motion

Angle of anteversion—The angle of inclination of the femoral neck relative to the femoral shaft in the transverse plane

Angular acceleration—The change in angular velocity of a body divided by the time over which change occurs, usually expressed in radians per second squared or degrees per second squared

Angular deformation—Internal structural change in an angular manner due to shear loading

Anisotropy—Variation of material property with direction

Antagonistic muscles—Muscles that oppose the actions of the agonistic muscles

Area moment of inertia—Quantity that takes into account the cross-sectional area and distribution of material around an axis during bending

Articular cartilage—A firm, highly hydrated hyaline connective tissue lining the bone ends within a diarthrodial joint

Asymptotic curve—A graph curve that reaches a certain point after which only one factor need remain constant for one end of the curve to parallel one axis to infinity

Axial rotation—Rotation about an axis

Axis of motion—Line about which all points move in a body in motion

Axis of rotation—Line about which all points in a rotating body describe circles

Bending—A loading mode in which a load is applied to a structure in a manner that causes it to bend about an axis, subjecting the structure to a combination of tension and compression (See *three-point bending* and *four-point bending)*

Bending moment—A quantity at a point in a structure equal to the product of the applied force and the perpendicular distance from the point to the force line, usually measured in newton meters

Biomechanics—The study of mechanical motion in biological systems

Biphasic material—A two-phase material; an example is articular cartilage, which is composed of a fluid and a solid phase

Bone remodeling—The ability of bone to adapt, by changing its size, shape, and structure, to the mechanical demands placed on it

Boundary lubrication—Mode of lubrication sustained by a monolayer of lubricant molecules adsorbed on the bearing surface (See *fluid film lubrication*)

Brittleness—The quality whereby a material exhibits little deformation before failure

Center of gravity—Equilibrium point of a supported body at which its total mass is considered concentrated (See *center of mass*)

Center of mass—That point at the exact center of an object's mass; often called the center of gravity (See *center of gravity*)

Center of rotation—A point around which circular motion is described

Chondroitin sulfate—A sulfated glycosaminoglycan (hexosamine) polymer chain composed of specific repeating disaccharide units; a basic constituent of proteoglycan

Chondrocyte—Cartilage cell occupying a lacuna within the cartilage matrix

Closed section—A cross section that has a continuous outer surface

Coefficient of friction—Ratio of the force required to slide one surface over another to the force acting perpendicular to the surfaces

Collagen—Basic fibrous protein of the body

Combined loading—Application of two or more loading modes to a structure

Compression—A loading mode in which equal and opposite loads are applied toward the surface of the structure, resulting in shortening and widening

Concentric work—Work produced by a muscle when it contracts and shortens
Concurrent—Meeting or intersecting in a point
Conservation of momentum—Maintenance of the relationship between velocity and mass
Contact area—Area of load support or direct surface-to-surface contact
Contact point—The junction point between two joint surfaces
Coplanar—Lying or acting in the same plane
Coronal plane—See *frontal plane*
Coxa valga—Condition in which the angle formed by the axes of the femoral head and shaft is greater than 125 degrees
Coxa vara—Condition in which the angle formed by the axes of the femoral head and shaft is less than 125 degrees
Creep—Progressive deformation of soft tissues due to constant low loading over an extended period
Creep displacement—The movement of a structure resulting from creep (See *creep*)
Cross-links—Interconnections between collagen molecules and between collagen fibers
Cross-sectional area—The area of a material on a plane perpendicular to its longitudinal axis
Deformation rate—The speed at which an applied load deforms a structure (See *speed of loading*)
Degrees of freedom—The number of ways in which a body can move
Density—The mass of matter in a given space
Diarthrodial joints—The freely moving joints of the body
Direction—The path along which motion takes place; with reference to a vector, direction includes line of application and sense
Direction of displacement—The direction of change in position of the contact points of two surfaces
Distraction—The movement of two surfaces away from each other
Donnan osmotic pressure—Pressure generated from counter ions associated with charged macromolecules trapped on one side of a semipermeable membrane
Dorsiflexion—Bending about the ankle and wrist joints in a dorsal direction
Ductility—The quality whereby a material exhibits extensive deformation before failure
Dynamics—The study of forces acting on a body in motion
Eccentric work—Work produced by a muscle when it contracts and lengthens
Elasticity—Property of a material that allows the material to return to its original shape and size after being deformed
Elastohydrodynamic lubrication—Mode of fluid film lubrication wherein the deformation of the bearing surfaces plays an important role
Equilibrium—State of a body at rest in which the sum of all forces and the sum of all moments are zero
Extension—The unbending of a joint whereby the angle between the bones is increased; the opposite of flexion
Fatigue—Mode of material damage caused by repeated cyclic deformation
Fatigue curve—A graph plotting the relationship of load and the frequency of loading that produces failure of a material
Fatigue fracture—A fracture typically produced by either infrequent repetition of high loads or frequent repetition of relatively normal loads
Fatigue wear—Removal of material from solid surfaces by mechanical action owing to deformation of the contacting bodies
Flexion—The bending of a joint whereby the angle between the bones is diminished; the opposite of extension
Fluid film lubrication—Mode of lubrication caused by a thin film of lubricant between the bearing surfaces
Force—A physical quantity that can accelerate and/or deform a body
Force couple—Two parallel forces of equal magnitude but opposite direction applied to a structure
Force-elongation curve—See *load-deformation curve*
Four-point bending—A type of bending that takes place when two force couples acting on a structure produce two equal moments (See *bending* and *three-point bending*)
Free body—A structure considered in isolation for the purpose of studying the effect of forces acting on it
Free body diagram—Diagram of an isolated portion of a structure used during free body analysis for the purpose of studying the effect of forces acting on the free body
Friction force—A tangential force opposing motion that acts between two bodies in contact
Frictional drag—Resistive force caused by the flow of the interstitial fluid through a porous permeable material
Frontal plane—Any plane passing longitudinally through the body from side to side

Glycosaminoglycan (GAG)—A long flexible chain of repeating disaccharide units that are the building blocks of proteoglycans

Gravitational force—Force produced by gravitational attraction of the earth on a body

Ground reaction force—A gravitational force produced by the weight of an object against the surface on which it lies

Helicoid—Spiral-like

Horizontal plane—See *transverse plane*

Hyaluronic acid—A linear nonsulfated glycosaminoglycan polymer chain composed of repeating disaccharide units; combines noncovalently with proteoglycan monomers to yield proteoglycan aggregates; the principal macromolecule in synovial fluid

Hydrodynamic lubrication—Mode of lubrication sustained by a layer of fluid film caused by the relative motion of the bearing surfaces and fluid viscosity

Hydrostatic lubrication—A mode of fluid film lubrication in which there is no relative sliding motion of the bearing surfaces; thus, pressure is usually generated by an external pressure supply

Inhomogeneity—Variation of material property with location within the piece of material

Instant center—See *instantaneous center of motion*

Instant center of rotation—See *instantaneous center of motion*

Instant center pathway—A pathway of the instant center for a joint in different positions throughout the range of motion in one plane

Instant center technique—A technique used to describe the relative uniplanar motion of two adjacent segments of a body and the direction of displacement of the contact points between these two segments

Instantaneous center of motion—The immovable point existing at an instant in time created by one segment (link) of a body rotating about an adjacent segment; all other points on the body rotate about this immovable point

Interfacial wear—Removal of material by either adhesion or abrasion due to interaction of the bearing surfaces

Interstitial fluid—Fluid within the porous collagen-proteoglycan solid matrix of articular cartilage

Intrinsic material properties—Material properties of the solid component of a biphasic material

Joint lubrication—A design feature of the joint that maintains the continuity of the thin film of synovial fluid between the joint surfaces, minimizing contact and wear of the cartilaginous surfaces

Joint reaction force—The internal reaction force acting at the contact surfaces when a joint in the body is subjected to external loads

Keratan sulfate—A sulfated glycosaminoglycan (hexosamine) polymer chain composed of specific repeating disaccharide units; a basic constituent of proteoglycan

Kinematics—The branch of mechanics that deals with motion of a body without reference to force or mass

Kinetics—The branch of mechanics that deals with the motion of a body under the action of given forces

Lever arm—The perpendicular distance from the line of application of a force to the center of motion in a rigid structure, also known as the moment arm of the force

Line of gravity—Line of application, or action line, of the force of gravity

Link—One of two adjacent segments of a body that move about an instantaneous center of motion in a joint

Load-deformation curve—A curve that plots the deformation of a structure when the structure is loaded in a known direction

Load relaxation—Decrease in load with time once the material under loading is deformed to a constant length (See *stress relaxation*)

Loading mode—The manner in which forces are applied to a structure (See *tension, compression, bending, shear, torsion,* and *combined loading*)

Longitudinal axis—A lengthwise line or plane about which a body or system rotates

Longitudinal plane—See *frontal plane*

Loss modulus—Component of dynamic response of a viscoelastic medium under steady sinusoidal deformation; energy loss per unit volume during one cycle of deformation

Mass moment of inertia—The measure of resistance to change in angular velocity, usually expressed in kilograms times meters squared or, equivalently, in newton meters times seconds squared

Matrix—The intercellular substance of a tissue

Modulus—A constant that expresses numerically the degree to which a property is possessed by a material

Modulus of elasticity (Young's modulus)—The ratio of stress to strain at any point in the elastic region of a load-deformation curve, yielding a value for stiffness

Moment—Torque; quantity necessary to angularly accelerate a body, usually expressed in newton meters

Moment arm—See *lever arm*

Motion segment—The functional unit of the spine consisting of two adjacent vertebrae and their intervening soft tissues

Neck-shaft angle—The angle of inclination of the femoral neck relative to the femoral shaft in the frontal plane

Neutral axis—The central plane on which the tensile and compressive stresses and strains produced by bending equal zero

Open section—A cross section of a hollow structure in which the surface of the material is no longer continuous

Open section defect—A large defect that disrupts the continuity of the outer surface of a cylindrical-type structure, roughly equal in its major diameter to the diameter of the cylinder

Osmotic pressure—Pressure generated from unequal distribution of solute particles on two sides of a semipermeable membrane

Permeability—Measure of the ease with which fluid can flow through a porous permeable medium

Perpendicular bisector—A line at right angles to a segment that divides the segment into two parts

Phase shift angle—Phase lag angle between imposed sinusoidal strain and the sinusoidal stress response of a viscoelastic medium

Physiological cross-sectional area—The amount of muscle fiber in a given cross section of a muscle

Planar—Relating to one plane

Plantar flexion—Bending about the ankle joint in the direction of the sole of the foot

Polar moment of inertia—A quantity that takes into account the cross-sectional area and the distribution of material around a neutral axis in torsional loading

Porosity—Ratio of fluid volume to the total volume in a porous material

Pressure—The surface stress acting perpendicular to a unit area

Prestress—Internal stresses in a material that counteract the stresses that result from an applied load

Proteoglycan—Protein-carbohydrate complex composed of a protein core to which one or more glycosaminoglycans are covalently bound; abundant in cartilaginous tissues

Range of motion—The range of translation and rotation of a joint for each of its six degrees of freedom

Repetitive loading—Repeated application of a load to a structure

Resiliency—The capacity of a strained body to recover its size and shape after deformation

Resultant (resultant force)—The single force that is the sum of a given set of forces with a common point of application

Rotation—Motion in which all points describe circular arcs about an immovable line or axis

Sagittal plane—The median plane of the body or any plane parallel thereto

Screw-home mechanism—A combination of knee extension and external rotation of the tibia

Shear—A loading mode in which a load is applied parallel to the surface of the structure, causing internal angular deformation

Shear modulus—An innate property of a deformable body that indicates how much resistance the body presents when an attempt is made to shear it, represented by the slope of the shear stress-strain curve; related to the modulus of elasticity

Shear strain—The amount of angular deformation of a structure under shear loading (See *strain*)

Speed of loading—The rate at which load is applied to a structure (See *deformation rate*)

Split line pattern—Pattern of elongated fissures on the articular surface induced by piercing the cartilage with a blunt round awl

Squeeze film lubrication—A mode of fluid lubrication in which the approaching surfaces generate a pressure field in the lubricant as it is forced out of the area of impending contact

Statics—The study of forces acting on a body in equilibrium

Storage modulus—Component of dynamic response of a viscoelastic medium under steady sinusoidal deformation; energy stored per unit volume during one cycle of deformation

Strain—Deformation (change in dimension) that develops within a structure in response to externally applied loads

Strain gauge—A device that permits strain to be measured

Strain rate—The speed at which a strain-producing load is applied

Stress—Load per unit area that develops on a plane surface within a structure in response to externally applied loads

Stress concentration effect—The increase in

stress that results from the concentration of stresses around a small defect in a structure

Stress raiser (stress riser)—Any geometric characteristic (such as a small hole or a sharp internal corner) in a loaded body that causes an abrupt increase in local stress

Stress relaxation—Decrease in stress with time once a material under loading has deformed to a constant length (See *load relaxation*)

Stress-strain curve—A curve generated by plotting the stress and the strain during compressive, tensile, or shear loading of a structure

Surface velocity—The surface speed of a body in a given direction

Synovial fluid—The fluid in a synovial joint

Tangential—Relating to a straight line that is the limiting position of a secant of a curve through a fixed point

Tensile strength—Maximum tensile stress or load sustained by a material in a uniaxial tension test

Tension—A loading mode in which equal and opposite loads are applied away from the surface of a structure, resulting in lengthening and narrowing

Three-point bending—A type of bending that takes place when three forces act on a structure (See *bending* and *four-point bending*)

Time-dependent (rate-dependent) or viscoelastic (recoverable) material behavior—Material behavior wherein deformation and recovery are influenced by rate and amount of time during loading and unloading

Time-independent (rate-dependent) or elastic (recoverable) material behavior—Material behavior wherein the structure deforms instantaneously when loaded and recovers instantaneously when unloaded

Torsion—A loading mode in which a load is applied to a structure in a manner that causes it to twist about an axis, subjecting the structure to a combination of shear, tensile, and compressive loads

Translation—Parallel motion of one surface across another

Transverse—Crosswise; in a horizontal direction

Transverse plane—Any plane that extends or lies in a crosswise, or horizontal, direction

Ultimate failure point—The point on the load-deformation curve past which complete failure of the structure occurs due to continued loading in the plastic region

Ultrafiltration—Filtration of a solution through a fine porous filter, which thereby removes the large solutes from the solvent

Ultrastructure—Organization of the collagen fiber network within connective tissues

Uniplanar—See *planar*

Vector—A quantity that has magnitude, sense, line of application, and point of application, commonly represented by a directed line segment

Velocity—The displacement of a body divided by the time over which displacement occurs

Viscoelasticity—Property whereby a material exhibits a change of stress or deformation when under constant deformation (stress relaxation) or under constant load (creep)

Viscosity—The resistance of a fluid to flowing

Wear—The removal of material from solid surfaces by mechanical action (See *interfacial wear* and *fatigue wear*)

Wolff's law—A law that states that bone is laid down where it is needed and resorbed where it is not needed

Yield point—The point of the load-deformation curve at which a material begins to yield; if loading continues, permanent deformation results

INDEX

INDEX

Page numbers in *italics* refer to figures; page numbers followed by *t* refer to tables.

B

C

D

M